Annals of Mathematics Studies

Number 145

# Surveys on Surgery Theory

# Volume 1

*Papers dedicated to C. T. C. Wall*

edited by

Sylvain Cappell, Andrew Ranicki,
and Jonathan Rosenberg

PRINCETON UNIVERSITY PRESS

PRINCETON, NEW JERSEY

2000

The Annals of Mathematics Studies are edited by
Luis A. Caffarelli, John N. Mather, and Elias M. Stein

Library of Congress Catalog Card Number 99-068088

ISBN 0-691-04937-8 (cloth)
ISBN 0-691-04938-6 (pbk.)

The publisher would like to acknowledge the authors of this
volume for providing the camera-ready copy from which this book was printed

The paper used in this publication meets the minimum requirements of
ANSI/NISO Z39.48-1992 (R 1997) (*Permanence of Paper*)

http://pup.princeton.edu

Printed in the United States of America

1  3  5  7  9  10  8  6  4  2

1  3  5  7  9  10  8  6  4  2
(Pbk.)

# Contents

# Surveys on Surgery Theory : Volume 1

*Papers dedicated to C. T. C. Wall*

## Preface

Surgery theory is now about 40 years old. The 60th birthday (on December 14, 1996) of C. T. C. Wall, one of the leaders of the founding generation of the subject, led us to reflect on the extraordinary accomplishments of surgery theory, and its current enormously varied interactions with algebra, analysis, and geometry. Workers in many of these areas have often lamented the lack of a single source surveying surgery theory and its applications. Indeed, no one person could write such a survey. Therefore we attempted to make a rough division of the areas of current interest, and asked a variety of experts to report on them. This is the first of two volumes which are the result of that collective effort. We hope that they prove useful to topologists, to other interested researchers, and to advanced students.

Sylvain Cappell
Andrew Ranicki
Jonathan Rosenberg

Surveys on Surgery Theory

Volume 1

# C. T. C. Wall's contributions to the topology of manifolds

Sylvain Cappell,* Andrew Ranicki,†
and Jonathan Rosenberg‡

**Note.** Numbered references in this survey refer to the Research Papers in Wall's Publication List.

## 1   A quick overview

C. T. C. Wall[1] spent the first half of his career, roughly from 1959 to 1977, working in topology and related areas of algebra. In this period, he produced more than 90 research papers and two books, covering

- cobordism groups,

- the Steenrod algebra,

- homological algebra,

- manifolds of dimensions 3, 4, $\geq 5$,

- quadratic forms,

- finiteness obstructions,

- embeddings,

- bundles,

- Poincaré complexes,

- surgery obstruction theory,

*Partially supported by NSF Grant # DMS-96-26817.
†This is an expanded version of a lecture given at the Liverpool Singularities Meeting in honor of Wall in August 1996.
‡Partially supported by NSF Grant # DMS-96-25336.
[1] Charles Terence Clegg Wall, known in his papers by his initials C. T. C. and to his friends as Terry.

- homology of groups,

- 2-dimensional complexes,

- the topological space form problem,

- computations of $K$- and $L$-groups,

- and more.

One quick measure of Wall's influence is that there are two headings in the *Mathematics Subject Classification* that bear his name:

- 57Q12 (Wall finiteness obstruction for $CW$ complexes).

- 57R67 (Surgery obstructions, Wall groups).

Above all, Wall was responsible for major advances in the topology of manifolds. Our aim in this survey is to give an overview of how his work has advanced our understanding of classification methods. Wall's approaches to manifold theory may conveniently be divided into three phases, according to the scheme:

1. All manifolds at once, up to cobordism (1959–1961).

2. One manifold at a time, up to diffeomorphism (1962–1966).

3. All manifolds within a homotopy type (1967–1977).

## 2   Cobordism

Two closed $n$-dimensional manifolds $M_1^n$ and $M_2^n$ are called *cobordant* if there is a compact manifold with boundary, say $W^{n+1}$, whose boundary is the disjoint union of $M_1$ and $M_2$. Cobordism classes can be added via the disjoint union of manifolds, and multiplied via the Cartesian product of manifolds. Thom (early 1950's) computed the cobordism ring $\mathfrak{N}_*$ of unoriented smooth manifolds, and began the calculation of the cobordism ring $\Omega_*$ of oriented smooth manifolds.

After Milnor showed in the late 1950's that $\Omega_*$ contains no odd torsion, Wall [1, 3] completed the calculation of $\Omega_*$. This was the ultimate achievement of the pioneering phase of cobordism theory. One version of Wall's main result is easy to state:

**Theorem 2.1** (Wall [3]) *All torsion in $\Omega_*$ is of order 2. The oriented cobordism class of an oriented closed manifold is determined by its Stiefel-Whitney and Pontrjagin numbers.*

For a fairly detailed discussion of Wall's method of proof and of its remarkable corollaries, see [Ros].

# 3   Structure of manifolds

What is the internal structure of a cobordism? Morse theory has as one of its main consequences (as pointed out by Milnor) that any cobordism between smooth manifolds can be built out of a sequence of *handle attachments*.

**Definition 3.1** Given an $m$-dimensional manifold $M$ and an embedding $S^r \times D^{m-r} \hookrightarrow M$, there is an associated *elementary cobordism* $(W; M, N)$ obtained by *attaching an $(r + 1)$-handle* to $M \times I$. The cobordism $W$ is the union

$$W = M \times I \cup_{\left(S^r \times D^{m-r}\right) \times \{1\}} D^{r+1} \times D^{m-r},$$

and $N$ is obtained from $M$ by deleting $S^r \times D^{m-r}$ and gluing in $S^r \times D^{m-r}$ in its place:

$$N = \left(M \smallsetminus \left(S^r \times D^{m-r}\right)\right) \cup_{S^r \times S^{m-r-1}} D^{r+1} \times S^{m-r-1}.$$

The process of constructing $N$ from $M$ is called *surgery on an $r$-sphere*, or *surgery in dimension $r$ or in codimension $m - r$*. Here $r = -1$ is allowed, and amounts to letting $N$ be the *disjoint* union of $M$ and $S^m$.

Any cobordism may be decomposed into such elementary cobordisms. In particular, any closed smooth manifold may be viewed as a cobordism between empty manifolds, and may thus be decomposed into handles.

**Definition 3.2** A cobordism $W^{n+1}$ between manifolds $M^n$ and $N^n$ is called an *h-cobordism* if the inclusions $M \hookrightarrow W$ and $N \hookrightarrow W$ are homotopy equivalences.

The importance of this notion stems from the *h-cobordism theorem* of Smale (ca. 1960), which showed that if $M$ and $N$ are simply connected and of dimension $\geq 5$, then every $h$-cobordism between $M$ and $N$ is a cylinder $M \times I$. The crux of the proof involves handle cancellations as well as Whitney's trick for removing double points of immersions in dimension $> 4$. In particular, if $M^n$ and $N^n$ are simply connected and $h$-cobordant, and if $n > 4$, then $M$ and $N$ are diffeomorphic (or $PL$-homeomorphic, depending on whether one is working in the smooth or the $PL$ category).

For manifolds which are not simply connected, the situation is more complicated and involves the fundamental group. But Smale's theorem was extended a few years later by Barden,[2] Mazur, and Stallings to give the *s-cobordism theorem*, which (under the same dimension restrictions) showed that the possible $h$-cobordisms between $M$ and $N$ are in natural bijection with the elements of the *Whitehead group* $\operatorname{Wh} \pi_1(M)$. The bijection sends

---

[2] One of Wall's students!

an $h$-cobordism $W$ to the *Whitehead torsion* of the associated homotopy
equivalence from $M$ to $W$, an invariant from algebraic $K$-theory that arises
from the combinatorics of handle rearrangements. One consequence of this
is that if $M$ and $N$ are $h$-cobordant and the Whitehead torsion of the $h$-
cobordism vanishes (and in particular, if $\operatorname{Wh} \pi_1(M) = 0$, which is the case
for many $\pi_1$'s of practical interest), then $M$ and $N$ are again diffeomorphic
(assuming $n > 4$).

The use of the Whitney trick and the analysis of handle rearrangements,
crucial to the proof of the $h$-cobordism and $s$-cobordism theorems, became
the foundation of Wall's work on manifold classification.

# 4    4-Manifolds

Milnor, following J. H. C. Whitehead, observed in 1956 that a simply connected 4-dimensional manifold $M\cdot$ is classified up to homotopy equivalence
by its *intersection form*, the non-degenerate symmetric bilinear from on
$H_2(M; \mathbb{Z})$ given by intersection of cycles, or in the dual picture, by the
cup-product

$$H^2(M; \mathbb{Z}) \times H^2(M; \mathbb{Z}) \to H^4(M; \mathbb{Z}) \xrightarrow{\cong} \mathbb{Z}.$$

Note that the isomorphism $H^4(M; \mathbb{Z}) \to \mathbb{Z}$, and thus the form, depends
on the orientation.

Classification of 4-dimensional manifolds up to homeomorphism or diffeomorphism, however, has remained to this day one of the hardest problems in topology, because of the failure of the Whitney trick in this dimension. Wall succeeded in 1964 to get around this difficulty at the expense of
"stabilizing." He used handlebody theory to obtain a stabilized version of
the $h$-cobordism theorem for 4-dimensional manifolds:

**Theorem 4.1** (Wall [19]) *For two simply connected smooth closed oriented
4-manifolds $M_1$ and $M_2$, the following are equivalent:*

1. *they are $h$-cobordant;*

2. *they are homotopy equivalent (in a way preserving orientation);*

3. *they have the same intersection form on middle homology.*

*If these conditions hold, then $M_1 \# k(S^2 \times S^2)$ and $M_2 \# k(S^2 \times S^2)$ are
diffeomorphic (in a way preserving orientation) for $k$ sufficiently large (depending on $M_1$ and $M_2$).*

Note incidentally that the converse of the above theorem is not quite
true: $M_1 \# k(S^2 \times S^2)$ and $M_2 \# k(S^2 \times S^2)$ are diffeomorphic (in a way

preserving orientation) for $k$ sufficiently large if and only if the intersection forms of $M_1$ and $M_2$ are *stably* isomorphic (where stability refers to addition of the hyperbolic form $\begin{pmatrix} 0 & 1 \\ 1 & 0 \end{pmatrix}$).

From the 1960's until Donaldson's work in the 1980's, Theorem 4.1 was basically the *only* significant result on the diffeomorphism classification of simply-connected 4-dimensional manifolds. Thanks to Donaldson's work, we now know that the stabilization in the theorem (with respect to addition of copies of $S^2 \times S^2$) is unavoidable, in that without it, nothing like Theorem 4.1 could be true.

# 5  Highly connected manifolds

The investigation of simply connected 4-dimensional manifolds suggested the more general problem of classifying $(n-1)$-connected $2n$-dimensional manifolds, for all $n$. The intersection form on middle homology again appears as a fundamental algebraic invariant of oriented homotopy type. In fact this invariant also makes sense for an $(n-1)$-connected $2n$-dimensional manifold $M$ with boundary a homology sphere $\partial M = \Sigma^{2n-1}$. If $\partial M$ is a homotopy sphere, it has a potentially *exotic* differentiable structure for $n \geq 4$.[3]

**Theorem 5.1** (Wall [10]) *For $n \geq 3$ the diffeomorphism classes of differentiable $(n-1)$-connected $2n$-dimensional manifolds with boundary a homotopy sphere are in natural bijection with the isomorphism classes of $\mathbb{Z}$-valued non-degenerate $(-1)^n$-symmetric forms with a quadratic refinement in $\pi_n(BSO(n))$.*

(The form associated to a manifold $M$ is of course the intersection form on the middle homology $H_n(M; \mathbb{Z})$. This group is isomorphic to $\pi_n(M)$, by the Hurewicz theorem, so every element is represented by a map $S^n \to M^{2n}$. By the Whitney trick, this can be deformed to an embedding, with normal bundle classified by an element of $\pi_n(BSO(n))$. The quadratic refinement is defined by this homotopy class.)

The sequence of papers [15, 16, 22, 23, 37, 42] extended this diffeomorphism classification to other types of highly-connected manifolds, using a combination of homotopy theory and the algebra of quadratic forms. These papers showed how far one could go in the classification of manifolds without surgery theory.

---

[3]It was the study of the classification of 3-connected 8-dimensional manifolds with boundary which led Milnor to discover the existence of exotic spheres in the first place. [Mil]

# 6   Finiteness obstruction

Recall that if $X$ is a space and $f : S^r \to X$ is a map, the space obtained from $X$ by *attaching an $(r + 1)$-cell* is $X \cup_f D^{r+1}$. A *CW complex* is a space obtained from $\emptyset$ by attaching cells. It is called *finite* if only finitely many cells are used. One of the most natural questions in topology is:

> When is a space homotopy equivalent to a *finite* $CW$ complex?

A space $X$ is called *finitely dominated* if it is a *homotopy* retract of a finite $CW$ complex $K$, i.e., if there exist maps $f : X \to K$, $g : K \to X$ and a homotopy $gf \simeq 1 : X \to X$. This is clearly a necessary condition for $X$ to be of the homotopy type of a finite $CW$ complex. Furthermore, for spaces of geometric interest, finite domination is much easier to verify than finiteness. For example, already in 1932 Borsuk had proved that every compact $ANR$, such as a compact topological manifold, is finitely dominated. So another question arises:

> Is a finitely dominated space homotopy equivalent to a finite $CW$ complex?

This question also has roots in the study of the free actions of finite groups on spheres. A group with such an action necessarily has periodic cohomology. In the early 1960's Swan had proved that a finite group $\pi$ with cohomology of period $q$ acts freely on an infinite $CW$ complex $Y$ homotopy equivalent to $S^{q-1}$, with $Y/\pi$ finitely dominated, and that $\pi$ acts freely on a *finite* complex homotopy equivalent to $S^{q-1}$ if and only if an algebraic $K$-theory invariant vanishes. Swan's theorem was in fact a special case of the following general result.

**Theorem 6.1** (Wall [26, 43]) *A finitely dominated space $X$ has an associated obstruction $[X] \in \widetilde{K}_0(\mathbb{Z}[\pi_1(X)])$. The space $X$ is homotopy equivalent to a finite $CW$ complex if and only if this obstruction vanishes.*

The obstruction defined in this theorem, now universally called the *Wall finiteness obstruction*, is a fundamental algebraic invariant of non-compact topology. It arises as follows. If $K$ is a finite $CW$ complex dominating $X$, then the cellular chain complex of $K$, with local coefficients in the group ring $\mathbb{Z}[\pi_1(X)]$, is a finite complex of finitely generated free modules. The domination of $X$ by $K$ thus determines a direct summand subcomplex of a finite chain complex, attached to $X$. Since a direct summand in a free module is projective, this chain complex attached to $X$ consists of finitely generated projective modules. The Wall obstruction is a kind of "Euler characteristic" measuring whether or not this chain complex is chain equivalent to a finite complex of finitely generated free modules.

The Wall finiteness obstruction has turned out to have many applications to the topology of manifolds, most notably the Siebenmann end obstruction for closing tame ends of open manifolds.

# 7 Surgery theory and the Wall groups

The most significant of all of Wall's contributions to topology was undoubtedly his development of the general theory of non-simply-connected surgery. As defined above, surgery can be viewed as a means of creating new manifolds out of old ones. One measure of Wall's great influence was that when other workers (too numerous to list here) made use of surgery, they almost invariably drew upon Wall's contributions.

As a methodology for classifying manifolds, surgery was first developed in the 1961 work of Kervaire and Milnor [KM] classifying homotopy spheres in dimensions $n \geq 6$, up to $h$-cobordism (and hence, by Smale's theorem, up to diffeomorphism). If $W^n$ is a parallelizable manifold with homotopy sphere boundary $\partial W = \Sigma^{n-1}$, then it is possible to kill the homotopy groups of $W$ by surgeries if and only if an obstruction

$$\sigma(W) \in P_n = \begin{cases} \mathbb{Z} & \text{if } n \equiv 0 \,(\mathrm{mod}\,4), \\ \mathbb{Z}/2 & \text{if } n \equiv 2 \,(\mathrm{mod}\,4), \\ 0 & \text{if } n \text{ is odd} \end{cases}$$

vanishes. Here

$$\sigma(W) = \begin{cases} \text{signature}(W) \in \mathbb{Z} & \text{if } n \equiv 0 \,(\mathrm{mod}\,4), \\ \text{Arf invariant}(W) \in \mathbb{Z}/2 & \text{if } n \equiv 2 \,(\mathrm{mod}\,4). \end{cases}$$

In 1962 Browder [Br] used the surgery method to prove that, for $n \geq 5$, a simply-connected finite $CW$ complex $X$ with $n$-dimensional Poincaré duality

$$H^{n-*}(X) \cong H_*(X)$$

is homotopy equivalent to a closed $n$-dimensional differentiable manifold if and only if there exists a vector bundle $\eta$ with spherical Thom class such that an associated invariant $\sigma \in P_n$ (the simply connected surgery obstruction) is 0. The result was proved by applying Thom transversality to $\eta$ to obtain a suitable degree-one map $M \to X$ from a manifold, and then killing the kernel of the induced map on homology. For $n = 4k$ the invariant $\sigma \in P_{4k} = \mathbb{Z}$ is one eighth of the difference between signature$(X)$ and the $4k$-dimensional component of the $\mathcal{L}$-genus of $-\eta$. (The minus sign comes from the fact that the tangent and normal bundles are stably the negatives of one another.) In this case, the result is a converse of the Hirzebruch signature theorem. In other words, $X$ is homotopy-equivalent to a differentiable manifold if and only if the formula of the theorem holds with $\eta$ playing the role of the stable normal bundle. The hardest step was to find enough embedded spheres with trivial normal bundle in the middle dimension, using the Whitney embedding theorem for embeddings $S^m \subset M^{2m}$ — this requires $\pi_1(M) = \{1\}$ and $m \geq 3$. Also in 1962, Novikov initiated the use of surgery in the study of the uniqueness of differentiable manifold structures in the homotopy type of a manifold, in the simply-connected case.

From about 1965 until 1970, Wall developed a comprehensive surgery obstruction theory, which also dealt with the non-simply-connected case. The extension to the non-simply-connected case involved many innovations, starting with the correct generalization of the notion of Poincaré duality. A connected finite $CW$ complex $X$ is called a *Poincaré complex* [44] of dimension $n$ if there exists a *fundamental homology class* $[X] \in H_n(X; \mathbb{Z})$ such that cap product with $[X]$ induces isomorphisms from cohomology to homology with local coefficients,

$$H^{n-*}(X; \mathbb{Z}[\pi_1(X)]) \xrightarrow{\cong} H_*(X; \mathbb{Z}[\pi_1(X)]).$$

This is obviously a *necessary* condition for $X$ to have the homotopy type of a closed $n$-dimensional manifold. A *normal map* or *surgery problem*

$$(f,b) \; : \; M^n \to X$$

is a degree-one map $f : M \to X$ from a closed $n$-dimensional manifold to an $n$-dimensional Poincaré complex $X$, together with a bundle map $b : \nu_M \to \eta$ so that

$$\begin{array}{ccc} \nu_M & \xrightarrow{b} & \eta \\ \downarrow & & \downarrow \\ M & \xrightarrow{f} & X \end{array}$$

commutes.

Wall defined ([41], [W1]) the *surgery obstruction groups* $L_*(A)$ for any ring with involution $A$, using quadratic forms over $A$ and their automorphisms. They are more elaborate versions of the Witt groups of fields studied by algebraists.

**Theorem 7.1** (Wall, [41], [W1]) *A normal map* $(f,b) : M \to X$ *has a surgery obstruction*

$$\sigma_*(f,b) \in L_n(\mathbb{Z}[\pi_1(X)]),$$

*and* $(f,b)$ *is normally bordant to a homotopy equivalence if (and for $n \geq 5$ only if)* $\sigma_*(f,b) = 0$.

One of Wall's accomplishments in this theorem, quite new at the time, was to find a way to treat both even-dimensional and odd-dimensional manifolds in the same general framework. Another important accomplishment was the recognition that surgery obstructions live in groups *depending on the fundamental group*, but not on any other aspect of $X$ (except for the dimension modulo 4 and the orientation character $w_1$. Here we have concentrated on the oriented case, $w_1 = 0$). In general, the groups $L_n(\mathbb{Z}[\pi_1(X)])$ are not so easy to compute (more about this below and elsewhere in this volume!), but in the simply-connected case, $\pi_1(X) = \{1\}$, they are just the Kervaire-Milnor groups, $L_n(\mathbb{Z}[\{1\}]) = P_n$.

Wall formulated various relative version of Theorem 7.1 for manifolds with boundary, and manifold $n$-ads. An important special case is often quoted, which formalizes the idea that the "surgery obstruction groups only depend on the fundamental group." This is the celebrated:

**Theorem 7.2** "$\pi$-$\pi$ Theorem" ([W1], 3.3) *Suppose one is given a surgery problem* $(f, b) : M^n \to X$, *where $M$ and $X$ each have connected non-empty boundary, and suppose* $\pi_1(\partial X) \to \pi_1(X)$ *is an isomorphism. Also assume that $n \geq 6$. Then $(f, b)$ is normally cobordant to a homotopy equivalence of pairs.*

The most important consequence of Wall's theory, which appeared for the first time in Chapter 10 of [W1], is that it provides a "classification" of the manifolds in a fixed homotopy type (in dimensions $\geq 5$ in the absolute case, $\geq 6$ in the relative case). The formulation, in terms of the "surgery exact sequence," was based on the earlier work of Browder, Novikov, and Sullivan in the simply connected case. The basic object of study is the *structure set* $\mathcal{S}(X)$ of a Poincaré complex $X$. This is the set of all homotopy equivalences (or perhaps simple homotopy equivalences, depending on the way one wants to formulate the theory) $M \xrightarrow{f} X$, where $M$ is a manifold, modulo a certain equivalence relation: $M \xrightarrow{f} X$ and $M' \xrightarrow{f'} X$ are considered equivalent if there is a diffeomorphism $\phi : M \to M'$ such that $f' \circ \phi$ is homotopic to $f$. One should think of $\mathcal{S}(X)$ as "classifying all manifold structures on the homotopy type of $X$."

**Theorem 7.3** (Wall [W1], Theorem 10.3 *et seq.*) *Under the above dimension restrictions, the structure set $\mathcal{S}(X)$ of a Poincaré complex $X$ is non-empty if and only if there exists a normal map $(f, b) : M \to X$ with surgery obstruction $\sigma_*(f, b) = 0 \in L_n(\mathbb{Z}[\pi_1(X)])$. If non-empty, $\mathcal{S}(X)$ fits into an exact sequence (of sets)*

$$L_{n+1}(\mathbb{Z}[\pi_1(X)]) \to \mathcal{S}(X) \to \mathcal{T}(X) \to L_n(\mathbb{Z}[\pi_1(X)]),$$

*where $\mathcal{T}(X) = [X, G/O]$ classifies "tangential data."*

Much of [W1] and many of Wall's papers in the late 1960's and early 1970's were taken up with calculations and applications. We mention only a few of the applications: a new proof of the theorem of Kervaire characterizing the fundamental groups of high-dimensional knot complements, results on realization of Poincaré (i.e., homotopy-theoretic) embeddings in manifolds by actual embeddings of submanifolds, classification of free actions of various types of discrete groups on manifolds (for example, free involutions on spheres), the classification of "fake projective spaces," "fake lens spaces," "fake tori," and more. The work on the topological space form problem (free actions on spheres) is particularly significant: the $CW$ complex version had already motivated the work of Swan and Wall on the finiteness obstruction discussed above, while the manifold version was one of the impulses for Wall's (and others') extensive calculations of the $L$-groups of finite groups.

# 8    PL and topological manifolds

While surgery theory was originally developed in the context of smooth manifolds, it was soon realized that it works equally well in the $PL$ category of combinatorial manifolds. Indeed, the book [W1] was written in the language of $PL$ manifolds. Wall's theory has the same form in both categories, and in fact the surgery obstruction groups are the same, regardless of whether one works in the smooth or in the $PL$ category. The only differences are that for the $PL$ case, vector bundles must be replaced by $PL$ bundles, and in theorem 7.3, $G/O$ should be replaced by $G/PL$.

Passage from the $PL$ to the topological category was a much trickier step (even though we now know that $G/PL$ more closely resembles $G/TOP$ than $G/O$). Wall wrote in the introduction to [44]:

> This paper was originally planned when the only known fact about topological manifolds (of dimension > 3) was that they were Poincaré complexes. Novikov's proof of the topological invariance of rational Pontrjagin classes and subsequent work in the same direction has changed this ....

Novikov's work introduced the torus $T^n$ as an essential tool in the study of topological manifolds. A *fake torus* is a manifold which is homotopy equivalent to $T^n$. The surgery theoretic classification of PL fake tori in dimensions $\geq 5$ by Wall and by Hsiang and Shaneson [HS] was an essential tool in the work of Kirby [K] and Kirby-Siebenmann [KS] on the structure theory of topological manifolds. (See also [KSW].) This in turn made it possible to extend surgery theory to the topological category.

# 9 Invariance properties of the signature

Wall made good use of the signature invariants of quadratic forms and manifolds. We pick out three particular cases:

1. One immediate (but non-trivial) consequence of the Hirzebruch signature theorem is that if $\widetilde{M}$ is a $k$-fold covering of a closed manifold $M$, then
$$\text{signature}\,(\widetilde{M}) \;=\; k \cdot \text{signature}\,(M) \in \mathbb{Z}\,.$$
   It is natural to ask whether this property is special to manifolds, or whether it holds for Poincaré complexes in general. But in [44], Wall constructed examples of 4-dimensional Poincaré complexes $X$ where this fails, and hence such $X$ are not homotopy equivalent to manifolds.

2. For finite groups $\pi$, Wall [W1] showed that the surgery obstruction groups $L_*(\mathbb{Z}[\pi])$ are finitely generated abelian groups, and that the torsion-free part of these groups is determined by a collection of signature invariants called the *multisignature*. This work led to a series of deep interactions between algebraic number theory and geometric topology.

3. The Novikov additivity property of the signature is that the signature of the boundary-connected union of manifolds is the sum of the signatures. In [54], Wall showed that this additivity fails for unions of manifolds along parts of boundaries which are not components, and obtained a homological expression for the non-additivity of the signature (which is also known as the Maslov index).

# 10 Homological and combinatorial group theory

Wall's work on surgery theory led to him to several problems in combinatorial group theory. One of these was to determine what groups can be the fundamental groups of aspherical Poincaré complexes. Such groups are called *Poincaré duality groups*. There are evident connections with the topology of manifolds:

([W2], problem G2, p. 391) Is every Poincaré duality group $\Gamma$ the fundamental group of a closed $K(\Gamma, 1)$ manifold? Smooth manifold? Manifold unique up to homeomorphism? (It will not be unique up to diffeomorphism.)

([W2], problem F16, p. 388) Let $\Gamma$ be a Poincaré duality group of dimension $\geq 3$. Is the 'fundamental group at infinity' of $\Gamma$ necessarily trivial? This is known in many cases, e.g. if $\Gamma$ has a finitely presented normal subgroup $\Gamma'$ of infinite index and either $\Gamma'$ or $\Gamma/\Gamma'$ has one end. In dimensions $\geq 5$ it is equivalent to having the universal cover of a compact $K(\Gamma, 1)$ manifold homeomorphic to euclidean space.

For more on the subsequent history of these problems, see [FRR] and [D].

# References

[Br]    W. Browder, Homotopy type of differentiable manifolds, in *Proc. Århus Topology Conference* (1962), reprinted in *Novikov Conjectures, Index Theorems and Rigidity*, Lond. Math. Soc. Lecture Notes **226**, Cambridge Univ. Press (1995), 97–100.

[D]     M. Davis, Poincaré duality groups, this volume.

[FRR]   S. Ferry, A. Ranicki and J. Rosenberg (editors), *Novikov Conjectures, Index Theorems and Rigidity*, Lond. Math. Soc. Lecture Notes **226**, **227**, Cambridge Univ. Press (1995).

[HS]    W.-C. Hsiang and J. Shaneson, Fake tori, in *Topology of Manifolds*, Proc. Conf. at Univ. of Georgia, 1969, Markham, Chicago (1970), 18–51.

[KM]    M. Kervaire and J. Milnor, Groups of homotopy spheres, I. Ann. of Math. **77** (1963), 504–537.

[K]     R. Kirby, Stable homeomorphisms and the annulus conjecture, Ann. of Math. **89** (1969), 575–583.

[KS]    R. Kirby and L. Siebenmann, *Foundational Essays on Topological Manifolds, Smoothings and Triangulations*, Ann. of Math. Studies **88**, Princeton Univ. Press, 1977.

[KSW]   R. Kirby, L. Siebenmann, and C. T. C. Wall, The annulus conjecture and triangulation, Abstract # 69T-G27, Notices Amer. Math. Soc. **16** (1969), 432.

[KT]    R. C. Kirby and L. R. Taylor, A survey of 4-manifolds through the eyes of surgery, vol. 2 of this collection.

[Mil]   J. Milnor, Classification of $(n-1)$-connected $2n$-dimensional manifolds and the discovery of exotic spheres, this volume.

[Ros] J. Rosenberg, Reflections on C. T. C. Wall's work on cobordism, vol. 2 of this collection.

[W1] C. T. C. Wall, *Surgery on Compact Manifolds*, (London Math. Soc. monographs, no. 1), Academic Press, London and New York, 1970. 2nd edition, A.M.S. Surveys and Monographs, A.M.S., 1999.

[W2] C. T. C. Wall (editor), *Homological Group Theory*, Lond. Math. Soc. Lecture Notes **36**, Cambridge Univ. Press (1979).

AUTHOR ADDRESSES:

S. C.: Courant Institute for the Mathematical Sciences
New York University
215 Mercer St.
New York, NY 10012–1110
USA
email: cappell@cims.nyu.edu

A. R.: Department of Mathematics
University of Edinburgh
The King's Buildings
Edinburgh EH9 3JZ
Scotland, UK
email: aar@maths.ed.ac.uk

J. R.: Department of Mathematics
University of Maryland
College Park, MD 20742–4015
USA
email: jmr@math.umd.edu

# C. T. C. Wall's publication list

## 1 Books

*A geometric introduction to topology*, vi, 168 pp., Addison Wesley, 1971; reprinted in paperback by Dover, 1993.

*Surgery on compact manifolds*, London Math. Soc. Monographs no. 1, x, 280 pp. Academic Press, 1970; 2nd edition (ed. A.A. Ranicki), Amer. Math. Soc. Surveys and Monographs **69**, A.M.S., 1999.

*Proceedings of Liverpool Singularities Symposium I., II.* (edited), Lecture Notes in Math. **192, 209**, Springer, 1971.

*Homological group theory* (edited), London Math. Soc. Lecture Notes **36**, Cambridge University Press, 1979.

*The geometry of topological stability* (with A.A. du Plessis), viii, 572 pp., London Math. Soc. Monographs, New Series, no. 9, Oxford University Press, 1995.

## 2 Research papers

1. Note on the cobordism ring, Bull. Amer. Math. Soc. **65** (1959) 329–331.
2. On a result in polynomial rings, Proc. Camb. Phil. Soc. **56** (1960) 104–108.
3. Determination of the cobordism ring, Ann. of Math. **72** (1960) 292–311.
4. Generators and relations for the Steenrod algebra, Ann. of Math. **72** (1960) 429–444.
5. Rational Euler characteristics, Proc. Camb. Phil. Soc. **57** (1961) 182–183.
6. Cobordism of pairs, Comm. Math. Helv. **35** (1961) 136–145.
7. Resolutions for extensions of groups, Proc. Camb. Phil. Soc. **57** (1961) 251–255.
8. On the cohomology of certain groups, Proc. Camb. Phil. Soc. **57** (1961) 731–733.

9. Killing the middle homotopy groups of odd dimensional manifolds, Trans. Amer. Math. Soc. **103** (1962) 421–433.
10. Classification of $(n-1)$-connected $2n$-manifolds, Ann. of Math. **75** (1962) 163–189.
11. Cobordism exact sequences for differential and combinatorial manifolds, Ann. of Math. **77** (1963) 1–15.
12. On the orthogonal groups of unimodular quadratic forms, Math. Ann. **147** (1962) 328–338.
13. The action of $\Gamma_{2n}$ on $(n-1)$-connected $2n$-manifolds, Proc. Amer. Math. Soc. **13** (1962) 943–944.
14. A characterisation of simple modules over the Steenrod algebra mod. 2, Topology **1** (1962) 249–254.
15. Classification problems in differential topology I. Classification of handlebodies, Topology **2** (1963) 253–261.
16. Classification problems in differential topology II. Diffeomorphisms of handlebodies, Topology **2** (1963) 263–272.
17. On the orthogonal groups of unimodular quadratic forms II., J. reine angew. Math. **213** (1963) 122–136.
18. Diffeomorphisms of 4-manifolds, J. London Math. Soc. **39** (1964) 131–140.
19. On simply-connected 4-manifolds, J. London Math. Soc. **39** (1964) 141–149.
20. Cobordism of combinatorial $n$-manifolds for $n \leq 8$, Proc. Camb. Phil. Soc. **60** (1964) 807–812.
21. Graded Brauer groups, J. reine angew. Math. **213** (1964) 187–199.
22. Classification problems in differential topology III. Applications to special cases, Topology **3** (1965) 291–304.
23. Quadratic forms on finite groups and related topics, Topology **2** (1963) 281–298.
24. Classification problems in differential topology VI. Classification of $(n-1)$-connected $(2n+1)$-manifolds, Topology **6** (1967) 273–296.
25. An obstruction to finiteness of $CW$-complexes, Bull. Amer. Math. Soc. **70** (1964) 269–270.
26. Finiteness conditions for $CW$-complexes, Ann. of Math. **81** (1965) 56–89.
27. (with W.-C. Hsiang) Orientability of manifolds for generalised homology theories, Trans. Amer. Math. Soc. **118** (1963) 352–359.
28. An extension of results of Novikov and Browder, Amer. Jour. of Math. **88** (1966) 20–32.
29. On the exactness of interlocking sequences, l'Enseignement Math. **111** (1966) 95–100.
30. Arithmetic invariants of subdivision of complexes, Canad. J. Math. **18** (1966) 92–96.

31. Formal deformations, Proc. London Math. Soc. **16** (1966) 342–352.
32. Open 3-manifolds which are 1-connected at infinity, Quart. J. Math. Oxford **16** (1963) 263–268.
33. Survey: topology of smooth manifolds, Jour. London Math. Soc. **40** (1965) 1–20.
34. Unknotting tori in codimension one and spheres in codimension two, Proc. Camb. Phil. Soc. **61** (1965) 659–664.
35. All 3-manifolds embed in 5-space, Bull. Amer. Math. Soc. **71** (1965) 564–567.
36. Piecewise linear normal microbundles, Bull. Amer. Math. Soc. **71** (1965) 638–641.
37. Classification problems in differential topology IV. Thickenings, Topology **5** (1966) 73–94.
38. (with A. Haefliger) Piecewise linear bundles in the stable range, Topology **4** (1965) 209–214.
39. Addendum to a paper of Conner and Floyd, Proc. Camb. Phil. Soc. **62** (1966) 171–176.
40. Locally flat $PL$-submanifolds with codimension two, Proc. Camb. Phil. Soc. **63** (1967) 5–8.
41. Surgery of non simply-connected manifolds, Ann. of Math. **84** (1966) 217–276.
42. Classification problems in differential topology V. On certain 6-manifolds, Invent. Math. **1** (1966) 355–374; corrigendum ibid. **2** (1967) 306.
43. Finiteness conditions for CW complexes II., Proc. Roy. Soc. **295A** (1966) 129–139.
44. Poincaré complexes I., Ann. of Math. **86** (1967) 213–245.
45. On bundles over a sphere with fibre euclidean space, Fund. Math. **61** (1967) 57–72.
46. Homeomorphism and diffeomorphism classification of manifolds, pp. 450–460 in *Proc. ICM, Moscow, 1966* Mir, 1968.
47. Graded algebras, anti-involutions, simple groups and symmetric spaces, Bull. Amer. Math. Soc. **74** (1968) 198–202.
48. Free piecewise linear involutions on spheres, Bull. Amer. Math. Soc. **74** (1968) 554–558.
49. On the axiomatic foundations of the theory of hermitian forms, Proc. Camb. Phil. Soc. **67** (1970) 243–250.
50. On groups consisting mostly of involutions, Proc. Camb. Phil. Soc. **67** (1970) 251–262.
51.a,b Geometric connectivity I., II., Jour. London Math. Soc. **3** (1971) 597–604, 605–608.
52. (with A. Fröhlich) Foundations of equivariant algebraic $K$-theory, pp. 12–27 in Lecture Notes in Math. **108**, Springer (1969).

53. On homotopy tori and the annulus theorem, Bull. London Math. Soc. **1** (1969) 95–97.

54. Non-additivity of the signature, Invent. Math. **7** (1969) 269–274.

55. (with W.-C. Hsiang) On homotopy tori II., Bull. London Math. Soc. **1** (1969) 341–342.

56. The topological space-form problem, pp. 319–351 in *Topology of manifolds* (ed. J.C. Cantrell & C.H. Edwards Jr.) Markham, 1970.

57. Pairs of relative cohomological dimension 1, J. Pure Appl. Alg. **1** (1971) 141–154.

58. (with W. Browder and T. Petrie) The classification of free actions of cyclic groups of odd order on homotopy spheres, Bull. Amer. Math. Soc. **77** (1971) 455–459.

59. (with C.B. Thomas) The topological spherical space-form problem I., Compositio Math. **23** (1971) 101–114.

60. Classification of Hermitian forms I. Rings of algebraic integers, Compositio Math. **22** (1970) 425–451.

61. Geometric topology: manifolds and structures, pp. 213–219 in *Proc. Internat. Cong. Math., Nice, 1970* Vol. 1 (Gauthier-Villars, 1971).

62.a Lectures on $C^\infty$-stability and classification, pp. 178–206

62.b Introduction to the preparation theorem, pp. 90–96

62.c Stratified sets: a survey, pp. 133–140 in *Proc. Liverpool Singularities Symposium I.*, ed. C.T.C. Wall, Lecture Notes in Math. **192**, Springer (1971).

62.d Remark on geometrical singularities, p. 121

62.e Reflections on gradient vector fields, pp. 191–195 in *Proc. Liverpool Singularities Symposium II.*, ed. C.T.C. Wall, Lecture Notes in Math. **209**, Springer (1971).

63. Classification of Hermitian forms II. Semisimple rings, Invent. Math. **18** (1972) 119–141.

64. Classification of Hermitian forms III. Complete semilocal rings, Invent. Math. **19** (1973) 59–71.

65. Classification of Hermitian forms V. Global rings, Invent. Math. **23** (1974) 261–288.

66. Quadratic forms on finite groups II., Bull. London Math. Soc. **4** (1972) 156–160.

67. (with A. Fröhlich) Equivariant Brauer groups in algebraic number theory, Bull. Math. Soc. France Mémoire **25** (1971) 91–96.

68. A remark on gradient dynamical systems, Bull. London Math. Soc. **4** (1972) 163–166.

69. On the commutator subgroups of certain unitary groups, J. Algebra **27** (1973) 306–310.

70. (with A. Fröhlich) Graded monoidal categories, Compositio Math. **28** (1974) 229–286.

71. (with F.E.A. Johnson) Groups satisfying Poincaré duality, Ann. of Math. **96** (1972) 592–598.

72. Equivariant algebraic $K$-theory, pp. 111–118 in *New developments in topology*, London Math. Soc. Lecture Notes **11** (1974).

73. Foundations of algebraic $L$-theory, pp. 266–300 in *Algebraic K-theory III. Hermitian K-theory and geometric applications*, ed. H. Bass, Lecture Notes in Math. **343**, Springer (1973).

74. Some $L$-groups of finite groups, Bull. Amer. Math. Soc. **79** (1973) 526–529.

75. On rationality of modular representations, Bull. London Math. Soc. **5** (1973) 199–202.

76. Periodicity in algebraic $L$-theory, pp. 57–68 in *Manifolds, Tokyo 1973*, Univ. of Tokyo Press, 1975.

77. Classification of Hermitian forms IV. Adèle rings, Invent. Math. **23** (1974) 241–260.

78. Norms of units in group rings, Proc. London Math. Soc. **29** (1974) 593–632.

79. Regular stratifications, pp. 332–344 in *Dynamical systems — Warwick 1974*, ed. A. Manning, Lecture Notes in Math. **468**, Springer (1975)

80. Classification of Hermitian forms VI. Group rings, Ann. of Math. **103** (1976) 1–80.

81. Formulae for surgery obstructions, Topology **15** (1976) 189–210; corrigendum ibid **16** (1977) 495–496.

82. (with C.B. Thomas and I. Madsen) The topological spherical space-form problem II. Existence of free actions, Topology **15** (1976) 375–382.

83. Nets of conics, Math. Proc. Camb. Phil. Soc. **81** (1977) 351–364.

84. Geometric properties of generic differentiable manifolds, pp. 707–774 in *Geometry and topology: III Latin American school of mathematics* ed. J. Palis & M.P. do Carmo, Lecture Notes in Math. **597**, Springer (1977).

85. Free actions of finite groups on spheres, pp. 115–124 in *Proc. Symp. in Pure Math.* **32i** (*Algebraic and Geometric Topology*) (ed. J. Milgram) Amer. Math. Soc. 1978.

86. Nets of quadrics and theta-characteristics of singular curves, Phil. Trans. Roy. Soc. **289A** (1978) 229–269.

87. Periodic projective resolutions, Proc. London Math. Soc. **39** (1979) 509–533.

88. Note on the invariant of plane cubics, Math. Proc. Camb. Phil. Soc. **85** (1979) 403–406.

89. (with G.P. Scott) Topological methods in group theory, pp. 137–204 in *Homological group theory*, ed. C.T.C. Wall and D. Johnson, London Math. Soc. Lecture Notes **36** (1979).

90. Affine cubic functions I. Functions on $\mathbb{C}^2$, Math. Proc. Camb. Phil. Soc. **85** (1979) 387–401.

91. Are maps finitely determined in general? Bull. London Math. Soc. **11** (1979) 151–154.

92. Singularities of nets of quadrics, Compositio Math. **42** (1981) 187–212.

93. (with J.W. Bruce) On the classification of cubic surfaces, Jour. London Math. Soc. **19** (1979) 245–256.

94. Affine cubic functions II. Functions on $\mathbb{C}^3$ with a corank 2 critical point, Topology **19** (1980) 89–98.

95. Affine cubic functions III. Functions on $\mathbb{R}^2$, Math. Proc. Camb. Phil. Soc. **87** (1980) 1–14.

96. Relatively 1-dimensional complexes, Math. Zeits. **172** (1980) 77–79.

97. The first canonical stratum, Jour. London Math. Soc. **21** (1980) 419–433.

98. (omitted)

99.a A note on symmetry of singularities, Bull. London Math. Soc. **12** (1980) 169–175.

99.b A second note on symmetry of singularities, Bull. London Math. Soc. **12** (1980) 347–354.

100. Affine cubic functions IV. Functions on $\mathbb{C}^3$, nonsingular at infinity, Phil. Trans. Roy. Soc. **302A** (1981) 415–455.

101. Stability, pencils and polytopes, Bull. London Math. Soc. **12** (1980) 401–421.

102. Finite determinacy of smooth map-germs, Bull. London Math. Soc. **13** (1981) 481–539.

103. On finite $C^k$ left determinacy, Invent. Math. **70** (1983) 399–405.

104. A splitting theorem for maps into $\mathbb{R}^2$, Math. Ann. **259** (1982) 443–453.

105. Classification of unimodal isolated singularities of complete intersections, pp. 625–640 in *Proc. Symp. in Pure Math.* **40ii** (*Singularities*) (ed. P. Orlik) Amer. Math. Soc., 1983.

106. Topological invariance of the Milnor number mod 2, Topology **22** (1983) 345–350.

107. (with C.B. Thomas and I. Madsen) The topological spherical space-form problem III. Dimensional bounds and smoothing, Pacific J. Math. **106** (1983) 135–143.

108. Geometric invariant theory of linear systems, Math. Proc. Camb. Phil. Soc. **93** (1983) 57–62.

109. Pencils of real binary cubics, Math. Proc. Camb. Phil. Soc. **93** (1983) 477–484.

110. Notes on the classification of singularities, Proc. London Math. Soc. **48** (1984) 461–513.

111. Periods of integrals and topology of algebraic varieties, Proc. Roy. Soc. **391A** (1984) 231–254.

112. (with W. Ebeling) Kodaira singularities and an extension of Arnol'd's strange duality, Compositio Math. **56** (1985) 3–77.

113. (with A.A. du Plessis) On $C^1$-stability and $\mathcal{A}^{(1)}$-determinacy, Publ. Math. I. H. E. S., **70** (1989) 5–46.

114. Equivariant jets, Math. Ann. **272** (1985) 41–65.

115. Infinite determinacy of equivariant map-germs, Math. Ann. **272** (1985) 67–82.

116. Determination of the semi-nice dimensions, Math. Proc. Camb. Phil. Soc. **97** (1985) 79–88.

117. Survey of recent results on equivariant singularity theory, Banach Centre Publ. **20** (1988) 457–474.

118. (with J.W. Bruce and A.A. du Plessis) Determinacy and unipotency, Invent. Math. **88** (1987) 521–554.

119. (omitted)

120. Geometries and geometric structures in real dimension 4 and complex dimension 2, pp. 268–292 in *Geometry and topology. Proceedings, University of Maryland 1983–1984* ed. J. Alexander and J. Harer, Lecture Notes in Math. **1167**, Springer (1985).

121. Geometric structures on compact complex analytic surfaces, Topology **25** (1986) 119–153.

122. Real forms of cusp singularities, Math. Proc. Camb. Phil. Soc. **99** (1986) 213–232.

123. Real forms of smooth del Pezzo surfaces, J. reine und angew. Math. **375/376** (1987) 47–66.

124. Functions on quotient singularities, Phil. Trans. Roy. Soc. **324A** (1987) 1–45.

125. Exceptional deformations of quadrilateral singularities and singular $K3$ surfaces, Bull. London Math. Soc. **19** (1987) 174–176.

126. Real forms of cusp singularities II., Math. Proc. Camb. Phil. Soc. **102** (1987) 193–201.

127. Deformations of real singularities, Topology **29** (1990) 441–460.

128. (with S. Edwards) Nets of quadrics and deformations of $\Sigma^{3(3)}$ singularities, Math. Proc. Camb. Phil. Soc. **105** (1989) 109–115.

129. Elliptic complete intersection singularities, pp. 340–372 in *Singularity theory and its applications. Warwick, 1989, I.,* ed. D. Mond and J. Montaldi, Lecture Notes in Math. **1462**, Springer (1991).

130. Pencils of cubics and rational elliptic surfaces, pp. 373–405 in *Singularity theory and its applications. Warwick, 1989, I.*, ed. D. Mond and J. Montaldi, Lecture Notes in Math. **1462**, Springer (1991).

131. Root systems, subsystems and singularities, Jour. Alg. Geom. **1** (1992) 597–638.

132. Is every quartic a conic of conics? Math. Proc. Camb. Phil. Soc. **109** (1991) 419–424.

133. (with A.A. du Plessis) Topological stability, pp. 351–362 in *Singularities, Lille 1991*, ed. J.-P. Brasselet, London Math. Soc. Lecture Notes **201**, Cambridge University Press, 1994.

134. Weighted homogeneous complete intersections, pp. 277–300 in *Algebraic geometry and singularities* (proceedings of conference at La Rabida 1991), ed. A. Campillo López and L. Narváez Macarro, Progress in Math. **134**, Birkhäuser, 1996.

135. Classification and stability of singularities of smooth maps, pp. 920–952 in *Singularity Theory* (Proceedings of College on Singularities at Trieste 1991), ed. D.T. Lê, K. Saito and B. Teissier, World Scientific, 1995.

136. Duality of singular plane curves, Jour. London Math. Soc., **50** (1994) 265–275.

137. Quartic curves in characteristic 2, Math. Proc. Camb. Phil. Soc. **117** (1995) 393–414.

138. Geometry of quartic curves, Math. Proc. Camb. Phil. Soc., **117** (1995) 415–423.

139. Highly singular quintic curves, Math. Proc. Camb. Phil. Soc., **119** (1996) 257–277.

140. Duality of real projective plane curves: Klein's equation, Topology, **35** (1996) 355–362.

141. Real rational quartic curves, pp. 1–32 in *Real and complex singularities*, ed. W.L. Marar, Pitman Research Notes in Math. **333**, Longman, 1995.

142. Pencils of binary quartics, Rend. Sem. Mat. Univ. Padova **99** (1998) 197–217.

143. (with A.A. du Plessis) Discriminants and vector fields, pp. 119–140 in *Singularities (Oberwolfach 1996)*, Progr. Math. **162**, Birkhäuser, 1998.

144. (with A.A. du Plessis) Versal deformations in spaces of polynomials with fixed weight, Compositio Math. **114** (1998) 113–124.

145. (with A.A. du Plessis) Applications of the theory of the discriminant to highly singular plane curves, Math. Proc. Camb. Phil. Soc. **126** (1999) 259–266.

146. Newton polytopes and non-degeneracy, J. reine angew. Math. **509** (1999) 1–19.

# Classification of $(n-1)$-connected $2n$-dimensional manifolds and the discovery of exotic spheres

## John Milnor

At Princeton in the fifties I was very much interested in the fundamental problem of understanding the topology of higher dimensional manifolds. In particular, I focussed on the class of $2n$-dimensional manifolds which are $(n-1)$-connected, since these seemed like the simplest examples for which one had a reasonable hope of progress. (Of course the class of manifolds with the homotopy type of a sphere is even simpler. However the generalized Poincaré problem of understanding such manifolds seemed much too difficult: I had no idea how to get started.) For a closed $2n$-dimensional manifold $M^{2n}$ with no homotopy groups below the middle dimension, there was a considerable body of techniques and available results to work with. First, one could easily describe the homotopy type of such a manifold. It can be built up (up to homotopy type) by taking a union of finitely many $n$-spheres intersecting at a single point, and then attaching a $2n$-cell $e^{2n}$ by a mapping of the boundary $\partial e^{2n}$ to this union of spheres, so that

$$M^{2n} \simeq (S^n \vee \cdots \vee S^n) \cup_f e^{2n} \ .$$

Here the attaching map $f$ represents a homotopy class in $\pi_{2n-1}(S^n \vee \cdots \vee S^n)$, a homotopy group that one can work with effectively, at least in low dimensions. Thus the homotopy theory of such manifolds is under control. We can understand this even better by looking at cohomology. The cohomology of such an $M^{2n}$, using integer coefficients, is infinite cyclic in dimension zero, free abelian in the middle dimension with one generator for each of the spheres, and is infinite cyclic in the top dimension where we have a cohomology class corresponding to this top dimensional cell; that is

$$H^0(M^{2n}) \cong \mathbf{Z} \ , \qquad H^n(M^{2n}) \cong \mathbf{Z} \oplus \cdots \oplus \mathbf{Z} \ , \qquad H^{2n}(M^{2n}) \cong \mathbf{Z} \ .$$

---

Taken from the lecture 'Growing Up in the Old Fine Hall' given on 20th March, 1996, as part of the Princeton 250th Anniversary Conference [9]. For accounts of exotic spheres, see [1]–[4], [6]. The classification problem for $(n-1)$-connected $2n$-dimensional manifolds was finally completed by Wall [10], making use of exotic spheres.

The attaching map $f$ determines a cup product operation: To any two co-homology classes in the middle dimension we associate a top dimensional cohomology class, or in other words (if the manifold is oriented) an integer. This gives a bilinear pairing from $H^n \otimes H^n$ to the integers. This pairing is symmetric if $n$ is even, skew-symmetric if $n$ is odd, and always has determinant $\pm 1$ by Poincaré duality. For $n$ odd this pairing is an extremely simple algebraic object. However for $n$ even such symmetric pairings, or equivalently quadratic forms over the integers, form a difficult subject which has been extensively studied. (See [7], and compare [5].) One basic invariant is the *signature*, computed by diagonalizing the quadratic form over the real numbers, and then taking the number of positive entries minus the number of negative entries.

So far this has been pure homotopy theory, but if the manifold has a differentiable structure, then we also have characteristic classes, in particular the Pontrjagin classes in dimensions divisible by four,

$$(1) \qquad\qquad p_i \in H^{4i}(M) \ .$$

This was the setup for the manifolds that I was trying to understand as a long term project during the 50's. Let me try to describe the state of knowledge of topology in this period. A number of basic tools were available. I was very fortunate in learning about cohomology theory and the theory of fiber bundles from Norman Steenrod, who was a leader in this area. These two concepts are combined in the theory of characteristic classes [8], which associates cohomology classes in the base space to certain fiber bundles. Another basic tool is obstruction theory, which gives cohomology classes with coefficients in appropriate homotopy groups. However, this was a big sticking point in the early 50's because although one knew very well how to work with cohomology, no one had any idea how to compute homotopy groups except in special cases: most of them were totally unknown. The first big breakthrough came with Serre's thesis, in which he developed an algebraic machinery for understanding homotopy groups. A typical result of Serre's theory was that the stable homotopy groups of spheres

$$\Pi_n \ = \ \pi_{n+k}(S^k) \qquad (k > n + 1)$$

are always finite. Another breakthrough in the early 50's came with Thom's cobordism theory. Here the basic objects were groups whose elements were equivalence classes of manifolds. He showed that these groups could be computed in terms of homotopy groups of appropriate spaces. As an immediate consequence of his work, Hirzebruch was able to prove a formula which he had conjectured relating the characteristic classes of manifolds to the signature. For any closed oriented $4m$-dimensional manifold, we can form the signature of the cup product pairing

$$H^{2m}(M^{4m}; \mathbf{R}) \otimes H^{2m}(M^{4m}; \mathbf{R}) \;\to\; H^{4m}(M^{4m}; \mathbf{R}) \;\cong\; \mathbf{R},$$

using real coefficients. If the manifold is differentiable, then it also has Pontrjagin classes (1). Taking products of Pontrjagin classes going up to the top dimension we build up various *Pontrjagin numbers*. These are integers which depend on the structure of the tangent bundle. Hirzebruch conjectured a formula expressing the signature as a rational linear combination of the Pontrjagin numbers. For example

$$(2) \qquad \text{signature}\,(M^4) \;=\; \frac{1}{3} p_1 [M^4]$$

and

$$(3) \qquad \text{signature}\,(M^8) \;=\; \frac{1}{45}(7p_2 - (p_1)^2)[M^8]\,.$$

Everything needed for the proof was contained in Thom's cobordism paper, which treated these first two cases explicitly, and provided the machinery to prove Hirzebruch's more general formula.

These were the tools which I was trying to use in understanding the structure of $(n-1)$-connected manifolds of dimension $2n$. In the simplest case, where the middle Betti number is zero, these constructions are not very helpful. However in the next simplest case, with just one generator in the middle dimension and with $n = 2m$ even, they provide quite a bit of structure. If we try to build up such a manifold, as far as homotopy theory is concerned we must start with a single $2m$-dimensional sphere and then attach a cell of dimension $4m$. The result is supposed to be homotopy equivalent to a manifold of dimension $4m$:

$$S^{2m} \cup e^{4m} \;\simeq\; M^{4m}\,.$$

What can we say about such objects? There are certainly known examples; the simplest is the complex projective plane in dimension four – we can think of that as a 2-sphere (namely the complex projective line) with a 4-cell attached to it. Similarly in dimension eight there is the quaternionic projective plane which we can think of as a 4-sphere with an 8-cell attached, and in dimension sixteen there is the Cayley projective plane which has similar properties. (We have since learned that such manifolds can exist only in these particular dimensions.)

Consider a smooth manifold $M^{4m}$ which is assumed to have a homotopy type which can be described in this way. What can it be? We start with a $2m$-dimensional sphere $S^{2m}$, which is certainly well understood. According to Whitney, this sphere can be smoothly embedded as a subset $S^{2m} \subset M^{4m}$ generating the middle dimensional homology, at least if $m > 1$. We look at a tubular neighborhood of this embedded sphere, or equivalently at its normal $2m$-disk bundle $E^{4m}$. In general this must be twisted as we go around the sphere — it can't be simply a product or the manifold wouldn't have the right properties. In terms of fiber bundle theory, we can look at it in the following way: Cut the $2m$-sphere into two hemispheres $D_+^{2m}$ and $D_-^{2m}$, intersecting along their common boundary $S^{2m-1}$. Over each of these hemispheres we must have a product bundle, and we must glue these two products together to form

$$E^{4m} \;=\; (D_+^{2m} \times D^{2m}) \cup_F (D_-^{2m} \times D^{2m}) \,.$$

Here the gluing map $F(x,y) = (x, f(x)y)$ is determined by a mapping $f : S^{2m-1} \to \mathrm{SO}(2m)$ from the intersection $D_+^{2m} \cap D_-^{2m}$ to the rotation group of $D^{2m}$. Thus the most general way of thickening the $2m$-sphere can be described by an element of the homotopy group $\pi_{2m-1}\mathrm{SO}(2m)$. In low dimensions, this group was well understood.

In the simplest case $4m = 4$, we start with a $D^2$-bundle over $S^2$ determined by an element of $\pi_1\mathrm{SO}(2) \cong \mathbf{Z}$. It is not hard to check that the only 4-manifold which can be obtained from such a bundle by gluing on a 4-cell is (up to orientation) the standard complex projective plane: This construction does not give anything new. The next case is much more interesting. In dimension eight we have a $D^4$-bundle over $S^4$ which is described by an element of $\pi_3(\mathrm{SO}(4))$. Up to a 2-fold covering, the group $\mathrm{SO}(4)$ is just a Cartesian product of two 3-dimensional spheres, so that $\pi_3\mathrm{SO}(4) \cong \mathbf{Z} \oplus \mathbf{Z}$. More explicitly, identify $S^3$ with the unit 3-sphere in the quaternions. We get one mapping from this 3-sphere to itself by left multiplying by an arbitrarily unit quaternion and another mapping by right multiplying by an arbitrary unit quaternion. Putting these two operations together, the most general $(f) \in \pi_3(\mathrm{SO}(4))$ is represented by the map $f(x)y = x^i y x^j$, where $x$ and $y$ are unit quaternions and where $(i, j) \in \mathbf{Z} \oplus \mathbf{Z}$ is an arbitrary pair of integers.

Thus to each pair of integers $(i, j)$ we associate an explicit 4-disk bundle over the 4-sphere. We want this to be a tubular neighborhood in a closed 8-dimensional manifold, which means that we want to be able to attach a 8-dimensional cell which fits on so as to give a smooth manifold. For that to work, the boundary $M^7 = \partial E^8$ must be a 7-dimensional sphere $S^7$. The question now becomes this: For which $i$ and $j$ is this boundary isomorphic to $S^7$? It is not difficult to decide when it has the right homotopy type: In fact $M^7$ has the homotopy type of $S^7$ if and only if $i + j$ is equal to $\pm 1$. To fix our ideas, suppose that $i + j = +1$. This still gives infinitely many choices of $i$. For each choice of $i$, note that $j = 1 - i$ is determined, and we get as boundary a manifold $M^7 = \partial E^8$ which is an $S^3$-bundle over $S^4$ having the homotopy type of $S^7$. Is this manifold $S^7$, or not?

Let us go back to the Hirzebruch-Thom signature formula (3) in dimension 8. It tells us that the signature of this hypothetical 8-manifold can be computed from $(p_1)^2$ and $p_2$. But the signature has to be $\pm 1$ (remember that the quadratic form always has determinant $\pm 1$), and we can choose the orientation so that it is $+1$. Since the restriction homomorphism maps $H^4(M^8)$ isomorphically onto $H^4(S^4)$, the Pontrjagin class $p_1$ is completely determined by the tangent bundle in a neighborhood of the 4-sphere, and hence by the integers $i$ and $j$. In fact it turns out that $p_1$ is equal to $2(i - j) = 2(2i - 1)$ times a generator of $H^4(M^8)$, so that $p_1^2[M^8] = 4(2i-1)^2$. We have no direct way of computing $p_2$, which depends on the whole manifold. However, we can solve equation (3) for $p_2[M^8]$, to obtain the formula

$$(4) \qquad p_2[M^8] \;=\; \frac{p_1{}^2[M^8] + 45}{7} \;=\; \frac{4(2i - 1)^2 + 45}{7} \, .$$

For $i = 1$ this yields $p_2[M^8] = 7$, which is the correct answer for the quaternion projective plane. But for $i = 2$ we get $p_2[M^8] = \frac{81}{7}$, which is impossible! Since $p_2$ is a cohomology class with integer coefficients, this Pontrjagin number $p_2[M^8]$, whatever it is, must be an integer.

What can be wrong? If we choose $p_1$ in such a way that (4) does not give an integer value for $p_2[M^8]$, then there can be no such differentiable manifold. The manifold $M^7 = \partial E^8$ certainly exists and has the homotopy type of a 7-sphere, yet we cannot glue an 8-cell onto $E^8$ so as to obtain a smooth manifold. What I believed at this point was that such an $M^7$ must be a counterexample to the seven dimensional Poincaré hypothesis: I thought that $M^7$, which has the homotopy type of a 7-sphere, could not be homeomorphic to the standard 7-sphere.

Then I investigated further and looked at the detailed geometry of $M^7$. This manifold is a fairly simple object: an $S^3$-bundle over $S^4$ constructed in an explicit way using quaternionic multiplication. I found that I could actually prove that it was homeomorphic to the standard 7-sphere, which made the situation seem even worse! On $M^7$, I could find a smooth real-valued function which had just two critical points: a non-degenerate maximum point and a non-degenerate minimum point. The level sets for this function are 6-dimensional spheres, and by deforming in the normal direction we obtain a homeomorphism between this manifold and the standard $S^7$.

(This is a theorem of Reeb: if a closed $k$-manifold possesses a Morse function with only two critical points, then it must be homeomorphic to the $k$-sphere.) At this point it became clear that what I had was not a counterexample to the Poincaré hypothesis as I had thought. This $M^7$ really was a topological sphere, but with a strange differentiable structure.

There was a further surprising conclusion. Suppose that we cut this manifold open along one of the level sets, so that

$$M^7 = D_+^7 \cup_f D_-^7 \, ,$$

where the $D_\pm^7$ are diffeomorphic to 7-disks . These are glued together along their boundaries by some diffeomorphism $g : S^6 \to S^6$. *Thus this manifold $M^7$ can be constructed by taking two 7-dimensional disks and gluing the boundaries together by a diffeomorphism.* Therefore, at the same time, the proof showed that there is a diffeomorphism from $S^6$ to itself which is essentially exotic: It cannot be deformed to the identity by a smooth isotopy, because if it could then $M^7$ would be diffeomorphic to the standard 7-sphere, contradicting the argument above.

## References

[1] M. Kervaire and J. Milnor, *Groups of homotopy spheres I*, Ann. Math. 77, 504–537 (1963)

[2] A. Kosinski, "Differential Manifolds", Academic Press (1993)

[3] T. Lance, *Differentiable structures on manifolds*, (in this volume)

[4] J. Milnor, *On manifolds homeomorphic to the 7-sphere*, Ann. of Math. 64, 399–405 (1956)

[5] —— *On simply connected 4-manifolds*, pp. 122–128 of "Symposium Internacional de Topologia Algebraica", UNAM and UNESCO, Mexico (1958)

[6] —— "Collected Papers 3, Differential Topology", Publish or Perish (in preparation)

[7] —— and D. Husemoller, "Symmetric Bilinear Forms", Springer (1973)

[8] —— and J. Stasheff, "Characteristic Classes", Ann. Math. Stud. 76, Princeton (1974)

[9] H. Rossi (ed.), *Prospects in Mathematics: Invited Talks on the Occasion of the 250th Anniversary of Princeton University*, Amer. Math. Soc., Providence, RI, 1999.

[10] C. T. C. Wall, *Classification of $(n-1)$-connected $2n$-manifolds*, Ann. of Math. 75, 163–189 (1962)

DEPARTMENT OF MATHEMATICS
SUNY AT STONY BROOK
STONY BROOK, NY 11794-3651
*E-mail address*: jack@math.sunysb.edu

# Surgery in the 1960's

S. P. Novikov

## Contents

## §1. Introduction

I began to learn topology in 1956, mostly from Mikhail Mikhailovich Postnikov and Albert Solomonovich Schwartz (so by the end of 1957 I already knew much topology). I remember very well how Schwartz announced a lecture in 1957 in the celebrated "Topological Circle" of Paul Alexandrov entitled something like "On a differentiable manifold which does not admit a differentiable homeomorphism to $S^7$". Some topologists even regarded this title as announcing the discovery of non-differentiable manifolds homeomorphic to $S^7$. In particular, this happened in the presence of my late father, who remarked that such a result contradicted his understanding of the basic definitions! I told him immediately that he was right, and such an interpretation was nonsense. In fact, it was an exposition of the famous discovery of the exotic spheres by Milnor ([6], [8]). At that time my interests were far from this subject. Until 1960 I worked on the calculations of the homotopy and cobordism groups. Only in 1960 did I begin to learn the classification theory of exotic spheres, from the recent preprint of Milnor [7] (which was probably given to me by Rochlin, who had already become my great friend). I was also very impressed by the new work of Smale [24] on the generalized Poincaré conjecture and the $h$-cobordism theorem, which provided the foundation of the classification. The beautiful work of Kervaire on a 10-dimensional manifold without a differentiable structure [4] completed the list of papers which stimulated me to work in this area. (Let me point out that the final work of Kervaire and Milnor [5] appeared later, in 1963.) Milnor, Hirzebruch and Smale personally influenced me when they visited the Soviet Union during the summer of 1961. They already knew my name from my work on cobordism.

## §2. The diffeomorphism classification of simply connected manifolds

I badly wanted to contribute to the diffeomorphism classification of manifolds. The very first conclusion I drew from Milnor's theory was that it was necessary to work with $h$-cobordisms instead of diffeomorphisms. My second guess was that there should exist some cobordism-type theory solving the diffeomorphism (or $h$-cobordism) classification problem. In the specific case of homotopy spheres these types of arguments had already been developed, using Pontrjagin's framed cobordism, reducing the problem to the calculation of the homotopy groups of spheres. However, it was not clear how to extend this approach to more complicated manifolds. Some time during the autumn of 1961, I observed some remarkable homotopy-theoretic properties of the maps of closed $n$-dimensional manifolds $f : M_1 \to M$ of degree 1: I saw that the manifold $M_1$ is really homologically bigger than $M$. The homology of $M_1$ splits as

$$H_*(M_1) = H_*(M) \oplus K_*$$

with the kernel groups

$$K_* = \ker(f_* : H_*(M_1) \to H_*(M))$$

satisfying ordinary Poincaré duality and a Hurewicz theorem connecting it with $\ker(f_* : \pi_*(M_1) \to \pi_*(M))$ in the first non-trivial dimension. It may well be that the idea of studying degree 1 maps came to me from my connections with my friends who were learning algebraic geometry at that time. Algebraic geometers call algebraic maps of degree 1 *birational equivalences*. It is well known to them that the manifold $M_1$ is bigger than $M$ in many senses. This kernel $K_*$ became geometrically analogous to an almost parallelizable manifold in the case when the stable tangent bundles $\tau_M \oplus 1$, $\tau_{M_1} \oplus 1$ of $M, M_1$ are in the natural agreement generated by the map $f$

$$f^*(\tau_M \oplus 1) = \tau_{M_1} \oplus 1 .$$

We call such maps (tangential) *normal maps* of degree 1. It is technically better and geometrically more clear to work with the stable normal $(N - n)$-bundles of manifolds embedded in high-dimensional Euclidean space $\mathbb{R}^N : M \subset E \subset \mathbb{R}^N \subset S^N$ with $E$ a small $\epsilon$-neighbourhood of $M$ in $\mathbb{R}^N$. Then $E$ is the total space of the normal bundle $\nu$, and that collapsing the complement of $E$ in $S^N$ to a point gives a mapping $S^N$ to the so-called Thom space

$$T(\nu) = E/\partial E = S^N/(S^N - E) .$$

Therefore, we have a preferred element of the homotopy group $\pi_N(T(\nu))$ associated with the manifold $M$. More generally, any normal map $f : M_1 \to M$ determines a specific element of the group $\pi_N(T(\nu))$ associated to $M$. This is a special case of the Thom construction. By Serre's theorem this homotopy group is

$$\pi_N(T(\nu)) = \mathbb{Z} \oplus A$$

where $A$ is a finite abelian group. The set of preferred elements of all the normal maps $f : M_1 \to M$ is precisely the finite subset

$$\{1\} \oplus A \subset \pi_N(T(\nu)) = \mathbb{Z} \oplus A .$$

It turned out that Milnor's surgery technique can be applied here. Instead of killing all the homotopy groups like in the case of the exotic spheres, one can try to kill just the kernels $K_*$. It worked well. Finally, I proved two theorems [10]:

(i) For dimensions $n \geq 5$ with $n \neq 4k+2$ any preferred element of $\pi_N(T(\nu))$ can be realized by a normal map $f : M_1 \to M$ with $M_1$ homotopy equivalent to $M$, and that for $n = 4k + 2$ there is an obstruction (Pontrjagin for $n = 2, 6, 14$, Kervaire-Arf invariant in general) and the realizable subset might be either the full set of preferred elements or subset of index 2.

(ii) If two (tangential) normal maps $f_1 : M_1 \to M$, $f_2 : M_2 \to M$ are already homotopy equivalences and represent the same element in $\pi_N(T(\nu))$ then $f_1, f_2$ are normal bordant by an $h$-cobordism, and for $n \geq 5$ the corresponding manifolds $M_1, M_2$ are canonically diffeomorphic for $n$ even, and diffeomorphic modulo special exotic Milnor spheres (those which bound parallelizable manifolds) for $n$ odd. This difference in behaviour between the odd and even dimensions disappears in the $PL$ category.

This gave some kind of diffeomorphism classification of a class of simply-connected manifolds normally homotopy equivalent to the given one. Of course, the automorphism and inertia groups (and so on) have to be taken into account, so the final classification will be given by the factorized set. One of the most important general consequences of this result was that the homotopy type and rational Pontrjagin classes determine the diffeomorphism class of a differentiable simply-connected manifold of dimension $\geq 5$ up to a finite number of possibilities. This is also true in dimension 4, replacing diffeomorphism by $h$-cobordism.

Soon afterwards I observed that it was not necessary to start with the normal bundle of $M \subset \mathbb{R}^N$ : this could have been replaced by any bundle such that the fundamental class of the Thom space is spherical for $n$ odd. For $n = 4k$ it is also necessary to ensure that the signature be given by the $\mathcal{L}$-genus, in accordance with the Hirzebruch formula expressing the signature in terms of the Pontrjagin classes. As before, for $n = 4k + 2$ it is necessary to deal with the Arf invariant, and such difficulties disappear in the *PL* category. This gave also a characterization of the normal bundles of manifolds homotopy equivalent to the given one ([9]). It turned out that completely independently, Bill Browder discovered the significance of maps of degree 1 ([1]). He had obtained the same characterization of the normal bundles of simply-connected manifolds, in a more general form. In his theorem the original manifold was replaced by a Poincaré complex, giving therefore the final solution to the problem of the recognition of the homotopy types of differentiable simply-connected manifolds.

## §3. The role of the fundamental group in homeomorphism problems

For a long time I thought that algebraic topology cannot deal with continuous homeomorphisms: the quantities such as homology, homotopy groups, Stiefel-Whitney classes whose topological invariance was already established, were in fact homotopy invariants. The only exception was Reidemeister torsion, which was definitely not a homotopy invariant, not even for 3-dimensional lens spaces. The topological invariance of Reidemeister torsion for 3-dimensional manifolds was obtained as a corollary of the Hauptvermutung proved by Moïse: every homeomorphism of 3-dimensional manifolds can be approximated by a piecewise linear homeomorphism, using an elementary approximating technique. It was unrealistic to expect something like that in higher dimensions. Indeed, Rochlin pointed out to me that he in fact had proved the topological invariance of the first Pontrjagin class $p_1$ of 5-dimensional manifolds in his work [22] preceding the well-known result of Thom [25] and Rochlin-Schwartz [23] on the combinatorial invariance of the rational Pontrjagin classes. I decided to analyze the codimension 1 problem, with the intention of proving that Rochlin's theorem is inessential, and that $p_1$ of a 5-dimensional manifold is in fact a homotopy invariant. This program was successfully realized: I found a beautiful formula for the codimension 1 Pontrjagin-Hirzebruch class $\mathcal{L}_k(p_1, \ldots, p_k)$ of a $4k + 1$-dimensional manifold $M$ involving the infinite cyclic covering $p : \widehat{M}_z \to M$ associated with an indivisible codimension 1 cycle $z \in H_{4k}(M) = H^1(M)$. Let $\widehat{z} \in H_{4k}(\widehat{M}_z)$ be the canonical cycle such that $p_*(\widehat{z}) = z$. Consider the symmetric scalar product on the infinite dimensional real vector space $H^{2k}(\widehat{M}_z; \mathbb{R})$

$$\langle x, y \rangle = \int_{\widehat{z}} x \wedge y .$$

The radical of this form has finite codimension, and therefore the signature $\tau(z)$ of this quadratic form is well-defined. The actual formula is

$$\langle \mathcal{L}_k(p_1, \ldots, p_k), z \rangle = \tau(z) .$$

Let me point out that this expression made essential use of the fundamental group and coverings. After finishing this theorem, a strange idea occurred to me, that a Grothendieck-type approach could be used in the study of continuous homeomorphisms. I had in mind the famous idea of Grothendieck defining the homology of algebraic varieties over fields of finite characteristic. He invented the important idea of the étale topology, using category of coverings over open sets rather than just the open sets themselves. I invented the specific category of *toroidal* open sets in the closed differentiable manifold:

$$M^{4k} \times T^{n-4k-1} \times \mathbb{R} \subset M^{4k} \times \mathbb{R}^{n-4k} \subset M^n .$$

Changing the differentiable structure in the manifold we still have the same open sets. However, I managed to prove that the rational Pontrjagin-Hirzebruch classes can be completely defined through the homotopy structure of these toroidal open sets, studying sequences of infinite cyclic coverings over them.

My first proof for codimension 2 was purely homological: it was already enough to establish the existence of homotopy equivalent but non-homeomorphic simply-connected closed manifolds (the Hurewicz problem) [11], [14]. I was not able to solve the general problem in this way: recently, this idea was completely realized by M. Gromov who added a beautiful homological argument of a new type ([3]). I found the full proof of the topological invariance of the rational Pontrjagin classes based on the technique of differential topology, including surgery theory, applied to non-simply-connected manifolds ([12], [15], [20]). To be more specific, it was necessary to deal with manifolds with a free abelian fundamental group, even for the proof of the topological invariance of the rational Pontrjagin classes for simply-connected manifolds.

It thus became clear for me that some special rational Pontrjagin classes (the so-called Pontrjagin-Hirzebruch classes $\mathcal{L}_k$) have a deep connection with the homotopy structure of non-simply-connected manifolds. The homeomorphism problems used only the specific case of the free abelian fundamental group. (This approach was extended in 1968 by Kirby, leading to the final solution of several fundamental problems of continuous topology.) This led to the so-called *higher signature conjecture*, elaborated in the process of interaction with V. Rochlin and G. Kasparov – one of my best students. (See [16], [17], [18]). The full formulation of the conjecture was found during the period 1965–1970, classifying all the homotopy invariant expressions from the Riemannian curvature tensor of a closed manifold $M$:

*if $x \in H^*(M; \mathbb{Q})$ is a cohomology class determined by the fundamental group $\pi_1(M)$ in the sense of Hopf-Eilenberg-MacLane then the integral of* *the Pontrjagin-Hirzebruch class* $\mathcal{L}_k(M)$ *along the cycle* $D(x)$ *is a homotopy invariant for all closed manifolds* $M$.

Important early results on this conjecture were obtained by Kasparov, Lusztig and Mishchenko. Much work has been done since then ([2]).

It became clear that the surgery technique can also be applied to the classification theory for non-simply-connected manifolds, extending the simply-connected case described in §1. However, in 1966 some important points were still unclear, especially in the odd dimensions. A breakthrough in this direction was made by C.T.C. Wall, whose preprint appeared in 1968 ([26]). I combined his results with my own ideas on the higher signature and with some algebraic and geometric material from the hamiltonian formalism of classical mechanics. Using this combination I started to construct the hermitian $K$-theory over any ring with involution. Ignoring 2-torsion this theory provided a nonstandard interpretation of Bott periodicity. My aim was also to formulate the higher signatures as some kind of Chern character in hermitian $K$-theory. It is interesting to point out that the standard algebraic $K$-theory of Quillen and others does not have Bott periodicity at all, and there are no analogs of the Chern character. This work was published in [17], and completed and developed further by Ranicki [21].

As a final remark I would like to mention that many years later, in 1981, I used my experience to work with free abelian coverings in the construction of the Morse-type theory on manifolds with closed 1-forms ([19]).

# Appendix

## Smooth manifolds of the same homotopy type
(Short communication to 1962 Stockholm ICM)
### Гладкие многообразия общего гомотопического типа.

Рассматриваются гладкие односвязные многообразия $M^n$, имеющие общий гомотопический тип и стабильный нормальный пучок. Устанавливается связь между этими многообразиями и гомотопической группой $\Pi_{N+n}(T_N)$, где $T_N$ – пространство Тома нормального пучка к многообразию. Извлекаются качественные следствия (конечностьчисла таких многообразий) и разбирается ряд примеров. Результаты применяются к следующим задачам:

1. Какой $SO(N)$-пучок над $M^n$ является нормальным для некоторого $M^n$, гомотопически эквивалентного $M^n$ и гладкого (быть может, $M^n$ комбинаторно).

2. В каком случае связная сумма $M^n \# \tilde{S}^n$ диффеоморфна (со степенью $+1$) многообразию $M^n$, где $\tilde{S}^n$ - нестандартная сфера Милнора.

Для установления связи между гомотопическими и дифференцируемыми задачами используются, с одной стороны, алгебраические методы, с другой стороны – $t$-регулярность, теория перестроек Морса и результаты Смейла-Уоллеса. При разборе конкретных примеров используется умножение в стабильных гомотопических группах сфер.

## English translation

Consider smooth simply-connected manifolds $M^n$, which have the same homotopy type and stable normal bundle. We establish a connection between such manifolds and the homotopy groups $\pi_{N+n}(T_N)$, where $T_N$ is the Thom space of the normal bundle of the manifold. Certain general corollaries are obtained (e.g. the finiteness of such manifold types) and a number of examples is discussed. The results are applied to the following problems:

1. Which $SO(N)$-bundles over $M^n$ (which could be combinatorial) come from the normal bundle of a homotopy equivalent smooth manifold?

2. In which case is the connected sum $M^n \# \tilde{S}^n$ diffeomorphic to $M^n$ (with degree 1), with $\tilde{S}^n$ an exotic Milnor sphere?

To establish the connection between homotopy theoretic and differentiable problems we use on the one hand algebraic methods, and on the other hand $t$-regularity, the theory of Morse surgery and the results of Smale-Wallace. In considering concrete examples, we use the multiplication in the stable homotopy groups of spheres.

## References

[1] W. Browder, *Homotopy type of differentiable manifolds*, in Proc. Århus Topology Conference (1962), reprinted in [2], Vol. 1, London Math. Soc. Lecture Notes 226, Cambridge Univ. Press 97–100 (1995)

[2] S. Ferry, A. Ranicki and J. Rosenberg (eds.), *Novikov Conjectures, Index Theorems and Rigidity*, Proceedings 1993 Oberwolfach Conference, Vols. 1, 2, London Math. Soc. Lecture Notes 226, 227, Cambridge Univ. Press (1995)

[3] M. Gromov, *Positive curvature, macroscopic dimension, spectral gaps*

*and higher signatures*, Functional Analysis on the Eve of the 21st Century, In Honor of the Eightieth Birthday of I.M. Gelfand, Vol. II, Progress in Mathematics 132, 1–213, Birkhäuser (1996)

[4] M. Kervaire, *A manifold which does not admit a differentiable structure*, Comm. Math. Helv. 34, 257–270 (1960)

[5] — — and J. Milnor, *Groups of homotopy spheres I.*, Ann. of Math. 77, 504–537 (1963)

[6] J. Milnor, *On manifolds homeomorphic to the 7-sphere*, Ann. of Math. 64, 399–405 (1956)

[7] — —, *A procedure for killing the homotopy groups of differentiable manifolds*, Proc. Symp. Pure Math. 3 (Differential Geometry), 39–55 (1961), Amer. Math. Soc.

[8] — —, *Classification of (n − 1)-connected 2n-dimensional manifolds and the discovery of exotic spheres*, in this volume.

[9] S. P. Novikov, *Smooth manifolds of common homotopy type*, Short Communication to Stockholm ICM (1962) (reprinted and translated in this article).

[10] — —, *Diffeomorphisms of simply-connected manifolds*, Dokl. Akad. Nauk SSSR 143 (1962), 1046–1049. English transl.: Soviet Math. Doklady 3, 540–543.

[11] — —, *Homotopic and topological invariance of certain rational classes of Pontrjagin*, Dokl. Akad. Nauk SSSR 162 (1965), 1248–1251. English transl.: Soviet Math. Doklady 6, 854–857.

[12] — —, *Topological invariance of rational Pontrjagin classes*, Dokl. Akad. Nauk SSSR 163 (1965), 298–300. English transl.: Soviet Math. Doklady 6, 921–923.

[13] — —, *Homotopy equivalent smooth manifolds I.*, Izv. Akad. Nauk SSSR, Ser. Mat. 28, 365–474 (1964). English transl.: Translations Amer. Math. Soc. 48, 271–396 (1965)

[14] — —, *Rational Pontrjagin classes. Homeomorphism and homotopy type of closed manifolds I.*, Izv. Akad. Nauk SSSR, Ser. Mat. 29, 1373–1388 (1965), English transl.: Translations Amer. Math. Soc. (2) 66, 214–230 (1968)

[15] — —, *On manifolds with free abelian fundamental group and applications (Pontrjagin classes, smoothings, high-dimensional knots)*, Izv. Akad. Nauk SSSR, Ser. Mat. 30, 208–246 (1966)

[16] — —, *Pontrjagin classes, the fundamental group and some problems of stable algebra*, Essays on Topology and Related Topics (Mémoires dédiés á Georges de Rham), 147–155, Springer (1970)

[17] — —, *The algebraic construction and properties of hermitian analogues of K-theory for rings with involution, from the point of view of the Hamiltonian formalism. Some applications to differential topology and the theory of characteristic classes.*, Izv. Akad. Nauk SSSR, Ser. Mat.

34, I. 253–288, II. 478–500 (1970)

[18] — —, *Analogues hermitiens de la K-théorie*, Proc. 1970 Nice I. C. M., Gauthier–Villars, Vol. 2, 39–45 (1971)

[19] — —, *The Hamiltonian formalism and a multivalued analogue of Morse theory*, Uspekhi Mat. Nauk 37, no. 5, 3–49, 248 (1982) English transl.: Russian Math. Surv. 37, no. 5, 1–56 (1982)

[20] — —, *Topology I*, Encyclopædia of Mathematical Sciences, Vol. 12, Springer (1996)

[21] A. Ranicki, *Algebraic L-theory*, Proc. Lond. Math. Soc. 27 (3), I. 101–125, II. 126–158 (1973)

[22] V. Rochlin, *On Pontrjagin characteristic classes*, Dokl. Math. Nauk SSSR 113, 276–279 (1957)

[23] — — and A. Schwartz, *The combinatorial invariance of Pontrjagin classes*, Dokl. Akad. Nauk SSSR 114, 490–493 (1957)

[24] S. Smale, *A survey of some recent developments in differential topology*, Bull. Amer. Math. Soc. 69, 131–145 (1963)

[25] R. Thom, *Les classes charactéristiques de Pontrjagin des variétés triangulées*, Symp. Int. Top. Alg. UNESCO, Mexico, 54–67 (1958)

[26] C. T. C. Wall, *Surgery on compact manifolds*, Academic Press (1970)

Institute for Physical Science and Technology
University of Maryland
College Park, MD 20742
email: novikov@ipst.umd.edu

# Differential topology
# of higher dimensional manifolds[*]

William Browder

Colloquium Lectures, January 1977

## Preface (written 1998)

These notes and the lectures they accompanied, were in no way intended
to be any sort of treatise, or even rough outline of the subject of surgery
theory at the time (1977). It was rather designed to give a little bit of the
flavor of the subject to as wide an audience as possible, while at the same
time giving an avenue, by way of the references, for someone to find their
way into the subject in a serious way.

The standard model for a colloquium lecture (or lecture series) in mathematics is that the first quarter is understandable by a general audience,
the second quarter by the specialists, the third quarter by only the speaker,
and the last quarter by absolutely no one.

While I have tried to avoid following this model too closely, there is
certainly an increasing level of technicality, as one proceeds through the
lectures. However, I hope that anyone with a first course in algebraic
topology under their belt, can follow the exposition and see the scenery, if
not the nuts and bolts of the detailed proofs (which are not given for the
most part).

In the first lecture, I gave proofs of very special cases of two theorems
from surgery theory, which are easy consequences of the general theorems
but have very simple direct proofs which illustrate much of the surgery
technique. In the second paragraph I tried to outline the contour of the
general machinery which had evolved, while in the third lecture I tried to
describe the important application to triangulation of manifolds, again in
the special simpler context of the Annulus Conjecture.

---

[*]These notes are a corrected reprinting of notes distributed in conjunction with the
Colloquium Lectures given at the 83rd annual meeting of the American Mathematical
Society, St. Louis, MO, January 26–30, 1977. Those notes were copyright © by the
American Mathematical Society, 1977, and are reprinted by permission of the Society.

My initial plan had been to use the final lecture for a discussion of the applications of surgery to the theory of compact transformation groups, but I did not write notes for this part. In the interim, between the writing of these notes and the delivery of the lectures, the work of Edwards on the "Double suspension theorem" took the limelight, and I devoted the last lecture instead to an account of a very special case, using a very simple argument found by Giffen, and of course no notes are given for this lecture.

The informal nature of these notes should not be forgotten, and let me emphasize again that the references constituted my attempt to offer a more detailed and formal picture of the subject.

I am very happy to be able to contribute this paper to a volume honoring Terry Wall. His central contributions to the theory of surgery are well known, and his name has justly found its way onto several of the most important objects of the theory.

# Introduction (written 1977)

The past twenty-five years have seen the birth and the flowering of the subject called "differential topology" which established a new point of view for solving geometric problems about manifolds and similar spaces. The new method reversed the point of view which had been so powerfully developed in the previous twenty years, in which the algebraic topological implications of geometric properties had been intensively investigated. The latter development had included the enormous development of homotopy theory, the theory of fibre bundles and characteristic classes, the theory of Whitehead torsion, and finally the cobordism theory of Thom, which marked the transition between the two eras.

The work of Thom studied a difficult geometrical question: given a closed manifold $M^m$, does there exist a manifold $W^{m+1}$ one dimension higher, whose boundary is $M$, $\partial W = M$? Thom showed that an algebraic condition on the characteristic classes in cohomology of the tangent bundle of $M$ was *equivalent* to the geometric condition, that one could pass from algebraic topological information to get existence of manifolds with certain properties. The principal geometric tool in his study was his Transversality Theorem, which we will describe soon.

The beginning of differential topology, as I will discuss it here, probably should be dated from the work of Milnor. In that work, on differential structures on spheres or on homotopy spheres, he began to ask the questions which created the point of view of this field, and introduced the methods of surgery which play such a central role. These lectures will be devoted to an outline of this point of view and method, with its applications to diverse problems in topology, and we will not try to encompass many other interesting developments in somewhat different directions, e.g., immersion theory and its developments, the Morse-Smale critical point theory and Cerf theory, foliation theory, etc. In fact, in four lectures it will only be possible to touch on a few chosen aspects of the surgery theory, and I hope to at least give the flavor of the subject, and suggest its scope.

# 1 Surgery theory and "classification" of manifolds

The objective of describing a complete set of invariants which determines a manifold up to homeomorphism, or diffeomorphism, is easily attained in dimensions less than three. For example, for closed 2-manifolds, the Euler Characteristic is a complete invariant. While this objective might still seem plausible for 3-manifolds, for manifolds of dimension 4 or more, the fundamental diversity of possibilities would make such a set of invariants inordinately complex. For example any finitely presented group occurs as the fundamental group of a 4-dimensional manifold, and the homeomorphism or diffeomorphism problem is at least as complicated as the isomorphism problem for such groups.

In fact let $K$ be a finite simplicial complex of dimension $k$. If we embed $K$ simplicially in $\mathbb{R}^n$, which is always possible for $n > 2k$, then a regular neighborhood $N$ of $K$ (which is a naturally defined closed simplicial neighborhood of $K$ homotopy equivalent to $K$) is a manifold with boundary of dimension $n$, and the homotopy groups of $\partial N$ in dimensions $< n - k$ are the same as those of $K$. Thus the homotopy type of an $n$-manifold is quite arbitrary in low dimensions, and it is reasonable to suppose that the "classification" of manifolds may be more difficult than homotopy classification of finite complexes.

The basic premise of surgery theory is that we fix a homotopy type, thus avoiding such questions, and study the manifolds of that homotopy type. We get two basic questions.

1. Given a space $X$, when is $X$ of the homotopy type of a manifold $M$ of dimension $n$?

2. How can one "classify" such manifolds $M$ with the given homotopy type?

The most obvious property of a compact closed oriented manifold from the point of view of algebraic topology is the Poincaré Duality theorem, and we make it into a definition.

*Definition.* Let $X$ be a finite complex, and suppose $H_n(X) \cong \mathbb{Z}$ with generator $[X] \in H_n(X)$. Also suppose $[X]$ defines a natural isomorphism $\mathcal{P} : H^q(X) \to H_{n-q}(X)$ for all $q$ (given by an algebraic map called "cap product"). We call such an $X$ a *Poincaré complex*, the choice of $[X] \in H_n(X)$ the *orientation*, $n =$ the *dimension* of $X$.

The explicit definition of the cap product defining the isomorphism is only relevant in the technical details so we omit it. Naturality is described by:

If $X, Y$ are Poincaré complexes of dimension $n$, $f : X \to Y$ a map such that $f_*([X]) = [Y]$, the diagram

$$
\begin{array}{ccc}
H^q(X) & \xleftarrow{\ f^* \ } & H^q(Y) \\
\downarrow{\scriptstyle \mathcal{P}} & & \downarrow{\scriptstyle \mathcal{P}} \\
H_{n-q}(X) & \xrightarrow{\ f_* \ } & H_{n-q}(Y)
\end{array}
$$

commutes, i.e., $\mathcal{P}(y) = f_* \mathcal{P} f^* y$ for $y \in H^q(Y)$.

Besides Poincaré duality, another necessary condition on a space $X$ to be homotopy equivalent to a closed manifold is the existence of a bundle which corresponds to the tangent bundle. It is more convenient to interpret the existence of this bundle in a different way, through its inverse bundle, that is, the normal bundle of an embedding of the manifold in Euclidean space, or the sphere.

If $M^m$ is a closed compact smooth manifold (dimension $m$), then Whitney proved in the 1930's an embedding theorem which implies that there is a smooth embedding $M^m \subset S^{m+k}$ ($S^{m+k}$ is the $(m+k)$-sphere in $\mathbb{R}^{m+k+1}$), provided $k \geq m$, and this embedding is unique up to isotopy if $k > m$. (An isotopy is a smooth family of such embeddings.)

For such a smooth embedding $M^m \subset S^{m+k}$, we can consider the tangent vectors to $S^{m+k}$ at a point of $M^m$, which are *normal* to $M^m$ in some chosen Riemannian metric. These vectors form a vector bundle, called the *normal bundle* of $M^m$ in $S^{m+k}$, denoted by $\nu_M$. If $\tau_M$ is the tangent bundle of $M$, then $\tau_M \oplus \nu_M$ is the bundle of tangents to $S^{m+k}$ at points of $M$, and is thus the product bundle (since $M \subset S^{m+k} - \{\infty\} = \mathbb{R}^{m+k}$). Now the set of normal vectors of length $< \varepsilon$ (for some small $\varepsilon > 0$) is diffeomorphic (using the exponential map) to an open neighborhood $U$ of $M^m$ in $S^{m+k}$. We can thus consider $U$ as the total space $E(\nu_M)$ of the vector bundle $\nu_M$, since the set of vectors of length $< \varepsilon$ is diffeomorphic to the set of all vectors. If one takes the one-point compactification of $E(\nu_M)$ we get a space $T(\nu_M)$ called the *Thom complex* of $\nu_M$. On the other hand the one-point compactification $U^*$ of $U$ is homeomorphic to $S^{m+k}/S^{m+k} - U$ (i.e., identifying the subspace to one point). Thus we have a natural map $\alpha \colon S^{m+k} \to S^{m+k}/S^{m+k} - U \cong T(\nu_M)$. One can calculate $H_*(T(\nu_M))$, and one sees that $H_{m+k}(T(\nu_M)) \cong \mathbb{Z}$ (if $M$ is oriented), and that $\alpha_*$ is an isomorphism $H_{m+k}(S^{m+k}) \to H_{m+k}(T(\nu_M))$.

The existence of such a bundle $\xi$ over $X$, with such a map $\alpha \colon S^{m+k} \to$

$T(\xi)$, is thus another necessary condition for $X$ to have the homotopy type of a smooth closed manifold.

These conditions are very close to being sufficient as well as necessary. For example, we have the following propositions developed by S. P. Novikov and myself:

**Proposition 1.1** *Let $X$ be a 1-connected Poincaré complex of dimension $n$, and $\xi^k$ a $k$-dimensional vector bundle over $X$, $k > n$, with a map $\alpha\colon S^{n+k} \to T(\xi)$ of degree 1.[1] If $n$ is odd, $n \geq 5$, then there is a homotopy of $\alpha$ to $\beta$ such that $\beta^{-1}(X) = M^m$ is a closed manifold, and $\beta|M\colon M \to X$ is a homotopy equivalence, and the normal bundle $\nu_M$ is mapped linearly into $\xi$ by $\beta$. If $n = 2k > 4$, there is an obstruction to achieving this conclusion, and if $k$ is even, this obstruction is calculable from the characteristic classes of $\xi$ and the index[2] of $M$.*

**Proposition 1.2** *Suppose $X$ is a 1-connected Poincaré complex of dimension $n$ and $\xi^k$ is a vector bundle over $X$, etc., as in Proposition 1.1. Then $X \times S^k$ for $k > 1$ is the homotopy type of a smooth manifold (and in fact the conclusion of Proposition 1.1 holds for $X \times S^k$).*

Examples of "uniqueness" theorems, analogous to these, can be easily given.

**Proposition 1.3** *Let $M$, $M'$ be closed simply connected manifolds of dimension $n \geq 5$. Let $f\colon M \to M'$ be a homotopy equivalence, $h\colon \nu_M \to \nu_{M'}$ a linear bundle map of the normal bundles of $M$, $M'$ in $S^{n+k}$, such that $h$ lies over the map $f$ of base spaces. Let $\alpha\colon S^{n+k} \to T(\nu_M)$, $\alpha'\colon S^{n+k} \to T(\nu_{M'})$ be the maps described above, and suppose $T(h)\alpha$ is homotopic to $\alpha'$. If $n$ is even, $f$ is homotopic to a diffeomorphism $f'\colon M \to M'$, while if $n$ is odd, $f$ is homotopic to $f'$ which is a diffeomorphism on the complement of a point.*

**Proposition 1.4** *Let $M$, $M'$ be as in Proposition 1.3. Then $f \times 1\colon M \times S^k \to M' \times S^k$ is homotopic to a diffeomorphism for $k > 1$.*

These four propositions give a hint of the quantities of geometric information contained in the normal bundle $\nu_M$ of $M \subset S^{n+k}$ and the map $\alpha\colon S^{n+k} \to T(\nu_M)$, and we will later describe the comprehensive theory that has emerged from this study. In the first lecture, I will outline some of the geometric tools used in this development, and outline the proof of a special case of Propositions 1.2 and 1.4.

---

[1]Editor's note: Degree 1 means that $\alpha_*$ sends the canonical generator of $H_{n+k}(S^{n+k})$ to the generator of $H_{n+k}(T(\xi))$ corresponding to $[X] \in H_n(X)$.

[2]Editor's note: This is also known as the signature. Its role will be explained in Lecture 2.

Let us consider the situation of Propositions 1.1 and 1.2, where $X$ is simply connected and satisfies Poincaré duality, $\xi^k$ is a $k$-plane bundle over $X$, and $\alpha: S^{n+k} \to T(\xi)$ is a map of degree 1. The first step is the:

**Thom Transversality Theorem.** *Let $\eta$ be a linear $k$-plane bundle over a compact space $X$, let $W^{n+k}$ be a smooth manifold and $g: W \to T(\eta)$. Then $g$ is homotopic (by an arbitrarily small homotopy) to a map $g'$ such that $g'$ is transversal to $X$, i.e., $g'^{-1}(X)$ is a smooth submanifold $M^n \subset W$ with normal bundle $\omega^k$, and $g'$ restricted to a bundle neighborhood of $X$ is a linear bundle map of $\omega$ onto $\eta$. Further, if $g|U$ is already transversal to $X$ for $U$ an open neighborhood of $A$ closed in $W$, the homotopy can be taken fixed on $A$.*

In our circumstances the theorem tells us that $\alpha: S^{n+k} \to T(\xi)$ is homotopic to $\alpha': S^{n+k} \to T(\xi)$ such that $\alpha'^{-1}(X) = N^n \subset S^{n+k}$ (with normal bundle $\nu_N$) and $\alpha'|E(\nu_N) = b$ is a linear bundle map of $\nu_N$ into $\xi$ covering $\alpha'|N = f: N \to X$ (where $E(\nu_N)$ is a bundle neighborhood of $N$ in $S^{n+k}$).

One can show, using Poincaré duality or the Thom isomorphism theorem, that $H_i(X) \overset{\iota}{\cong} H_{i+k}(T(\xi))$, and that $f_*[N] = \iota[X]$ ($N$ is oriented in a natural way).

We take this to be a definition:

A *smooth normal map* is a pair of maps $(f, b)$, $f: M \to X$, $M$ a smooth closed manifold, $X$ satisfying Poincaré duality, $b: \nu_M \to \xi$ is a linear bundle map lying over $f$, where $\nu_M$ is the normal bundle of $M \subset S^{n+q}$, $q > n = $ dimension $M$, $\xi$ a linear $q$-plane bundle over $X$, and $f_*[M] = \iota[X]$.

The map $f: N \to X$ created by the Thom Transversality Theorem is far from being a homotopy equivalence, so starting from $(f, b)$ we must get closer to a homotopy equivalence, by a *normal bordism*: a pair $(F, B)$ where $F: W \to X \times [0, 1]$, with $\partial W = M_0 \cup M_1$, $F|M_i \subset X \times \{i\}$, $B: \omega \to \xi \times [0, 1]$ (covering $F$), a linear bundle map, where $\omega$ is the normal bundle of $W \subset S^{n+q} \times [0, 1]$ with $M_i \subset S^{n+q} \times \{i\}$. If $(F, B)|(M_i, \nu_i) = (f_i, b_i)$, we say $(F, B)$ is a normal bordism of $(f_0, b_0)$ to $(f_1, b_1)$.

The Thom Transversality Theorem shows that normal maps and normal bordisms correspond to maps $\alpha: S^{n+q} \to T(\xi)$ and homotopies of such maps.

Now we try to construct a normal bordism of $(f, b)$ to make a map as close to a homotopy equivalence as possible. This construction is the process called *surgery*:

Let $M^m$ be a manifold (dimension $m$), and suppose we have an embedding $S^k \times D^{m-k} \subset$ interior $M$. Let $M_0 = M - (\text{interior } S^k \times D^{m-k})$ and define $M' = M_0 \cup (D^{k+1} \times S^{m-k-1})$ with $S^k \times S^{m-k-1}$ identified together in $M_0$ (as boundary of the embedded $S^k \times D^{m-k} \subset M$) and in $D^{k+1} \times S^{m-k-1}$ (as its boundary). The process of passing from $M$ to $M'$

is called surgery.

We note that if we define $W = (M \times [0,1]) \cup (D^{k+1} \times D^{m-k})$, identifying the $S^k \times D^{m-k} \subset$ boundary $(D^{k+1} \times D^{m-k})$ with $S^k \times D^{m-k} \times \{1\} \subset M \times [0,1]$, then $W$ is a manifold with boundary $\partial W = M \cup M' \cup (\partial M) \times [0,1]$, and is a bordism from $M$ to $M'$ if $M$ is closed $(\partial M = \emptyset)$. We call $W$ the *trace* of the surgery.

If $M$ is smooth, and $S^k \times D^{m-k}$ is smoothly embedded, then $W$ can be made smooth, etc.

We note that shrinking the $D^{m-k}$ down to a point in $D^{k+1} \times D^{m-k}$, or shrinking $M \times [0,1]$ down to $M \times \{0\}$, are deformation retractions, and we get

**Lemma 1.5** *$W$ is homotopy equivalent to $M \cup D^{k+1}$, with $D^{k+1}$ attached along $S^k \times \{0\} \subset M$.*

Now $M'$ has embedded in it $D^{k+1} \times S^{m-k-1} \subset \partial(D^{k+1} \times D^{m-k})$, so that one could do a surgery on $M'$ using this embedding, and clearly one would arrive back at $M$ ("the inverse surgery"). In fact it is not hard to see that the trace of this surgery is the same $W$, only considered from the opposite point of view, as a bordism from $M'$ to $M$. Thus it is also clear that

**Lemma 1.6** *$W$ is homotopy equivalent to $M' \cup D^{m-k}$, etc.*

Thus for surgery on a sphere of dimension $k < m - k - 1$, we get:

**Lemma 1.7** *$H_i(M') \cong H_i(W) \cong H_i(M)/(\varphi)$ for $i \leq k < m-k-1$, where $(\varphi)$ is the subgroup of $H_i(M)$ generated by the homology of the embedded manifold $S^k \times \{0\}$, so is non-zero only in dimension $k$. The same is true in homotopy.*

Thus surgery on low dimensional spheres produces bordisms which may make homotopy and homology smaller. Now consider a normal map $(f, b)$, $f: M^n \to X$, etc.

**Lemma 1.8** *If $\varphi: S^k \times D^{n-k} \to M^n$ is a smooth embedding, the map $f: M \to X$ extends to the bordism $F: W \to X \times [0,1]$ if and only if $f\varphi: S^k \times \{0\} \to X$ extends to $D^{k+1}$.*

This follows immediately from Lemma 1.5, and in that case it can be arranged for $F(M') \subset X \times \{1\}$, $F(M) \subset X \times \{0\}$.

Similarly:

**Lemma 1.9** *An extension $F: W \to X \times [0,1]$, as in Lemma 1.8, can be covered by a bundle map $B: \omega \to \xi \times [0,1]$ if and only if $b|(\nu_M|S^k)$ extends to $\omega|D^{k+1}$, where $\omega$ is the normal bundle to $W \subset S^{n+q} \times [0,1]$.*

Now the relative homotopy groups of the map $f: M \to X$ are represented by commutative diagrams:

$$
\begin{array}{ccc}
S^k & \xrightarrow{\;\alpha_0\;} & M^n \\
\cap\Big\downarrow & & \Big\downarrow f \\
D^{k+1} & \xrightarrow{\;\alpha_1\;} & X.
\end{array}
$$

The embedding theorem of Whitney tells us that if $k < \frac{n}{2}$, $\alpha_0$ is homotopic to an embedding, and $f \cup \alpha_1$ defines a map $g: M \cup D^{k+1} \to X$. Then $\nu_M$ extends to the pullback bundle $g^*(\xi)$ over $M \cup D^{k+1}$, and the embedding $E(\nu_M) \subset S^{n+q} \times \{0\}$ extends to an embedding of $g^*(\xi) \subset S^{n+q} \times [0,1]$, and the normal vectors to this bundle will define the trace $W$ of a surgery on $S^k \subset M$, which will define a normal bordism of $(f, b)$.

Now we can prove:

**Theorem 1.10** *Let* $(f, b)$ *be a normal map,* $f: M \to X$, *dimension* $M = n$. *Then* $(f, b)$ *is normally bordant to* $(f', b')$, *where* $f'$ *is* $\left[\frac{n}{2}\right]$-*connected* $\left(\left[\frac{n}{2}\right] = \right.$ *greatest integer* $\left. \leq \frac{n}{2}\right)$. *That is,* $f'_*: \pi_i(M') \to \pi_i(X)$ *is an isomorphism for* $i < \left[\frac{n}{2}\right]$ *and onto for* $i = \left[\frac{n}{2}\right]$.

Proof proceeds by induction on $k < \left[\frac{n}{2}\right]$, assuming $f: M \to X$ is $k$-connected. Consider $\pi_{k+1}(f)$, which fits in the exact sequence

$$
\ldots \to \pi_{k+1}(M) \xrightarrow{f_*} \pi_{k+1}(X) \to \pi_{k+1}(f) \to \pi_k(M) \xrightarrow{f_*} \pi_k(X) \to \ldots;
$$

elements of $\pi_{k+1}(f)$ are represented by homotopy classes of commutative diagrams:

$$
\begin{array}{ccc}
S^k & \xrightarrow{\;\alpha_0\;} & M \\
\cap\Big\downarrow & & \Big\downarrow f \\
D^{k+1} & \xrightarrow{\;\alpha_1\;} & X.
\end{array}
$$

By the Whitney embedding theorem and the above remarks we can find a normal cobordism $(F, B)$, $F: W \to X \times [0,1]$ of $(f, b)$ with $W$ homotopy equivalent to $M \cup_{\alpha_0} D^{k+1}$ (Lemma 1.5). Then $\pi_k(M') \cong \pi_k(M)/(\alpha_0)$ and one can easily show that $\pi_{k+1}(f') \cong \pi_{k+1}(f)/(\alpha_0, \alpha_1)$. From the fact that $X$ and $M$ are finite complexes one may deduce that $\pi_{k+1}(f)$ is finitely generated,[3] and thus in a finite number of such surgeries we may reduce $\pi_{k+1}(f)$ to 0.

---

[3] Editor's note: The author is assuming here that $X$ is simply connected. The problem if this assumption is dropped is that there are finite complexes with homotopy groups that are not finitely generated, e.g., $S^1 \vee S^2$.

We note that several things go wrong with this technique if $k > \left[\frac{n}{2}\right] - 1$. First, the Whitney embedding theorem will fail in this form in case $\pi_1 M \neq 0$, and the relation between homology and homotopy of $M$ and of $M'$ is no longer so simple. However, from Theorem 1.10, the most elementary result of surgery, we can deduce Proposition 1.2 in the special case where $n$ is even.

First we note some consequences of Poincaré duality:

**Lemma 1.11** *Let $f: M \to X$ be a map of degree 1 $(f_*[M] = [X])$, where $M$, $X$ satisfy Poincaré duality in dimension $n$. Then there are maps $\beta_*$ : $H_*(X) \to H_*(M)$, $\beta^* : H^*(M) \to H^*(X)$, such that $f_*\beta_* = 1_{H_*(X)}$ and $\beta^*f^* = 1^{H_*(X)}$, so that*

$$H_*(M) \cong H_*(X) \oplus K_*(f),$$
$$H^*(M) \cong H^*(X) \oplus K^*(f),$$

$K_*(f) = \ker f_*$, $K^*(f) = \ker \beta^*$, *Further, $K^*$ and $K_*$ satisfy Poincaré duality algebraically.*

This follows immediately from naturality of Poincaré duality, using the commutative diagram:

$$
\begin{array}{ccc}
H^*(M) & \xleftarrow{\;f^*\;} & H^*(X) \\
\scriptstyle\mathcal{P}\downarrow & & \downarrow\scriptstyle\mathcal{P} \\
H_*(M) & \xrightarrow[\;f_*\;]{} & H_*(X).
\end{array}
$$

Define $\beta_* = \mathcal{P}f^*\mathcal{P}^{-1}$, $\beta_* = \mathcal{P}^{-1}f_*\mathcal{P}$.

If $(f, b)$ is a normal map, $f: M \to X$, dimension $= 2\ell$, and $f$ is $\ell$-connected (which we may assume by Theorem 1.10), then $K_i(f) = 0$ for $i < \ell$, so that $K^i(f) = 0$ for $i > \ell$, by Poincaré duality. Since $H_*$ and $H^*$ satisfy the Cohomology Universal Coefficient Theorem:

$$0 \to \mathrm{Ext}\,(H_{i-1}(X); \mathbb{Z}) \to H^i(X; \mathbb{Z}) \to \mathrm{Hom}\,(H_i(X); \mathbb{Z}) \to 0,$$

it follows that $K_i$ and $K^i$ are related by such an exact sequence, so that $K^i = 0$ for $i = \ell + 1$ implies that $\mathrm{Ext}\,(K_\ell; \mathbb{Z}) = 0$, which implies $K_\ell$ has no torsion, and is a free $\mathbb{Z}$-module. Similarly, we get $K_i = 0$ for $i > \ell$, so $K_i \neq 0$ only for $i = \ell$, and is free over $\mathbb{Z}$.

**Lemma 1.12** *Let $(f, b)$, $f: M \to X$, be a normal map, $\dim M = \dim X = n$, $M$, $X$ simply connected, and suppose $K_i(f) \neq 0$ only for $i = q$, $n-q$, and $K_q(f)$ is free, with $q < \frac{n}{2}$. Then $(f, b)$ is normally bordant to a homotopy equivalence.*

Assume Lemma 1.12 for the moment and we will complete the proof of Proposition 1.2 in case $n = 2\ell$. We have seen that $K_i(f) = 0$ for $i \neq \ell$, and that $K_\ell$ is free. Multiplying by $S^k$, $(f \times 1, b \times 1)$, $f \times 1 : M \times S^k \to X \times S^k$ is a normal map, and now

$$K_i(f \times 1_{S^k}) \cong K_i(f) \otimes H_0(S^k) \oplus K_{i-k}(f) \otimes H_k(S^k)$$

from the Künneth formula relating the homology of a product to the homologies of the factors. Thus $K_i(f \times 1_{S^k}) \neq 0$ only for $i = \ell$ and $i = \ell + k$, $K_\ell(f \times 1_{S^k}) \cong K_\ell(f)$ is free over $\mathbb{Z}$, and $\ell < \frac{1}{2}(n + k) = \frac{1}{2}(2\ell + k) = \ell + \frac{k}{2}$. Then Lemma 1.12 applies provided $X \times S^k$ is still simply connected, i.e., $k > 1$, so that $f \times 1_{S^k}$ is normally bordant to a homotopy equivalence, and Proposition 1.2 is proved.

The case where $n = 2\ell + 1$ can be proved similarly with an extra step involving the torsion in $K_\ell(f)$.

Proposition 1.4 can be proved in a similar way, starting from the normal bordism $W$ of $f : M \to M'$ to $1_M : M' \to M'$, which comes from applying the Transversality Theorem to the homotopy of $\alpha$ to $\alpha'$. One does surgery on the interior of the normal bordism $W$ to make $F$ $\left[\frac{n+1}{2}\right]$-connected by Theorem 1.10 ($n = \dim M = \dim M'$, so $n + 1 = \dim W$), $F : W \to M' \times [0, 1]$. If $n + 1 = 2\ell$, we follow the above line of argument to get $K_i(F) = 0$ for $i \neq \ell$, $K_\ell(F)$ free, and again use Lemma 1.12 to get $F' : W' \to M' \times S^k \times [0, 1]$ a homotopy equivalence, $\partial W' = (M \times S^k) \cup (M' \times S^k)$, $F|M = f \times 1_{S^k}$, $F|M' = 1_{M' \times S^k}$. Now Proposition 1.4 follows from the

**h-bordism Theorem of Smale.** *Let $U$ be a compact manifold, simply connected with dimension $U \geq 6$, $\partial U = N \cup N'$. If the inclusion maps $N \to U$, $N' \to U$ are homotopy equivalences, then the identity map $1_N : N \to N$ extends to a diffeomorphism $e : N \times [0, 1] \to U$, so in particular $N \times \{1\}$ is diffeomorphic to $N'$.*

The theorem of Smale, and its generalization to non-simply connected manifolds due to Barden, Mazur, and Stallings, is the key to translating homotopy and bordism information into diffeomorphism information.

We end this lecture with a sketch proof of Lemma 1.12.

Note first that we have the homotopy exact sequence mapping into the homology exact sequence by the Hurewicz homomorphism:

$$\longrightarrow \pi_i(M) \xrightarrow{f_*} \pi_i(X) \longrightarrow \pi_i(f) \longrightarrow \pi_{i-1}(M) \xrightarrow{f_*} \pi_{i-1}(X) \longrightarrow$$
$$\downarrow \qquad \downarrow \qquad \downarrow \qquad \downarrow \qquad \downarrow$$
$$\longrightarrow H_i(M) \xrightarrow{f_*} H_i(X) \longrightarrow H_i(f) \longrightarrow H_{i-1}(M) \xrightarrow{f_*} H_{i-1}(X) \longrightarrow \cdot$$
$$\underset{\beta_*}{\smile}$$

The splitting $\beta_*$ shows that $K_{i-1}(f) \cong H_i(f)$ and the relative Hurewicz theorem implies $\pi_i(f) \cong H_i(f) \cong K_{i-1}(f) = 0$ for $i < q+1$ and $\pi_{q+1}(f) \cong H_{q+1}(f) \cong K_q(f)$ is free over $\mathbb{Z}$. Choose a basis for $\pi_{q+1}(f)$. We can do surgery on an embedded sphere $\varphi\colon S^q \to M$ representing an element of this basis. The exact sequence of the pair $W$, $M$ ($W$ = trace of the surgery) gives

$$\cdots \longrightarrow H_{q+1}(W,M) \longrightarrow H_q(M) \longrightarrow H_q(W) \longrightarrow H_q(W,M) \longrightarrow \cdots$$

$$\begin{array}{ccc} & j^* \big\uparrow \cong & \varphi_* \big\downarrow \\ H_{q+1}(D^{q+1}, S^q) & \xrightarrow{\cong} & H_q(S^q) \end{array} \qquad\qquad 0 \qquad ,$$

where $j_*$ is an isomorphism by the Excision Theorem of homology (see Lemma 1.5). Now $\varphi_*$ carries a generator of $H_q(S^q)$ into a basis element of $K_q(f) \subseteq H_q(M)$, and it follows that $H_q(W) \cong H_q(M)/(\varphi)$ and $H_{q+1}(W) \cong H_{q+1}(M)$. From Lemma 1.6 it follows that $H_i(M') \cong H_i(W)$ for $i < n-q-1$ so that if $n-q-1 > q+1$ (i.e., $q < \frac{n}{2}-1$), we get $K_i(f') = 0 = K_i(f)$ (where $f'\colon M' \to X$) for $i \neq q$, and $K_q(f') \cong K_q(f)/(\varphi)$, which is a free module over $\mathbb{Z}$ with rank $K_q(f') = \operatorname{rank} K_q(f) - 1$. Thus for $q < \frac{n}{2} - 1$, the proof follows by induction.

For $\frac{n}{2} - 1 \leq q < \frac{n}{2}$, more use of Poincaré duality must be made to calculate $K_q(f')$.

Much of this lecture could have been done without the assumption of simple connectivity (for example Theorem 1.10), but at the cost of using homology with local coefficients, or considering some slightly more delicate homological algebra over $\mathbb{Z}\pi$, the group ring of the fundamental group $\pi = \pi_1(X)$. But if one removes the dimension restrictions, and tries to do surgery in the middle dimension, a whole new set of difficulties arises, which leads to a theory, in place of simple theorems.

# 2 Surgery invariants, classifying spaces, and exact sequences

In this lecture I will try to describe the various parts of the theory of surgery and the relations between them. This theory arises inevitably from trying to refine theorems such as the rather elementary ones of the first lecture.

First recall the invariant of the *index* of a manifold or Poincaré space. Let $X$ satisfy Poincaré duality in dimension $4k$, and consider the form $H^{2k}(X) \otimes H^{2k}(X) \xrightarrow{(\;,\;)} \mathbb{Z}$ defined by $(x,y) = (x \cup y)[X]$ for $x$, $y \in H^{2k}(X)$, $[X] \in H_{4k}(X)$ the orientation class. This is a symmetric pairing and we define

$$I(X) = \text{signature } (\;,\;) = n_+ - n_-,$$

where $n_+$ $(n_-)$ is the number of $+$ $(-)$ signs in a diagonal matrix representing $(\ ,\ )$ over the real numbers.

It follows from relative Poincaré duality that if $M^{4k} = \partial W^{4k+1}$, $W$ orientable, then $I(M) = 0$. For any $z \in H^{2k}(W)$, $(i^*z, i^*z) = (i^*z \cup i^*z)[M] = i^*(z^2)[M] = z^2(i_*[M]) = 0$, so that $i^*H^{2k}(W)$ is a self-annihilating subspace of $H^{2k}(X)$. Poincaré duality implies that with $\mathbb{Q} = $ rational coefficients, $i^*H^{2k}(W;\mathbb{Q})$ has rank $= \frac{1}{2}$rank $H^{2k}(M;\mathbb{Q})$, which implies signature $(\ ,\ ) = 0$.

Therefore, if $F: W \to X \times [0,1]$ is a bordism of $f: M \to X \times \{0\}$ to $f': M' \to X \times \{1\}$, it follows that $I(\partial W) = I(M) - I(M') = 0$ so $I(M) = I(M')$, and if $f'$ is a rational homology equivalence, then $I(M') = I(X)$. Hence if $(f, b)$, $f: M \to X$, is a normal map and $I(M) \neq I(X)$, then $(f, b)$ could not be bordant to a homotopy equivalence. In fact:

**Theorem 2.1** *Let $(f, b)$, $f: M \to X$, be a normal map, $X$ simply connected, $f_*[M] = [X]$ and $\dim M = \dim X = n = 4k > 4$. Then $(f, b)$ is normally bordant to a homotopy equivalence if and only if $I(M) = I(X)$. Furthermore, $I(M) - I(X)$ is divisible by 8.*

This description of a "surgery obstruction" in terms of index was first found by Milnor in his initial paper on surgery, in the special context with $X$ a disk.

One first interprets $I(M) - I(X)$ as the signature of $(\ ,\ )$ restricted to $K^{2k}(f) \cong \operatorname{coker} f^*: H^{2k}(X) \to H^{2k}(M)$ (see Lecture 1), and then shows that the problem of finding embedded spheres $S^{2k} \subset M$ on which to do surgery is related to $(x, x)$, where $x \in H^{2k}(M)$ is Poincaré dual to $[S^{2k}] \in H_{2k}(M)$. In particular, if $X$ is $2k$-connected and $\mathcal{P}x \in K_{2k}(f)$, $k > 1$, then $\mathcal{P}x$ is represented by an embedding $S^{2k} \times D^{2k} \subset M$ if and only if $(x, x) = 0$. If one could find half a basis for $K_{2k}(f)$ which annihilated itself under $(\ ,\ )$, then an induction argument could be carried out to "kill" $K_{2k}(f)$ and obtain a homotopy equivalence. If sign $(\ ,\ )|K_{2k}(f) = 0$, then such a half-basis can be found.

Thus we can say in this case that the "surgery obstruction" is $\frac{1}{8}(I(M) - I(X))$, and in fact all integer values are assumed for this obstruction for various problems.

This situation is the first example and the prototype for the theory of the "surgery obstruction" in general. We shall now describe the general surgery theory as it has emerged and then show how it relates to algebraic $K$-theory on the one hand, and to homotopy theory, bundle theory, etc., on the other.

Let us define a $\pi$-*Poincaré space* to be a finite complex $X$, connected, with $\pi_1 X = \pi$, and such that Poincaré duality holds for all coefficient systems (which is equivalent to holding for the coefficient system $\mathbb{Z}\pi$).

If $f_i: M_i \to X$, $i = 0, 1$ are homotopy equivalences, $M_i$ a closed smooth manifold, we say that $(M_0, f_0)$ is *concordant* to $(M_1, f_1)$ if there is a smooth manifold $W$ with boundary $\partial W = M_0 \cup M_1$ and a *homotopy equivalence* $F: W \to X \times [0,1]$ with $F|M_0 = (f_0, 0)$, $F|M_1 = (f_1, 1)$.

We note that a similar definition of concordance could be made for other kinds of equivalences — e.g., replace homotopy equivalence in the definition by simple homotopy equivalence, or by homology equivalence, or by normal maps and normal bordism.

If $X$ is a $\pi$-Poincaré space, define $\mathcal{S}(X) = \{$concordance classes of pairs $(M, f)$, $M$ is a smooth closed manifold, $f: M \to X$ a homotopy equivalence$\}$. We call $\mathcal{S}(X)$ the set of homotopy smoothings of $X$. We could define similar sets using simple homotopy equivalence, or other relations, but the technical complication of the exposition would increase greatly, so we will confine ourselves to homotopy equivalence in these lectures.

If $f: M \to X$ is a homotopy equivalence, then we can make $f$ into a normal map by taking $\xi = g^*(\nu_M)$, $g$ a homotopy inverse to $f$, so that there is a linear bundle map $b: \nu_M \to \xi$ covering $f$. The choice of $b$ is not unique; we could compose it with any linear automorphism of $\xi$.

Let us then define $\mathcal{N}(X) = \{$normal bordism classes of normal maps $(f, b)$, $f: M \to X$, $b: \nu_M \to \xi$, for all $M$, $\xi$, with the relation of identifying $(f, b)$ with $(f, cb)$ if $c: \xi \to \xi$ is a linear automorphism$\}$.

Then we have a natural map $\eta : \mathcal{S}(X) \to \mathcal{N}(X)$. Now the existence and uniqueness questions of Lecture 1 can be interpreted as questions about the map $\eta$. How big is the image of $\mathcal{S}(X)$? How big is the inverse image $\eta^{-1}(x)$ for $x \in \mathcal{N}(X)$?

One of the central formulations of surgery theory is an *exact sequence* in which $\eta$ is embedded, with "surgery obstruction" groups on both sides.

**Theorem 2.2** *Let $\pi$ be a finitely presented group and $X$ a $\pi$-Poincaré space of dimension $n \geq 5$. There are abelian groups $L_n(\pi)$, $L_{n+1}(\pi)$, and an exact sequence of sets:*

$$L_{n+1}(\pi) \overset{\omega}{\dashrightarrow} \mathcal{S}(X) \overset{\eta}{\longrightarrow} \mathcal{N}(X) \overset{\sigma}{\longrightarrow} L_n(\pi).$$

More specifically, for $x \in \mathcal{N}(X)$, $\sigma(x) = 0$ if and only if $x \in \eta(\mathcal{S}(X))$, and $\omega$ denotes an *action* $L_{n+1}(\pi) \times \mathcal{S}(X) \to \mathcal{S}(X)$ such that if $a, b \in \mathcal{S}(X)$, $\eta(a) = \eta(b)$ if and only if there is an $n \in L_{n+1}(\pi)$ such that $\omega(n, a) = b$.

Note that if $\omega$ were simply a *map*, that would imply $\mathcal{S}(X)$ is non-empty, which is not always true.

The groups $L_n(\pi)$ are called the *surgery obstruction groups* or *Wall groups*, and they have many good properties.

**Theorem 2.3** $L_i(\pi)$ *is a covariant functor from finitely presented groups to abelian groups, periodic of order 4 in $i$, i.e., $L_{i+4}(\pi) \cong L_i(\pi)$. Further,* *if $\varphi: \pi \to \pi'$ is a homomorphism, we have another functor $L_i(\varphi)$, periodic of order 4, and an exact sequence*

$$\cdots \to L_i(\pi) \xrightarrow{\varphi_*} L_i(\pi') \to L_i(\varphi) \to L_{i-1}(\pi) \to \cdots .$$

The groups $L_n(\varphi)$ are associated to a sequence, similar to that of Theorem 2.2, for the situation of manifolds with boundary, with $\pi = \pi_1(\partial X)$, $\pi' = \pi_1(X)$.

The periodicity of $L_n(\pi)$ can be represented geometrically by taking the product with $\mathbb{CP}^2$, i.e.,

$$
\begin{array}{ccccccc}
L_{n+1}(\pi) & \xdashrightarrow{\omega} & S(X) & \xrightarrow{\eta} & N(X) & \xrightarrow{\sigma} & L_n(\pi) \\
\Big\downarrow{\cong} & & \Big\downarrow{\times 1 \cdots 2} & & \Big\downarrow{\times 1 \cdots 2} & & \Big\downarrow{\cong} \\
L_{n+5}(\pi) & \xdashrightarrow{\omega} & S(X \times \mathbb{CP}^2) & \xrightarrow{\eta} & N(X \times \mathbb{CP}^2) & \xrightarrow{\sigma} & L_{n+4}(\pi).
\end{array}
$$

This is an often used method for raising the dimension of surgery problems to put calculations into higher dimension, where often more methods are available. We will see examples of this later.

Replacing our general problems of Lecture 1, we may now set some general problems involved in the calculation of $S(X)$:

**(A)** Calculate $L_n(\pi)$.

**(B)** Calculate $N(X)$.

**(C)** Calculate the map $\sigma$.

**(D)** Calculate the image of $\eta$.

**(E)** Calculate the action $\omega$.

None of these problems is completely understood in general, but important special cases lead to a very interesting variety of theorems.

In particular, if we take $\pi = 1$, the trivial group, we have the result of Kervaire-Milnor:

**Proposition 2.4**

$$
L_n(1) = \begin{cases} 0 & n \text{ odd} \\ \mathbb{Z} & n = 4k \\ \mathbb{Z}/2 & n = 4k+2. \end{cases}
$$

The case $n = 4k$ is Theorem 2.1 of this Lecture, and Propositions 1.1 and 1.3 of Lecture 1 follow readily from this proposition.

The group $L_{4k}(1)$ as we saw is simply detected by the index and is a Grothendieck group of non-singular, symmetric, even-valued quadratic forms over $\mathbb{Z}$. The group $L_{4k+2}(1)$ is interpreted as an analogous group of $\mathbb{Z}/2$-valued quadratic forms, detected by the so-called Kervaire-Arf invariant, which is analogous to the index. The general Wall group $L_n(\pi)$ has a similar algebraic interpretation, where $L_{4k}(\pi)$ is a group of non-singular, symmetric quadratic forms over the group ring $\mathbb{Z}\pi$, $L_{4k+2}(\pi)$ is a similar group of antisymmetric forms over $\mathbb{Z}\pi$, while $L_{2n+1}(\pi)$ is a Grothendieck group of automorphisms of the trivial class of forms in $L_{2n}(\pi)$.

Besides the case $\pi = 1$, the case $\pi$ finite can be very effectively studied by using this algebraic definition, which has been done by Wall, R. Lee, H. Bass, A. Bak, and others. But even in this case the picture is not complete.[4]

Another approach to the calculation of $L_n(\pi)$ is based on using the geometry of a manifold $M$ with $\pi_1 M = \pi$, which gives a calculation for $\pi$ free abelian, or free, and many other cases.[5] We will say more about this in the next lecture.

The set $\mathcal{N}(X)$ of bordism classes of normal maps into $X$, modulo bundle equivalences over $X$, can be interpreted as homotopy classes of sections of a certain bundle over $X$, with fibre a space called $G/O$. This space is essentially a homogeneous space, the quotient of $G = $ limit as $k \to \infty$ of the space $G_k$ of homotopy equivalences of $S^{k-1}$, by $O = $ limit as $k \to \infty$ of $O(k)$, the orthogonal group. If $\mathcal{N}(X) \neq \emptyset$, i.e., if there is one section of this bundle, then as in the case of a principal bundle, the total space is homotopy equivalent to $X \times G/O$, and homotopy classes of sections are given by $\mathcal{N}(X) = [X, G/O]$, the set of homotopy classes of maps of $X$ into $G/O$.

From the form of $G/O$ as a "homogeneous space" we get an exact homotopy sequence:

$$\cdots \to \pi_i(O) \xrightarrow{J} \pi_i(G) \to \pi_i(G/O) \to \pi_{i-1}(O) \to \cdots .$$

The periodicity theorem of Bott calculates

$$\pi_i(O) = \begin{cases} \mathbb{Z}/2 & i = 0, 1 \bmod 8 \\ \mathbb{Z} & i = 3, 7 \bmod 8 \\ 0 & \text{otherwise.} \end{cases}$$

The homotopy groups of $G$ are the stable homotopy groups of spheres $\pi_i(G) \cong \pi_{i+n}(S^n)$, $n$ large, which can be seen by using the fibration

$$\Omega^k S^k \to G_{k+1} \xrightarrow{e} S^k,$$

---

[4]Editor's note: This was written in 1977. For the present understanding of the Wall groups of finite groups, see the survey by Hambleton and Taylor in this volume.

[5]Editor's note: Again, for the present status of this approach, see the surveys by J. Davis and C. Stark in this volume.

where $e$ is the fibre map which evaluates the homotopy equivalence of $S^k$ at a fixed base point $x_0 \in S^k$. The map $J: \pi_i(O) \to \pi_i(G)$ is the famous $J$-homomorphism of G. W. Whitehead, which has now been calculated completely, through the work of J. F. Adams and many others, and one gets $\pi_i(G/O) = \text{cokernel } J \oplus A_i$, where

$$A_i = \begin{cases} 0 & i \not\equiv 0 \bmod 4 \\ \mathbb{Z} & i \equiv 0 \bmod 4, \end{cases}$$

$A_i \cong \text{kernel } J$ on $\pi_{i-1}(O)$.

Putting together these facts with the surgery exact sequence (Theorem 2.2) we can get the results of Kervaire-Milnor: a long exact sequence in which all terms are groups:

$$\cdots \to L_{n+1}(1) \to \mathcal{S}(S^n) \to \pi_n(G/O) \to L_n(1) \to \cdots \qquad (n \geq 5).$$

Using the relation of the surgery obstruction with the index if $n = 4k$ and the Hirzebruch Index Theorem to calculate the obstruction for elements of $\pi_n(G/O)$, $n = 4k$, together with the results on $J$ above, the sequence breaks down into short exact sequences:

$$0 \to bP_{n+1} \to \mathcal{S}(S^n) \to C_n \to 0,$$

where $bP_{n+1}$ is a finite cyclic group which is 0 for $n$ even, a subgroup of $\mathbb{Z}/2$ is $n = 4k + 1$, and for $n = 4k - 1$ calculable from numerators of Bernoulli numbers $B_m$.[6] The group $C_n \subseteq \text{cokernel } J: \pi_n(O) \to \pi_n(G)$, and is the kernel of a homomorphism coker $J \to \mathbb{Z}/2$, given by using the Kervaire invariant. This homomorphism is trivial for $n \neq 2^i - 2$, is non-trivial for $n = 6, 14, 30, 62$, but is not known to be non-trivial for other values of $n$.[7]

The group of exotic spheres of Milnor, $\mathcal{S}(S^n)$ (denoted also by $\Theta_n$ or $\Gamma^n$) is thus in fact $\mathcal{S}(S^n) = bP_{n+1} \oplus \text{coker } J$, for $n \neq 2^i - 2$, where $bP_{n+1}$ is finite cyclic and calculated for $n \neq 2^i - 3$. (That the exact sequence actually splits was shown by Brumfiel.)

# 3 Applications to topological manifolds

One of the most striking applications of surgery theory is in the applications to the problems of existence and uniqueness of piecewise linear manifold structures on topological manifolds, i.e., the triangulation problem and the "Hauptvermutung".

First we note that the surgery theory developed in Lectures 1 and 2 can be developed also for piecewise linear manifolds instead of smooth manifolds. (A simplicial complex $K$ is a piecewise linear (PL) manifold if each

---

[6] Editor's note: See the paper by Lance in this volume for more details.

[7] Editor's note: See the paper by E. Brown in this volume for more details.

point has a neighborhood $U$ which in some linear subdivision of $K$ is linearly isomorphic to an open set in the linear space $\mathbb{R}^n$.) For linear bundles, one substitutes PL bundles or so-called "block bundles". Transversality theorems can be proved in this context, and with the aid of the smoothing theory of PL manifolds, the surgery exact sequence of Lecture 2, Theorem 2.2 can be derived:

$$L_{n+1}(\pi) \xrightarrow{\ \omega\ } \mathcal{S}_{\mathrm{PL}}(X) \xrightarrow{\ \eta\ } \mathcal{N}_{\mathrm{PL}}(X) \xrightarrow{\ \sigma\ } L_n(\pi),$$

where $X$ is a $\pi$-Poincaré space of dimension $n \geq 5$, $\mathcal{S}_{\mathrm{PL}}$, $\mathcal{N}_{\mathrm{PL}}$ are the analogs of $\mathcal{S}$ and $\mathcal{N}$ using PL manifolds and bundles in place of smooth manifolds and linear bundles, but $L_n(\pi)$ are the *same* groups as in the smooth case.

While the problem (A) of Lecture 2, that of calculating $L_n(\pi)$, remains identical, problem (B) becomes easier. If $\mathcal{N}_{\mathrm{PL}}(X) \neq \emptyset$, then $\mathcal{N}_{\mathrm{PL}}(X) \cong [X, G/PL]$, and the space $G/PL$ has much simpler properties. In particular

$$\pi_i(G/PL) = \begin{cases} 0 & i \text{ odd} \\ \mathbb{Z} & i = 4k \\ \mathbb{Z}/2 & i = 4k+2, \end{cases}$$

and in fact Sullivan showed that except for a slight complication in dimension 4, $[X, G/PL]$ can be calculated from $H^*(X; \mathbb{Z}_{(2)})$, $H^*(X; \mathbb{Z}/2)$ and $KO(X)_{(p)}$, $p$ odd, where the subscript $(p)$ denotes localization at the prime $p$. Thus $\mathcal{N}_{\mathrm{PL}}(X)$ is very calculable, when it is non-empty.

Sullivan, in his version of the "Hauptvermutung", used this result together with a study of the map $G/PL \to G/Top$.

The "generalized Poincaré conjecture" of Smale and Stallings tells us that a PL manifold $M^n$ of the homotopy type of $S^n$, $n \geq 5$, is PL equivalent to $S^n$. This shows that $\mathcal{S}_{\mathrm{PL}}(S^n)$ has exactly one element. One may then interpret this in the surgery exact sequence to show that the action $\omega$ of $L_{n+1}(1)$ on $\mathcal{S}_{\mathrm{PL}}(X)$ is trivial, and one gets the statement:

If $X$ is a 1-connected closed PL manifold of dimension $\geq 5$, then $\mathcal{S}_{\mathrm{PL}}(X) \cong [X_0, G/PL]$, where $X_0 = $ complement of a point in $X$.

Thus Sullivan's result on the homotopy type of $G/PL$ makes possible extensive "classification" theorems for simply connected PL manifolds, or manifolds whose fundamental group has relatively simple Wall groups. In particular, for $X = T^n$ a very beautiful PL classification theorem can be proved, as we shall see.

Perhaps the most dramatic application of surgery theory has been to the study of triangulations of topological manifolds developed by Kirby and Siebenmann. We will describe in some detail the first step in this development, Kirby's proof of the Annulus Conjecture for dimensions $> 4$.

An equivalent form of this conjecture can be stated as follows:

**Conjecture $S_n$.** *If $f: \mathbb{R}^n \to \mathbb{R}^n$ is a homeomorphism, then there exist homeomorphisms $g, h: \mathbb{R}^n \to \mathbb{R}^n$ and open sets $U, V \subset \mathbb{R}^n$ such that $h$ is PL, $f|U = g|U$ and $g|V = h|V$. (Such a homeomorphism is called stable. If $f$ is stable and $g = f$ on an open set, then $g$ is stable.)*

If two homeomorphisms of $\mathbb{R}^n$ agree on an open set, then they are isotopic, i.e., one can be deformed to the other through homeomorphisms. Thus $(S_n)$ implies any homeomorphism of $\mathbb{R}^n$ is isotopic to a PL homeomorphism, and this can be thought of as the 0-dimensional step in a proof of the triangulability of topological manifolds.

One starts with Connell's observation that if $f: \mathbb{R}^n \to \mathbb{R}^n$ is *bounded*, i.e., if $\|f(x) - x\| < K$ for all $x \in \mathbb{R}^n$ and fixed $K$, then $f$ is stable. For, if we treat $\mathbb{R}^n$ as interior $D^n$, the closed unit ball in $\mathbb{R}^n$, then $f$ bounded implies $f$ extends to $\overline{f}: D^n \to D^n$ with $\overline{f}|S^{n-1} = 1_{S^{n-1}}$. The identification of $\mathbb{R}^n$ with int $D^n$ can be made to be the identity on a small ball, so the result follows.

Kirby then observed that if $f_0: T^n \to T^n$ is a homeomorphism of the torus, $T^n = S^1 \times \cdots \times S^1$ ($n$ times), the induced homeomorphism of the universal cover, $\widetilde{f}_0: \mathbb{R}^n \to \mathbb{R}^n$, is the product of a linear homeomorphism $g_1 \in GL(n, \mathbb{Z})$ (which gives the effect of $f_0$ on homology) and $g_2$ which is periodic, preserving the integral lattice. Hence $g_2$ is bounded by by $\max \|g_2(x) - x\|$, for $x \in$ fundamental domain in $\mathbb{R}^n$, and hence $\widetilde{f}_0$ is stable.

One then asks if one can show that a homeomorphism $f: \mathbb{R}^n \to \mathbb{R}^n$ agrees on some open set with $\widetilde{f}_0: \mathbb{R}^n \to \mathbb{R}^n$ for some homeomorphism $f_0: T^n \to T^n$.

Using the Smale-Hirsch immersion theory, or by a direct argument, one can find an immersion $u : T_0^n \to \mathbb{R}^n$, where $T_0^n = T^n - \{\text{point}\}$. ($u$ an immersion means $u$ is a smooth map of maximal rank everywhere.) One can "pull back" the smooth structure on $\mathbb{R}^n$ by the composite $fu$ to get a new smooth structure $\overline{T}_0^n$ on the topological manifold $T_0^n$, so that $\overline{T}_0^n$ is immersed in $\mathbb{R}^n$ (by $\overline{u}$). Then we have a commutative diagram

$$
\begin{array}{ccc}
\mathbb{R}^n & \xrightarrow{\ f\ } & \mathbb{R}^n \\[4pt]
{\scriptstyle u}\big\uparrow & & \big\uparrow{\scriptstyle \overline{u}} \\[4pt]
T_0^n & \xrightarrow[\ h_0\ ]{} & \overline{T}_0^n,
\end{array}
$$

where $h_0$ is the (identity) homeomorphism of the topological manifolds $T_0^n$ and $\overline{T}_0^n$. Note that on a small open ball in $T_0^n$, $fu$ agrees with $\overline{u}h_0$. Since $u$, $\overline{u}$ are smooth immersions, if $h_0$ were the restriction of a homeomorphism of $T^n$ to itself, the conjecture $(S_n)$ would follow. But $\overline{T}_0^n$ may not be smoothly or PL equivalent to $T_0^n$, so the argument must be more complicated.

First extend $h_0$ to $h \colon T^n \to \overline{T}^n$, where $\overline{T}^n$ is the one-point compactification of $\overline{T}_0^n$. One shows that $\overline{T}^n$ is a PL manifold by a codimension-one surgery argument, which we will not elaborate. The map $h$ is a homeomorphism. Now we have the result of Hsiang-Shaneson and Wall:

**Theorem 3.1** *Let* $f \colon M^n \to T^n$ *be a homotopy equivalence,* $M$ *a PL manifold,* $T^n$ *the torus,* $n \geq 5$. *Then a finite* $2^n$-*sheeted cover of* $f$, $f_c \colon M_c^n \to T_c^n = T^n$, *is homotopic to a PL equivalence.*

This follows from a classification theorem for homotopy PL structures on $T^n$, which we will discuss later.

Thus the $2^n$-fold cover $\overline{T}_c^n$ of $\overline{T}^n$ is PL equivalent to $T^n$, and let $g \colon \overline{T}_c^n \to T^n$ be the equivalence. Now we describe Kirby's "Main Diagram":

Let $B^n$ be an $n$-dimensional ball in $\mathbb{R}^n$, so small that the inclusion $B^n \subset \mathbb{R}^n$ can be factored though an embedding in $T_0^n$ and the immersion $u$, and let $\overline{B}^n$ be the image of $B^n$ under $h$, which has similar properties, and let $h_1 = f|B^n$. Let $\pi \colon T^n \to T^n$, $\overline{\pi} \colon \overline{T}_c^n \to \overline{T}^n$ be the $2^n$-fold covers (2-fold along each $S^1$ factor).

**Main Diagram.**

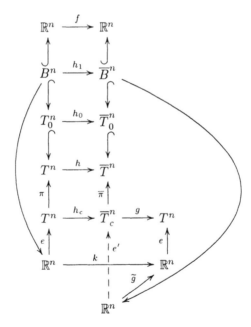

The balls $B^n$, $\overline{B}^n$ are so small that the large curved arrows exist, lifting the embeddings. Since $g$ is PL, $\widetilde{g}$ is PL, hence stable, and $k$ is stable, as the lift of a homeomorphism of $T^n$ to itself, where $k$ is the induced map from $gh_c$. Hence $\widetilde{g}^{-1}k$ is stable and $f$ agrees with $\widetilde{g}^{-1}k$ on the small ball $B^n$.

Now we return to a discussion of Theorem 3.1 and homotopy PL structures on $T^n$. The surgery exact sequence in that case gives us:

$$L_{n+1}(\mathbb{Z}^n) \xdashrightarrow{\omega} \mathcal{S}_{\mathrm{PL}}(T^n) \xrightarrow{\eta} [T^n, G/PL] \xrightarrow{\sigma} L_n(\mathbb{Z}^n).$$

Note that we have identified $\mathcal{N}_{\mathrm{PL}}(T^n)$ with $[T^n, G/PL]$ using the identity $T^n \to T^n$ as the "base point" which determines the isomorphism of sets $\mathcal{N}_{\mathrm{PL}}(T^n) \to [T^n, G/PL]$.

First we will calculate $L_n(\mathbb{Z}^n)$, then $[T^n, G/PL]$, and show $\sigma$ is an injection. Then we show that the action of $L_{n+1}(\mathbb{Z}^n)$ on $\mathcal{S}_{\mathrm{PL}}(T^n)$ is mostly trivial, and that the non-trivial part of this action is through a number of $\mathbb{Z}/2$'s, which become trivial by taking the $2^n$-fold covers. The arguments rely heavily on the fibering theorem of Farrell, generalizing the Browder-Levine Theorem. We give a special case:

**Theorem 3.2** *Let $W^{n+1} \to S^1$ be a fibre bundle (smooth or PL) with fibre a closed manifold $M^n$, $n \geq 5$, and suppose $\pi_1(W)$ is free abelian. If $U \xrightarrow{h} W$ is a homotopy equivalence, then $U$ is also a fibre bundle over $S^1$, and $h$ is homotopic to a fibre-preserving homotopy equivalence. Similarly, if $\partial M \neq \emptyset$, and if $\partial U \to \partial W$ is already a homotopy equivalence of bundles, we get the same result preserving the given structure on $\partial U$, i.e., the fibering and the maps extend from $\partial U$ to $U$.*

The proof of Theorem 3.2 is another "codimension 1" surgery argument, which we will not present.

Considering the iterated fibrations $T^n \to T^{n+1} \to S^1$, we could apply Theorem 3.2 to a homotopy equivalence $M \to T^{n+1}$ to get similar fibrations for $M$ over $S^1$, provided that the dimensions were kept $> 5$. Multiplying both sides by $\mathbb{CP}^2 \times \mathbb{CP}^2$ ensures that the dimensions involved will remain large enough to apply Theorem 3.2, while the periodicity theorem of surgery will show that the surgery obstruction information is preserved. Inductive application of "cutting" along fibres will finally reduce a surgery problem with range $T^{n+1} \times \mathbb{CP}^2 \times \mathbb{CP}^2$ to a collection of simply connected problems. For example for $T^2$, we consider $T^2$ as the identification space of $[0,1] \times [0,1]$, the rectangle with opposite sides identified:

Using transversality to cut up a surgery problem

$$M \to T^2 \times \mathbb{CP}^2 \times \mathbb{CP}^2$$

in a similar way, we get maps into $V \times \mathbb{CP}^2 \times \mathbb{CP}^2$, $a \times \mathbb{CP}^2 \times \mathbb{CP}^2$ and $b \times \mathbb{CP}^2 \times \mathbb{CP}^2$, and if we can solve each surgery problem (inductively), the solutions can be "glued" back together to solve the problem for $M \to T^2 \times \mathbb{CP}^2 \times \mathbb{CP}^2$. The fibering theorem (Theorem 3.2) shows that this is the *only* way to solve this problem.

This enables us to interpret $L_n(\mathbb{Z}^n)$ as the collection of simply connected surgery obstructions along the pieces of the "cut up" torus, and enables us to show that $\sigma \colon [T^n, G/PL] \to L_n(\mathbb{Z}^n)$ is injective.

The action $\omega \colon L_{n+1}(\mathbb{Z}^n) \dashrightarrow \mathcal{S}_{\mathrm{PL}}(T^n)$ can be similarly interpreted *after* multiplying with $\mathbb{CP}^2 \times \mathbb{CP}^2$, but to show the action is trivial, this information is not adequate. Most of the action of $L_{n+1}(\mathbb{Z}^n)$ can be interpreted on $T^n$ itself, cut up, except for the part of $L_{n+1}(\mathbb{Z}^n)$ which comes from the 3-dimensional pieces of the carved $T^n$. This part of the action is seen to be non-trivial as a consequence of Rohlin's theorem on the index of 4-dimensional spin manifolds, but this non-triviality is of order 2, and disappears on taking 2-fold covers in each direction.

# 4 Selected References

## 4.1 Books and expository articles

- G. E. Bredon, *Introduction to Compact Transformation Groups*, Pure and Applied Mathematics, vol. 46, Academic Press, New York, 1972.

- W. Browder, *Surgery on Simply Connected Manifolds*, Ergebnisse der Mathematik und ihrer Grenzgebiete, vol. 65, Springer-Verlag, Berlin, New York, 1972.

- ——, Manifolds and homotopy theory, *Manifolds—Amsterdam 1970*, Lecture Notes in Math., no. 197, Springer-Verlag, Berlin, 1971, pp. 17–35.

- M. W. Hirsch, *Differential Topology*, Graduate Texts in Mathematics, vol. 33, Springer-Verlag, New York, 1976.

- J. Milnor, *Morse Theory*, Annals of Math. Studies, no. 51, Princeton Univ. Press, Princeton, NJ, 1963.

- ——, *Lectures on the h-Cobordism Theorem*, notes by L. Siebenmann and J. Sondow, Princeton Math. Notes, no. 1, Princeton Univ. Press, Princeton, NJ, 1965.

- ──, Whitehead torsion, *Bull. Amer. Math. Soc.* **72** (1966), pp. 358–426.

- J. Milnor and J. Stasheff, *Characteristic Classes*, Annals of Math. Studies, no. 76, Princeton Univ. Press, Princeton, NJ, 1974.

- E. Spanier, *Algebraic Topology*, McGraw-Hill Series in Higher Math., McGraw-Hill, New York, 1966.

- C. T. C. Wall, *Surgery on Compact Manifolds*, London Math. Soc. Monographs, vol. 1, Academic Press, London, New York, 1971.

## 4.2 References from the lectures and other references in surgery

- S. E. Cappell and J. L. Shaneson, On four-dimensional surgery and applications, *Comment. Math. Helv.* **46** (1971), pp. 500–528.

- J.-C. Hausmann, Homological surgery, *Ann. of Math.* **104** (1976), pp. 573–584.

- M. W. Hirsch, Immersions of manifolds, *Trans. Amer. Math. Soc.* **93** (1959), pp. 242–276.

- W. C. Hsiang and J. L. Shaneson, Fake tori, *Topology of Manifolds* (Proc., Athens, GA, 1969), Markham, Chicago, IL, 1970, pp. 18–51.

- ──, Fake tori, the annulus conjecture, and the conjectures of Kirby, *Proc. Nat. Acad. Sci. U.S.A.* **62** (1969), pp. 687–691.

- M. Kervaire and J. Milnor, Groups of homotopy spheres, I, *Ann. of Math.* (2) **77** (1963), pp. 504–537.

- R. C. Kirby, Stable homeomorphisms and the annulus conjecture, *Ann. of Math.* (2) **89** (1969), pp. 575–582.

- R. C. Kirby and L. C. Siebenmann, On the triangulation of manifolds and the Hauptvermutung, *Bull. Amer. Math. Soc.* **75** (1969), pp. 742–749.

- ──, Foundations of TOPology, *Notices Amer. Math. Soc.* **16** (1969), Abstract #69T-G101, p. 848.

- R. Lashof and M. Rothenberg, Triangulation of manifolds, I, II, *Bull. Amer. Math. Soc.* **75** (1969), pp. 750–754 and 755–757.

- B. Mazur, Differential topology from the point of view of simple homotopy theory, *Inst. Hautes Études Sci. Publ. Math.*, no. 15, 1963. Corrections and further comments, *ibid.*, no. 22 (1964), 81–91.

- J. Milnor, A procedure for killing homotopy groups of differentiable manifolds, *Differential geometry*, Proc. Sympos. Pure Math., vol. 3, Amer. Math. Soc., Providence, RI, 1961, pp. 39–55.

- ――, On simply connected 4-manifolds, *Symposium internacional de topología algebraica*, Univ. Nacional Autónoma de México, Mexico City, 1958, pp. 122–128.

- ――, *Differentiable manifolds which are homotopy spheres*, mimeographed notes, Princeton Univ., Princeton, NJ, 1958.

- ――, Microbundles, I, *Topology* **3** (1964), suppl. 1, pp. 53–80.

- S. P. Novikov, Homotopically equivalent smooth manifolds, I, *Izv. Akad. Nauk SSSR, Ser. Mat.* **28** (1964), pp. 365–474; English translation in *Amer. Math. Soc. Transl.* (2) **48** (1965), pp. 271–396.

- F. Quinn, *A geometric formulation of surgery*, Thesis, Princeton Univ., Princeton, NJ, 1969.

- V. A. Rohlin, New results in the theory of four-dimensional manifolds, *Doklady Akad. Nauk SSSR* (N.S.) **84** (1952), pp. 221–224.

- C. P. Rourke and B. J. Sanderson, Block bundles, I, *Ann. of Math.* (2) **87** (1968), pp. 1–28.

- R. W. Sharpe, Surgery on compact manifolds: The bounded even dimensional case, *Ann. of Math.* (2) **98** (1973), pp. 187–209.

- S. Smale, Generalized Poincaré's conjecture in dimensions greater than four, *Ann. of Math.* (2) **74** (1961), pp. 391–406.

- ――, On the structure of manifolds, *Amer. J. Math.* **84** (1962), pp. 387–399.

- ――, The classification of immersions of spheres in Euclidean spaces, *Ann. of Math.* (2) **69** (1959), pp. 327–344.

- D. Sullivan, *Triangulating homotopy equivalences*, Thesis, Princeton Univ., Princeton, NJ, 1965.

- ――, On the Hauptvermutung for manifolds, *Bull. Amer. Math. Soc.* **73** (1967), pp. 598–600.

- ――, Triangulating and smoothing homotopy equivalences and homeomorphisms: *Geometric Topology Seminar Notes*, Princeton Univ., 1967. Reprinted in *The Hauptvermutung book*, *K*-Monogr. Math., vol. 1, Kluwer Acad. Publ., Dordrecht, 1996, pp. 69–103.

- ——, *Smoothing homotopy equivalences*, mimeographed notes, Univ. of Warwick, Coventry, England, 1966.

- ——, *Geometric Topology, I: Localization, periodicity, and Galois symmetry*, mimeographed notes, M.I.T., Cambridge, MA, 1970.

- R. Thom, Quelques propriétés globales des variétés différentiables, *Comment. Math. Helv.* **28** (1954), pp. 17–86.

- C. T. C. Wall, An extension of results of Novikov and Browder, *Amer. J. Math.* **88** (1966), pp. 20–32.

- ——, Surgery of non-simply-connected manifolds, *Ann. of Math.* (2) **84** (1966), pp. 217–276.

## 4.3   The Adams conjecture

- J. F. Adams, On the groups $J(X)$, I, *Topology* **2** (1963), pp. 181–195.

- ——, *ibid.*, II, *Topology* **3** (1965), pp. 137–171.

- ——, *ibid.*, III, *Topology* **3** (1965), pp. 193–222.

- ——, *ibid.*, IV, *Topology* **5** (1966), pp. 21–71. Correction, *Topology* **7** (1968), p. 331.

- J. C. Becker and D. H. Gottlieb, The transfer map and fiber bundles, *Topology* **14** (1975), pp. 1–12.

- E. Friedlander, Fibrations in étale homotopy theory, *Inst. Hautes Études Sci. Publ. Math.*, no. 42 (1973), pp. 5–46.

- D. Quillen, Some remarks on étale homotopy theory and a conjecture of Adams, *Topology* **7** (1968), pp. 111–116.

- ——, The Adams conjecture, *Topology* **10** (1971), pp. 67–80.

- D. Sullivan, Genetics of homotopy theory and the Adams conjecture, *Ann. of Math.* (2) **100** (1974), pp. 1–79.

## 4.4   Calculating surgery obstruction groups

- A. Bak, Odd dimension surgery groups of odd torsion groups vanish, *Topology* **14** (1975), no. 4, pp. 367–374.

- H. Bass, $L_3$ of finite abelian groups, *Ann. of Math.* (2) **99** (1974), pp. 118–153.

- W. Browder, Manifolds with $\pi_1 = \mathbb{Z}$, *Bull. Amer. Math. Soc.* **72** (1966), pp. 238–244.

- S. E. Cappell, On homotopy invariance of higher signatures, *Invent. Math.* **33** (1976), no. 2, pp. 171–179.

- F. X. Connolly, Linking numbers and surgery, *Topology* **12** (1973), pp. 389–409.

- F. T. Farrell and W.-C. Hsiang, A formula for $K_1 R_\alpha [T]$, *Applications of Categorical Algebra* (New York, 1968), Proc. Sympos. Pure Math., vol. 17, Amer. Math. Soc., Providence, RI, 1970, pp. 192–218.

- ——, Manifolds with $\pi_1 = G \times_\alpha T$, *Amer. J. Math.* **95** (1973), pp. 813–848.

- R. Lee, Computation of Wall groups, *Topology* **10** (1971), pp. 149–176.

- G. Lusztig, Novikov's higher signature and families of elliptic operators, *J. Differential Geometry* **7** (1972), pp. 229–256.

- T. Petrie, The Atiyah-Singer invariant, the Wall groups $L_n(\pi, 1)$, and the function $(te^x + 1)/(te^x - 1)$, *Ann. of Math.* (2) **92** (1970), pp. 174–187.

- J. L. Shaneson, Wall's surgery obstruction groups for $G \times \mathbb{Z}$, *Ann. of Math.* (2) **90** (1969), pp. 296–334.

- C. T. C. Wall, Classification of Hermitian Forms, VI: Group rings, *Ann. of Math.* (2) **103** (1976), no. 1, pp. 1–80.

## 4.5 Codimension one surgery

- W. Browder, Structures on $M \times \mathbb{R}$, *Proc. Cambridge Philos. Soc.* **61** (1965), pp. 337–345.

- W. Browder and J. Levine, Fibering manifolds over a circle, *Comment. Math. Helv.* **40** (1966), pp. 153–160.

- W. Browder and G. R. Livesay, Fixed point free involutions on homotopy spheres, *Tôhoku Math. J.* (2) **25** (1973), pp. 69–87.

- W. Browder, J. Levine, and G. R. Livesay, Finding a boundary for an open manifold, *Amer. J. Math.* **87** (1965), pp. 1017–1028.

- S. E. Cappell, A splitting theorem for manifolds, *Invent. Math.* **33** (1976), no. 2, pp. 69–170.

- F. T. Farrell, The obstruction to fibering a manifold over a circle, *Indiana Univ. Math. J.* **21** (1971/1972), pp. 315–346.

- S. P. Novikov, On manifolds with free abelian fundamental group and their application, *Izv. Akad. Nauk SSSR, Ser. Mat.* **30** (1966), pp. 207–246; English translation in *Amer. Math. Soc. Transl.* (2) **71** (1968), pp. 1–42.

- ——, Pontrjagin classes, the fundamental group and some problems of stable algebra, *Essays on Topology and Related Topics* (Mémoires dédiés à Georges de Rham), Springer-Verlag, New York, 1970, pp. 147–155.

- L. C. Siebenmann, *The obstruction to finding the boundary of an open manifold*, Thesis, Princeton Univ., Princeton, NJ, 1965.

## 4.6 Group actions

- J. P. Alexander, G. C. Hamrick, and J. W. Vick, Involutions on homotopy spheres, *Invent. Math.* **24** (1974), pp. 35–50.

- ——, Cobordism of manifolds with odd order normal bundle, *Invent. Math.* **24** (1974), pp. 83–94.

- W. Browder, Surgery and the theory of differentiable transformation groups, *Proc. Conf. on Transformation Groups* (New Orleans, LA, 1967), Springer-Verlag, New York, 1968, pp. 1–46.

- W. Browder and T. Petrie, Semi-free and quasi-free $S^1$ actions on homotopy spheres, *Essays on Topology and Related Topics* (Mémoires dédiés à Georges de Rham), Springer-Verlag, New York, 1970, pp. 136–146.

- ——, Diffeomorphisms of manifolds and semifree actions on homotopy spheres, *Bull. Amer. Math. Soc.* **77** (1971), pp. 160–163.

- W. Browder and F. Quinn, A surgery theory for $G$-manifolds and stratified sets, *Manifolds—Tokyo 1973* (Proc. Internat. Conf., Tokyo, 1973), Univ. of Tokyo Press, Tokyo, 1975, pp. 27–36.

- W.-C. Hsiang and W.-Y. Hsiang, Differentiable actions of compact connected classical groups, I, *Amer. J. Math.* **89** (1967), pp. 705–786. *Ibid.*, II, *Ann. of Math.* (2) **92** (1970), pp. 189–223.

- ——, The degree of symmetry of homotopy spheres, *Ann. of Math.* (2) **89** (1969), pp. 52–67.

- W.-Y. Hsiang, On the unknottedness of the fixed point set of differentiable circle group actions on spheres—P. A. Smith conjecture, *Bull. Amer. Math. Soc.* **70** (1964), pp. 678–680.

- ——, On the degree of symmetry and the structure of highly symmetric manifolds, *Tamkang J. Math.* **2** (1971), no. 1, 1–22.

- L. E. Jones, The converse to the fixed point theorem of P. A. Smith, I, *Ann. of Math.* (2) **94** (1971), pp. 52–68.

- ——, *ibid.*, II, *Indiana Univ. Math. J.* **22** (1972/73), pp. 309–325. Erratum, *ibid.* **24** (1974/75), pp. 1001–1003.

- R. Lee, Non-existence of free differentiable actions of $S^1$ and $\mathbb{Z}_2$ on homotopy spheres, *Proc. Conf. on Transformation Groups* (New Orleans, LA, 1967), Springer-Verlag, New York, 1968, pp. 208–209.

- ——, Semicharacteristic classes, *Topology* **12** (1973), pp. 183–199.

- J. Levine, Semi-free circle actions on spheres, *Invent. Math.* **22** (1973), pp. 161–186.

- S. López de Medrano, *Involutions on manifolds*, Ergebnisse der Mathematik und ihrer Grenzgebiete, vol. 59, Springer-Verlag, New York, Heidelberg, 1971.

- R. Oliver, Fixed-point sets of group actions on finite acyclic complexes, *Comment. Math. Helv.* **50** (1975), pp. 155–177.

- T. Petrie, Free metacyclic group actions on homotopy spheres, *Ann. of Math.* (2) **94** (1971), pp. 108–124.

- ——, Equivariant quasi-equivalence, transversality and normal cobordism, *Proc. International Cong. Math.* (Vancouver, B. C., 1974), vol. 1, Canad. Math. Congress, Montreal, Que., 1975, pp. 537–541.

- F. Quinn, Semifree group actions and surgery on PL homology manifolds, *Geometric topology* (Proc. Conf., Park City, Utah, 1974), Lecture Notes in Math., vol. 438, Springer-Verlag, Berlin, 1975, pp. 395–414.

- M. Rothenberg, Differentiable group actions on spheres, *Proc. Advanced Study Inst. on Algebraic Topology* (Aarhus, 1970), Mathematical Inst., Aarhus Univ., Aarhus, Denmark, 1970, pp. 455–475.

- M. Rothenberg and J. Sondow, *Nonlinear smooth representations of compact Lie groups*, mimeographed notes, Univ. of Chicago, Chicago, IL, 1969.

- R. Schultz, Differentiable group actions on homotopy spheres, I: Differential structure and the knot invariant, *Invent. Math.* **31** (1975), no. 2, pp. 105–128.

- ——, The nonexistence of free $S^1$ actions on some homotopy spheres, *Proc. Amer. Math. Soc.* **27** (1971), pp. 595–597.

- ——, Improved estimates for the degree of symmetry of certain homotopy spheres, *Topology* **10** (1971), pp. 227–235.

- ——, Semifree circle actions and the degree of symmetry of homotopy spheres, *Amer. J. Math.* **93** (1971), pp. 829–839.

- ——, Composition constructions on diffeomorphisms of $S^p \times S^q$, *Pacific J. Math.* **42** (1972), pp. 739–754.

- ——, Circle actions on homotopy spheres bounding plumbing manifolds, *Proc. Amer. Math. Soc.* **36** (1972), pp. 297–300.

- ——, Circle actions on homotopy spheres bounding generalized plumbing manifolds, *Math. Ann.* **205** (1973), pp. 201–210.

- ——, Homotopy sphere pairs admitting semifree differentiable actions, *Amer. J. Math.* **96** (1974), pp. 308–323.

- ——, Differentiable $\mathbb{Z}_p$ actions on homotopy spheres, *Bull. Amer. Math. Soc.* **80** (1974), pp. 961–964.

- ——, Circle actions on homotopy spheres not bounding spin manifolds, *Trans. Amer. Math. Soc.* **213** (1975), pp. 89–98.

- M. Sebastiani, Sur les actions à deux points fixes de groupes finis sur les sphères, *Comment. Math. Helv.* **45** (1970), pp. 405–439.

## 4.7   Embedding manifolds, knots, etc.:

- W. Browder, Embedding 1-connected manifolds, *Bull. Amer. Math. Soc.* **72** (1966), pp. 225–231; Erratum: *ibid.* **72** (1966), p. 736.

- ——, Embedding smooth manifolds, *Proc. Internat. Congr. Math.* (Moscow, 1966), Mir, Moscow, 1968, pp. 712–719.

- S. E. Cappell and J. L. Shaneson, The codimension two placement problem and homology equivalent manifolds, *Ann. of Math.* (2) **99** (1974), pp. 277–348.

- ——, Piecewise linear embeddings and their singularities, *Ann. of Math.* (2) **103** (1976), no. 1, pp. 163–228.

- ——, There exist inequivalent knots with the same complement, *Ann. of Math.* (2) **103** (1976), no. 2, pp. 349–353.

- A. Haefliger, Knotted $(4k-1)$-spheres in $6k$-space, *Ann. of Math.* (2) **75** (1962), pp. 452–466.

- ——, Differential embeddings of $S^n$ in $S^{n+q}$ for $q > 2$, *Ann. of Math.* (2) **83** (1966), pp. 402–436.

- ——, Knotted spheres and related geometric problems, *Proc. Internat. Congr. Math.* (Moscow, 1966), Mir, Moscow, 1968, pp. 437–445.

- ——, Enlacements de sphères en codimension supérieure à 2, *Comment. Math. Helv.* **41** (1966/1967), pp. 51–72.

- L. E. Jones, Topological invariance of certain combinatorial characteristic classes, *Bull. Amer. Math. Soc.* **79** (1973), pp. 981–983.

- M. Kervaire, Les noeuds de dimensions supérieures, *Bull. Soc. Math. France* **93** (1965), pp. 225–271.

- J. Levine, On differentiable imbeddings of simply-connected manifolds, *Bull. Amer. Math. Soc.* **69** (1963), pp. 806–809.

- ——, A classification of differentiable knots, *Ann. of Math.* (2) **82** (1965), pp. 15–50.

- ——, Knot cobordism groups in codimension two, *Comment. Math. Helv.* **44** (1969), pp. 229–244.

- ——, Invariants of knot cobordism, *Invent. Math.* **8** (1969), pp. 98–110; addendum, *ibid.* **8** (1969), p. 355.

- ——, Unknotting spheres in codimension two, *Topology* **4** (1965), pp. 9–16.

## 4.8 The Kervaire invariant

- W. Browder, The Kervaire invariant of framed manifolds and its generalization, *Ann. of Math.* (2) **90** (1969), pp. 157–186.

- ——, Cobordism invariants, the Kervaire invariant and fixed point free involutions, *Trans. Amer. Math. Soc.* **178** (1973), pp. 193–225.

- E. H. Brown, Jr., Generalizations of the Kervaire invariant, *Ann. of Math.* (2) **95** (1972), pp. 368–383.

- E. H. Brown, Jr., and F. P. Peterson, The Kervaire invariant of $(8k+2)$-manifolds, *Amer. J. Math.* **88** (1966), pp. 815–826.

- M. Kervaire, A manifold which does not admit any differentiable structure, *Comment. Math. Helv.* **34** (1960), pp. 257–270.

- C. P. Rourke and D. P. Sullivan, On the Kervaire obstruction, *Ann. of Math.* (2) **94** (1971), pp. 397–413.

## 4.9 Classifying spaces of surgery

- G. Brumfiel, On the homotopy groups of BPL and PL/O, *Ann. of Math.* (2) **88** (1968), pp. 291–311. *Ibid.*, II, *Topology* **8** (1969), pp. 305–311. *Ibid.*, III, *Michigan Math. J.* **17** (1970), pp. 217–224.

- G. Brumfiel and I. Madsen, Evaluation of the transfer and the universal surgery classes, *Invent. Math.* **32** (1976), no. 2, pp. 133–169.

- G. Brumfiel, I. Madsen, and R. J. Milgram, PL characteristic classes and cobordism, *Ann. of Math.* (2) **97** (1973), pp. 82–159.

- I. Madsen and R. J. Milgram, The universal smooth surgery class, *Comment. Math. Helv.* **50** (1975), no. 3, pp. 281–310.

- R. J. Milgram, Surgery with coefficients, *Ann. of Math.* (2) **100** (1974), pp. 194–248.

- J. W. Morgan and D. P. Sullivan, The transversality characteristic class and linking cycles in surgery theory, *Ann. of Math.* (2) **99** (1974), pp. 463–544.

## 4.10 Poincaré spaces

- W. Browder, Poincaré spaces, their normal fibrations and surgery, *Invent. Math.* **17** (1972), pp. 191–202.

- ——, The Kervaire invariant, products and Poincaré transversality, *Topology* **12** (1973), pp. 145–158.

- L. E. Jones, Patch spaces: a geometric representation for Poincaré spaces, *Ann. of Math.* (2) **97** (1973), pp. 306–343. Corrections, *ibid.* **102** (1975), no. 1, pp. 183–185.

- N. Levitt, Poincaré duality cobordism, *Ann. of Math.* (2) **96** (1972), pp. 211–244.

- F. Quinn, Surgery on Poincaré and normal spaces; *Bull. Amer. Math. Soc.* **78** (1972), pp. 262–267.

- M. Spivak, Spaces satisfying Poincaré duality, *Topology* **6** (1967), pp. 77–101.

- C. T. C. Wall, Poincaré complexes, I, *Ann. of Math.* (2) **86** (1967), pp. 213–245.

Department of Mathematics, Fine Hall
Princeton University
Princeton, NJ 08544-0001
email: browder@math.princeton.edu

# Differentiable structures on manifolds

## Timothy Lance

### Prologue.

Suppose $M$ is a closed smooth manifold. Is the "smoothness" of the underlying topological manifold unique up to diffeomorphism?

The answer is no, and the first, stunningly simple examples of distinct smooth structures were constructed for the 7-sphere by John Milnor as 3-sphere bundles over $S^4$.

THEOREM 1.1. (Milnor [52]) *For any odd integer $k = 2j + 1$ let $M_k^7$ be the smooth 7-manifold obtained by gluing two copies of $D^4 \times S^3$ together via a map of the boundaries $S^3 \times S^3$ given by $f_j : (u, v) \to (u, u^{1+j} v u^{-j})$ where the multiplication is quaternionic. Then $M_k^7$ is homeomorphic to $S^7$ but, if $k^2 \not\equiv 1 \bmod 7$, is not diffeomorphic to $S^7$.*

This paper studies smooth structures on compact manifolds and the role surgery plays in their calculation. Indeed, one could reasonably claim that surgery was created in the effort to understand these structures. Smooth manifolds homeomorphic to spheres, or homotopy spheres, are the building blocks for understanding smoothings of arbitrary manifolds. Milnor's example already hints at surgery's role. $M_k^7$ is the boundary of the 4-disk bundle over $S^4$ constructed by gluing two copies of $D^4 \times D^4$ along $S^3 \times D^4$ using the same map $f_j$. Computable invariants for the latter manifold identify its boundary as distinct from $S^7$.

Many homotopy spheres bound manifolds with trivial tangent bundles. Surgery is used to simplify the bounding manifold so that invariants such as Milnor's identify the homotopy sphere which is its boundary. We will encounter obstructions lying in one of the groups $0$, $\mathbf{Z}/2$, or $\mathbf{Z}$ (depending on dimension), to simplifying the bounding manifold completely to a contractible space, so that its boundary will be the usual sphere. We call these groups the Wall groups for surgery on simply connected manifolds.

Except in the concluding §7, no advanced knowledge of topology is required. Some basic definitions are given below, and concepts will be introduced, intuitively or with precision, as needed, with many references to the literature. Expanded presentations of some of this material are also available, e.g. [40] or Levine's classic paper [45].

## §1 Topological and smooth manifolds.

A topological $n$-manifold (perhaps with boundary) is a compact Hausdorff space $M$ which can be covered by open sets $V_\alpha$, called coordinate neighborhoods, each of which is homeomorphic to $\mathbf{R}^n$ (or $\mathbf{R}^{n-1} \times [0, \infty)$) via some "coordinate map" $\varphi_\alpha : V_\alpha \to \mathbf{R}^n$, with any points of the boundary $\partial M$ carried to $\mathbf{R}^{n-1} \times 0$ via the maps $\varphi_\alpha$ ($M$ is closed if no such points exist). $M$ is a smooth manifold if it has an "atlas" of coordinate neighborhoods and maps $\{(V_\alpha, \varphi_\alpha)\}$ such that the composites $\varphi_\alpha \circ \varphi_{\alpha'}^{-1}$ are smooth bijections between subsets of Euclidean space where defined (i.e., on the sets $\varphi_{\alpha'}(V_\alpha \bigcap V_{\alpha'})$.)

Similarly, $M$ is piecewise linear, or $PL$, if an atlas exists such that the composites $\varphi_\alpha \circ \varphi_{\alpha'}^{-1}$, when defined, are piecewise linear. For any $PL$ manifold there is a polyhedron $P \subset \mathbf{R}^q$ for some large $q$ and a homeomorphism $T : P \to M$, called a triangulation, such that each composite $\varphi_\alpha \circ T$ is piecewise linear. Any smooth manifold $M$ may be triangulated and given the structure of a $PL$ manifold, and the underlying $PL$-manifold is unique up to a $PL$-isomorphism.

The triangulation $T$ may be chosen so that the restriction to each simplex is a smooth map. Any $PL$ manifold clearly has an underlying topological structure. A deep result of Kirby and Siebenmann [39] (see also §7) shows that most topological manifolds may be triangulated.

We assume that all manifolds are also orientable. If $M$ is smooth this means that coordinate maps $\varphi_\alpha$ can be chosen so that the derivatives of the composites $\varphi_\alpha \circ \varphi_{\alpha'}^{-1}$ have positive determinants at every point. The determinant condition ensures the existence, for each coordinate neighborhood $V_\alpha$, of a coherent choice of orientation for $\varphi_\alpha(V_\alpha) = \mathbf{R}^n$. Such a choice is called an orientation for $M$, and the same manifold with the opposite orientation we denote $-M$. Orient the boundary (if non-empty) by choosing the orientation for each coordinate neighborhood of $\partial M$ which, followed by the inward normal vector, yields the orientation of $M$.

The sphere $S^n$, consisting of all vectors in $\mathbf{R}^{n+1}$ of length 1, is an example of an orientable smooth $n$-manifold. $S^n$ has a smooth structure with two coordinate neighborhoods $V_n$ and $V_s$ consisting of all but the south (north) pole, with $\varphi_n$ carrying a point $x \in V_n$ to the intersection with $\mathbf{R}^n \times 0$ of the line from the south pole to $x$, and similarly for $\varphi_s$. $S^n$ is the boundary of the smooth $(n+1)$-manifold $D^{n+1}$.

If $M$ is a closed, smooth, oriented manifold, then the question regarding the uniqueness of "smoothness" means the following: given another set of coordinate neighborhoods $U_\beta$ and maps $\psi_\beta$, does there exist a homeomorphism $\Phi$ of $M$ such that the composites $\varphi_\alpha \circ \Phi \circ \psi_\beta^{-1}$ and their inverses are smooth bijections of open subsets of $\mathbf{R}^n$ which preserve the chosen orientations?

One might also ask whether a topological or *PL* manifold has at least one smooth structure. The answer is again no, with the first examples due to Kervaire [37] and Milnor [52]. In this paper we assume that all manifolds have a smooth structure. But we shall see in 4.5 and again in 4.8 examples (including Kervaire's and Milnor's) of topological manifolds which have smooth structures everywhere except a single point. If a neighborhood of that point is removed, the smooth boundary is a homotopy sphere.

## §2 The groups of homotopy spheres.

Milnor's example inspired intensive study of the set $\Theta_n$ of $h$-cobordism classes of manifolds homotopy equivalent to the $n$-sphere, culminating in Kervaire and Milnor's beautiful *Groups of homotopy spheres: I* [38]. Two manifolds $M$ and $N$ are homotopy equivalent if there exist maps $f : M \to N$ and $g : N \to M$ such that the composites $g \circ f$ and $f \circ g$ are homotopic to the identity maps on $M$ and $N$, respectively. They are $h$-cobordant if each is a deformation retraction of an oriented $(n + 1)$-manifold $W$ whose boundary is the disjoint union $M \sqcup (-N)$.

For small values of $n \neq 3$ the set $\Theta_n$ consists of the $h$-cobordism class of $S^n$ alone. This is clear for $n = 1$ and 2 where each topological manifold has a unique smooth structure, uniquely determined by its homology. The triviality of $\Theta_4$, due to Cerf [24], is much harder, requiring a meticulous study of singularities. The structure of $\Theta_3$ is unknown, depending as it does on the Poincaré conjecture. But each topological 3-manifold has a unique differentiable structure ([65], [98]), so if a homotopy 3-sphere is homeomorphic to $S^3$ it is diffeomorphic to it. The vanishing of $\Theta_5$ and $\Theta_6$ will use surgery theory, but depends as well on the $h$-cobordism theorem of Smale.

THEOREM 2.1. (Smale [79]) *Any $n$-dimensional simply connected $h$-cobordism $W$, $n > 5$, with $\partial W = M \sqcup (-N)$, is diffeomorphic to $M \times [0, 1]$.*

Smale's proof is a striking demonstration of reflecting geometrically the algebraic simplicity of the triple $(W, M, N)$, that is, $H_*(W, M) \cong H_*(W, N) \cong 0$. One can find a smooth real valued function $f : (W, M, N) \to ([a, b], \{a\}, \{b\})$ such that, around each point $x \in W$ where the derivative of $f$ vanishes, there is a coordinate neighborhood $(V_\alpha, \varphi_\alpha)$ such that the composite $f \circ \varphi_\alpha^{-1} : \mathbf{R}^n \to \mathbf{R}$ equals $(x_1, \ldots x_n) \to -x_1^2 - x_2^2 - \ldots - x_\lambda^2 + x_{\lambda+1}^2 + \ldots + x_n^2$. We call $x$ a non-degenerate singularity of index $\lambda$, and $f$ a Morse function for $W$. The singularities are necessarily isolated, and f can be adjusted so that $[a, b] = [-1/2, n + 1/2]$ and $f(x) = \lambda$ for any singularity of index $\lambda$. Morse functions for $W$ not only exist, but are plentiful ([56], [57]). If $f$ could be found with no singularities, then the integral curves of this function (roughly, orthogonal trajectories to the level sets of $f$, whose existence and uniqueness follow by standard differential equations arguments) yield a diffeomorphism $W \cong M \times [0, 1]$. This is always possible given the above assumptions about trivial homology of $(W, M)$ and $(W, N)$.

To check this, let $W_\lambda = f^{-1}((-\infty, \lambda + 1/2])$, and let $M_{\lambda-1}$ be the level set $f^{-1}(\lambda - 1/2)$ (the level set of any value between $\lambda - 1$ and $\lambda$ would be equivalent). Let $x_\alpha$ be an index $\lambda$ critical point. Then $x_\alpha$ together with the union of all integral curves beginning in $M_{\lambda-1}$ and approaching $x_\alpha$ form a disk $D^\lambda_{\alpha,L}$, called the left-hand disk of $x_\alpha$, with bounding left-hand sphere $S^{\lambda-1}_{\alpha,L} \subset M_{\lambda-1}$. $W_\lambda$ is homotopy equivalent to the union of $W_{\lambda-1}$ and all left hand disks associated to critical points of index $\lambda$, so that $C_\lambda = H_\lambda(W_\lambda, W_{\lambda-1})$ is a free abelian group with a generator for each such singularity. We can similarly define, for any index $(\lambda - 1)$ critical point $y_\beta$, the right-hand disk $D^{n-\lambda+1}_{\beta,R}$ and right-hand sphere $S^{n-\lambda}_{\beta,R} \subset M_{\lambda-1}$.

If the intersection number $S^{\lambda-1}_{\alpha,L} \cdot S^{n-\lambda}_{\beta,R} = \pm 1$, we can move $S^{\lambda-1}_{\alpha,L}$ by a homotopy so that it intersects $S^{n-\lambda}_{\beta,R}$ transversely in a single point, and change $f$ to a new Morse function $g$ with the same critical points and the newly positioned left hand sphere for $x_\lambda$. (The dimension restriction $n > 5$ is critical here, providing enough room to slide $S^{\lambda-1}_{\alpha,L}$ around to remove extraneous intersection points.) With this new Morse function there is a single integral curve from $y_\beta$ to $x_\alpha$. By a result of Morse, $g$ can be further altered in a neighborhood of this trajectory to eliminate both critical points $x_\alpha$ and $y_\beta$.

This cancellation theorem is the key tool in proving the $h$-cobordism theorem. The groups $C_\lambda$ form a chain complex with $\partial_\lambda : C_\lambda \rightarrow C_{\lambda-1}$, the boundary map of the triple $(W_\lambda, W_{\lambda-1}, W_{\lambda-2})$, given explicitly by intersection numbers: the $y_\beta$ coefficient of $\partial_\lambda(x_\alpha)$ equals $S^{\lambda-1}_{\alpha,L} \cdot S^{n-\lambda}_{\beta,R}$. But $H_*(C) \cong H_*(W, M) \cong 0$. Thus for each $\lambda$, $kernel(\partial_\lambda)$ is the isomorphic image under $\partial_{\lambda+1}$ of some subgroup of $C_{\lambda+1}$. Thus the matrices for the boundary maps $\partial_\lambda$ corresponding to bases given by critical points can, by elementary operations, be changed to block matrices consisting of identity and trivial matrices. These operations can be reflected by correspondingly elementary changes in the Morse function. By the above cancellation theorem all critical points can thus be removed, and $W \cong M \times [0, 1]$.

As an immediate consequence of 2.1, two homotopy spheres are $h$-cobordant if and only if they are orientation preserving diffeomorphic. The $h$-cobordism theorem also fixes the topological type of a homotopy sphere in dimensions $\geq 6$. If $\Sigma$ is a homotopy $n$-sphere, $n \geq 6$, and $W$ equals $\Sigma$

with the interiors of two disks removed, what remains is an $h$-cobordism which, by 2.1, is diffeomorphic to $S^{n-1} \times [0,1]$. Since this product may be regarded as a boundary collar of one of the two disks, it follows that $\Sigma$ may be obtained by gluing two disks $D^n$ via some diffeomorphism $f$ of the boundaries $S^{n-1}$ of the two disks. If $\Sigma'$ is constructed by gluing $n$-disks via a diffeomorphism $f'$ of $S^{n-1}$, we may try to construct a diffeomorphism $\Sigma \to \Sigma'$ by beginning with the identity map of the "first" disk in each sphere. This map induces a diffeomorphism $f' \circ f^{-1}$ of the boundaries of the second disks, which extends radially across those disks. Such an extension is clearly a homeomorphism, and smooth except perhaps at the origin. If $n = 5$ or $6$, then $\Sigma$ bounds a contractible 6- or 7-manifold [38], and by the above argument is diffeomorphic to $S^5$ or $S^6$.

COROLLARY 2.2. ([79], [81], [100]) *If $n \geq 5$, any two homotopy $n$-spheres are homeomorphic by a map which is a diffeomorphism except perhaps at a single point.*

$\Theta_n$ has a natural group operation #, called connected sum, defined as follows. If $\Sigma_1$ and $\Sigma_2$ are homotopy $n$-spheres, choose points $x_i \in \Sigma_i, i = 1, 2$, and let $D_i$ be a neighborhood of $x_i$ which maps to the disk $D^n$ under some coordinate map $\varphi_i$ which we may assume carries $x_i$ to 0. Define $\Sigma_1 \# \Sigma_2$ as the identification space of the disjoint union $(\Sigma_1 - x_1) \sqcup (\Sigma_2 - x_2)$ in which we identify $\varphi_1^{-1}(tu)$ with $\varphi_2^{-1}((1-t)u)$ for every $u \in S^{n-1}$ and $0 < t < 1$.

Give $\Sigma_1 \# \Sigma_2$ an orientation agreeing with those given on $\Sigma_1 - x_1$ and $\Sigma_2 - x_2$ (which is possible since the map of punctured disks $tu \to (1-t)u$ induced by the gluing is orientation preserving). Intuitively, we are cutting out the interiors of small disks in $\Sigma_1$ and $\Sigma_2$ and gluing along the boundaries, appropriately oriented.

Connected sum is well defined. By results of Cerf [23] and Palais [69], given orientation preserving embeddings $g_1, g_2 : D^n \to M$ into an oriented $n$-manifold, then $g_2 = f \circ g_1$ for some diffeomorphism $f$ of $M$. (One may readily visualize independence of the choice of points $x_i$. Given $x_1$ and $x_1'$ in $\Sigma_1$, there is an $n$-disk $D \subset \Sigma_1$ containing these points in the interior and a diffeomorphism carrying $x_1$ to $x_1'$ which is the identity on $\partial D$.) Connected sum is clearly commutative and associative, and $S^n$ itself is the identity.

The inverse of any homotopy sphere $\Sigma$ is the oppositely oriented $-\Sigma$. If we think of $\Sigma \# (-\Sigma)$ as two disks $D^n$ glued along their common boundary $S^{n-1}$, then we may intuitively visualize a contractible $(n+1)$-manifold $W$ bounding $\Sigma \# (-\Sigma)$ by rotating one of the disks 180° around the boundary $S^{n-1}$ till it meets the other — rather like opening an $n$-dimensional awning with $S^{n-1}$ as the hinge. Removing the interior of a disk from the interior of $W$ yields an $h$-cobordism from $\Sigma \#(-\Sigma)$ to $S^n$.

THEOREM 2.3. (Kervaire, Milnor [38]) *For $n \neq 3$ the group $\Theta_n$ is finite.*

We shall see below that in almost all dimensions, $\Theta_n$ is a direct sum of two groups: one is a cyclic group detected, much as in Milnor's example, from invariants of manifolds which the spheres bound; the second is a quotient group of the stable $n^{th}$ homotopy of the sphere.

The above definition of $\#$ applies to arbitrary closed $n$-manifolds $M_1$ and $M_2$. Though not a group operation in this case, it does define a group action of $\Theta_n$ on $h$-cobordism classes of $n$-manifolds. For bounded manifolds the analogous operation, connected sum along the boundary, is defined as follows.

Suppose $M_i = \partial W_i$, $i = 1, 2$. Choose a disk $D^{n+1}$ in $W_i$ such that the southern hemisphere of the bounding sphere lies in $M_i$. Remove the interior of $D^{n+1}$ from $W_i$, and the interior of the southern hemisphere from $\partial W_i = M_i, i = 1, 2$. What remains of these $(n + 1)$-disks are the northern hemispheres of their bounding spheres. Glue the two resulting manifolds together along these hemispheres $D^n$ to form $W_1 \# W_2$. Restricted to the boundaries this operation agrees with $\#$ defined above, and again respects $h$-cobordism classes.

### §3 An exact sequence for smoothings.

To compute the group $\Theta_n$, we consider the tangent bundles of homotopy spheres and the manifolds they bound. Let $M$ be any compact smooth $m$-manifold. We may suppose $M$ is a differentiable submanifold of $\mathbf{R}^k$ via a differentiable inclusion $\Phi : M \to \mathbf{R}^k$ for some $k$ sufficiently large. (In fact, this is a fairly direct consequence of the definition of smooth manifold). For any $x \in M$, coordinate neighborhood $V_\alpha$ containing $x$, and coordinate map $\varphi : V_\alpha \to \mathbf{R}^m$, define the tangent space to $M$ at $x$, $\tau(M)_x$, to be the image of the derivative of $\Phi \circ \varphi_\alpha^{-1}$ at $\varphi_\alpha(x)$. Change of variables in calculus shows that the $m$-dimensional subspace $\tau(M)_x$ of $\mathbf{R}^k$ is independent of the choice of $V_\alpha$ and $\varphi_\alpha$. Define the tangent bundle $\tau(M)$ to be the set $\{(x, v) \in \mathbf{R}^k \times \mathbf{R}^k | v \in \tau(M)_x\}$ together with the map $p : \tau(M) \to M$ induced by projection to the first coordinate. The fiber $p^{-1}(x)$ is the $m$-dimensional vector space $\{x\} \times \tau(M)_x$.

The tangent bundle is a special case of an $n$-dimensional vector bundle $\xi$ consisting of a total space $E$, base space $B$, and map $p : E \to B$ which locally is a projection map of a product. Thus we assume there are open sets $U_\beta$ covering $B$ (or $M$ in the case of $\tau(M)$) and homeomorphisms : $\psi_\beta : U_\beta \times \mathbf{R}^n \to p^{-1}(U_\beta)$ such that $\psi_\beta^{-1} \circ \psi_{\beta'}$ is a linear isomorphisms on $x \times \mathbf{R}^n$ for every $x \in U_\beta \cap U_{\beta'}$, and $p \circ \psi_\beta$ is the projection onto $U_\beta$. When the base space is a smooth manifold, we will assume the maps $\psi_\beta$ are diffeomorphisms. Any operation on vector spaces defines a corresponding operation on bundles. For example, using direct sum of spaces we define the Whitney sum $\oplus$ as follows.

If $\xi_1$ and $\xi_2$ are $m$- and $n$-plane bundles, with total spaces $E_1$ and $E_2$ and common base $B$, then $\xi_1 \oplus \xi_2$ is the $(m+n)$-plane bundle with base $B$ and total space the fiber product $\{(x_1, x_2) | p_1(x_1) = p_2(x_2)\}$. Bundles over a manifold "stabilize" once the fiber dimension exceeds that of the manifold. That is, if $\xi_1$ and $\xi_2$ are bundles over an $m$-manifold $M$ of fiber dimension $k > m$, and if $\xi_1 \oplus \epsilon_M^j \cong \xi_2 \oplus \epsilon_M^j$, where $\epsilon_M^j$ is the product bundle $M \times \mathbf{R}^j \to M$, then $\xi_1 \cong \xi_2$.

We will need other vector bundles associated with $M$. If $M$ is embedded as a submanifold of an $n$-dimensional smooth manifold $N$, where for simplicity we assume both $M$ and $N$ are closed, contained in $\mathbf{R}^k$ for some $k$, and $m < n$, the $(n-m)$-dimensional normal bundle of $M$ in $N$, $\nu(M, N)$, has as fiber at $x \in M$ the elements of $\tau(N)_x$ which are orthogonal to $\tau(M)_x$. Here orthogonality can be defined using dot product in $\mathbf{R}^k$. We denote the normal bundle of $M$ in $\mathbf{R}^k$ by $\nu(M)$.

We call a manifold $M$ parallelizable if $\tau(M)$ is trivial, that is, isomorphic to $M \times \mathbf{R}^m \to M$. The sphere $S^n$ is parallelizable precisely when $n$ equals 1, 3, or 7, a magical fact proved by Bott and Milnor [10] (and independently by Kervaire) who also show that these are the only spheres which support multiplications (complex, quaternionic, and Cayley). Recall that Milnor used the quaternionic multiplication on $S^3$ in his first construction of homotopy spheres.

A somewhat weaker condition on $\tau(M)$ is stable parallelizability, that is, the bundle $\tau(M) \oplus \epsilon_M^1 \cong \epsilon_M^{n+1}$. More generally, two vector bundles $\xi_1$ and $\xi_2$ over a base $B$ are stably isomorphic if $\xi_1 \oplus \epsilon_B^j \cong \xi_2 \oplus \epsilon_B^k$ where, if $B$ is a complex of dimension $r$, the total fiber dimension of these Whitney sums exceeds $r$. Such bundles are said to be in the "stable range". A connected, compact $m$-manifold $M$ with non-trivial boundary is parallelizable iff it is stably parallelizable, since it has the homotopy type of an $(m-1)$-complex and thus $\tau(M)$ is already in the stable range.

Though few spheres are parallelizable, all are stably parallelizable. In fact, if we envision the fiber of $\nu(S^n)$ at $x$, in the usual embedding of $S^n \subset \mathbf{R}^{n+1}$, as generated by $x$, then $\nu(S^n) \cong S^n \times \mathbf{R}$. Thus $\tau(S^n) \oplus \epsilon_{S^n}^1$ is isomorphic to the restriction of the trivial tangent bundle of $\mathbf{R}^{n+1}$ to $S^n$. Far less obvious is the following result of Kervaire and Milnor ([38], 3.1), which follows from obstruction theory and deep computations of Adams about the $J$-homomorphism:

<u>THEOREM 3.1.</u> *Every homotopy sphere $\Sigma^n$ is stably parallelizable.*

As an immediate corollary, if $\Sigma^n$ is embedded in $\mathbf{R}^k$ where $k > 2n+1$, then $\nu(\Sigma^n) \cong \epsilon_{\Sigma^n}^{k-n}$. For $\tau(\Sigma^n) \oplus \epsilon_{\Sigma^n}^1 \oplus \nu(\Sigma^n)$ equals the restriction of $\mathbf{R}^k \oplus \epsilon_{\mathbf{R}^k}^1$ restricted to $\Sigma^n$. But $\tau(\Sigma^n) \oplus \epsilon^1 \cong \epsilon_{\Sigma^n}^{n+1}$ since the tangent bundle is stably parallelizable, so $\nu(\Sigma^n)$ is trivial by stability.

Given an isomorphism $\varphi : \nu(\Sigma^n) \cong \Sigma \times \mathbf{R}^{k-n}$ we define a continuous map $S^k \to S^{k-n}$ as follows. Regard $\Sigma \times \mathbf{R}^{k-n}$ as a subset of $S^k$, and $S^{k-n}$ as the disk $D^{k-n}$ with its boundary identified to a point $*$. Then send the pair $\varphi^{-1}(x,y)$, where $(x,y) \in \Sigma \times D^{k-n}$ to the point in $S^{k-n}$ corresponding to $y$, and send all other points of $S^k$ to $*$. Following [38], let $p(\Sigma^n, \varphi)$ denote the homotopy class of this map in the stable homotopy group of the sphere $\Pi_n(S) = \pi_k(S^{k-n})$.

Generally, if $(M, \varphi)$ is any $n$-manifold with framing $\varphi : \nu(M) \overset{\cong}{\to} (M \times \mathbf{R}^{k-n})$ of the normal bundle in $\mathbf{R}^k$, the same definition yields a map $p(M, \varphi) \in \Pi_n(S)$. This is the Pontrjagin-Thom construction. If $(M_1, \varphi_1) \sqcup (M_2, \varphi_2) \subset \mathbf{R}^k$ form the framed boundary of an $(n+1)$-manifold $(W, \partial W, \Phi) \subset (\mathbf{R}^k \times [0, \infty), \mathbf{R}^k \times 0)$, we say that they are framed cobordant.

<u>THEOREM 3.2.</u> (Pontrjagin [72], Thom [89]) *For any manifold $M$ with stably trivial normal bundle with framing $\varphi$, there is a homotopy class $p(M, \varphi)$ dependent on the framed cobordism class of $(M, \varphi)$. If $p(M) \subset \Pi_n(S)$ is the set of all $p(M, \varphi)$ where $\varphi$ ranges over framings of the normal bundle, it follows that $0 \in p(M)$ iff $M$ bounds a parallelizable manifold.*

The set $p(S^n)$ has an explicit description. Any map $\alpha : S^n \to SO(r)$ induces a map $J(\alpha) : S^{n+r} \to S^r$ by writing $S^{n+r} = (S^n \times D^r) \cup (D^{n+1} \times S^{r-1})$, sending $(x,y) \in S^n \times D^r$ to the equivalence class of $\alpha(x)y$ in $D^r/\partial D^r = S^r$, and sending $D^{n+1} \times S^{r-1}$ to the (collapsed) $\partial D^r$. Let $J : \pi_n(SO) \to \Pi_n(S)$ be the stable limit of these maps as $r \to \infty$. Then $p(S^n) = \text{image}(J(\pi_n(SO)) \subseteq \Pi_n(S)$.

Let $bP_{n+1}$ denote the set of those $h$-cobordism classes of homotopy spheres which bound parallelizable manifolds. In fact, $bP_{n+1}$ is a subgroup of $\Theta_n$. If $\Sigma_1, \Sigma_2 \in bP_{n+1}$, with bounding parallelizable manifolds $W_1, W_2$, then $\Sigma_1 \# \Sigma_2$ bounds the parallelizable manifold $W_1 \# W_2$ where the latter operation is connected sum along the boundary.

<u>THEOREM 3.3.</u> *For $n \neq 3$, there is a split short exact sequence*

$$0 \to bP_{n+1} \to \Theta_n \to \Theta_n/bP_{n+1} \to 0$$

where the left hand group is finite cyclic and $\Theta_n/bP_{n+1}$ injects into

$$\Pi_n(S)/J(\pi_n(SO))$$

via the Pontrjagin-Thom construction. The right hand group is isomorphic to $\Pi_n(S)/J(\pi_n(SO))$ when $n \neq 2^j - 2$.

Injectivity of $\Theta_n/bP_{n+1} \to \Pi_n(S)/J(\pi_n(SO))$ follows from 3.2; see [38] for details. Since the stable homotopy groups are finite, so is $\Theta_n/bP_{n+1}$. In the next two sections we examine how surgery is used to calculate $bP_{n+1}$ and show that the sequence splits. In particular, we will get an exact order for the group $bP_{n+1}$ for most $n$, and verify the finiteness asserted in 2.1.

## §4 Computing $bP_{n+1}$ using surgery.

Suppose $\Sigma^n \in bP_{n+1}$ bounds a parallelizable manifold $W$ whose homotopy groups $\pi_i(W)$ vanish below dimension $j$ for some $j < n/2$. With this latter restriction, any element of $\pi_j(W)$ may be represented by an embedding $f : S^j \to \text{interior}(W)$. Since $W$ is parallelizable, the restriction of $\tau(W)$ to $f(S^j)$ is trivial and hence, by stability, so is the normal bundle $\nu(f(S^j), W)$. Let $F : S^j \times D^{n+1-j} \to \text{interior}(W)$ be an embedding which extends $f$ and frames the normal bundle. Let $W(F)$ denote the quotient space of the disjoint union $(W \times [0,1]) \sqcup (D^{j+1} \times D^{n+1-j})$ in which $(x, y) \in S^j \times D^{n+1-j}$ is identified with $(F(x,y), 1) \in W \times 1$. Think of the $(n+2)$-manifold $W(F)$ as obtained from $W \times [0,1]$ by attaching a $(j+1)$-handle $D^{j+1} \times D^{n+1-j}$ via $F$. This manifold seems to have non-smooth corners near the gluing points $S^j \times S^{n-j}$, but a straightforward argument shows how to smoothly straighten the angle on this set. The resulting manifold has boundary $(W \times \{0\}) \cup (\Sigma^n \times [0,1]) \cup W'$ where $W'$, the "upper boundary" of $W(F)$, is obtained by cutting out the interior of $F(S^j \times D^{n+1-j})$, leaving a boundary equal to $S^j \times S^{n-j}$, and gluing $D^{j+1} \times S^{n-j}$ to it along its boundary.

We say that $W'$ is obtained from $W$ by doing surgery via the framed embedding $F$. Since this process attaches a $(j+1)$-disk via $f$ and $j < n/2$, it follows that $\pi_i(W') \cong \pi_i(W)$ for $i < j$, and $\pi_j(W') \cong \pi_j(W)/\Lambda$ for some group $\Lambda$ containing the homotopy class of $f$. The surgery can be done in such a way that the tangent bundle $\tau(W')$ is again trivial. The restriction of $\tau(W(F))$ to the image of $f$ has two trivializations, one coming from the parallelizability of $W \times [0,1]$, the other from the triviality of any bundle over $D^{j+1} \times D^{n+1-j}$, a contractible space. Comparing them gives a map $\alpha : S^j \to SO(n+2)$. Since $j < n - j$, this factors as a composite $S^j \xrightarrow{\beta} SO(j+1) \hookrightarrow SO(n+2)$, where the second map is the natural inclusion. (This is an elementary argument using exact homotopy sequences of fibrations $SO(r) \to SO(r+1) \to S^r$ for $r \geq j$.) It follows that the $(n+2)$-manifold $W(F_{\beta^{-1}})$ is parallelizable, where $\beta^{-1} : S^j \to SO(n-j)$ carries $x$ to $(\beta(x))^{-1}$, and $F_{\beta^{-1}}(x,y) = F(x, \beta^{-1}(x)y)$. The restriction of the tangent bundle of the "upper boundary" $W'_{\beta^{-1}}$ of $W(F_{\beta^{-1}})$ is isomorphic to $\tau(W'_{\beta^{-1}}) \oplus \epsilon^1_{W'_{\beta^{-1}}}$, with the trivial subbundle $\epsilon^1_{W'_{\beta^{-1}}}$ generated by the inward normal vectors along the boundary. Thus $W'_{\beta^{-1}}$ is stably parallelizable and, since $\partial W'_{\beta^{-1}} = \Sigma \neq \emptyset$, parallelizable.

Though surgery kills the homotopy class represented by $f$, it opens up an $(n-j)$-dimensional "hole" represented by the homotopy class of $F|_{x \times S^{n-j}}$ for any $x \in D^{j+1}$. But no matter. Our strategy is to start with a generator $g$ of the lowest non-zero homotopy group $\pi_j(W)$. As long as $j < n/2$ we can do surgery to kill $g$, adding no new homotopy in dimension $j$ or lower, and leaving $\partial M = \Sigma$ fixed. Thus working inductively on the finite number of generators in a given dimension $j$, and on the dimension, we obtain:

PROPOSITION 4.1. *If $\Sigma^n$ bounds a parallelizable manifold, it bounds a parallelizable manifold $W$ such that $\pi_j(W) = 0$ for $j < n/2$.*

Suppose that $n = 2k$. The first possible non-zero homotopy (and hence homology) group of the manifold $W$ of 4.1 occurs in dimension $k$. By Poincaré duality, all homology and cohomology of $W$ is concentrated in dimensions $k$ and $k+1$. If by surgery we can kill $\pi_k(W)$, the resulting manifold $W'$ will have trivial homology. Removing a disk from the interior of $W'$ thus yields an $h$-cobordism between $\Sigma$ and $S^n$.

But if we do surgery on $W$ using a framed embedding $F : S^k \times D^{k+1} \to$ interior$(W)$ to kill the homotopy class of $f = F|_{S^k \times 0}$, it is possible that the homotopy class of $f' = F|_{0 \times S^k}$ might be a "new" non-zero element of $\pi_k(W')$. If there were an embedding $g : S^{k+1} \to W$ whose image intersected that of $f$ transversely in a single point, then $f'$ would be null-homotopic. For we may suppose that image$(g) \cap$ image$(F) = F(x \times D^{k+1})$ for some $x \in S^k$. Then $f'$ is homotopic to $\tilde{f} = F|_{x \times S^k}$, and $\tilde{f}$ deforms to a constant in the disk formed by the image of $g$ lying outside $F(S^n \times int(D^{n+1}))$.

By moving to homology, we get criteria which are easier to fulfill and insure the triviality of $f'$. Let $\lambda \in H_k(W)$ and $\lambda' \in H_k(W')$ be the homology classes corresponding to $f$ and $f'$ under the Hurewicz isomorphism, and suppose also that $\lambda$ generates a free summand in $H_k(W)$. By Poincaré duality there is $\mu \in H_{k+1}(M)$ such that $\lambda \cdot \mu = 1$ where $\cdot$ denotes intersection number. The element $\mu$ plays the role of the transverse sphere. A straightforward argument involving homology exact sequences of the pairs $(W, W_0)$ and $(W', W_0)$, where $W_0 = W \backslash int(F(S^k \times D^{k+1}))$ shows that

$\lambda' = 0$, even when the framing $F$ is replaced by $F_{\beta-1}$ to ensure paralleliz-ability. Thus, by a sequence of surgeries, we can reduce $H_k(W)$ to a torsion group $T$.

Here the argument becomes more technical and delicate. Kervaire and Milnor show that if $k$ is even, surgery always changes the free rank of $H_k(W)$, so if $\lambda$ is a generator of $T$, surgery on $\lambda$ reduces $|T|$ at the cost of introducing non-zero $\mathbf{Z}$ summands, which are then killed by subsequent surgeries. If $k$ is odd, special care must be taken to choose a framing which both reduces the size of $T$ and preserves parallelizability. In both cases, $H_k(W)$ can be eliminated by surgery and we obtain:

THEOREM 4.2. (Kervaire-Milnor [38]) *For any* $k \geq 1, bP_{2k+1} = 0$.

## §5 The groups $bP_{2k}$.

Suppose $W$ is a parallelizable $2k$-manifold with boundary the homo-topy sphere $\Sigma^{2k-1}$, $k > 2$. As in §4, we may assume, after performing surgery on $W$ leaving $\partial W$ fixed, that $W$ is $(k-1)$-connected. By Poincaré duality the homology of $W$ is free and concentrated in dimension $k$. Once again, the homotopy class of an embedding $f : S^k \to W$ can be killed by surgery without adding new non-trivial homotopy classes if the geometry near it is nice — i.e., if $f(S^k)$ has a trivial normal bundle and there is an embedding $g : S^k \to W$ whose image intersects $f(S^k)$ transversely in a single point. Of course, we have no assurance that such transverse spheres exists and, since $\nu(f(S^k), W)$ is just below the stable range, parallelizabil-ity of $W$ does not guarantee triviality of this normal bundle. But there is a simple criterion for ensuring homological intersection conditions which enable elimination of $H_k(W)$ by surgery.

Let $k = 2m$. The intersection number defines a symmetric bilinear map $H_{2m}(W) \times H_{2m}(W) \to \mathbf{Z}$. Since $\partial W$ is homeomorphic to a sphere, we can view $W$ as a topologically closed manifold and hence, by Poincaré du-ality, the intersection pairing is non-singular. If this pairing is diagonalized over $\mathbf{R}$, define the signature $\sigma(W)$ to be the number of positive diagonal entries minus the number of negative ones.

THEOREM 5.1. ([38], [54]) *The homotopy (and hence homology) groups of* $W$ *can be killed by surgery if and only if* $\sigma(W) = 0$.

The intersection form is also even (for any homology class $\lambda \in H_{2m}(W)$, the self-intersection number $\lambda \cdot \lambda$ is an even integer), and hence the signature must be divisible by 8 ([15], [76]). A $(2m-1)$-connected parallelizable $4m$-manifold $W_0^{4m}$ with boundary a homotopy sphere and signature $\sigma(W_0^{4m})$ precisely equal to 8 can be constructed as follows. Let $E_1, E_2, \ldots, E_8$ be disjoint copies of the subset of $\tau(S^{2m})$ of vectors of length $\leq 1$. We glue $E_1$ to $E_2$ as follows. The restriction of $E_i$ to a $2m$-disk in the base is diffeomorphic to $D_{b,i}^{2m} \times D_{f,i}^{2m}$ where the subscripts $b$ and $f$ denote base and fiber (i.e., for $x \in D_{b,i}^{2m}$, an element of the base of $E_i$, $x \times D_{f,i}^{2m}$ is the fiber over $x$). Then we identify $(x, y) \in D_{b,1}^{2m} \times D_{f,1}^{2m}$ with $(y, x) \in D_{b,2}^{2m} \times D_{f,2}^{2m}$. Thus the disk in the base of $E_1$ maps onto a fiber in $E_2$, transversely crossing the base in a single point. We say that we have attached these disks by "plumbing". As before, there are corners, but these can be easily smoothed by straightening the angle. We similarly attach $E_2$ to $E_3$, $E_3$ to $E_4, \ldots$, $E_6$ to $E_7$, and $E_8$ to $E_5$. The resulting manifold $W_0^{4m}$ has boundary a homotopy sphere $\Sigma_0^{4m-1}$ and homology intersection form given by the following matrix with determinant 1 and signature 8 (see [15] or [60] for nice expositions):

$$
A = \begin{pmatrix}
2 & 1 & & & & & & \\
1 & 2 & 1 & & & & & \\
& 1 & 2 & 1 & & & & \\
& & 1 & 2 & 1 & & & \\
& & & 1 & 2 & 1 & 0 & 1 \\
& & & & 1 & 2 & 1 & 0 \\
& & & & 0 & 1 & 2 & 0 \\
& & & & 1 & 0 & 0 & 2
\end{pmatrix}
$$

where all omitted entries are 0. The 2's on the main diagonal are the self-intersections of the 0-section of $\tau(S^{2m})$ in $E_1, \ldots E_8$. Note that even though the 0-section of each $E_i$ has a sphere intersecting transversely in a single point, it cannot be killed by surgery since its normal bundle in $W_0^{4m}$ is non-trivial.

By taking connected sums along the boundary (as described in §2) we obtain, for any $j$, a parallelizable manifold with signature $8j$ and boundary a homotopy sphere. If $(W_0^{4m})^{\#j}$ equals the $j$-fold sum $W_0 \# \ldots \# W_0$ if $j > 0$, and the $(-j)$-fold sum $(-W_0) \# \ldots \# (-W_0)$ if $j < 0$, then $\sigma((W_0^{4m})^{\#j}) = 8j$. By 2.2, the boundary $(\Sigma_0^{4m-1})^{\#j}$ is a homotopy sphere, homeomorphic to $S^{4m-1}$. We use this construction to compute the cyclic group $bP_{4m}$.

THEOREM 5.2. *For $m > 1$ the homomorphism $\sigma : \mathbf{Z} \to bP_{4m}$ given by*

$$
\sigma(j) = \partial(W_0^{4m})^{\#j} = (\Sigma_0^{4m-1})^{\#j}
$$

*is a surjection with kernel all multiples of*

$$
\sigma_m = a_m 2^{2m-2}(2^{2m-1} - 1)\text{numerator}(B_m/4m)
$$

Here $a_m = 2$ or $1$ depending on whether $m$ is odd or even, and the rational Bernoulli numbers $B_m$ are defined by the power series

$$\frac{z}{e^z - 1} = 1 - \frac{z}{2} + \frac{B_1}{2!}z^2 - \frac{B_2}{4!}z^4 + \frac{B_3}{6!}z^6 - + \dots$$

This lovely result, announced in *Groups of Homotopy Spheres I* , is a confluence of earlier work of Kervaire and Milnor, the signature theorem of Hirzebruch ([31], [62]) and *J*-homomorphism computations of Adams ([1] – [4]). We sketch a proof.

Suppose $W^{4m}$ is an oriented, closed, smooth manifold with a framing $\varphi$ of the stable tangent bundle in the complement of a disc. By the signature theorem,

$$\sigma(W) = \left\langle \frac{2^{2m}(2^{2m-1} - 1)B_m}{(2m)!} p_m(W), [W] \right\rangle$$

where $p_m(W)$ is the $m^{th}$ Pontrjagin class and $[W] \in H_{4m}(W)$ is the orientation class. There is an obstruction $O(W, \varphi) \in \pi_{4m-1}(SO) \cong \mathbf{Z}$ to extending to all $W$ the given framing on $W$ less a disk. Milnor and Kervaire [61] showed that the Pontrjagin number $\langle p_m(W), [W] \rangle \in \mathbf{Z}$ corresponds to $\pm a_m(2m - 1)! O(W, \varphi)$ under this identification of groups. This shows that $O(W, \varphi)$ is independent of the choice of $\varphi$, and that an almost parallelizable $W$ is stably parallelizable iff $\sigma(W) = 0$. A straightforward argument using the Pontrjagin-Thom construction shows that an element $\gamma \in \pi_{j-1}(SO)$ occurs as an obstruction $O(W)$ to framing an almost parallelizable $W^j$ iff $J(\gamma) = 0$. A hard computation of Adams [4] showed that the order$(J(\pi_{4m-1}(SO))) =$ denominator$(B_m/4m)$, up to (perhaps) multiplication by 2 in half the dimensions. In their solutions to the Adams conjecture, Quillen [73] and Sullivan [87] showed that this multiplication by 2 is unnecessary, completing the proof.

COROLLARY 5.3. *Let $\widehat{W}_j^{4m}$ denote the space obtained from $(W_0^{4m})^{\#j}$ by attaching a cone on the boundary. If $j \not\equiv 0 \mod \sigma_m$, then $\widehat{W}_j^{4m}$ is a closed topological 4m-manifold with a smooth structure in the complement of a point, but no smooth structure overall.*

A second application of Adams' *J*-homomorphism computation yields the exact order of $\Theta_{4n-1}$. By the Pontrjagin-Thom construction, any element of the stable homotopy group $\Pi_{4n-1}(S)$ corresponds uniquely to a framed cobordism class of $(4n - 1)$-manifolds. From the same argument that showed that $bP_{4m-1} = 0$, any such class is represented by a homotopy sphere. Thus the injection of $\Theta_{4m-1}/p(S^{4m-1}) \to \Pi_{4n-1}(S)/J(\pi_{4m-1}(SO))$ of 3.3 is a bijection and we have:

THEOREM 5.4. [38] *For $m > 1$, $\Theta_{4m-1}$ has order*

$$a_m 2^{2m-4}(2^{2m-1} - 1)B_m(\text{order}(\Pi_{4m-1}(S)))/m.$$

Brieskorn ([11], [12]), Hirzebruch [33], and others ([34], [71]) have studied these homotopy spheres and their bounding manifolds in a very different context. Let $a = (a_0, a_1, \ldots, a_n)$ be an $(n + 1)$-tuple of integers $a_j \geq 2$, $n \geq 3$. Define a complex polynomial $f_a(z_0, \ldots, z_n) = z_0^{a_0} + \ldots + z_n^{a_n}$. The intersection $f_a^{-1}(0) \cap S^{2n+1}$ of the affine variety $f_a^{-1}(0) \subset \mathbf{C}^{n+1}$ with the sphere is a smooth $(2n - 1)$-manifold $M_a$. For small $\epsilon > 0$, $M_a$ is diffeomorphic to $f_a^{-1}(\epsilon) \cap S^{2n+1}$, and this in turn bounds the parallelizable $2n$-manifold $f_a^{-1}(\epsilon) \cap D^{2n+2}$. Brieskorn [12] shows that if $a = (3, 6j - 1, 2, \ldots, 2)$, with $2$ repeated $2m - 1$ times (so that $n = 2m$), then $\sigma(f_a^{-1}(\epsilon) \cap D^{2n+2}) = (-1)^m 8j$. In particular, $M_a$ is diffeomorphic to $(\Sigma_0^{4m-1})^{\#(-1)^m j}$. It follows from the work above, and is shown directly in [12] or [34], that $\sigma(f_a^{-1}(\epsilon) \cap D^{2n+2})$ is diffeomorphic to $(W_0^{4m})^{\#(-1)^m j}$.

Finally, we consider $bP_{n+1}$ when $n = 4m + 1$, even more delicate and still not computed for all $m$. Suppose $\Sigma = \partial W^{4m+2}$ where $W$ is parallelizable with framing $\varphi$. By surgery, we may assume $W$ is $2m$-connected. The obstruction to continuing this framed surgery to obtain a contractible space, the Kervaire invariant $c(W, \varphi) \in \mathbf{Z}/2$, derives from the Arf-invariant for non-singular $\mathbf{Z}/2$ quadratic forms. If $V$ is a $\mathbf{Z}/2$ vector space, we say that $\xi : V \to \mathbf{Z}/2$ is a quadratic form if $\xi(x + y) - \xi(x) - \xi(y) = (x, y)$ is bilinear, and $\xi$ is non-singular if the associated bilinear form is. Suppose $V$ is finite dimensional, and choose a symplectic basis $\{\alpha_i, \beta_i | i = 1 \ldots r\}$ where $(\alpha_i, \alpha_i) = (\beta_i, \beta_i) = 0$ and $(\alpha_i, \beta_j) = \delta_{i,j}$. Define the Arf invariant of $\xi$ by $A(\xi) = \sum_{i=1}^r \xi(\alpha_i)\xi(\beta_i) \in \mathbf{Z}/2$. A theorem of Arf [6] (see also [15], pp 54-55) states that two non-singular quadratic forms on a finite dimensional $\mathbf{Z}/2$ vector space are equivalent iff their Arf invariants agree. The Arf invariant has been used by Kervaire [37], Kervaire and Milnor [38], Browder [16], Brown [17], Brown and Peterson [18], and others to study surgery of spheres and other simply-connected $(4m + 2)$-manifolds, and by Wall ([95], [96]) to extend this work to the non-simply connected case.

THEOREM 5.5. *There is a non-singular quadratic form*

$$\psi : H^{2k+1}(W, \partial W; \mathbf{Z}/2) \to \mathbf{Z}/2$$

*with associated quadratic form* $(x, y) \to \langle x \cup y, [W] \rangle$. *Let* $c([W, \partial W], \varphi)$, *the Kervaire invariant, be the Arf invariant of* $\psi$, *which depends on the framed cobordism class of* $([W, \partial W], \varphi)$. *Then* $([W, \partial W], \varphi)$ *is framed cobordant to a contractible manifold iff* $c([W, \partial W], \varphi) = 0$. *In particular,* $bP_{4k+2}$ *is isomorphic to* $\mathbf{Z}/2$ *or* $0$.

In [38], $\psi$ is defined as a cohomology operation which detects $[\iota, \iota](x \cup y)$, where $[,]$ is the Whitehead product; Browder [16] defines $\psi$ using functional cohomology operations. Using the Poincaré duality isomorphism $H^{2k+1}(W, \partial W; \mathbf{Z}/2) \cong H_{2k+1}(W; \mathbf{Z}/2)$, an alternative description of $\psi$ may be given using homology. From the Hurewicz theorem, any integral homology class reducing to $w \in H_{2k+1}(W, \mathbf{Z}/2)$ can be represented by a map $\omega : S^{2k+1} \to W$. By [28] or [77], this map is homotopic to a framed immersion. Define $\psi_0(\omega)$ to be the self-intersection number of this immersion mod 2. Then $A(\psi) = A(\psi_0') = c(W, \varphi)$.

If $\dim(W) = 6$ or $14$, we can find a symplectic basis represented by framed embeddings, and $bP_6 = bP_{14} = 0$. Thus we suppose that $m \neq 1, 3$, and that $W^{4m+2}$ is a framed manifold with boundary a homotopy sphere. As in the case of the signature, $c(W, \varphi)$ is independent of the choice of framing. We write $c(W)$ for the Kervaire invariant, which vanishes iff the quadratic form vanishes on more than half the elements of $H^{2m+1}(W, \partial W; \mathbf{Z}/2)$ (see, e.g. [15]). But computing this invariant has proved extraordinarily hard.

By plumbing together two copies of $\tau(S^{2m+1})$, we obtain a $(4m + 2)$-manifold $W_0$ with $c(W_0) = 1$ and $\partial W_0$ a homotopy sphere. If $\partial W_0 = S^{4m+1}$, then by attaching a disk we obtain a closed, almost framed $(4m+2)$-manifold of Kervaire invariant 1. Adams [2] showed that if $m \not\equiv 3 \mod 4$, the $J$-homomorphism $\pi_m(SO) \to \Pi_m(S)$ is injective, so the almost framed manifold can be framed. Thus $bP_{4m+2} = \mathbf{Z}/2$ precisely when the Kervaire invariant vanishes for framed, closed $(4m + 2)$-manifolds — that is, when $\partial W_0^{4m+2}$ is non-trivial. Kervaire [37] showed this for dimensions 10 and 18, Brown and Peterson [18] in dimensions $8k + 2$. Browder extended this to show that the Kervaire invariant vanishes in all dimensions $\neq 2^i - 2$, and is non-zero in one of those dimensions precisely when a certain element in the Adams spectral sequence survives to $E^\infty$. Combining this with calculations of Mahowald and Tangora [48], and of Barratt, Jones, and Mahowald [8], we have:

THEOREM 5.6. $bP_{4m+2} = \mathbf{Z}/2$ *if* $4m + 2 \neq 2^i - 2$, *and vanishes if* $4m + 2 = 6, 14, 30,$ *and* $62$.

It is interesting to compare the results of this section with the $h$-cobordism theorem. Suppose, for example, that $j > 0$ is chosen so that $\partial((W_0^{4m})^{\# j})$ is the usual sphere $S^{4m-1}$. Removing a disk from the interior, we obtain a cobordism $W$ of $S^{4m-1}$ to itself. Giving $W$ a Morse function $f$ and following the proof of 2.1, it is possible to replace $f$ with a new Morse function $f'$ on $W$ with critical points of index $2m$ only, with each left hand disk corresponding to one of the $2m$-spheres used to construct $(W_0^{4m})^{\#j}$. These disks are embedded in $W$, but their bounding spheres $S^{2m-1}$ in the "lower" boundary component $S^{4m-1}$ link according to the same rules for intersections of the spheres plumbed together in constructing $(W_0^{4m})^{\#j}$.

## §6 Computation of $\Theta_n$ and number theory.

Let $\Omega_k^{\text{framed}}$ denote the family of framed cobordism classes of $k$-manifolds with a framing $\varphi$ of the stable trivial normal bundle in Euclidean space. The Pontrjagin-Thom construction gives an equivalence $\Omega_k^{\text{framed}} \cong \Pi_k(S)$ which generates the injection $\Theta_n/bP_{n+1} \to \Pi_n(S)/J(\pi_n(SO))$ of §3. In particular, $\Omega_k^{\text{framed}}$ is finite group, with disjoint union as the group operation.

By placing different restrictions on the normal bundle, we obtain other cobordism groups. For example, $\Omega_k^U$ denotes the class of manifold where the stable normal bundle has the structure of a complex vector bundle. Milnor [53] showed that the groups $\Omega_k^U$ are torsion free, so that the canonical map $\Omega_k^{\text{framed}} \to \Omega_k^U$ must be trivial. Thus for any $\Sigma \in \Theta_k$ there is a $U$-manifold $W^k$ with $\partial W = \Sigma$, even though $\Sigma$ may not bound a parallelizable manifold. When $k = 4m - 1$, Brumfiel [19] shows that $W^{4m}$ may be chosen with all decomposable Chern numbers vanishing. In this case, $\sigma(W)$ is again divisible by 8, and independent mod $8\sigma_m$ of the choice of such $W$. Define a homomorphism $\alpha_m : \Theta_{4m-1} \to \mathbf{Z}/\sigma_m$ by sending the $h$-cobordism class of $\Sigma^{4m-1}$ to $\sigma(W)/8$ mod $\sigma_m$. Then $\alpha_m$ is a splitting map for the exact sequence 3.3 in dimension $n = 4m - 1$. By similar arguments, Brumfiel ([20], [21]) defines splittings in all dimensions $n = 4m + 1$ not equal to $2^j - 3$. Combining this, the bijection $\Theta_n/bP_{n+1} \to \Pi_n(S)/J(\pi_n(SO))$, and the calculations of $bP_n$ for $n$ even we obtain:

THEOREM 6.1. *If $n = 4m + 1 \neq 2^j - 3$, then*

$$\Theta_{4m+1} \cong \mathbf{Z}/2 \oplus \Pi_{4n+1}/J(\pi_{4m+1}(SO)).$$

*If $n = 4m - 1 \geq 7$, then $\Theta_{4m-1} \cong \mathbf{Z}/\sigma_m \oplus \Pi_{4n+1}/J(\pi_{4m+1}(SO))$, where $\sigma_m = a_m 2^{2m-2}(2^{2m-1} - 1)\text{numerator}(B_m/4m)$.*

The calculation of $\Theta_n$ is thus reduced to determination of

$$\Pi_n(S)/J(\pi_n(SO)),$$

the cokernel of the $J$-homomorphism, a hard open problem in stable homotopy theory, and calculating $bP_{n+1}$. Surgery techniques yield $bP_{n+1} = 0$

when n is even, $\mathbf{Z}/2$ most of the time when $n + 1 \equiv 2 \bmod 4$ (and a hard open homotopy theory problem if $n + 1 = 2^j - 2$), and the explicit formula $bP_{4m} = \mathbf{Z}/\sigma_m$ for $m > 1$.

Even for the latter, there are intricacies and surprises. For any given $m$, it is possible (with patience) to display $bP_{4m}$ explicitly. When $m = 25$, for example, we get a cyclic group of order

$$62,514,094,149,084,022,945,360,663,993,469,995,647,144,254,828,014,731,264,$$

generated by the boundary of the parallelizable manifold $W_0^{100}$ of signature 8.

The integer $\sigma_m$ increases very rapidly with $m$, with the fastest growing contribution made by numerator$(B_m)/m$. For all $m > \pi e$,

$$\text{numerator}\left(\frac{B_m}{m}\right) > \frac{B_m}{m} > \frac{4}{\sqrt{e}}\left(\frac{m}{\pi e}\right)^{2m-\frac{1}{2}} > 1$$

where the first three terms are asymptotically equal as $m \to \infty$ (see [62], Appendix B, or [66]). As noted in §5, denominator$(B_m/4m)$ equals the image of the $J$-homomorphism ([4], [47]). Unlike the numerator, it is readily computable. In 1840 Clausen [26] and von Staudt [83] showed that denominator$(B_m)$ is the product of all primes $p$ with $(p-1)$ dividing $2m$, and the next year von Staudt showed that $p$ divides the denominator of $B_m/m$ iff it divides the denominator of $B_m$. Thus for any such prime $p$, if $p^\mu$ is the highest power dividing $m$, then $p^{\mu+1}$ is the highest power of $p$ dividing the denominator of $B_m/m$.

Such results suggest that it might be better to compute one prime $p$ at a time. Let $\mathbf{Z}_{(p)}$, the integers localized at $p$, denote the set of rational numbers with denominators prime to $p$. Then for any finite abelian group $G$, $G \otimes \mathbf{Z}_{(p)}$ is the $p$-torsion of $G$. We investigate the $p$-group $bP_{4m} \otimes \mathbf{Z}_{(p)}$.

Let $p$ be a fixed odd prime (the only 2-contribution in $\sigma_m$ comes from the factor $a_m 2^{2m-4}$), and suppose $k \in \mathbf{Z}$ generate the units in $\mathbf{Z}/p^2$. Define sequences $\{\eta_m\}$, $\{\zeta_m\}$, $\{\tilde{\sigma}_m\}$, and $\{\beta_m\}$ by $\eta_m = (-1)^{m+1}(k^{2m}-1)B_m/4m$, $\zeta_m = 2^{2m-1} - 1$, $\tilde{\sigma}_m = (-1)^m 2^{2m}(k^{2m} - 1)(2^{2m-1} - 1)B_m/2m$, and $\beta_m = (-1)^m B_m/m$ if $m \not\equiv 0 \bmod (p-1)/2$ and 0 otherwise. The first three are sequences in $\mathbf{Z}_{(p)}$ since, for any generator $k$ of the units in $\mathbf{Z}/p^2$, $\nu_p(k^{2m} - 1) = \nu_p(\text{denominator}(B_m/4m))$, where $\nu_p(x)$ denotes the exponent $p$ in a prime decomposition of the numerator of $x \in Q$ ([2], §2 or [62], Appendix B). The last sequence lies in $\mathbf{Z}_{(p)}$ from Clausen's and von Staudt's description of the denominator of $B_m/m$.

The sequence $\eta$ isolates $p$-divisibility in the numerator of $B_m/m$: $\nu_p(\eta_m) = \nu_p(\text{numerator}(B_m/m))$, and one is a unit in $\mathbf{Z}_{(p)}$ times the other.

Similarly, $\nu_p(\sigma_m) = \nu_p(\tilde{\sigma}_m)$ where $\sigma_m$ from 5.2 is the order of $bP_{4m}$. These sequences come from maps, described in §7, of the classifying space $BO$ for stable bundles. The homology of these maps yields congruences between terms of the sequences, and descriptions of the growth of $p$-divisibility of those terms satisfied for many primes $p$:

THEOREM 6.2. [42] *Let* $\lambda = (\lambda_1, \lambda_2, \ldots)$, *denote any of the above sequences.*

1) $\lambda_m \equiv \lambda_n$ *mod* $p^{k+1}$ *whenever* $m \equiv n$ *mod* $p^k(p-1)/2$.

2) *Suppose* $m \equiv n$ *mod* $(p-1)/2$ *are prime to* $p$, *and* $j$ *is minimal such that* $\nu_p(\lambda_{mp^j}) \leq j$. *Then* $\nu_p(np^i) = j$ *for all* $i \geq j$.

For the sequence $\beta$, the congruences are the familiar congruences of Kummer: $(-1)^m B_m/m \equiv (-1)^n B_n/n$ mod $p^{k+1}$ if $m \equiv n$ mod $p^k(p-1)$ and $m, n \not\equiv 0$ mod $(p-1)/2$.

For $\lambda = \eta, \zeta$, or $\tilde{\sigma}$, 6.2 gives tools for mapping out the $p$-torsion in the groups $bP_m$. If $\nu_p(\lambda_m) = 0$, that is, $\lambda_m$ is a unit in $\mathbf{Z}_{(p)}$, the same is true for any $\lambda_n$ where $n \equiv m$ mod $(p-1)/2$. Applying this to $\tilde{\sigma}$, it follows that a given group $bP_{4m}$ has $p$-torsion iff $bP_{4n}$ does for every $n \equiv m$ mod $(p-1)/2$. Thus to map out where all $p$-torsion occurs, it suffices to check $p$-divisibility of the coefficients $\tilde{\sigma}_1, \tilde{\sigma}_2, \ldots, \tilde{\sigma}_{(p-1)/2}$ Furthermore, the growth of $p$-torsion is likely quite well behaved.

CONJECTURES 6.3. *Let* $\lambda = \eta, \zeta$, *or* $\tilde{\sigma}$, *and suppose* $m_0 \in \{1, 2, \ldots (p-1)/2\}$ *is such that* $p$ *divides* $\lambda_{m_0}$ *(there could be several such* $m_0$*).*

1. *For any* $n \equiv m_0$ *mod* $(p-1)/2$ *which is prime to* $p$, *the exponents of* $p$ *in the subsequence* $\lambda_n, \lambda_{pn}, \lambda_{p^2n}, \ldots$ *are given by* $j, \nu_p(\lambda_{m_0}), \nu_p(\lambda_{m_0}), \ldots,$ *where* $j$ *may be any integer* $\geq \nu_p(\lambda_{m_0})$.

2. $\nu_p(\lambda_{m_0} - \lambda_{m_0+(p-1)/2}) = \nu_p(\lambda_{m_0})$.

3. $\eta_{m_0}$ *and* $\eta_{pm_0}$ *are non-zero mod* $p^2$.

These have been verified by computer for many primes. By the congruences in 6.2, conjectures 1 and 2 are actually equivalent. When $\lambda = \zeta$, 1 and 2 are not conjectures but true globally and easily proved. The statement for $\zeta$ analogous to 3 fails, however. There exist primes $p$, albeit not many, such that $p^2$ divides $2^{p-1} - 1$. For primes less than a million, $p = 1093$ and 3511 satisfy this. However, there is no 1093 torsion in the groups $bP_{4m}$, since $2^j \not\equiv 1$ mod 1093 for any odd exponent $j$. The prime 3511 is more interesting, with possible values of $m_0$ equal to 708 and 862 (where 3511 but not $3511^2$ divides $\eta_{m_0}$ and $\tilde{\sigma}_{m_0}$), and $m_0 = 878$ (where $3511^2$ but not $3511^3$ divides $2^{1755} - 1$ and $\tilde{\sigma}_{m_0}$).

The $p$-divisibility of these sequences have long been of interest because of their relationship to Fermat's Last Theorem, recently proved by Andrew Wiles ([99], [88]).

THEOREM 6.4. *1.* (Kummer [41]) *If $p$ does not divide the numerator of $B_m/m$ for $m = 1, 2, \ldots, (p-3)/2$, then there is no integral solution to $x^p + y^p = z^p$.*

*2.* (Wieferich [97]) *If $2^{p-1} \not\equiv 1 \bmod p^2$, then there is no integral solution to $x^p + y^p = z^p$ where $xyz$ is prime to $p$.*

A prime $p$ satisfying the condition in 1 is said to be regular. Thus $p$ is regular iff it is prime to the sequence $\eta$. The smallest irregular prime is 37, which divides the numerator of $B_{16}$. There are infinitely many irregular primes, with the same statement unknown for regular primes. Extensive computations suggest rough parity in the number of each (about 40% are irregular).

The condition in 2 is almost equivalent to $p^2$ not dividing the sequence $\zeta$ — almost, but not quite. The prime $p = 1093$ is prime to $\zeta$ even though $1093^2$ divides $(2^{1092} - 1)$ because $2^j \not\equiv 1 \bmod 1093$ for any odd factor of 1092. Vandiver [91], Miramanoff [63], and others (see [92] for an extensive summary) have shown that for primes $r \leq 43$, $r^{p-1} \not\equiv 1 \bmod p^2$ implies that $x^p + y^p = z^p$ has no integral solutions with $xyz$ prime to $p$.

All this work attempted to verify Fermat's last theorem. It would be wonderful to know whether Wiles's result could be used to establish any of the conjectures 6.3, potentially giving complete information about the $p$-torsion in the groups $bP_{4m}$. Other fairly recent algebraic results have yielded partial information. For example, by translating Ferrero and Washington's proof of the vanishing of the Iwasawa invariant [27] into the equivalent formulation using Bernoulli numbers [36], it follows that $p^2$ does not divide $(B_n/n) - B_{n+(p-1)/2}/(n + (p-1)/2)$.

## §7 Classifying spaces and smoothings of manifolds.

By comparing the linear structures on a "piecewise linear" bundle (discussed below), we are able to define a space $PL/O$ whose homotopy groups equal $\Theta_n$. Specifically, $PL/O$ is the fiber of a map $BO \to BPL$ where $BO$ and $BPL$, the spaces which classify these bundle structures, are defined as follows.

For any positive integers $n$ and $k$, let $G_n(\mathbf{R}^{n+k})$ be the compact Grassmann $nk$-manifold $O(n+k)/O(n) \times O(k)$ where $O(j)$ is the orthogonal group. We may think of this as the space of $n$-planes in $(n+k)$-space. There are natural maps $G_n(\mathbf{R}^{n+k}) \to G_n(\mathbf{R}^{n+k+1})$, and we write $BO(n) = \lim_{k \to \infty} G_n(\mathbf{R}^{n+k})$. The elements of the individual $n$-planes in

$G_n(\mathbf{R}^{n+k})$ form the fibers of a canonical $\mathbf{R}^n$-bundle $\gamma^n$, and given any $n$-bundle $\xi$ over a compact base $B$, there is a map $g : B \to BO(n)$ unique up to homotopy such that $\xi$ is isomorphic to the pullback bundle $f^*(\gamma^n)$. Set $BO = \lim_{n\to\infty} BO(n) \cong \lim_{n\to\infty} G_n(\mathbf{R}^{2n})$. The set of homotopy classes $[M, BO(n)]$ and $[M, BO]$ then correspond to $n$-dimensional and stable bundles over the compact manifold $M$.

For $PL$ manifolds the object corresponding to the vector bundle is the block bundle [74]. (Alternatively, one may use Milnor's microbundles [55]). We omit the definition, but note that the vector bundle tools used for surgery on a smooth manifold are also available for block bundles. Given any embedding $M \to N$ of $PL$ manifolds, for example, there is a normal block bundle of $M$ in $N$. One may construct a classifying space $BPL(n)$ for $n$-dimensional block bundles, which is the base space for a universal block bundle $\gamma^n_{PL}$. (We abuse notation slightly; $BPL$ is often denoted $\widetilde{BPL}$ in the literature, with $BPL$ used for its equivalent in the semisimplicial category.)

Set $BPL = \lim_{n\to\infty} BPL(n)$. Piecewise differentiable triangulation of the canonical vector bundles $\gamma^n$ yields $\gamma^n_{PL}$, classifying maps $BO(n) \to BPL(n)$, and the limit map $BO \to BPL$. Regarding this map as a fibration, we define $PL/O$ to be its fiber.

Products and Whitney sums of block bundles are defined analogously to $\times$ and $\oplus$ for vector bundles. Using these constructions, we obtain commutative $H$-space structures $\mu^\oplus : BO \times BO \to BO$ and $\mu^\oplus_{PL} : BPL \times BPL \to BPL$ under which the map $BO \to BPL$ is an $H$-map, and defines an $H$-space structure on the fiber $PL/O$ as well.

Let $\mathcal{S}(M)$ denote the set of concordance classes of smoothings of a $PL$ manifold $M$, where two smoothings of $M$ are concordant if there is a smoothing of $M \times [0, 1]$ which restricts to the given smoothings on $M \times 0$ and $M \times 1$. If $M$ is the smooth triangulation of a smooth manifold $M_\alpha$, we think of $\mathcal{S}(M_\alpha)$ as the concordance classes of smoothings of $M$ with a given preferred smoothing $M_\alpha$. Note that $\mathcal{S}(S^n) = \Theta_n$ except for $n = 3$. The unique smooth and $PL$ structure on a topological $S^3$ dictates that $\mathcal{S}(S^3)$ consists of a single element.

If a $PL$ manifold $M$ has a smooth structure, then the normal block bundle of the diagonal $\Delta$ in $M \times M$ is actually the normal vector bundle. In fact, such a linearization is sufficient for existence of a smoothing. For any $PL$ manifold $M$ and submanifold $K \subset M$, a linearization of $(M, K)$ is a piecewise differentiable vector bundle $p : E \to K$ where $E$ is a neighborhood of $K$ in $M$, and $M$ induces a compatible $PL$ structure on $E$. Let $\mathcal{L}(M, K)$ denote the set of all equivalence classes of such linearizations, and $\mathcal{L}_s(M, K)$

the classes of stable linearizations (i.e., the direct limit of $\mathcal{L}(M, K) \to \mathcal{L}(M \times \mathbf{R}^1, K) \to \mathcal{L}(M \times \mathbf{R}^2, K) \to \ldots$, where the maps are defined by Whitney sum with a trivial bundle).

THEOREM 7.1. ([55], [30], [44]) *A closed PL-manifold $M$ has a smooth structure iff $\mathcal{L}(M \times M, \Delta) \neq \emptyset$ iff $\mathcal{L}_s(M \times M, \Delta) \neq \emptyset$, and there is a bijection $\mathcal{S}(M) \to \mathcal{L}_s(M \times M, \Delta)$.*

This description uses block bundles and follows the notation of [30], but essentially identical results using microbundles are true. The theorem suggests that "smoothability" is a stable phenomenon. This is true; the natural map $\mathcal{S}(M) \to \mathcal{S}(M \times \mathbf{R}^m)$ is a bijection (see [29], [50] for smoothing products with $\mathbf{R}$, or [90], [65] for the $M \times [0, 1]$ analogue). By 7.1, a $PL$ manifold $M$ supports a smooth structure iff the classifying map $M \to BPL$ for the stable normal block bundle of the diagonal $\Delta \subset M \times M$ lifts to $BO$. But the homotopy classes of such lifts are in turn classified by maps into the fiber of $BO \to BPL$:

THEOREM 7.2. ([30], [44]) *Let $M$ be closed $PL$-manifold which can be smoothed, and let $M_\alpha$ be some fixed smooth structure on it. Then there is a bijection $\Psi_\alpha : \mathcal{S}(M) \to [M, PL/O]$ which carries the concordance class of $M_\alpha$ to the trivial homotopy class.*

Despite apparent dependence on a particular smooth structure and, given $M_\alpha$, on a choice of smooth triangulation, there is a great deal of naturality in the bijection $\Psi_\alpha$. If $N$ is another smoothable $PL$ manifold with chosen smooth structure $N_\beta$, and if $f : M_\alpha \to N_\beta$ is both a diffeomorphism and $PL$-homeomorphism, then $\Psi_\beta \circ f^* = f^\# \circ \Psi_\alpha$ where $f^* : \mathcal{S}(N) \to \mathcal{S}(M)$ and $f^\# : [N, PL/O] \to [M, PL/O]$ are the natural maps. The bijections $\Psi_\alpha$ can be used to reformulate 7.2 as a well defined homotopy functor defined on "resmoothings" of a smooth manifold [30].

The $H$-multiplication $PL/O \times PL/O \to PL/O$ gives $[M, PL/O]$ the structure of an abelian group. Given a smoothing $M_\alpha$ of $M$, the bijection $\Psi_\alpha$ gives $\mathcal{S}(M)$ the structure of a group, which we denote $\mathcal{S}(M_\alpha)$. For any smoothing $M_\beta$ of $M$, let $[M_\beta]$ denote its concordance class, a group element of $\mathcal{S}(M_\alpha)$

THEOREM 7.3. *The group operation $*$ in $\mathcal{S}(M_\alpha)$ is given by $[M_\beta] * [M_\gamma] = [M_\omega]$ where $M_\omega$ is the unique (up to concordance) smoothing such that the germ of the smooth manifold $M_\alpha \times M_\omega$ along the diagonal equals that of $M_\beta \times M_\gamma$. In particular, $[M_\alpha]$ is the identity element. If $M = S_0^n$ denotes the $n$-sphere regarded as a $PL$ manifold given the usual smoothing, the resulting bijection $\Psi_0$ is a group isomorphism $\Theta_n \to \pi_n(PL/O)$ for $n \neq 3$.*

Using 7.2, the $H$-space structure on $PL/O$, induced by Whitney sum of vector- and block-bundles, allowed us to define (isomorphic) group structures on $\mathcal{S}(M)$ via the bijections $\Psi_\alpha$.

It is interesting to note that the finite group structures on $\Theta_n$ (with 0 substituted for the unknown $\Theta_3$) can in turn be used to describe the $H$-multiplication on $PL/O$. See [30] for a proof. Theorem 7.2 also provides a homotopy theoretic description of smoothings for an arbitrary smoothable $PL$ manifold, one which recasts the obstruction theories for smoothings of Munkres [65] and Hirsch [29] in terms of classical obstruction theory.

We examine the homotopy theory of $PL/O$, studying it one prime at a time, just as we did for the coefficients in §6. For any prime $p$ and well-behaved space $X$ (for example, any $CW$ complex or any $H$-space), there is a space $X_{(p)}$, the localization of $X$ at $p$, and map $X \to X_{(p)}$ which on homotopy groups is the algebraic localization $\pi_n(X) \to \pi_n(X) \otimes \mathbf{Z}_{(p)}$. Similarly, $H_*(X_{(p)}) \cong H_*(X) \otimes \mathbf{Z}_{(p)}$. We will see below that the localization $PL/O_{(p)}$ is a product, reflecting homotopy theoretically the splitting of the exact sequence of 3.3.

Suppose first that $p$ is an odd prime. The sequences $\eta, \zeta, \tilde{\sigma}$, and $\beta$ of §6 all arise from self-maps of the $p$-localizations of $BO$ and $BU$ (the analogue of $BO$ which classifies stable complex vector bundles) which are reflections of geometric operations on bundles. An important example is the Adams map $\psi^k : BU_{(p)} \to BU_{(p)}$, which arises from the $K$-theory operation

$$\psi^k(x) = \sum_{w(\alpha)=k} (-1)^{|\alpha|+k}(k/|\alpha|)\{\alpha\}(\wedge^1(x)^{\alpha_1} \ldots \wedge^j(x)^{\alpha_j}),$$

where $x \in K(X)$, $\wedge^i$ is the exterior power, the sum is taken over all $j$-tuples of non-negative integers $\alpha = (\alpha_1, \ldots, \alpha_j)$ of weight $w(\alpha) = \alpha_1 + 2\alpha_2 + \ldots + j\alpha_j = k$, and $\{\alpha\}$ is the multinomial coefficient $(\alpha_1 + \ldots + \alpha_j)!/\alpha_1! \ldots \alpha_j!$. The reader may recognize this as the Newton polynomial applied to exterior operators. The induced map on the homotopy group $\pi_{2m}(BU_{(p)}) = \mathbf{Z}_{(p)}$ is multiplication by $k^m$. The Adams map on $BU_{(p)}$ induces one on $BO_{(p)}$ by the following:

THEOREM 7.4. (Adams [5], Peterson [70]).  *There are $H$-space equivalences*

$$BU_{(p)} \to W \times \Omega^2 W \times \ldots \Omega^{2p-4}W \quad and \quad BO_{(p)} \to W \times \Omega^4 W \times \ldots \Omega^{2p-6}W$$

*where $\pi_{2j(p-1)}(W) = \mathbf{Z}_{(p)}, j = 1, 2, \ldots,$ and $\pi_i(W) = 0$ otherwise. In particular, $BU_{(p)} \cong BO_{(p)} \times \Omega^2 BO_{(p)}$. Any $H$-map $f : BO_{(p)} \to BO_{(p)}$*

induces $H$-maps $f_{4j}$ of $\Omega^{4j}W$, and under this equivalence $f$ becomes a product $f_0 \times f_4 \times \ldots \times f_{2p-6}$, with an analogous decomposition for a self $H$-map of $BU_{(p)}$.

In fact, these are infinite loop space equivalences. Peterson constructs $W$ as the bottom space of a spectrum associated to a bordism theory with singularities. The maps $f_{4j}$ allow us to write the fiber $F$ of $f$ as a product $F_0 \times F_4 \times \ldots \times F_{2p-6}$, where $F_{4j}$ can be seen to be indecomposable by examining the action of the Steenrod algebra on it.

Returning to the Adams map, associated to the $K$-theory operation $x \to \psi^k(x) - x$ is an $H$-map $\psi^k - 1$ of $BU_{(p)}$ and, by 7.4, $\psi^k - 1 : BO_{(p)} \to BO_{(p)}$. The induced homomorphism on $\pi_{4m}(BO_{(p)}) \cong \mathbf{Z}_{(p)}$ is multiplication by $k^{2m} - 1$. If $k$ generates the units in $\mathbf{Z}/p^2$, the fiber $J$ of $\psi^k - 1$ (sometimes called "Image $J$") is independent, up to infinite loop space equivalence, of the choice of $k$, and has homotopy groups equal to the $p$ component of the image of the $J$-homomorphism. Since $k^{2j} \not\equiv 1 \mod p$ unless $j \equiv 0 \mod (p-1)/2$, using 7.4 we may also describe $J$ as the fiber of $(\psi^k - 1)_0 : W \to W$.

The numbers $k^{2j} - 1$ are an example of a "characteristic sequence". If $f : BO_{(p)} \to BO_{(p)}$ induces multiplication by $\lambda_j \in \mathbf{Z}_{(p)}$ on $\pi_{4m}(BO_{(p)})$, it is determined up to homotopy by the characteristic sequence $\lambda = (\lambda_1, \lambda_2, \ldots)$ (with a similar statement for self-maps of $BU$). The elements of $\lambda$ satisfy the congruences and $p$-divisibility conditions of 6.2 for any self-map of $BO_{(p)}$, and analogous statements for almost any map (including all $H$-maps) of $BU_{(p)}$ ([42], [25]).

We can use the Adams map and other bundle operations to realize the sequences of §6 as characteristic sequences. To construct these maps we must depart from the more self-contained material of the first six sections, and in particular require $p$-local versions of oriented bundle theory and the Thom isomorphism $\Phi$. We refer the reader to May [49] or the Adams $J(X)$ papers [1] - [4] for beautiful presentations of this material, and present the constructions without greater detail simply to show that these sequences, so rich with number theoretic information, all arise geometrically.

$\zeta$. Since $p$ is an odd prime, $\psi^2$ is a homotopy equivalence. Then

$$(1/2)(\psi^2)^{-1} \circ (\psi^4 - 2\psi^2) : BO_{(p)} \to BO_{(p)}$$

has characteristic sequence $\zeta$, where $\zeta_m = 2^{2m-1} - 1$.

$\eta$. For an oriented bundle $\xi$ define $\rho^k(\xi) = (\Phi)^{-1}\psi^k\Phi(1).\in KO(B)$ where $B$ is the base of the bundle $\xi$ and $k$ generates the units in $\mathbf{Z}/p^2$ as above. The resulting $H$-map $\rho^k : BO^{\oplus}_{(p)} \to BO^{\otimes}_{(p)}$, the so called Adams-Bott cannibalistic class ([2], [9], [49]), has characteristic sequence $\eta$ given by $\eta_m = (-1)^{m+1}(k^{2m} - 1)B_m/4m$. The superscripts $\oplus$ and $\otimes$ indicate that $BO_{(p)}$ carries the $H$-multiplication coming from Whitney sum and tensor product, respectively.

$\tilde{\sigma}$. In the $J(X)$ papers ([1] to [4]) Adams conjectured, and Sullivan [87] and Quillen [73] proved, that for any $x \in K(X)$, where $X$ is a finite complex, the underlying spherical fiber space of $k^q(\psi^k(x) - x)$ is stably trivial for large enough $q$. Localized at $p$, this means that the map $BO_{(p)} \overset{\psi^k - 1}{\longrightarrow} BO_{(p)} \to BG_{(p)}$ is null-homotopic, where $BG$ is the classifying space for stable spherical fiber spaces. (Its loop space $G = \Omega BG$ has homotopy equal to the stable homotopy of spheres.) Thus $\psi^k - 1$ lifts to a map $\gamma^k : BO_{(p)} \to (G/O)_{(p)}$, where $G/O$ is the fiber of $BO \to BG$. In his thesis [84] Sullivan showed that, when localized at an *odd* prime $p$, the fiber $G/PL$ of $BPL \to BG$ is $H$-space equivalent to $BO_{(p)}^\otimes$. Let $\theta^k$ denote the composite $BO_{(p)}^\oplus \overset{\gamma^k}{\to} (G/O)_{(p)} \to (G/PL)_{(p)} \overset{\approx}{\to} BO_{(p)}^\otimes$. Then $\theta^k$ has characteristic sequence $\tilde{\sigma}$ where $\tilde{\sigma}_m = (-1)^m 2^{2m}(k^{2m} - 1)(2^{2m-1} - 1)B_m/2m$.

$\beta$. For $j = 1, 2, \ldots, (p-3)/2$, define $b_j : \Omega^{4j}W \to \Omega^{4j}W$ to be $\rho^k \circ (\psi^k - 1)^{-1}$, and let $b_0$ be the constant map on $W$. Taking the product of these maps and applying 7.4, the resulting map $b : BO_{(p)} \to BO_{(p)}$ has characteristic sequence $\beta$ where $\beta_m = (-1)^m B_m/m$.

Let $bP$ denote the fiber of $\theta^k$. Clearly there a map $\iota : bP \to (PL/O)_{(p)}$. There is also a $p$-local space $C$, the so-called "cokernel of $J$", whose homotopy is the $p$-component of the cokernel of the $J$-homomorphism in $\Pi_*(S)$ (see e.g. [49]). These spaces yield a $p$-local splitting of $PL/O$ (see [49] for a beautiful presentation).

THEOREM 7.5. *There is a map* $\kappa : C \to (PL/O)_{(p)}$ *such that the composite*

$$bP \times C \overset{\iota \times \kappa}{\longrightarrow} (PL/O)_{(p)} \times (PL/O)_{(p)} \overset{mult}{\longrightarrow} (PL/O)_{(p)}$$

*is an equivalence of $H$-spaces (indeed, of infinite loop spaces).*

The homotopy equivalence $(PL/O)_{(p)} \approx bP \times C$ now yields a $p$-local isomorphism $\Theta_n \cong bP_{n+1} \oplus \Pi_n(S)/J(\pi_n(SO))$, not just when $n = 4m - 1$ (6.1), but in all dimensions $\neq 3$ (where $\pi_3(PL/O)$ vanishes). Hidden in this is the representability of framed cobordism classes by homotopy spheres. In dimension $4m$ this is a consequence of the signature theorem and finiteness of $\Pi_*(S)$. In dimension $4m + 2$ there may be a Kervaire invariant obstructions to such a representative, but the obstruction lies in $\mathbf{Z}/2$ and hence vanishes when localized at odd $p$.

Let $L$ and $M$ denote the ($p$-local) fibers of the maps $2\psi^2 - \psi^4$ and $\rho$. Comparing characteristic sequences, it follows that $\theta^k$ is homotopic to $\rho^k \circ (2\psi^2 - \psi^4)$, and there is a fibration $L \to bP \to M$. If the factor spaces $L_{4j}$ and $M_{4j}$ are both non-trivial for some $j = 1, \ldots, (p-3)/2$, the induced fibration $L_{4j} \to bP_{4j} \to M_{4j}$ cannot be a product. Otherwise, we do have a homotopy equivalence $bP \approx L \times M$ by 7.4.

This is usually the case; $bP$ can be written this way, for example, for all primes $p < 8000$ except $p = 631$. When $p = 631$, both $\rho_{452}^k$ and $(2\psi^2 - \psi_{452}^4)$ are non-trivial, and the indecomposable space $bP_{452}$ cannot be written as a product.

In sections 4 and 5, we saw that it was possible to kill a framed homotopy class $x$ by surgery with no new homotopy introduced if there was a framed sphere crossing a representing sphere for $x$ transversely at a single point. The same kind of criterion provides a tool for computing the group $[M, bP]$ of smoothings classified by $bP$. Suppose $M$ is a smoothable $PL$ $n$-manifold, with smooth handle decomposition $\emptyset \subseteq M_0 \subseteq M_1 \subseteq \ldots \subseteq M_n = M$, where each $M_j$ is obtained from $M_{j-1}$ by attaching handles $D^j \times D^{n-j}$ via embeddings $\varphi_\alpha : S^{j-1} \times D^{n-j} \to \partial M_{j-1}$. Thus with each attachment of a $j$-handle $h_\alpha^j$ we are performing surgery on a homotopy class in $\partial M_{j-1}$. For any $j$-handle, we refer to the image of $D^j \times 0$ as the left hand disk of the handle.

Handles give us a way of trying to build new smoothings. Given a $j$-handle $h_\alpha^j$, and a homotopy $j$-sphere $\Sigma$ regarded as the union of two $j$-disks attached by a diffeomorphism $f_{j,\beta} : S^{j-1} \to S^{j-1}$, form a new smooth manifold $M_j \# \Sigma$ by attaching the handle using the map $\varphi_\alpha \circ (f_{j,\beta} \times 1)$. We describe circumstances under which such resmoothings extend to all of $M$ and give a tool for explicitly calculating smoothings.

Suppose that the homology of $M$ and of its suspension $\Sigma M$ is $p$-locally Steenrod representable. Thus given any $x$ in the $p$-local homology of $M$ or of $\Sigma M$, there is an orientable smooth manifold $X$ and map $X \to M$ which carries the orientation class of $X$ to $x$. Suppose in addition that the odd prime $p$ satisfies the conjecture 6.3. Then there is a set of manifolds $\{X_\alpha\}$, and maps $X_\alpha \to M$ with the top handle $D^{j_\alpha}$ of $X_\alpha$ mapped homeomorphically onto the left hand disk of some $j_\alpha$-handle $h_\alpha^{j_\alpha}$ (the rest mapping to $M_{j_\alpha - 1}$) satisfying the following: any resmoothing of $M$ corresponding to a homotopy class in $[M, bP]$ is formed by extensions to $M$ of smoothings of $M_{j_\alpha}$ of the form $M_{j_\alpha} \# \Sigma_\beta^{j_\alpha}$. This is a sort of "characteristic variety theorem" for smoothings classified by $bP$. See [43] for details.

We conclude with a few remarks about the prime 2, where life is very different. At the outset we have the problem posed by Kervaire in 1960, and still not completely settled, on the existence of a framed, closed $(4m + 2)$-manifolds $W$ with Kervaire invariant 1. This prevents the algebraic splitting $\Theta_{4m+1} \cong \mathbf{Z}/2 \oplus \Theta_{4m+1}/bP_{4m+2}$ for some values of $m$. Furthermore, not all parallelizable manifolds are representable by homotopy spheres, so we may not in general identify $\Theta_{4m+1}/bP_{4m+2}$ with the cokernel of the $J$-homomorphism.

Many of the results above at odd primes depend on the solution of the Adams conjecture — a lift $\gamma^k : BSO_{(2)} \to (G/O)_{(2)}$ of $\psi^k - 1$. Such a solution exists at 2, but cannot be an $H$-map, a condition needed to define the $H$-space structure for $bP$.

Finally, at odd $p$ Sullivan defined an equivalence of $H$-spaces

$$(G/PL)_{(p)} \to (BO^{\otimes})_{(p)}.$$

At 2, $G/PL$ is equivalent to a product

$$S \times \prod_{j \geq 1} (K(\mathbf{Z}/2, 4j + 2) \times K(\mathbf{Z}_{(2)}, 4j + 4)),$$

where $K(G, n)$ denotes the Eilenberg-Maclane space with a single homotopy group $G$ in dimension $n$, and where $S$ is a space with two non-zero homotopy groups, $\mathbf{Z}/2$ in dimension 2, and $\mathbf{Z}_{(2)}$ in dimension 4. There is a non-trivial obstruction in $\mathbf{Z}/2$ to $S$ being a product, the first $k$-invariant of $S$. For the 2-localization of the analogous space $G/TOP$ (the fiber of $BTOP \to BG$) that obstruction vanishes. This is a consequence of the extraordinary work of Kirby and Siebenmann:

THEOREM 7.6. [39] *The fiber $TOP/PL$ of the fibration $G/PL \to G/TOP$ is an Eilenberg-MacLane space $K(\mathbf{Z}/2, 3)$, and the following homotopy exact sequence does not split:*

$$0 \to \pi_4(G/PL) \to \pi_4(G/TOP) \to \pi_3(TOP/PL) \to 0.$$

## Epilogue.

Surgery techniques, first developed to study smooth structures on spheres, have proved fruitful in an extraordinary array of topological problems. The Browder-Novikov theory of surgery on normal maps of simply connected spaces, for example, attacked the problem of finding a smooth manifold within a homotopy type. This was extended by Wall to the non-simply connected space. Surgery theory has been used to study knots and links, to describe manifolds with special restrictions on their structure (e.g., almost complex manifolds, or highly connected manifolds), and to under-stand group actions on manifolds. The articles in this volume expand on some of these topics and more, and further attest to the rich legacy of surgery theory.

## References

1. Adams, J. F., *On the groups J(X). I*, Topology **2** (1963), 181–195.

2. Adams, J. F., *On the groups J(X). II*, Topology **3** (1965), 136–171.

3. Adams, J. F., *On the groups J(X). III*, Topology **3** (1965), 193–222.

4. Adams, J. F., *On the groups J(X). IV*, Topology **5** (1966), 21–71.

5. Adams, J. F., *Lectures on generalized cohomology*, Homology Theory and Applications, III, Lecture Notes in Mathematics No. 99, Springer-Verlag, New York, 1969.

6. Arf, C., *Untersuchungen über quadratische Formen in Körpern der Charakteristik 2*, Jour. für reine u. angew. Math. **183** (1941), 148–167.

7. Barden, D., *The structure of manifolds*, Ph.D. thesis, Cambridge, 1963.

8. Barratt, M., Jones, J., and Mahowald, M., *Relations amongst Toda brackets and the Kervaire invariant in dimension 62*, J. London Math. Soc. (Series 2) **30** (1984), 533–550

9. Bott, R., *Lectures on K(X)*, W. A. Benjamin, Inc., 1969.

10. Bott, R., and Milnor, J.W., *On the parallelizability of the spheres*, Bull. Amer. Math. Soc. **64** (1958), 87–89.

11. Brieskorn, E. V., *Examples of singular normal complex spaces which are topological manifolds*, Proc. Nat. Acad. Sci. U.S.A. **55** (1966), 1395–1397.

12. Brieskorn, E. V., *Beispiele zur Differentialtopologie von Singularitäten*, Invent. Math. **2** (1966), 1–14.

13. Browder, W., *Homotopy type of differentiable manifolds*, Proceedings of the Aarhus Symposium, (1962), 42–46; reprinted in "Proc. 1993 Oberwolfach Conference on the Novikov conjectures, Rigidity and Index Theorems, Vol. 1." London Math. Soc. Lecture Notes **226**, Cambridge, 1995, 97–100.

14. Browder, W., *Diffeomorphisms of 1-connected manifolds*, Trans. Amer. Math. Soc. **128** (1967), 155–163.

15. Browder, W., *Surgery on Simply Connected Manifolds*, Springer-Verlag, Berlin, Heidelberg, New York, 1972

16. Browder, W., _The Kervaire invariant of a framed manifold and its generalization_, Annals of Math. **90** (1969), 157–186.

17. Brown, E. H., _Generalizations of the Kervaire invariant_, Annals of Math. **96** (1972), 368–383.

18. Brown, E. H., and Peterson, F. P., _The Kervaire invariant of $(8k+2)$-manifolds_, Amer. J. Math. **88** (1966), 815–826.

19. Brumfiel, G., _On the homotopy groups of BPL and PL/O_, Annals of Math. (2) **88** (1968), 291–311.

20. Brumfiel, G., _On the homotopy groups of BPL and PL/O, II_, Topology **8** (1969), 305–311.

21. Brumfiel, G., _On the homotopy groups of BPL and PL/O, III_, Michigan Math. Journal **17** (1970), 217–224.

22. Brumfiel, G., _Homotopy equivalence of almost smooth manifolds_, Comment. Math. Helv. **46** (1971), 381–407.

23. Cerf, J., _Topologie de certains espaces de plongements_, Bull. Soc. Math. France **89** (1961), 227–380.

24. Cerf, J., _Sur les difféomorphismes de la sphère de dimension trois $(\Gamma_4 = 0)$_, Lecture Notes in Mathematics No. 53, Springer-Verlag, New York, 1968.

25. Clarke, F. W., _Self-maps of BU_, Math. Proc. Cambridge Philos. Soc. **89** (1981), 491–500.

26. Clausen, T., _Auszug aus einem Schreiben an Herrn Schumacher_, Astronomische Nachr. **17** (1840), 351–352.

27. Ferrero, B., and Washington, L., _The Iwasawa invariant $\mu_p$ vanishes for algebraic number fields_, Annals of Math. (2) **109** (1979), 377–395.

28. Hirsch, M., _Immersions of manifolds_, Trans. Amer. Math. Soc. **93** (1959), 242–276.

29. Hirsch, M., _Obstructions to smoothing manifolds and maps_, Bull. Amer. Math. Soc. **69** (1963), 352–356.

30. Hirsch, M., and Mazur, B., _Smoothings of Piecewise Linear Manifolds_, Annals of Math. Studies **80**, Princeton University Press, 1974.

31. Hirzebruch, F., _New topological methods in algebraic geometry_, Springer-Verlag, Berlin, Heidelberg, New York, 1966

32. Hirzebruch, F., _Differentiable manifolds and quadratic forms_, Lecture Notes, Marcel Decker, New York, 1962.

33. Hirzebruch, F., _Singularities and exotic spheres_, Séminaire Bourbaki **314**, 1966/67.

34. Hirzebruch, F., and Mayer, K., *O(n)-Mannigfaltigkeiten, exotische Sphären und Singularitäten*, Lecture Notes in Mathematics No. 57, Springer-Verlag, Berlin, Heidelberg, New York, 1968.

35. Hudson, J.F.P., *Piecewise Linear Topology*, W. A. Benjamin, Inc., New York, 1969.

36. Iwasawa, K., *On some invariants of cyclotomic fields*, Amer. J. Math. **80** (1958), 773–783; erratum **81** (1959), 280.

37. Kervaire, M. A., *A manifold which does not admit any differentiable structure*, Comment. Math. Helv. **34** (1960), 257–270.

38. Kervaire, M.A., and Milnor, J.W., *Groups of homotopy spheres: I*, Annals of Math. **77** (1963), 504–537.

39. Kirby, R., and Siebenmann, L., *Foundational Essays on Topological Manifolds, Smoothings, and Triangulations*, Annals of Math. Studies 88, Princeton University Press, 1977.

40. Kosinski, A., *Differentiable Manifolds*, Academic Press, New York, 1993.

41. Kummer, E. E., *Allgemeiner Beweis des Fermat'schen Satzes dass die Gleichung $x^\lambda + y^\lambda = z^\lambda$ durch ganze Zahlen unlösbar ist, für alle diejenigen Potenz-Exponenten $\lambda$, welche ungerade Primzahlen sind und in die Zählern der ersten $\frac{1}{2}(\lambda - 3)$ Bernoulli'schen Zahlen als Factoren nicht vorkommen*, Jour. für reine u. angew. Math. **40** (1850), 130–138.

42. Lance, T., *Local H-maps of classifying spaces*, Trans. Amer. Math. Soc. **107** (1979), 195–215.

43. Lance, T., *Local H-maps of BU and applications to smoothing theory*, Trans. Amer. Math. Soc. **309** (1988), 391–424.

44. Lashof, R., and Rothenberg, M., *Microbundles and smoothing*, Topology **3** (1965), 357–380.

45. Levine, J. *Lectures on groups of homotopy spheres*, Algebraic and Geometric Topology, Lecture Notes in Mathematics No. 1126, Springer-Verlag (1985), 62–95.

46. Madsen, I., and Milgram, J., *The Classifying Spaces for Surgery and Cobordism of Manifolds*, Annals of Math. Studies 92, Princeton University Press, 1979.

47. Mahowald, M., *The order of the image of the J-homomorphism*, Bull. Amer. Math. Soc. **76** (1970), 1310–1313.

48. Mahowald, M., and Tangora, M., *Some differentials in the Adams spectral sequence*, Topology **6** (1967), 349–369.

49. May, J. P. (with contributions by F. Quinn, N. Ray, and J. Tornehave), *E∞ Ring Spaces and E∞ Ring Spectra*, Lecture Notes in Mathematics No. 577, Springer-Verlag, New York, 1977.

50. Mazur, B., *Séminaire de topòlogie combinatoire et differentielle*, Inst. des Hautes Etudes Scien., 1962.

51. Mazur, B., *Relative neighborhoods and a theorem of Smale*, Annals of Math. **77** (1963), 232–249.

52. Milnor, J., *On manifolds homeomorphic to the 7-sphere*, Annals of Math. **64** (1956), 399–405.

53. Milnor, J., *On the cobordism ring $\Omega^*$ and a complex analogue*, Amer. J. Math. **82** (1960), 505–521.

54. Milnor, J., *A procedure for killing the homotopy groups of differentiable manifolds*, A. M. S. Symposia in Pure Mathematics, Vol. III, 1961, 39–55.

55. Milnor, J., *Microbundles, part I*, Topology **3** supplement 1 (1964), 53–80.

56. Milnor, J., *Morse Theory*, Annals of Math. Studies No. 51, Princeton University Press, 1963.

57. Milnor, J., *Lectures on the h-cobordism theorem*, Princeton University Press, Princeton, 1965.

58. Milnor, J., *Whitehead torsion*, Bull. Amer. Math. Soc. **72** (1966), 358–426.

59. Milnor, J., *Singular Points of Complex Hypersurfaces*, Annals of Math. Studies No. 61, Princeton University Press, 1968.

60. Milnor, J., and Husemoller, D,. *Symmetric Bilinear Forms*, Springer-Verlag, New York, 1973

61. Milnor, J., and Kervaire, M., *Bernoulli numbers, homotopy groups, and a theorem of Rohlin*, Proc. Int. Congress of Mathematicians, Edinburgh, 1958, pp. 454–458.

62. Milnor, J., and Stasheff, J., *Characteristic Classes*, Annals of Math. Studies No. 76, Princeton University Press, Princeton, 1974.

63. Miramanoff, D., *Sur le dernier théorème de Fermat et la critérium de M. A. Wieferich*, Enseignement Math. **11** (1909), 455–459.

64. Munkres, J.R., *Obstructions to the smoothing of piecewise-linear homeomorphisms*, Bull. Amer. Math. Soc. **65** (1959), 332–334.

65. Munkres, J.R., *Obstructions to the smoothing of piecewise-differential homeomorphisms*, Annals of Math. **72** (1960), 521–554.

66. Nielsen, N., *Traité élémentaire des nombres de Bernoulli*, Gauthier-Villars, Paris, 1923.

67. Novikov, S., *Diffeomorphisms of simply connected manifolds*, Dokl. Akad. Nauk. SSSR **143** (1962), 1046–1049.

68. Novikov, S., *Homotopy equivalent smooth manifolds I*, Translations Amer. Math. Soc. **48** (1965), 271–396.

69. Palais, R., *Extending diffeomorphisms*, Proc. Amer. Math. Soc. **11** (1960), 274–277.

70. Peterson, F. P., *The mod p homotopy type of BSO and F/PL*, Bol. Soc. Mat. Mexicana **14** (1969), 22–27.

71. Pham, F., *Formules de Picard-Lefschetz généralisées et des ramifications des intégrales*, Bull. Soc. Math. France **93** (1965), 333–367.

72. Pontrjagin, L., *Smooth manifolds and their applications in homotopy theory*, Trudy Mat. Inst. im. Steklov **45** (1955), and AMS Translations, Series 2, **11**, 1–114.

73. Quillen, D., *The Adams conjecture*, Topology **10** (1971), 67–80.

74. Rourke, C.P., and Sanderson, B.J., *Block bundles I, II, and III*, Annals of Math. **87** (1968), 1–28, 256–278, and 431–483.

75. Rourke, C.P., and Sanderson, B.J., *Introduction to Piecewise Linear Topology*, Springer-Verlag, Berlin, Heidelberg, New York, 1972

76. Serre, J. P., *Formes bilinéaires symmétriques entières à discriminant ±1*, Séminaire Cartan **14** (1961–62).

77. Smale, S., *Immersions of spheres in Euclidean space*, Annals of Math. **69** (1959), 327–344.

78. Smale, S., *Generalized Poincaré conjecture in dimensions greater than four*, Annals of Math **74** (1961), 391–406.

79. Smale, S., *On the structure of manifolds*, Amer. J. Math. **84** (1962), 387–399.

80. Spivak, M., *Spaces satisfying Poincaré duality*, Topology **6** (1967), 77–104.

81. Stallings, J., *Polyhedral homotopy spheres*, Bull. Amer. Math. Soc. **66** (1960), 485–488.

82. Stasheff, J., *The image of J as an H-space mod p*, Conf. on Algebraic Topology, Univ. of Illinois at Chicago Circle, 1968.

83. von Staudt, C., *Beweis eines Lehrsatzes, die Bernoullischen Zahlen betreffend*, Jour. für reine u. angew. Math. **21** (1840), 372–374.

84. Sullivan, D., *Triangulating homotopy equivalences*, Ph.D. dissertation, Princeton University, 1965.

85. Sullivan, D., *On the Hauptvermutung for manifolds*, Bull. Amer. Math. Soc. **73** (1967), 598–600.

86. Sullivan, D., *Geometric topology, Part I, Localization, periodicity, and Galois symmetry*, mimeo. lecture notes, M.I.T., 1970

87. Sullivan, D., *Genetics of homotopy theory and the Adams conjecture*, Annals of Math. **100** (1974), 1–79.

88. Taylor, R., and Wiles, A., *Ring-theoretic properties of certain Hecke algebras*, Annals of Math. **141** (1995), 553–572.

89. Thom, R., *Quelques propriétés globales des variétés différentiables*, Comment. Math. Helv. **28** (1954), 17–86.

90. Thom, R., *Des variétés triangulées aux variétés différentiables*, Proc. Internat. Congr. Math. Edinburgh, 1958, 248–255.

91. Vandiver, H. S., *Extensions of the criteria of Wieferich and Miramanoff in connection with Fermat's last theorem*, Jour. für reine u. angew. Math. **144** (1914), 314–318.

92. Vandiver, H. S., *Fermat's last theorem: its history and the nature of the known results concerning it*, Amer. Math. Monthly **53** (1946), 555–578.

93. Wall, C. T. C., *Classification of $(n-1)$-connected $2n$-manifolds*, Annals of Math. **75** (1962), 163–189.

94. Wall, C. T. C., *An extension of a result of Novikov and Browder*, Amer. J. Math. **88** (1966), 20–32.

95. Wall, C. T. C., *Surgery of non-simply-connected manifolds*, Annals of Math. **84** (1966), 217–276.

96. Wall, C. T. C., *Surgery on Compact Manifolds*, Academic Press, London, New York, 1970

97. Wieferich, A., *Zum letzten Fermat'schen Theorem*, Jour. für reine u. angew. Math. **136** (1909), 293–302.

98. Whitehead, J. H. C., *Manifolds with transverse fields in Euclidean space*, Annals of Math. **73** (1961), 154–212.

99. Wiles, A., *Modular elliptic curves and Fermat's Last Theorem*, Annals of Math. **142** (1995), 443–551.

100. Zeeman, C., *The generalized Poincaré conjecture*, Bull. Amer. Math. Soc. **67** (1961), 270.

Mathematics Department
University at Albany
Albany, NY 12222

*Email:* lance@math.albany.edu

# The Kervaire invariant and surgery theory

## Edgar H. Brown, Jr.

ABSTRACT. We give an expository account of the development of the Kervaire invariant and its generalizations with emphasis on its applications to surgery and, in particular, to the existence of stably parallelizable manifolds with Kervaire invariant one.

## 1. INTRODUCTION

As an expository device we describe the development of this subject in chronological order beginning with Kervaire's original paper ([10]) and Kervaire-Milnor's Groups of Homotopy Spheres ([11]) followed by Frank Peterson's and my work using Spin Cobordism ([5], [7]), Browder's application of the Adams spectral sequence to the Kervaire invariant one problem ([3]), Browder-Novikov surgery ([16]) and finally an overall generalization of mine ([6]). In a final section we describe, with no detail, other work and references for these areas. We do not give any serious proofs until we get to the "overall generalization" sections where we prove the results about the generalized Kervaire invariant and Browder's Kervaire invariant one results.

## 2. COBORDISM PRELIMINARIES.

We make $\mathbb{R}^N \subset \mathbb{R}^{N+1}$ by identifying $x \in \mathbb{R}^N$ with $(x, 0)$. Then $BO_k = \bigcup G_{k,l}$, where $G_{k,l}$ is the space of $k$ dimensional linear subspaces of $\mathbb{R}^{k+l}$ and the universal bundle $\zeta_k \to BO_k$ is the space of all pairs $(P, v)$, where $v \in P \in G_{k,l}$ for some $l$. A vector bundle is assumed to have a metric on its fibres. Hence if $\xi$ is a $k$-plane bundle over $X$, it has associated disc and sphere bundles, $D\xi$ and $S\xi$, a Thom space $T\xi = D\xi/S\xi$, and a Thom class $U_k \in H^k(T\xi)$ (coefficients $\mathbb{Z}$ or $\mathbb{Z}/2\mathbb{Z}$ as appropriate).

Throughout this paper "$m$-manifold" means a smooth, compact manifold of dimension $m$, equipped with a smooth embedding into Euclidean space, $\mathbb{R}^{m+k}$, $k$ large ($k > 2m + 1$). If $M$ is such a manifold, its tangent and normal bundles are given by

$$\tau(M) = \{(x,v) \in M \times \mathbb{R}^{m+k} \mid v \text{ is tangent to } M \text{ at } x\},$$
$$\nu(M) = \{(x,v) \in M \times \mathbb{R}^{m+k} \mid v \text{ is perpendicular to } M \text{ at } x\}.$$

One associates to $M$ a map $t : S^{m+k} = \mathbb{R}^{m+k} \cup \{\infty\} \longrightarrow T(\nu)$, the Thom construction, as follows: For $\epsilon > 0$ sufficiently small, $e : D_\epsilon(\nu(M)) \to \mathbb{R}^{m+k}$, given by $e(x,v) = x + v$, is an embedding. Let $t(u) = (x, v/\epsilon)$ if $u = e(x,v)$ and $= \{S(\nu(M))\}$ otherwise. If $\xi$ is a $k$-plane bundle over $X$, a $\xi$-structure on a manifold $M$ is a bundle map $f : \nu(M) \longrightarrow \xi$; $f_M : M \longrightarrow X$ denotes the underlying map.

We define the $m^{\text{th}}$ $\xi$-cobordism group, $\Omega_m(\xi)$, to be the set of pairs $(M, f)$, where $M$ is a closed $m$-manifold with a $\xi$-structure $f$, modulo the equivalence relation generated by the following to relations. If $i : M \subset \mathbb{R}^{m+k}$ and $j : \mathbb{R}^{m+k} \subset \mathbb{R}^{m+k+1}$ are their given inclusions, then $M$ equipped with $i$ is equivalent to $M$ equipped with $ji$. Also $(M_1, f_1)$ and $(M_2, f_2)$ are equivalent if they are $\xi$-cobordant, that is, there is a $(m+1)$-manifold $N$ with a $\xi$-structure $F$ and an embedding $N \subset \mathbb{R}^{m+k} \times [0,1]$ such that $\partial N = M_1 \cup M_2$, $N$ is perpendicular to $\mathbb{R}^{m+k} \times \{0,1\}$, $(N, F) \cap \mathbb{R}^{m+k} \times \{i - 1\} = (M_i, f_i)$. Disjoint union of pairs $(M_1, f_1)$ and $(M_2, f_2)$ makes $\Omega_m(\xi)$ into an abelian group.

**Theorem 2.1 (Thom [21]).** *Sending $(M, f)$ to $T(f)t : S^{m+k} \longrightarrow T(\xi)$ induces an isomorphism, $\Psi : \Omega_m(\xi) \longrightarrow \pi_{m+k}(T(\xi))$.*

Sometimes when $\xi$ is a bundle over a particular $X$, we denote $T\xi$ by $TX$ and $\Omega_m(\xi)$ by $\Omega_m(X)$, or when $X = BG_k$ by $\Omega_m(G)$.

## 3. GROUPS OF HOMOTOPY SPHERES

Kervaire and Milnor ([11]) defined the group of homotopy $m$-spheres, $\theta_m$, to be the set of closed, oriented $m$-manifolds homotopy equivalent to $S^m$ (for $m > 4$, by Smale's Theorem, homeomorphic to $S^m$) modulo the relation of $h$-cobordism (for a cobordism $N$ between $M_1$ and $M_2$, the inclusions of $M_i$ into $N$ are required to be homotopy equivalences). Addition is defined using the connected sum operation. Using Bott's computation of $\pi_*(BO)$ and results of Adams concerning the $J$-homomorphism, they prove:

**Theorem 3.1.** *If $M$ is a homotopy $m$-sphere, $\nu(M)$ is trivial ($k$ large).*

For the remainder of this section we assume $m > 4$. Let $0^k$ denote the vector bundle $\mathbb{R}^k \longrightarrow pt$. Thus an $0^k$-structure on $M$ is a framing of $\nu(M)$. If $M$ is a homotopy $m$-sphere, choosing a framing of $\nu(M)$ gives an element of $\Omega_m(0^k) \approx \pi_{m+k}(S_k)$ and a simple argument shows that changing the framing adds to this element an element in the image of $J$. Thus we have an induced map $\Psi : \theta_m \longrightarrow \pi_{m+k}(S_k)/\operatorname{Im} J$. The Kervaire-Milnor paper is mainly devoted to computing the kernel and cokernel of this map, both of which turn out to be finite cyclic groups; the kernel is denoted by $bP_{m+1}$ = homotopy spheres bounded by stably parallelizable $(m+1)$-manifolds.

## 4. Surgery

The process of doing surgery on an $m$-manifold with respect to an embedding $g : S^n \times D^{m-n} \longrightarrow M$ ($S^{q-1}$ and $D^q$ are the unit sphere and disc in $\mathbb{R}^q$) consists of producing a new manifold $M'$ and a cobordism $N$ between $M$ and $M'$ as follows. Let $N$ be a smoothed version of the identification space formed from $M \times I \cup D^{n+1} \times D^{m-n}$ by identifying $(x,y) \in S^n \times D^{m-n}$ with $(g(x,y),1)$. The boundary of $N$ consists of three parts, $M = M \times \{0\}, M - g(S^n \times D^{m-n}) \cup D^{n+1} \times S^{m-n-1}$ and $(\partial M) \times I$. In what follows $M$ is closed or its boundary is a homotopy $(m-1)$-sphere which we can cone off to form a topological closed manifold. Henceforth we ignore $\partial M$. Note $M'$ has an embedding $D^{n+1} \times S^{m-n-1} \longrightarrow M'$, namely the inclusion, and applying surgery to it gives $M$. We define a $\xi$-surgery to be one in which $M, M'$ and $N$ have $\xi$-structures making a $\xi$-cobordism. For our applications to $bP_{m+1}$ we use $\xi = 0^k$. The change in homology going from $M$ to $M'$ can be easily computed from the homology exact sequences of the pairs $(N, M)$ and $(N, M')$ and the observation that $H_q(N, M) \approx H_q(D^{n+1}, S^n)$. If $2n < m - 1$, $H_q(M) \approx H_q(M')$ for $q < n$ and $H_n(M') \approx H_n(M)/\{u\}$, where $u$ is the homology class represented by $g(S^n, 0)$. If $2n = m - 1$ or $m$ the outcome is more complicated. If $m = 2n$, $M$ is oriented and if there is a $v \in H_n(M)$ such that the intersection number $u \cdot v = 1$, then $H_n(M') \approx H_n(M)/\{u,v\}$.

Suppose $M$ has a $\xi$-structure $f$ and $u$ is in the kernel of

$$(f_M)_* : H_n(M) \longrightarrow H_n(X)$$

($X$ is the base of $\xi$). If $X$ is simply connected, $M$ can be made simply connected by a sequence of $n = 0$ and $1$ $\xi$-surgeries. The standard procedure for killing $u$ by $\xi$-surgery proceeds through the following steps:

(i) Represent $u$ by a map $g : S^n \longrightarrow M$ such that $f_M g : S^n \longrightarrow X$ is homotopic to zero. If $M$ and $X$ are simply connected and $\pi_q(X, M)$ is zero for $q \leq n$, $g$ exists. (If $X = \{pt\}$, $M$ is $(n-1)$-connected.)

(ii) Choose $g$ so that it is a smooth embedding. If $2n < m$ or $M$ is simply connected and $2n = m$, such a $g$ exists.

(iii) Extend $g$ to an embedding $g : S^n \times D^{m-n} \longrightarrow M$. Such an extension exists if the normal bundle of $S^n$ in $M$, $\nu$, is trivial. Since $f_M g$ is homotopic to zero, $\nu$ is stably trivial. Hence $\nu$ is trivial if $2n < m$, or $2n = m = 4a$ and the Euler class of $\nu$ is zero, or $2n = m = 4a+2$ and from the two possibilities for $\nu$, trivial or $\tau(S^n)$, it is trivial. This last case is what this paper is all about.

(iv) Extend the $\xi$ structure over the cobordism $N$. This follows from the hypotheses in (ii), except when $m = 2n$ and $n = 1, 3, 7$.

## 5. Application of surgery to the calculation of $bP_m$

Suppose $M$ has an $0^k$-structure and the boundary of $M$ is null or a homotopy sphere. Starting in dimension zero, one can make it $([m/2] - 1)$-connected by a sequence of $0^k$-surgeries. When $m = 2n + 1$, delicate arguments show that $0^k$-surgery can be applied to produce an $n$-connected manifold and hence, by Poincaré duality, an $(m - 1)$-connected manifold which is either an $m$-disc or a homotopy $m$-sphere. Hence $bP_{2n+1} = 0$ and coker $\Psi = 0$ in odd dimensions.

Suppose $m = 2n$ and $M$ is $(n-1)$-connected. We first consider the case $n$ even, which provides techniques and results which one tries to mimic when $n$ is odd. As we described above, we can kill $u \in H_n(M; \mathbb{Z})$ by $0^k$-surgery if there is a class $u \in H_n(M)$ such that $u \cdot v = 1$ and, when $u$ is represented by an embedded $n$-sphere, its normal bundle, $\nu$, is trivial; $\nu$ is trivial if and only if its Euler class is zero if and only if $u \cdot u = 0$. Thus one can kill $H_n(M)$ by surgery if and only if $H_n(M)$ has a basis $u_i, v_i, i = 1, 2, \ldots, r$, such that $u_i \cdot u_j = v_i \cdot v_j = 0$ and $u_i \cdot v_j = \delta_{i,j}$, that is, a symplectic basis. From this one can deduce that $M$ can be made $n$-connected (and hence $m - 1$ connected) if and only if the index of $M$, that is, the signature of the quadratic form on $H_n(M; \mathbb{Z})$ given by the intersection pairing, is zero ([15]). For $M$ closed, the Hirzebruch index theorem expresses the index of $M$ as polynomial in the Pontrjagin classes. But since $M$ has an $0^k$-structure, the Pontrjagin classes are zero. Thus the cokernel of $\Psi$ is zero in dimensions $4a, a > 1$. Kervaire-Milnor also use the Hirzebruch index theorem to calculate $bP_{4a}$.

Now suppose $M$ is as above with $n$ odd, $n \neq 1, 3$ or $7$. In the $1, 3, 7$ cases there is an obstruction to extending the $0^k$-structure over the cobordism $N$. Although our function $\phi$ measures this obstruction, we do not treat this case because of the difficulty of the surgery details required. Suppose $M$ is $(n - 1)$-connected. Let $\phi : H_n(M; \mathbb{Z}) \longrightarrow \mathbb{Z}_2 = \mathbb{Z}/2\mathbb{Z}$ be defined as follows: For $u \in H_n(M; \mathbb{Z})$, represent $u$ by an embedded $n$-sphere and let $\nu_u$ be its normal bundle in $M$. Let $\phi(u) = 0$ or $1$ according as $\nu_u$ is trivial or isomorphic to $\tau(S^n)$ (the only two possibilities for $\nu_u$).

**Lemma 5.1.** $\phi$ is well defined and satisfies:

$$\phi(u + v) = \phi(u) + \phi(v) + u \cdot v.$$

Since $n$ is odd, $u \cdot u = 0$, and therefore $\phi(2u) = 0$. Hence we do not lose anything by taking $u$ in $H_n(M; \mathbb{Z}_2)$. For the remainder of this paper $H_n(M) = H_n(M; \mathbb{Z}_2)$. As above, we may make $M$ $(m-1)$-connected by $0^k$-surgery if $H_n(M)$ has a symplectic basis $u_i, v_i$ such that $\phi(u_i) = 0$ for all $i$. Arf associated to quadratic functions such as $\phi$ a $\mathbb{Z}_2$ invariant given by

$$A(\phi) = \sum \phi(u_i)\phi(v_i).$$

He also proved that given the pairing, $A(\phi)$ classifies such quadratic functions and $H_n(M)$ has a symplectic basis $u_i, v_i$ such that $\phi(u_i) = 0$, for all $i$, if and only if $A(\phi) = 0$. Thus, if $A(\phi) = 0$, $M$ can be made $(2n-1)$-connected. Starting with a closed $2n$-manifold, $n$ odd $\neq 1, 3, 7$ and $0^k$-structure $f$ on $M$, by a sequence of surgeries one can produce an $(n-1)$ connected $(M', f')$ and then a $\phi_{M'}$. Then $A(\phi_{M'})$ is the Kervaire invariant of $M$ and the following was proved in [10]:

**Theorem 5.2.** *Sending $(M, f)$ to $A(\phi_{M'})$ induces a homomorphism*

$$\alpha : \Omega_{2n}(0^k) \longrightarrow \mathbb{Z}_2 .$$

*An element $z \in \Omega_{2n}(0^k)$ can be represented by a homotopy sphere, if and only if $\alpha(z) = 0$.*

**Corollary 5.3.** *In dimensions $4a + 2$, the kernel and cokernel of $\Psi$ are $0$ or $\mathbb{Z}_2$; $\ker \Psi = 0$ if and only if $\alpha = 0$ and $\operatorname{coker} \Psi = 0$ if and only if $\alpha \neq 0$.*

The present state of knowledge on $\alpha$ is:

**Theorem 5.4.** *$\alpha \neq 0$ for $2n = 2, 6, 14$ ([11]), $30$ ([24]), $62$ ([2]) and $\alpha = 0$ for $2n = 10$ ([10]), $8a + 2$ ([7]), $\neq 2^j - 2$ ([3]). $\alpha \neq 0$ if and only if $h_j^2$ lives to $E_\infty$ in the Adams spectral sequence for stable homotopy groups of spheres. ([3]) (Should such an element exist it is called $\theta_j$.)*

We prove Browder's results in section 8.

An $(n-1)$-connected $2n$-manifold $M$ with $0^k$-structure, boundary a homotopy sphere and $A(\phi_M) = 1$ may be constructed as follows: One plumbs together two copies of $D\tau(S^n)$ as follows: Let $h : D^n \longrightarrow S^n$ be a homeomorphism onto the upper hemisphere of $S^n$ given by $h(x) = (x, \sqrt{1 - |x|^2})$ and $r : D^n \times D^n \longrightarrow \tau(S^n)$ be a bundle map covering $h$. Let $M$ be a smoothed version of $D\tau(S^n) \times \{0\} \cup D\tau(S^n) \times \{1\}$ with $(h(x, y), 0)$ identified with $(h(y, x), 1)$ for all $x, y$. An easy cell decomposition of $M$ shows that its boundary is two $n$-discs glued together along their boundaries and hence the boundary of $M$ is a homotopy sphere $\Sigma$. Let $N$ be $M$ with $\Sigma$ coned off.

**Theorem 5.5.** *If $\Sigma$ is diffeomorphic to $S^{2n-1}$, $N$ is smoothable and has Kervaire invariant one; otherwise $N$ is a topological manifold which does not admit a differentiable structure ([10]) and $\Sigma$ generates $bP_{2n} \approx \mathbb{Z}_2$.*

The proofs of the results cited in theorem 5.4 follow a very orderly path which we now outline. We switch from homology to cohomology, $H^n(M) = H^n(M; \mathbb{Z}_2)$. Composing with Poincaré duality, $\phi$ becomes $\phi$ : $H^n(M) \longrightarrow \mathbb{Z}_2$. One wants to associate to a closed $2n$-manifold $M$, $n$ odd, $\phi$ satisfying:

(5.6) $\phi(u + v) = \phi(u) + \phi(v) + uv([M])$ where $uv$ denotes cup product.

(5.7) If the Poincaré dual of $u$ can be represented by an embedded $n$-sphere with stably trivial normal bundle $\nu$ and $n \neq 1, 3, 7$, then $\phi(u) = 0$ if and only if $\nu$ is trivial.

In [10] Kervaire defines $\phi$ for an $(n - 1)$-connected $M$ as follows. Recall the loop space $\Omega = \Omega(S^{n+1})$ has cohomology generators $e_i$ in dimensions $ni$ and under the multiplication $\Omega \times \Omega \longrightarrow \Omega$, $e_1$ goes to $e_1 \otimes 1 + 1 \otimes e_1$ and $e_2$ goes to $e_2 \otimes 1 + 1 \otimes e_2 + e_1 \otimes e_1$. For $u \in H^n(M)$ there is a map $f_u : M \longrightarrow \Omega$ such that $f_u^*(e_1) = u$; $\phi$ is then defined by $\phi(u) = f_u^*(e_2)$ and satisfies 5.6 and 5.7. Then it is shown that $\alpha : \pi_{2n+k}(S^k) = \Omega_{2n} \longrightarrow \mathbb{Z}_2$, as above, is a well defined homomorphism. Kervaire proves that $\alpha$ is zero for $n = 5$ using knowledge of $\pi_{10+k}(S^k)$, namely, this group has a unique element $a$ of order two and $a = bc, b \in \pi_{k+1}(S^k)$ which represents a manifold $S^1 \times M'$ which then can be surgered to a homotopy sphere. We remark that an equivalent way of defining $\phi$ would be to represent $u$ by a map $F : SM \longrightarrow S_{n+1}$ ($S$ = suspension) and define $\phi(u)$ to be the functional squaring operation $Sq_F^{n+1}(s_{n+1})$, $s_{n+1}$ the cohomology generator.

In [5] $\alpha : \Omega_{8a+2}(\text{Spin}) \longrightarrow \mathbb{Z}_2$ was defined as follows: $\Omega_m(\text{Spin}) = \Omega_m(\xi)$, where $\xi \longrightarrow B\text{Spin}_k$ is the universal $\text{Spin}_k$ vector bundle. For $n = 4a + 1$, the Adem relation,

$$Sq^{n+1} = Sq^2 Sq^{n-1} + Sq^1 Sq^2 Sq^{n-2}$$

gives a relation on $H^n(M)$,

$$Sq^2 Sq^{n-1} + Sq^1 Sq^2 Sq^{n-2} = 0$$

which in turn gives a secondary cohomology operation on $M$ with a Spin structure,

$$\phi' : H^n(M) \longrightarrow H^{2n}(M) .$$

Define $\phi(u) = \phi'(u)(M)$. The Spin structure is used to ensure that $\phi'$ is defined on all of $H^n(M)$ with zero indeterminacy. For example, $Sq^1$ : $H^{2n-1}(M) \longrightarrow H^{2n}(M)$ is given by $Sq^1(v) = w_1 v$. Then $\phi$ satisfies 5.6

and 5.7 and defines $\alpha$. In [7] we showed that $\alpha$ was zero on the image of $\Omega_{8k+2}(0^k)$ (framed cobordism) in $\Omega_{8a+2}(\text{Spin})$ using the result of Conner and Floyd and Lashof and Rothenberg that if $A \in \Omega_{8k+2}(SU)$ goes to zero in $\Omega_{8k+2}(U)$, then $A = B^2 C, B \in \Omega_{8k}(SU)$.

In [3] Browder developed $\alpha : \Omega_{2n}(\xi) \longrightarrow \mathbb{Z}_2$, $n$ even or odd, $\xi \longrightarrow X$, as follows: One may assume that $X$ is a smooth closed $N$-manifold, $N$ large, with normal bundle in $\mathbb{R}^{N+k}$ equal to $\xi$. Suppose $M$ is a smooth closed $2n$-manifold with $\xi$-structure $f$. One may assume $f_M$ is an embedding. Then the normal bundle of $M$ in $X$ is trivial and trivializing and using the Thom construction one obtains a map $F : X \longrightarrow S^k M$ ($S^k M$, the $k$-fold suspension of $M$). Then $\phi$ is defined by

$$\phi(u) = Sq_F^{n+1}(S^k u).$$

In order for $\phi(u)$ to be defined, $F^*(S^k u)$ must equal zero, and for there to be no indeterminacy, $Sq^{n+1} : H^{N-n-1}(X) \longrightarrow H^N(X)$ must be zero. $Sq^{n+1}(w) = v_{n+1} w$ for $w \in H^{N-n-1}(X)$, where $v_{n+1}$ is the Wu class of $\xi$. One restricts the choice of $\xi$ to bundles with $v_{n+1} = 0$. Then $\phi : P \longrightarrow \mathbb{Z}_2$ where $P = \{u \in H^n(M) | F^*(S^k u) = 0\}$. Where defined $\phi$ satisfies 5.6 and 5.7. One restricts $\alpha$ to those $(M, f)$ such that $\phi$ is zero on all $u \in P$ such that $uv(M) = 0$ for all $v \in P$. The Arf invariant algebra then works to give an integer mod 2. Then $\alpha$ is related to the Adams spectral sequence by computing a Postnikov system, up to the relevant dimension of $MO[v_{n+1}]$. We give the details in section 8.

## 6. GENERALIZED GROUPS OF HOMOTOPY SPHERES

Several people, most notably Novikov, discovered that the Groups of Homotopy Spheres paper ([11]) could be generalized by the following two step process. Replace $0^k$, the bundle $\mathbb{R}^k \longrightarrow pt$, by $O_m^k$ the bundle $S^m \times \mathbb{R}^k \longrightarrow S^m$. Then the coker $\Psi$ question asks: "Which elements $\{M, f\}$, where $f_M$ has degree one, can be represented by $\{M, f\}$, where $f_M$ is a homotopy equivalence?" (For there to be an $(M, f)$ with $f_M$ of degree one, the top homology class of $T(\xi)$ must be spherical.) The $bP_{m+1}$ question asks, "If $\{M, f\} = \{S^m, id\}$, can the cobordism between them be chosen to be an $h$-cobordism?" Now replace all occurrences of $O_m^k$ with $\xi \longrightarrow X$, where $X$ is a simply connected CW complex of finite type satisfying Poincaré duality in dimension $m$, that is there a class $x \in H_m(X; \mathbb{Z})$ such that cap product with $x$ gives an isomorphism, $H^n(X) \longrightarrow H_{m-n}(X)$ for all $n$. Then everything in Kervaire-Milnor goes through. Suppose $\{M, f\} \in \Omega_m(\xi)$ and $f_M$ has degree one. The trick is to consider the commutative diagram:

$$H^n(M) \xleftarrow{\quad f_M{}^* \quad} H^n(X)$$

$$\Big\downarrow {}_{[M]\cap} \qquad\qquad \Big\downarrow {}_{x\cap}$$

$$H_{m-n}(M) \xrightarrow{\quad f_{M*} \quad} H_{m-n}(X)$$

By Poincaré duality $f_{M*}$ is an epimorphism and $f_M{}^*$ is a monomorphism. For $2n < m$ one can kill elements in the kernel of $f_{M*}$ just as in [11]. This material is thoroughly described in Browder's book "Surgery on Simply Connected Manifolds" ([4]). This material, when $X$ is not simply connected, is the subject of Wall's book "Surgery on Compact Manifolds" ([22]), where the general pattern of the above is followed but surgery in the middle dimensions is much more complicated and leads to Wall's $L$-groups, in which the obstructions to doing the middle dimension surgery lie. Ranicki develops in "Exact Sequences in the Algebraic Theory of Surgery" ([19]) a completely algebraic approach to these surgery obstructions, replacing manifolds by their chain complex analogs. Michael Weiss refines this algebraic approach to surgery obstructions and makes it more calculable ([23], [24]).

## 7. A FURTHER GENERALIZATION OF $\phi$

We state the main theorems of this section and then prove them.

In the remainder of this section most spaces have base points, $M^+$ is $M$ with a disjoint base point, $[X, Y]$ is the homotopy classes of maps from $X$ to $Y$, $\{X, Y\} = \lim[S^k X, S^k Y]$ and $\eta : [X, Y] \longrightarrow \{X, Y\}$ sends $[f]$ to $\{f\}$. Let $s : S^n \longrightarrow K_n$ be the $\pi_*$ generator.

**Lemma 7.1.** *The Hopf construction* $h(\lambda) : S^{2n+1} \longrightarrow SK_n$ *on*

$$\lambda : S^n \times S^n \xrightarrow{s \times s} K_n \times K_n \xrightarrow{\mu} K_n$$

*with $\mu$ the multiplication, gives a generator of* $\{S^{2n}, K_n\} \approx \mathbb{Z}_2$.

Suppose $M$ is a closed $2n$-manifold ($n$ even or odd). We form an abelian group $G(M) = H^n(M) \times H^{2n}(M)$ with addition

$$(u, v) + (u', v') = (u + u', v + v' + uu').$$

Let $j : \mathbb{Z}_2 \longrightarrow \mathbb{Z}_4$ be the homomorphism sending 1 to 2. Then functions $\phi : H^n(M) \longrightarrow \mathbb{Z}_4$ satisfying $\phi(u + v) = \phi(u) + \phi(v) + j(uv([M]))$ are in one to one correspondence with homomorphisms $h : G(M) \longrightarrow \mathbb{Z}_4$ such that $h(0, v) = j(v([M]))$. We will see that such functions occur in nature.

**Theorem 7.2.** *Let* $F : G(M) \longrightarrow \{M^+, K_n\}$, *given by* $F(u,v) = \eta(u) + h(\lambda)g_v$, *where* $g_v \longrightarrow S^{2n}$ *has degree one. Then* $F$ *is an isomorphism.*

Let $\nu$ be the normal bundle of $M$ in $\mathbb{R}^{2n+k}$ and $\Delta : T\nu \longrightarrow T\nu \wedge M^+$ be the diagonal map sending $v$ to $(v, p(v)), p : \nu \longrightarrow M$. Then $S^{2n+k} \xrightarrow{t} T\nu \xrightarrow{\Delta} T\nu \wedge M^+$ is an $S$ (Spanier-Whitehead) duality map ([20]). Then $\{M^+, K_n\} \approx \{S^2n, T\nu \wedge K_n\}$ under the map sending $S^l M \longrightarrow S^l K_n$ to

$$S^{2n+k+l} \longrightarrow S^l T\nu \longrightarrow S^l(T\nu \wedge M^+) = T\nu \wedge S^l M^+ \longrightarrow T\nu \wedge S^l K_n .$$

Let $q(\lambda) \in \{S^{2n}, T\nu \wedge K_n\}$ be the image of $h(\lambda)$ under this map.

**Lemma 7.3.** *If* $f : \nu \longrightarrow \xi$, *the image of* $f_*(q(\lambda))$ *is non-zero if and only if* $v_{n+1}(\xi) = 0$ *and it is at most divisible by 2.*

We call $\xi$ a Wu-$n$ spectrum if $v_{n+1}(\xi) = 0$ in which case we can choose a homomorphism $\omega : \{S^{2n}, T\xi \wedge K_n\} \longrightarrow \mathbb{Z}_4$ such that $\omega((f_*(q(\lambda))) = 2$. Let $\phi = \phi(M, f, \omega)$ be the composition $H^n(M) = [M^+, K_n] \longrightarrow \{M^+, K_n\} \longrightarrow \{S^{2n}, T\nu \wedge K_n\} \longrightarrow \{S^{2n}, T\xi \wedge K_n\} \longrightarrow \mathbb{Z}_4$. Hence,

**Theorem 7.4.** $\phi = \phi(M, f, \omega)$ *satisfies*

$$\phi(u + v) = \phi(u) + \phi(v) + j(uv([M])).$$

Let $BO_k[v_{n+1}] \longrightarrow BO_k$ be the fibration with fibre $K_n$ and $k$-invariant $v_{n+1}$ and let $\xi[v_{n+1}]$ be the pull back of the universal bundle over $BO_k$. Suppose $M$ has a $\xi[v_{n+1}]$-structure, $S^n \subset M$ has normal bundle $\mu$ and $\nu|S^n$ is trivial. Then $S^n \longrightarrow M \longrightarrow BO_k[v_{n+1}]$ factors through $K_n$. Let $\epsilon_1 = 0$ or $1 =$ degree of this map. Let $\epsilon_2 = 0$ or $1$ according to whether $\mu$ is trivial or $\tau(S^n)$. Let $u$ be the Poincaré dual of the homology class represented by $S^n \longrightarrow M$.

**Lemma 7.5.** *If* $n$ *is odd and* $\neq 1, 3, 7$, *then* $\phi(u) = 2(\epsilon_1 + \epsilon_2)$.

**Remark.** *If* $n$ *is even,* $\epsilon_2 =$ *Euler number of* $\mu$ *mod 4. If* $n = 1, 3,$ *or 7 and* $\epsilon_2 = 0$, *the element in* $\pi_{2n+k}(T\xi \wedge K_n)$ *involves a map* $g : S^{2n+k} \longrightarrow S^k$ *such that* $\epsilon_1 =$Hopf *invariant of* $g$.

Let $S^n \times S^n \longrightarrow BO_k[v_{n+1}]$ be the composition of $K_n \longrightarrow BO_k[v_{n+1}]$ and $s \otimes 1 + 1 \otimes s$. This lifts to a $\xi[v_{n+1}]$-structure $q$. Then 7.5 gives

**Lemma 7.6.** *The* $\phi$ *associated to* $(S^n \times S^n, q)$. *satisfies* $\phi(s \otimes 1) = \phi(1 \otimes s) = 2$.

The following gives an analog of the Arf invariant for these quadratic functions.

**Definition 7.7.** *Let $V$ be a finite dimensional vector space over $\mathbb{Z}_2$. A function $\phi : V \longrightarrow \mathbb{Z}_4$ is a (nonsingular) quadratic if it satisfies $\phi(u + v) = \phi(u) + \phi(v) + jt(u,v)$ where $j : \mathbb{Z}_2 \longrightarrow \mathbb{Z}_4$ sends 1 to 2 and $t$ is a nonsingular bilinear pairing. If $\phi_i : V_i \longrightarrow \mathbb{Z}_4, i = 1, 2$, are such functions $\phi_1 \approx \phi_2$ if there is an isomorphism $T : V_1 \approx V_2$ such that $\phi_1 = \phi_2 T$. $(\phi_1 + \phi_2) : V_1 + V_2 \longrightarrow \mathbb{Z}_4, -\phi$ and $(\phi_1\phi_2) : V_1 \otimes V_2 \longrightarrow \mathbb{Z}_4$ are defined by $(\phi_1 + \phi_2)(u,v) = \phi_1(u) + \phi_2(v), (-\phi)(u) = -\phi(u)$ and $(\phi_1\phi_2)(u \otimes v) = \phi_1(u)\phi_2(v)$.*

A proof of the following appears in [6] and is straightforward. The first part of the theorem is proved by showing that the Grothendieck group of these functions is cyclic of order eight.

**Theorem 7.8.** *There is a unique function $\sigma$ from quadratic functions as in 7.1 to $\mathbb{Z}_8$ satisfying:*

(i) *If $\phi_1 \approx \phi_2$ , then $\sigma(\phi_1) = \sigma(\phi_2)$*
(ii) *$\sigma(\phi_1 + \phi_2) = \sigma(\phi_1) + \sigma(\phi_2)$*
(iii) *$\sigma(-\phi) = -\sigma(\phi)$*
(iv) *$\sigma(\gamma) = 1$, where $\gamma : \mathbb{Z}_2 \longrightarrow \mathbb{Z}_4$ by $\gamma(0) = 0$ and $\gamma(1) = 1$.*

*Furthermore $\sigma$ satisfies:*

(v) *If $\phi = j\phi', \sigma(\phi) = 4 \operatorname{Arf}(\phi')$.*
(vi) *If $\phi : V \longrightarrow \mathbb{Z}_4, \sigma(\phi) = \dim V \mod 2$.*
(vii) *$\sigma(\phi_1\phi_2) = \sigma(\phi_1)\sigma(\phi_2)$.*
(viii) *If $U$ is a finitely generated abelian group, $\tau : U \otimes U \longrightarrow \mathbb{Z}$ is a symmetric unimodular form, $\Psi(u) = \tau(u,u)$ and $\phi : U/2U \longrightarrow \mathbb{Z}_4$ is defined by $\phi(u) = \Psi(u) \mod 4$, then $\phi$ is quadratic and $\sigma(\phi) =$ (signature $\Psi$) $\mod 8$.*
(ix) *Suppose $t$ is the bilinear form of $\phi : V \longrightarrow \mathbb{Z}_4, V_1 \xrightarrow{\nu} \nu V \xrightarrow{\delta} V_2$ is an exact sequence and $t' : V_1 \otimes V_2 \longrightarrow \mathbb{Z}_2$ is a nonsingular bilinear form such that $t'(u, \delta v) = t(\nu u, v)$. If $\phi\nu = 0$, then $\sigma(\phi) = 0$.*
(x) *With $i = \sqrt{-1}$,*

$$\sigma(\phi) = (4i/\pi)\ln(2^{(\dim V)/2}/(\sum_{u \in V} i^{\phi(u)})).$$

**Theorem 7.9.** *Sending $(M, f)$ to $\sigma(\phi(M, f, \omega))$ induces a homomorphism $\sigma_\omega : \Omega_{2n}(\xi[v_{n+1}]) \longrightarrow \mathbb{Z}_8$ such that $\sigma_\omega$ composed with*

$$\Omega_{2n}(0^k) \longrightarrow \Omega_{2n}(\xi[v_{n+1}])$$

*gives the Kervaire invariant.*

We can apply $\sigma_\omega$ to the general, simply connected surgery problem as follows.

**Definition 7.10.** *For $n$ odd, a $2n$-Poincaré quadruple $(X, \xi, \beta, \omega)$ is a connected finite CW complex $X$, a $k$-plane bundle $\xi$ over $X$, $\omega$ a homomorphism as above and $\beta \in \pi_{m+k}(T\xi)$ such that*

$$S^{m+k} \xrightarrow{\beta} T\xi \xrightarrow{\Delta} T\xi \wedge X_+$$

*is an S duality (which makes $\xi$ a Wu bundle).*

**Theorem 7.10.** *Then by the Thom Theorem, $\beta$ gives a $2n$-manifold $M$ with a $\xi$-structure $f$ such that $f_M$ has degree one and the surgery obstruction to making $f_M$ a homotopy equivalence is $\sigma(\phi(X, id_X, \omega)) - \sigma(\phi(M, f, \omega))$.*

*Proof of 7.1 and 7.2.* Let $\iota$ be the generator of $H^n(K_n)$. Let $E \longrightarrow K_{n+1}$ be the fibration with fibre $K_{2n+1}$ and $k$-invariant $\iota_{n+1}^2$. Then $S\iota_n \longrightarrow K_{n+1}$ lifts $g : S\iota_n \longrightarrow E$ and on homotopy groups $\pi_i \, \pi_i(g)$ is an isomorphism for $i \leq 2n + 1$. Then since $M$ is $2n$ dimensional $\{M, K_n\} = [SM, SK_n] \approx [SM, E] \approx [M, \Omega E] = [M, K_n \times K_{2n}] = H^n(M) \times H^{2n}(M)$. The additive structure comes from the fact that under multiplication $\Omega E \times \Omega E \longrightarrow \Omega E$, $\iota_{2n}$ goes to $\iota_{2n} \otimes 1 + 1 \otimes \iota_{2n} + \iota_n \otimes \iota_n$. Applying this to $M = S^n \times S^n$ gives 7.1.

*Proof of 7.3.* Let $V(X) = \{S^{2n+k+l}, X \wedge S^l(K_n)\}$. We want to know the image of $V(S^k)$ in $V(T\xi)$ and how divisible it is. We can assume $T\xi$ is a finite $(k-1)$-connected CW complex and $S^l(K_n) \longrightarrow K_{n+l}$ is a fibration with fiber $K_{2n+l}$ and $k$-invariant $Sq^{n+1}\iota_{n+l}$. Applying $\{S^{2n+k+l}, T\xi \wedge (\ )\}$, this gives an exact sequence

$$\longrightarrow H_{k+n+1}(T\xi) \xrightarrow{Sq^{n+1}} H_k(T\xi) \longrightarrow V(T\xi) \longrightarrow H_{k+n}(T\xi) \longrightarrow$$

and the same with $T\xi$ replaced by $S^k$. For $x \in H_{k+n+1}(T\xi)$ and $U$ the Thom class,

$$U(Sq^{n+1}(x)) = \chi(Sq^{n+1})(U)(x) = v_{n+1}U(x) .$$

The two exact sequences make a ladder from which the desired result can be read off.

*Proof of 7.5.* Note, applying the Thom construction and the Thom class give maps $S^n \times S^n \longrightarrow T\tau(S^n) \longrightarrow K_n$ and the Hopf construction gives $S^{2n+1} \longrightarrow ST\tau(S^n) \longrightarrow SK_n$. Also the element $a \in \pi_{2n+k}(T\xi \wedge K_n)$ such that $\phi(u)$ comes from it is given by $S^{2n+k} \longrightarrow T\nu_{S^n} \longrightarrow S^k(S^{n+}) \wedge T\mu \longrightarrow S^k(K_n) \wedge K_n = S^k K_n \vee S^k K_n \wedge K_n \longrightarrow T\xi \wedge K_n$. Combining these two gives the desired result.

*Proof of 7.9.* To show that $\sigma(\omega)$ is well defined suppose $(M, f) = \partial(N, F)$. By virtue of 7.8(ix) and the exact sequence

$$H^n(N) \xrightarrow{j^*} H^n(M) \longrightarrow H^{n+1}(N, M),$$

it is sufficient to show that $\phi(j^*(u) = 0$. The element $a \in \pi_{2n+k}(T\xi \wedge K_n)$ such that $\phi(j^*(u) = \omega(a)$ is

$$S^{2n+k+1} \longrightarrow T\nu_N / T\nu_M \longrightarrow T\nu_N \wedge (N/M) \longrightarrow$$
$$T\nu_N \wedge SM^+ \longrightarrow T\nu_N \wedge SN^+ \longrightarrow T\xi \wedge SK_n$$

But this is zero because $N/M \longrightarrow SM^+ \longrightarrow SN^+$ is zero.

## 8. PROOF OF THEOREM 5.4

Let $B = BO_k$, $K_n = K(\mathbb{Z}_2, n)$, $v_{n+1} : B \longrightarrow K_{n+1}$ represent the $(n+1)$-th Wu class, $C = \{(b, a) \in B \times K_{n+1}^I \mid a(1) = v_{n+1}(b)\}$, $D = BO_k[v_{n+1}] = \{(b, a) \in C \mid a(0) = *\}$, $i : B \longrightarrow C$ by $i(b) = (b, a_b)$ where $a_b(t) = v_{n+1}(b)$ and $\pi : (C, D) \longrightarrow (B \times K_{n+1}, B \times \{*\})$ and $\pi' : C \longrightarrow B \times K_{n+1}$ by $(b, a) \to (b, a(0))$. The following is easily verified:

**Lemma 8.1.** *The map $i$ is a homotopy equivalence. $\pi$ is a fibre map with fibre $K_n$ and in the Serre cohomology spectral sequence for $\pi$, $E_2^{p,q} = 0$ for $q > 0$ and $p+q \leq 2n+2$ except $E_2^{n+1,n} \approx \mathbb{Z}_2$. Hence $\pi^* : H^p(B \times K_{n+1}, B \times \{*\}) \approx H^p(C, D)$ for $p \leq 2n+2$ except it may have a kernel isomorphic to $\mathbb{Z}_2$ for $p = 2n+2$.*

**Lemma 8.2.** *The kernel of*

$$\pi^* : H^{2n+2}(B \times K_{n+1}, B \times \{*\}) \longrightarrow H^{2n+2}(C, D)$$

*is generated by $v_{n+1} \otimes \iota_{n+1} + 1 \otimes \iota_{n+1}^2$.*

*Proof.* Recall that via the inclusion map $j : X \longrightarrow (X, A)$, $H^*(X)$ acts on $H^*(X, A)$, and for $u \in H^*(X, A)$, $u^2 = j^*(u)u$. Note $i\pi'$ sends $b$ to $(b, v_{n+1}(b))$ and hence $(i\pi')^*(1 \otimes \iota_{n+1}) = v_{n+1}$. Thinking of $1 \otimes \iota_{n+1} \in H^{n+1}(B \times K_{n+1}, B \times \{*\})$,

$$\pi^*(1 \otimes \iota_{n+1}^2) = \pi'^* j^*(1 \otimes \iota_{n+1})(1 \otimes \iota_{n+1})$$
$$= \pi^*(v_{n+1} \otimes 1)(1 \otimes \iota_{n+1})$$
$$= \pi^*(v_{n+1} \otimes \iota_{n+1}).$$

Under the map $C \longrightarrow B$ sending $(b, a)$ to $b$, the universal bundle over $B$ pulls back to bundles over $C$ and $D$ whose Thom spaces we denote by $TD, TC$, and $TB$ and whose Thom classes we denote by $U$. The map $\pi$ induces a map $T\pi : TC \wedge TD \longrightarrow TB \wedge K_{n+1}$. And,

**Lemma 8.3.** *The kernel of* $H^q(T\pi)$ *for* $q \le 2n + 2 + k$ *is generated by*

$$v_{n+1}U \otimes \iota_{n+1} + U \otimes \iota_{n+1}^2 = \chi(Sq^{n+1})U \otimes \iota_{n+1} + 1 \otimes \iota_{n+1}^2$$

$$= \sum_{i>0} \chi(Sq^i)(U \otimes Sq^{n+1-i}\iota_{n+1}).$$

The same Steenrod square manipulations yield:

**Lemma 8.4.** *For* $j < n + 1$, $TB \longrightarrow TC \longrightarrow TC/TB \longrightarrow TC \wedge K_{n+1}$ *sends* $(U \otimes Sq^j\iota_{n+1})$ *to*

$$\sum_{k>0} (Sq^k(v_{n+1}v_{j-k}U) + \chi(Sq^k)((Sq^{j+1-k}v_j)U)).$$

We very briefly describe the portion of Adams' work on cohomology operations ([1]) relevant to this proof. All the spaces we will deal with will be approximately $k$-connected, and the results we state will be correct in a range of dimensions up to about $2k$. Suppose $L$ and $K$ are spaces and $F : L \longrightarrow K$ is a map, $E$ is the space of paths in $L$ starting at a base point and $P : E \longrightarrow L$ sends a path to its end point. Note $E$ is contractible and $P$ is a fibre map with fibre the loops on $K, \Omega(K)$. Let $p : E_F = F^*E \longrightarrow L$ be the induced fibration. Suppose $G_1 : L_1 \longleftarrow L_2$ is an inclusion and $F : L_2 \longrightarrow L_2/L_1$ is the quotient map. Then $L_1 = E_F$ and $p = G$. Suppose $G_2 : L_2 \longrightarrow L_3$ and $G_2G_1$ is homotopic to the constant map. Then $G_1$ lifts to $G_1' : L_1 \longrightarrow E_{G_2}$ and we form $E_{G_1'}$. There is a map $G_2' : E_{G_1} \longrightarrow \Omega(L_2)$ such that $E_{G_2'} = E_{G_1'}$. Call this space $E(G_1, G_2)$. We apply this to our situation by taking $L_1 = TB$, $L_2 = TB \wedge K_{n+1}, L_3 = K_{2n+2+k}, G_1 : TB \longrightarrow TC/TD \longrightarrow TB \wedge K_{n+1}$ and $G_2 : TB \wedge K_{n+1} \longrightarrow K_{2n+2+k}$ representing the cohomology element in Lemma 8.3. The map $TD = T(BO_k[v_{n+1}]) \longrightarrow T(BO_k) = TB$ lifts to $h : TD \longrightarrow E(G_1, G_2)$, and

**Lemma 8.5.** $\pi_q(h)$ *is an isomorphism for* $q \le 2n + k$.

*Proof.* $TC/TD \longrightarrow TB \wedge K_{n+1}$ lifts to $r : TC/TD \longrightarrow E_{G_2}$. $H^q(r)$ is an isomorphism for $q \le 2n + 2 + k$, hence $\pi_q(r)$ is an isomorphism for $q \le 2n + 1 + k$ and hence $\pi_q(h)$ is as an isomorphism for $q \le 2n + k$.

The cohomology of both $TC = TB$ and $TB \wedge K_{n+1}$ are free modules over the mod two Steenrod algebra $A$. If $\{u_i\}$ is a basis for $H^*(TB)$ over $A$ and $\{v_i\}$ is a basis for $H^*(K_{n+1})$, $\{u_i \otimes v_j\}$ is a free $A$-basis for $H^*(TB \wedge K_{n+1})$. Up to homotopy type $TB = \prod K_{|u_i|}$ and similarly for $TB \wedge K_{n+1}$.

Let $s : S^k \longrightarrow TD$ represent the generator of $\pi_k(TD)$. Then the map of $2n$-framed cobordism to $2n_BO_k[v_{n+1}]$ cobordism corresponds to $s_* :$

$\pi_{2n+k}(S^k) \longrightarrow \pi_{2n+k}(TD) \approx \pi_{2n+k}(E(G_1, G_2))$. Let $V : K_k \longrightarrow TB$ be the map such that $U$, the Thom class pulls back to $\iota_k$. Then $S^k \longrightarrow TD \longrightarrow TB$ factors through $V : K_k \longrightarrow TB$ and hence $S^k \longrightarrow TD \longrightarrow E(G_1, G_2)$ factors through $E(G_1 V, G_2)$ giving a map $t : S^k \longrightarrow E(G_1 V, G_2)$. In [1] Adams (with some refinements) that one may of viewing $E(G_1 V, G_2)$ is as the beginning of a tower building $S^k$. This tower gives a spectral sequence with $E_2 = Ext_A(\mathbb{Z}_2, \mathbb{Z}_2)$ and a map $S^{2n+k} \longrightarrow E(G_1 V, G_2)$ representing an element of $E_2$ is said to live to $E_\infty$ if it lifts all the way up the tower giving a map $S^{2n+k} \longrightarrow S^k$ (more or less). At this two stage level, the relevant elements of $E_2$ have names "$h_i h_j$", $i \leq j$. Adams proves that if a map $S^{2n+k} \longrightarrow S^k \longrightarrow E(G_1 V, G_2)$ is non zero, then $G_1 V$ and $G_2$ satisfy the following condition: The algebra $A$ is generated by elements $Sq^{2^i}$. Let $h_i : A \longrightarrow \mathbb{Z}_2$ be the linear map which is zero on decomposables and $h_i(Sq^{2^j} = \delta_{i,j}$. Let $x$, $\{y_i\}$ and $z$ be $A$ generators of $H^*(K_k)$, $H^*(TB \wedge K_{n+1})$ and $H^*(K_{k+2n+2})$. Then $G_2^*(z) = \sum a_i y_i$, and $G_1^*(y_i) = b_i x$. (Since $G_2 G_1 V$ is homotopic to zero, $\sum a_i b_i = 0$.)

**Theorem 8.6 (Adams).** *If a map $S^{2n+k} \longrightarrow S^k \longrightarrow E(G_1 V, G_2)$ is non-zero, then for some $s$ and $t \leq s$, $\sum h_s(a_i) h_t(b_i) = 0$.*

Using the fact that $\chi(Sq^{2^i}) = Sq^{2^i} +$ decomposables, and inspecting the elements in 8.2 and 8.4, one sees that the condition in 8.6 is satisfied exactly when $n$ is of the form $2^i - 1$. If a framed $2n$-manifold has Kervaire invariant one, it will be non-zero in $\Omega_{2n}(BO[v_{n+1}])$. Conversely, if $h_i^2$ lives to $E_\infty$, there is a non-zero map $s : S^{2^i - 2 + k} \longrightarrow S^k \longrightarrow E(G_1, G_2)$. By an easy Hopf invariant one type argument, it goes to zero in $E_{G_1}$ and hence must be the $\pi_*$ generator of the fiber of $E(G_1, G_2) \longrightarrow E_{G_1}, K_{2^i - 2 + k}$. But this generator corresponds to $(S^{2^i - 1} \times S^{2^i - 1}, q)$ since by 7.6 this manifold has Kervaire invariant one and the underlying map of $q$ factors through $K_n$. Hence,

**Corollary 8.7 (Browder).** *There is a framed $2n$-manifold with Kervaire invariant one, if and only if $n = 2^i - 1$ and $h_i^2$ lives to $E_\infty$ in the Adams spectral sequence for $\pi_*(S^0)$.*

## 9. OTHER WORK

An amusing low dimensional application of the generalized Kervaire invariant is afforded by immersions of surfaces = closed, compact smooth 2-manifolds in $\mathbb{R}^3$. If $f : S \longrightarrow \mathbb{R}^3$ is such an immersion, associate to it $\phi : H_1(S) = H_1(S; \mathbb{Z}_2) \longrightarrow \mathbb{Z}_4$ as follows. Represent $u \in H_1(S)$ by an embedded circle and let $T$ be a tubular neighborhood in $S$ of this circle. Define $\phi(u)$ to be the number of half twists (by $180°$) of the twisted strip

$f(T)$. This makes sense mod 4 and $\phi$ has the quadratic property with respect to the intersection pairing on $H_1(S)$. Then the quadratic functions associated to the intersection pairing are in one to one correspondence with the regular homotopy classes of immersion of $S$ in $\mathbb{R}^3$ and the Kervaire invariant gives an isomorphism of the cobordism group of such immersions onto $\mathbb{Z}_8$ ([6]).

Ochanine has generalized the above to surfaces immersed in 3- manifolds and to a $(8a + 2)$-manifold $V$ immersed in a $(8a + 4)$-manifold and dual to $w_2(M)$. He also related $KO$ characteristic classes for Spin $(8a + 2)$-manifolds to these issues ([18]).

A variant of the above is to take $S$ with boundary $S^1$ and $f : S \longrightarrow \mathbb{R}^3$ an embedding. Then $f(\partial S)$ is a knot. In this connection the Kervaire invariant appears in a number of knot and link theory papers. For example, Levine expresses the Kervaire invariant of a knot in terms of its Alexander polynomial ([12]).

There are a number of papers in homotopy theory studying the existence of framed manifolds having Kervaire invariant one, for example [13]. The existence of such manifolds in dimensions 30 and 62 was first proved by homotopy groups of spheres calculations ([2], [14]). In [8] Jones constructed a stably framed 30-manifold with Kervaire invariant one and also proved that a similar construction does not work in dimension 62. In [9] Browder's results for $2n \neq 2^j - 2$ are deduced from the Kahn-Priddy theorem.

## REFERENCES

[1] J. F. Adams, The non-existence of elements of Hopf invariant one, Ann. of Math. (2) 72 (1960), 20–104.

[2] M. G. Barratt, J. D. S. Jones, and M. E. Mahowald, Relations amongst Toda brackets and the Kervaire invariant in dimension 62, J. London Math. Soc. (2) 30 (1984), no. 3, 533–550.

[3] W. Browder, The Kervaire invariant of framed manifolds and its generalization, Ann. of Math. (2) 90 (1969), 157–186.

[4] W. Browder, Surgery on simply connected manifolds, Springer-Verlag, New York-Heidelberg, 1972.

[5] E. H. Brown, Note on an invariant of Kervaire, Michigan Math. J. 12 (1965), 23–24.

[6] E. H. Brown, Generalizations of the Kervaire invariant, Ann. of Math. (2) 95 (1972), 368–383.

[7] E. H. Brown and F. P. Peterson, The Kervaire invariant of $(8k + 2)$-manifolds, Amer. J. Math. 88 (1966), 815–826.

[8] J. D. S. Jones, The Kervaire invariant of extended power manifolds, Topology 17 (1978), 249–266.

[9] J. D. S. Jones and E. Rees, A note on the Kervaire invariant, Bull. London Math. Soc. 7 (1975) 279–282.

[10] M. A. Kervaire, A manifold that does not admit any differential structure, Comment. Math. Helv. 34 (1960), 257–270.

[11] M. A. Kervaire and J. W. Milnor, Groups of homotopy spheres I, Ann. of Math. (2) 77 (1963), 504–537.

[12] J. Levine, Polynomial invariants of knots of codimension two, Ann. of Math. (2) 83 (1966), 537–554.

[13] M. E. Mahowald, Some remarks on the Kervaire invariant problem from a homotopy point of view, Proc. Symp. Pure Math., Vol. XXII, Amer. Math. Soc., Providence, 1971.

[14] M. E. Mahowald and M. Tangora, Some differentials in the Adams spectral sequence, Pacif. J. Math. 60 (1975), 235–275.

[15] J. W. Milnor, A procedure for killing the homotopy groups of differentiable manifolds, Proc. Symposia in Pure Math., vol. III (1961), Amer. Math. Soc., Providence, pp. 39–55.

[16] S. P. Novikov, Homotopy equivalent manifolds I, Izv. Akad. Nauk SSSR, Ser. Mat. 28 (1964), 365–474.

[17] S. Ochanine, Signature modulo 16, invariants de Kervaire generalisés et nombres caractéristiques dans la $K$-théorie réelle, Mem. Soc. Math. France 81, no. 5 (1980).

[18] S. Ochanine, Formules "à la Hirzebruch" pour les invariants de Kervaire generalisés, C.R. Acad. Sci. Paris Ser. A–B 289 (1979), no. 9, 487–490.

[19] A. Ranicki, Exact sequences in the algebraic theory of surgery, Mathematical Notes 26, Princeton University Press, 1981.

[20] E. H. Spanier, Algebraic Topology, McGraw-Hill, 1966.

[21] R. Thom, Quelques propriétés globales des variétés différentiables, Com. Math. Helv. 28 (1954), 17–86.

[22] C. T. C. Wall, Surgery on compact manifolds, Academic Press, London and New York, 1970.

[23] M. Weiss, Surgery and the generalized Kervaire invariant, I, Proc. Lond. Math. Soc. 51 (1985), 146–192; II, *ibid.*, 193–230.

[24] M. Weiss, Visible $L$-theory, Forum Math. 4 (1992), 465–498.

DEPARTMENT OF MATHEMATICS
BRANDEIS UNIVERSITY
WALTHAM, MA 02254-9110
*E-mail address:* brown@math.brandeis.edu

# A guide to the classification of manifolds

Matthias Kreck

## 1. INTRODUCTION

The purpose of this note is to recall and to compare three methods for classifying manifolds of dimension $\geq 5$.

The author has the impression that these methods are only known to a rather small group of insiders. This is related to the fact that the literature is not in good shape. If somebody gets interested in the classification problem, he has to go through a vast amount of literature until he perhaps finds out that the literature he has studied does not solve his specific problem. By presenting the basic principles of classification methods, the author hopes to provide a little guide addressing non-experts in the field.

The first method shows that a smooth manifold $M$ is diffeomorphic to a certain explicit model $\mathcal{M}$ by decomposing the model into pieces $P_i$: $\mathcal{M} = \bigcup P_i$ and decomposing the manifold $M$ into diffeomorphic pieces by embedding them appropriately. Then $M$ is also the union of the same pieces but perhaps glued together in a different way. The final step is to study the different glueing diffeomorphisms. Wall has in the early sixties applied this method to classify highly connected manifolds at least up to homeomorphism, in particular $(n-1)$-connected $2n$-manifolds [W1].

The second method is the surgery program which classifies manifolds together with a homotopy equivalence to a given Poincaré complex. The first important application of this method is the classification of homotopy spheres, or better, its reduction to the stable homotopy groups of spheres by Kervaire and Milnor [KM]. A systematic treatment of the theory in the 1-connected case was independently developed by Browder [B1] and Novikov [N]. The much harder general case was done by Wall in his famous book [W2]. This book contains also a lot of applications. Later on, together with Madsen and Thomas, he solved the famous spherical space form problem, asking for those fundamental groups of smooth manifolds whose universal coverings are spheres [MTW]. Other fundamental results and problems dealt by Wall's surgery theory include the study of manifolds which are $K(\pi, 1)$-complexes. The key words here are the Novikov and Borel conjectures.

The third method is a modification of the second. In both cases one tries to find an $s$-cobordism between two given manifolds. For this, one assumes the existence of some cobordism between the given manifolds and tries to replace it via a sequence of surgeries by an $s$-cobordism. Then, for dimension of the given manifolds $\geq 5$, the $s$-cobordism theorem implies that the manifolds are diffeomorphic [Ke]. To carry the surgeries out, one needs some control. In the classical surgery theory, the control comes from assuming that the homotopy equivalences from the manifolds to a given Poincaré complex extend to the bordism in such a way that an appropriate stable vector bundle over the Poincaré complex pulls back to the normal bundle of the cobordism. In the early eighties the author modified this theory by replacing the Poincaré complex by something weaker, which roughly controls the homotopy type of the given manifolds up to half the dimension. This allows one sometimes to classify manifolds where the homotopy types are unknown or manifolds where the passage from the diffeomorphism types of pairs $(M, f)$ of manifolds $M$ together with a homotopy equivalence $f$ to a given Poincaré complex $X$ to the diffeomorphism types of the manifolds $M$ homotopy equivalent to $X$ alone is not known. This theory was successfully applied to classify certain complete intersections [K], [T], and to many 7-dimensional homogeneous spaces [KS1], [KS2].

We don't say anything about the classification problem in dimensions 3 and 4. For the homeomorphism classification of 4-manifolds one can (thanks to Freedman's $s$-cobordism theorem) apply similar surgery methods, if the fundamental groups are for example finite. Results in this direction are contained in [HK1], [HK2]. For the diffeomorphism classification of 4-manifolds the picture is completely different, as follows from the breakthroughs of Donaldson and later on the Seiberg-Witten theory. In dimension 3, again the surgery methods don't help. There Thurston's geometrization program is the most promising method.

## 2. THE EMBEDDING METHOD

I will explain this method by discussing a simple example. Consider the problem of how to decide if a smooth manifold $M$ is diffeomorphic to $S^n \times S^n$ for $n \geq 3$. The starting point is to observe that $S^n \times S^n$ can be decomposed into a regular neighbourhood of the two factors $S^n \times \{*\}$ and $\{*\} \times S^n$ and $D^{2n} = D^n \times D^n$. The regular neighbourhood of the union of the factors can itself be constructed by a plumbing construction from two copies of $S^n \times D^n$ identified along $D^n \times D^n$ where $D^n \times D^n$ is embedded into the first copy as $S^n_+ \times D^n$, $S^n_+$ the upper hemisphere, and into the second copy as $D^n \times S^n_+$:

If we call the result of this glueing of two copies of $D^n \times D^n$ by $X$, then $S^n \times S^n = X \cup_{S^{2n-1}} D^{2n}$. For this, note that the boundary of the plumbed manifold $X$ is $D^n \times S^{n-1} \cup S^{n-1} \times D^n = S^{2n-1}$.

Now one can decide if $M$ is diffeomorphic to $S^n \times S^n$ by trying to reconstruct $M$ as $X \cup_{S^{2n-1}} D^{2n}$. For this, one first tries to embed two copies of $S^n$ into $M$ meeting in one point transversally. This can be achieved if $M$ is 1-connected and there are two maps $\alpha \colon S^n \to M$ and $\beta \colon S^n \to M$ with homological intersection number 1. For then by Whitney's embedding theorem [H] one can approximate $\alpha$ and $\beta$ by embeddings and then by the Whitney trick [Mil] choose these embeddings such that they meet transversally in one point.

The next step is to look at a thickening of the union of these two embedded $n$-spheres and to decide if it is diffeomorphic to the plumbed manifold $X$. For this, it is enough to check if the normal bundles of the two embedded spheres are trivial, since then the union of small tubular neighbourhoods is $X$. The normal bundles are $n$-dimensional vector bundles over $S^n$. For $n \equiv 0 \bmod 4$, an $n$-dimensional vector bundle $E$ over $S^n$ is by Bott periodicity stably trivial if and only if its Pontrjagin class $p_{n/4}(E) = 0$. If $n$ is even a stably trivial bundle $E$ over $S^n$ is trivial if and only if its Euler class $e(E) = 0$. This is in turn equivalent to the vanishing of the homological self intersection number of the zero section in $E$. Recall that the hyperbolic plane $\mathcal{H}$ over $\mathbb{Z}$ is the symmetric bilinear form $S$ over $\mathbb{Z}^2$ with $S(a,a) = S(b,b) = 0$ and $S(a,b) = S(b,a) = 1$, where $a$ and $b$ are a basis. Thus, if $H \colon \pi_n(M) \to H_n(M;\mathbb{Z})$ denotes the Hurewicz homomorphism, the existence of $\alpha$ and $\beta$ with the desired properties is equivalent to the existence of an isometric embedding of the hyperbolic plane $\mathcal{H}$ into the image of $H$ in $H_n(M;\mathbb{Z})$ equipped with the intersection form, such that for each element $x \in \mathcal{H}$, the following holds:

$$\langle p_{n/4}(TM), x \rangle = 0.$$

Thus we have proved:

**Lemma 1.** *For $n \equiv 0 \bmod 4$ and $M$ a 1-connected smooth manifold of dimension $2n$, $X$ can be embedded into $M$ if and only if there is an isometric embedding of the hyperbolic plane $\mathcal{H}$ into the image of $H$ in $H_n(M;\mathbb{Z})$ equipped with the intersection form, such that for each element $x \in \mathcal{H}$, the following relation holds: $\langle p_{n/4}(TM), x \rangle = 0$.*

With this information one can proceed (for $n \equiv 0 \bmod 4$) with the embedding method as follows. Once one knows that $X$ can be embedded into $M$, one has as a next step to decide if the complement $M - X^\circ$ is diffeomorphic to the ball $D^{2n}$. By Smale's $h$-cobordism theorem [S] this is for $n \geq 3$ equivalent to: $M - X^\circ$ is 1-connected and $\tilde{H}_i(M - X^\circ; \mathbb{Z}) = 0$ for all $i$.

A simple exercise in algebraic topology shows that this condition is for $n \geq 3$ equivalent to: $M$ is $(n-1)$-connected and $H_n(X; \mathbb{Z}) \to H_n(M; \mathbb{Z})$ is an isomorphism. Summarizing the conditions of Lemma 1 and these considerations, we see that for $n \equiv 0 \bmod 4$ $M$ is diffeomorphic to $X \cup_f D^{2n}$ for some diffeomorphism $f$ if and only if $M$ is $(n-1)$-connected, $p_{n/4}(M) = 0$ and the intersection form $S$ on $H_n(M; \mathbb{Z})$ is isometric to the hyperbolic plane $\mathcal{H}$. Every diffeomorphism $f \colon \partial X = S^{2n-1} \to \partial D^{2n} = S^{2n-1}$ determines a homotopy sphere $\Sigma_f = D^{2n} \cup_f D^{2n}$ and $X \cup_f D^{2n}$ is diffeomorphic to $X \cup_{Id} D^{2n} \# \Sigma_f$ or to $S^n \times S^n \# \Sigma_f$. Thus $M$ is diffeomorphic to $S^n \times S^n \# \Sigma$ for some homotopy sphere $\Sigma$. In particular, $M$ is homeomorphic to $S^n \times S^n$.

The remaining question is to decide when $S^n \times S^n \# \Sigma$ is diffeomorphic to $S^n \times S^n$. This question has no simple answer. It turns out that this is equivalent to the question of when $\Sigma$ is diffeomorphic to $S^{2n}$. Using the results of Kervaire-Milnor on the classification of homotopy spheres, one knows that $\Sigma$ is diffeomorphic to $S^{2n}$ if and only if $\Sigma$ bounds a parallelizable manifold [KM]. Combining this with the considerations above leads to the following result.

**Theorem 1.** *Assume $n \equiv 0 \bmod 4$. A $2n$-dimensional smooth manifold $M$ is homeomorphic to $S^n \times S^n$ if and only if $M$ is $(n-1)$-connected, $p_{n/4}(M) = 0$ and the intersection form $S$ on $H_n(M; \mathbb{Z})$ is isometric to the hyperbolic plane $\mathcal{H}$.*

*$M$ is diffeomorphic to $S^n \times S^n$ if, in addition, $M$ bounds a parallelizable manifold.*

It is not difficult to generalize the homeomorphism classification to arbitrary $(n-1)$-connected $2n$-manifolds $M$, $n > 2$. The Pontrjagin class which measures the stable normal bundle of embedded $n$-spheres in $M$ has to be replaced by a more complicated invariant and, for $n$ odd and $n$ not equal to 3 or 7, the intersection form has to be refined by a quadratic form $q$ with values in $\mathbb{Z}/2$ defined on those homology classes which can be represented by spheres with stably trivial normal bundle. All this is carried out in [W1]. There is also some strong information by S. Stolz [St] about the passage to the diffeomorphism classification, which again boils down to the question of when $M \# \Sigma$ is diffeomorphic to $M$ for a homotopy sphere $\Sigma$.

How far can one further generalize the embedding method? If one looks closer at the case $S^n \times S^n$, one observes that the decomposition as $X \cup D^{2n}$ is a special case of a handle body decomposition:

$$M = D^k \cup H_1 \cup H_2 \cup \ldots$$

where $H_i$ are handles of the form $D^s \times D^{k-s}$.

Thus there is a chance only if one can (in principle) decide if for an embedding of $D^k \cup H_1 \cdots \cup H_r$ into $M$ one can extend this to an embedding of $D^k \cup H_1 \cdots \cup H_r \cup H_{r+1}$ into $M$. There are general results reducing this question to homotopy theory, which work rather nicely in the so called metastable range, i.e., if $M$ is roughly $k/3$-connected. Thus for such manifolds one can, if the homological picture is not too complicated, try to apply the embedding methods.

## 3. The surgery program

The starting point here is a finite connected Poincaré complex $X$. For simplicity we assume it to be oriented and $\pi_1(X)$ finite. Then by definition, $X$ is a finite $CW$-complex together with a class $[X] \in H_n(X; \mathbb{Z})$ such that if we call $[\tilde{X}]$ the transfer of $[X]$ in $H_n(\tilde{X}; \mathbb{Z})$, $\tilde{X}$ the universal covering, we assume that $[\tilde{X}]\cap: H^k(\tilde{X}; \mathbb{Z}) \to H_{n-k}(\tilde{X}; \mathbb{Z})$ is an isomorphism. Each closed (oriented with finite fundamental group) smooth manifold has a $CW$-structure, and Poincaré duality implies that it is a Poincaré complex.

Wall reserves the name finite Poincaré complex for the case where the map induced by $[\tilde{X}]\cap$ on the chain level,

$$C^k(\tilde{X}) \to C_{n-k}(\tilde{X}),$$

is a simple homotopy equivalence, i.e., its Whitehead torsion [W2] vanishes. Here $C^k(\tilde{X})$ and $C_k(\tilde{X})$ mean the cellular chain complexes with respect to the $CW$-composition of $X$ (pulled back to $\tilde{X}$), and the Whitehead torsion is taken with respect to the basis given by the cells. Again, a closed smooth manifold (oriented with finite fundamental group) is a simple finite Poincaré complex, and for those who are not used to Poincaré complexes, it is convenient to think of this case.

The surgery program aims to classify all smooth manifolds simply homotopy equivalent to a given finite simple Poincaré complex $X$ of dimension $\geq 5$. Unfortunately, it studies something even more complicated, the set of so called homotopy smoothings $S(X)$. $S(X)$ consists of diffeomorphism classes of pairs $(M, f)$, where $f: M \to X$ is a simple (again with respect to the cellular basis) homotopy equivalence of a smooth closed manifold $M$ to $X$. Two such homotopy smoothings are diffeomorphic if there is a diffeomorphism between the manifolds commuting up to homotopy with the simple homotopy equivalences.

The group of homotopy classes of simple self equivalences of $X$, $Aut_s(X)$, acts on $S(X)$ by composition, and the orbit space is what we are looking for, the set of diffeomorphism classes of manifolds simply homotopy equivalent to $X$. We denote this set by $\mathcal{M}(X)$. Then

$$\mathcal{M}(X) = S(X)/Aut_s(X).$$

One can make similar definitions in the topological category, replacing smooth manifolds by topological ones and looking for classification up to homeomorphisms. The whole program can be carried out in this context and the answers are in a certain sense even simpler. But there are deep facts behind this, the smoothing theory of Kirby and Siebenmann [KS]. On the other hand, the smooth surgery theory of Wall played an important role in smoothing theory, in particular in connection with the classification of homotopy tori. I decided to concentrate in this article on the smooth category and refer to the literature for the topological case.

Now, if one wants to classify the manifolds simply homotopy equivalent to $X$, one has to compute $S(X)$, $Aut_s(X)$ and the action of $Aut_s(X)$ on $S(X)$. None of these steps is simple. The group $Aut_s(X)$ fits into an exact sequence

$$1 \to Aut_s(X) \to Aut(X) \to Wh(\pi_1(X))$$

The middle term is the group of homotopy classes of self equivalences, an object of classical homotopy theory. Although there are general methods for attacking this group, typically based on obstruction theory, its computation is in general hopeless. The map $Aut(X) \to Wh(\pi_1(X))$ is Whitehead torsion and the kernel is $Aut_s(X)$.

For $S(X)$ has to distinguish the case where $S(X)$ is empty, i.e. there is no smooth manifold homotopy equivalent to $X$, from the other case, on which we want to concentrate. Then $S(X)$ can also be computed in terms of an exact sequence. We can assume that $X$ is itself a manifold. In this situation there is a fundamental invariant on $S(X)$, the normal invariant. Let $f: M \to X$ be a simple homotopy equivalence. Let $g$ be a homotopy inverse of $f$. Then $E := g^*\nu(M)$, $\nu(M)$ the stable normal bundle of $M$, is a stable vector bundle over $X$, and there is an isomorphism $\alpha$ between $\nu(M)$ and $f^*E$. The set of bordism classes of quadruples $(M, f, E, \alpha)$, where $f: M \to X$ is a degree 1 map, $E$ a stable vector bundle over $X$ and $\alpha$ an isomorphism between $\nu(M)$ and $f^*E$, is denoted by $\Omega d(X)$, and the normal invariant is given by the quadruple defined above, giving a map

$$\eta : S(X) \to \Omega d(X).$$

The next step in computing $S(X)$ is the determination of the fibres of $\eta$. This is the central result of [W2]. The basic idea is to look at a bordism between $(M, f, E, \alpha)$ and $(M', f', E', \alpha')$ and to ask if this bordism is itself bordant rel. boundary to an $s$-cobordism between $M$ and $M'$. If this is the case, the $s$-cobordism theorem implies that $(M, f, E, \alpha)$ and

$(M', f', E', \alpha')$ are diffeomorphic (assuming dim $M$ greater than 4). Since the $s$-cobordism theorem is so central for the classification of manifolds, let me say a few words about it. Let's first assume that all manifolds are simply connected. An $h$-cobordism is a bordism such that the inclusions from the two boundary components to the bordism is a homotopy equivalence. Thus, up to homotopy, the bordism is a cylinder. The surprising deep result of Smale [S] states that the bordism is actually diffeomorphic to a cylinder. If the manifolds are not simply connected, this is not always true. There is an additional obstruction to be taken care of, the Whitehead torsion, which assigns to a homotopy equivalences between finite CW-complexes (e.g., compact manifolds) an invariant in the Whitehead group of the fundamental group. This is how algebraic $K$-theory (the Whitehead group is essentially the algebraic $K$-group $K_1$) comes naturally into the game.

Let's return to the surgery program. Wall defines, in a purely algebraic way, groups denoted $L_m(\pi_1(X))$, depending on $\pi_1(X)$ and the dimension $m$ of $X$ (modulo 4). The groups are quadratic form analogs of algebraic $K$-groups. For $m$ even they are roughly Grothendieck groups of unimodular symmetric, if $m \equiv 0 \mod 4$, and antisymmetric, if $m \equiv 2 \mod 4$, hermitian forms over free $\mathbb{Z}[\pi_1(X)]$-modules. For details I refer to [W2]. Then Wall defines a map

$$\theta: \Omega d(X) \to L_m(\pi_1(X)),$$

where $m$ is the dimension of the smooth manifold $X$, and an action

$$L_{m+1}(\pi_1(X)) \times S(X) \to S(X)$$

such that the orbit space injects under $\theta$ into the "kernel" of $\eta$. This is the content of the famous surgery exact sequence.

**Theorem 2.** *Let $X$ be a closed smooth manifold of dimension $m \geq 5$. There is an exact sequence (in the sense described above)*

$$L_{m+1}(\pi_1(X)) \xrightarrow{\eta} S(X) \to \Omega d(X) \xrightarrow{\theta} L_m(\pi_1(X)).$$

Thus, to determine $S(X)$, one has as a first step to compute the group $L_{m(+1)}(\pi_1(X))$ and the set $\Omega d(X)$. For the first, Wall [W3] has developed deep algebraic techniques which allow very powerful computations. For trivial fundamental groups, Kervaire and Milnor, without having the language of $L$-groups available, indirectly compute these groups in [KM]. For an overview of computation of Wall groups compare [HT]. Although $\Omega d(X)$ has — as described — no group structure, there is another interpretation by Sullivan which not only gives a group structure (where $\theta$ is in general not a homomorphism) but, in addition, shows that the functor $\Omega d(X)$ is a generalized cohomology theory. Let $G$ be the limit as $k$ goes to infinity of the monoid of self equivalences on $S^k$. The homomorphism $O \to G$ induces a map between classifying spaces $BO \to BG$ whose fibre is denoted $G/O$. Sullivan identifies:

$$\Omega d(X) \cong [X, G/O]$$

and proves several deep facts concerning $G/O$ and the topological analogue $G/TOP$. Using this determination of $\Omega d$ leads to the final form of the surgery exact sequence:

**Theorem 3.** *Let $X$ be a closed smooth manifold of dimension $m \geq 5$. Then there is an exact sequence*

$$[\Sigma X, G/O] \overset{\psi}{\to} L_{m+1}(\pi_1(X)) \to S(X)$$
$$\overset{\eta}{\to} [X, G/O] \overset{\theta}{\to} L_m(\pi_1(X)).$$

Here $\psi$ is the analog of $\theta$ on the manifold $X \times I$ (rel. boundary). $\Sigma X$ is the suspension of $X$. It turns out that, in contrast to $\theta$, the map $\psi$ is a homomorphism. Thus, up to an extension problem, the computation of $S(X)$ is reduced to the computation of $L_{m(+1)}(\pi_1(X))$, of $[(\Sigma)X, G/O]$, and of the maps $\psi$ and $\theta$. The computation of $[X, G/O]$ and of $[\Sigma X, G/O]$ is very much simplified by the fact that $X \mapsto [X, G/O]$ is a generalized cohomology theory.

To make use of the fact that $[X, G/O]$ is a generalized cohomology theory, one has to understand the coefficients $[S^n, G/O] = \pi_n(G/O)$. For this one uses the homotopy exact sequence of the fibering $BO \to BG$:

$$\pi_{n+1}(BO) \to \pi_{n+1}(BG) \to \pi_n(G/O) \to \pi_n(BO) \to \pi_n(BG).$$

Then one uses the standard isomorphisms $\pi_n(BO) \cong \pi_{n-1}(O)$ and

$$\pi_n(BG) \cong \pi_{n-1}(G) \cong \pi_{n-1}^s,$$

the stable $(n-1)$-stem. It turns out the map $\pi_n(BO) \to \pi_n(BG)$ translates into the $J$-homomorphism $J_n$, implying a short exact sequence:

$$0 \to \text{cok } J_n \to \pi_n(G/O) \to \text{ker } J_{n-1} \to 0.$$

According to [A], ker $J_{n-1}$ is zero unless $n \equiv 0 \mod 4$, when it is isomorphic to $\mathbb{Z}$. Thus we obtain

$$\pi_n(G/O) \cong \begin{cases} \mathbb{Z} \times \text{cok } J_n & n \equiv 0 \mod 4, \\ \text{cok } J_n & n \not\equiv 0 \mod 4. \end{cases}$$

We want to apply Theorem 2 to compute the group of homotopy spheres of dimension $n$, which is denoted by $\Theta_n$. We first note that $Aut_s(S^n) = 1$, since the degree determines the homotopy class, and the Whitehead torsion is trivial for 1-connected spaces. Thus $\Theta_n \cong S(S^n)$.

Using the computation of the $L$-groups of a trivial fundamental group [W2] (Theorem 13 A.1) (indirectly contained in [KM]):

$$\begin{cases} L_{2k+1}\{1\} & = & \{0\} \\[2mm] L_{4k}\{1\} & \cong & \mathbb{Z} \\[2mm] L_{4k+2}\{1\} & \cong & \mathbb{Z}/2, \end{cases}$$

the fact that for $n = 0 \mod 4$ the map from $\pi_n(G/O)$ to $L_n(\{1\}) \cong \mathbb{Z}$ is given by the signature and thus injective, and that by construction [W2] the map $L_{m+1}\{1\} \to S(S^n)$ maps to those pairs $(\Sigma, f)$ where $\Sigma$ bounds a parallelizable manifold, one obtains the result originally proved by Kervaire and Milnor [KM]:

**Theorem 4.** *There is an exact sequence of groups*

$$0 \to bP_{n+1} \to \Theta_n \to \operatorname{cok} J_n,$$

*where $bP_{n+1}$ is the group of homotopy spheres which bound a parallelizable manifold.*

Kervaire and Milnor prove in addition that $bP_{n+1}$ is a cyclic group and determine its order, and show that the map $\Theta_n \to \operatorname{cok} J_n$ is surjective, unless perhaps $n \equiv 2 \mod 4$, where its cokernel is either trivial or isomorphic to $\mathbb{Z}/2$. This is equivalent to studying the map $\psi$ in Theorem 3. To decide when the map $\Theta_n \to \operatorname{cok} J_n$ is surjective is the famous Arf-invariant 1 manifold problem: It is surjective if and only if there is no closed parallelizable manifold of Arf invariant 1. By work of Browder, if such a manifold exists, $n$ is of the form $2^i - 2$ [B2].

There are not many other homotopy types of manifolds $X$ where a similar computation for $\mathcal{M}(X)$ is known. Wall himself studies the case where $X$ is a lens space and calls it "the best example for application of our techniques" [W2]. To my knowledge, even for lens spaces a complete answer analogous to the case for $S^n$ is up to now not known.

We have concentrated on the case where the Poincaré complex is homotopy equivalent to a manifold, i.e. $S(X) \neq \emptyset$. Wall's techniques also deal with the question of when $S(X) \neq \emptyset$. As mentioned in the introduction, he, together with Madsen and Thomas, solves with his techniques and a lot of computational work on $L$-groups the famous spherical space form problem. I consider this the best application of Wall's surgery theory. Of course, there is a vast literature making use of Wall's book. In particular, the computation of $L$-groups leads in many cases to a reduction of the computation of $S(X)$ to stable homotopy (i.e. the computation of $[X, G/O]$). My impression is that the final step, the investigation of the action of $Aut_s(X)$ on $S(X)$, is almost never studied. One has best chances if the manifolds are $K(\pi, 1)$'s or at least close to such spaces. This is one of the reasons why the method was so successfully applied to such important problems like the space form problem, the Novikov and Borel conjectures. To make surgery theory accessible to some other natural contexts where the action of $Aut_s(X)$ on $S(X)$ is too difficult is one of the motivations for the third classification method described below. Of course, this is only a modification of Wall's method.

## 4. Surgery and Duality

This is the name of a program which is a modification of the surgery program. The idea is to weaken the homotopy theoretical input, hoping that Poincaré duality or stronger duality forces allow one to obtain full classification results. I will demonstrate the method by an example, the diffeomorphism classification of complete intersections. Let $f_1, \cdots, f_r$ be homogeneous polynomials on $\mathbb{C}P^{n+r}$ of degrees $d_1, \cdots, d_r$. If the gradients of these polynomials are linearly independent, the set of common zeros is a smooth complex manifold of dimension $n$, a non-singular complete intersection. As was noted by Thom, the diffeomorphism type of non-singular complete intersections depends only on the unordered tuple $\delta = (d_1, \cdots, d_r)$, called the multidegree. We denote this diffeomorphism type by $X_\delta^n$. It is natural to ask for a diffeomorphism classification of this very interesting class of algebraic manifolds.

Except under some restrictive assumptions ([LW1], [LW2]), even the homotopy classification of the $X_\delta^n$'s, which is the first step in the ordinary surgery theory, is unknown. On the other hand the topology of $X_\delta^n$ up to half the dimension is known. According to Lefschetz the inclusion

$$i \colon X_\delta^n \longrightarrow \mathbb{C}P^\infty$$

is an $n$-equivalence.

Moreover, it is easy to see that the normal bundle of $X_\delta^n$ is isomorphic to

$$\nu(X_\delta^n) \cong i^*(\nu(\mathbb{C}P^{n+r}) \oplus H^{d_1} \oplus \cdots \oplus H^{d_r})$$
$$\cong i^*(-(n+r+1) \cdot H \oplus H^{d_1} \oplus \cdots \oplus H^{d_r})$$

where $H$ is the Hopf bundle and $H^{d_i}$ means the $d_i$-fold tensor product. Denote $-(n+r+1) \cdot H \oplus H^{d_1} \oplus \cdots \oplus H^{d_r}$ by $\xi(n, \delta)$.

To obtain numerical invariants we choose a generator

$$x \in H^2(X_\delta^n; \mathbb{Z}).$$

Then the Pontrjagin classes are expressed as $p_i(X_\delta^n) = \alpha_i(n, \delta) \cdot x^{2i}$. We say that two complete intersections $X_\delta^n$ and $X_{\delta'}^n$ have equal Pontrjagin classes if $\alpha_i(n, \delta) = \alpha(n, \delta')$ for all $i \leq 2n$. Obviously diffeomorphic complete intersections have equal Pontrjagin classes.

Now we are looking for a space $B$ in which we can compare two complete intersections. $B$ will play the role of the Poincaré complex $X$ in the standard surgery theory. As topological space $B$ will be $\mathbb{C}P^\infty \times BO\langle n+1\rangle$, where the second factor denotes the $(n+1)$-connected cover over $BO$. This is a space admitting a fibration over $BO$ whose homotopy groups vanish up to dimension $n+1$ and are mapped isomorphically in higher dimensions. For $n = 0$ one obtains of course the usual universal cover. Denote by $E \to BO\langle n+1\rangle$ the stable vector bundle obtained as pullback of the universal bundle over $BO$. Then $\xi(n,\delta) \times E$ is a stable vector bundle over $B$. Obviously the inclusion into the first factor gives a map $f : X_\delta^n \to B$ under which $\xi(n,\delta) \times E$ pulls back to the normal bundle of $X_\delta^n$. This map is an $n$-equivalence. If $X_{\delta'}^n$ is a second complete intersection with equal Pontrjagin classes, an easy obstruction theory argument shows that there is another $n$-equivalence $f' = X_{\delta'}^n \to B$ under which $\xi(n,\delta) \times E$ pulls back to the normal bundle of $X_{\delta'}^n$.

Then we are in a similar situation as in the surgery program. The only difference is that the space $B$ in which we compare our two manifolds is not a Poincaré complex.

At this point we generalize for a moment from complete intersection to arbitrary manifolds. Let $B$ be a $CW$-complex and $E$ a stable vector bundle over $B$. Let $(M, f, \alpha)$ be a triple of a closed oriented smooth manifold $M$ of dimension $k = 2n$ or $2n + 1$ together with an $n$-equivalence $f : M \to B$ and an isomorphism $\alpha$ between $f^*E$ and $\nu(M)$. This is called a normal $n$-smoothing. Denote the bordism group of such triples, where we do not require that $f$ is an $n$-equivalence, by $\Omega_k(B, E)$. Denote the set of diffeomorphism classes of normal $n$-smoothings $(M, f, \alpha)$ by $S_k(B, E)$. The main result of [K] is the construction of a monoid $l_{k+1}(\pi_1(B))$ and of an invariant $\theta(W, g, \beta) \in l_{k+1}(\pi_1(B))$. Here $(W, g, \beta)$ is a bordism between two normal $n$-smoothings $(M, f, \alpha)$ and $(M', f', \alpha')$, and $(W, g, \beta)$ is bordant rel. boundary to an $s$-cobordism if and only if $\theta(W, g, \beta)$ is elementary, which is a purely algebraic property of elements in $l_{k+1}(\pi_1(B))$.

**Theorem 5.** ([K], Theorem B) *Assume $k \geq 5$. The forgetful map $S_k(B, E) \to \Omega_k(B, E)$ is surjective. If $(W, g, \beta)$ is a bordism between two elements in $S_k(B, E)$ with the same Euler characteristic, then this is bordant to an $s$-cobordism if and only if $\theta(W, g, \beta) \in l_{k+1}(\pi_1(B))$ is elementary.*

The obstruction $l$-monoids are very complicated and algebraically — in contrast to the $L$-groups — not understood. Thus it is a surprise that for $k$ even, at least for finite fundamental groups, one can under some mild conditions forget the obstruction.

**Theorem 6.** ([K], Theorem 5) *Let $(W, g, \beta)$ be a bordism between two normal $n$-smoothings of $2n$-dimensional manifolds $M$ and $M'$ with same Euler characteristic. Then there is another bordism $(W', g', \beta')$ between the same manifolds such that $\theta(W', g', \beta') \in l_{2n+1}(\pi_1(B))$ is elementary if one of the following conditions is fulfilled:*

**i):** *q odd and B 1-connected*

**ii):** *q even, B 1-connected and* $\ker(\pi_q(M) \to \pi_q(B))/\mathrm{rad}$ *splits off* $\mathcal{H}$.

**iii):** $\pi_1(B)$ *is finite and* $\ker(\pi_q(M) \to \pi_q(B))/\mathrm{rad}$ *splits off* $H_\epsilon(\Lambda^2)$.

*Here* $\Lambda$ *is the group ring* $\mathbb{Z}[\pi_1(B)]$ *and* $H_\epsilon(\Lambda^2)$ *is the hyperbolic form over* $\Lambda^2$.

In §2 we have found necessary and sufficient conditions for fulfilling ii). One can show that these conditions are fulfilled for all complete intersections (of complex dimension $\geq 3$) except for $X_1^n, X_2^n, X_{(2,2)}^n$, in which case the total degree $d = \prod d_i$ and the Pontrjagin class determine the $d_i$'s and thus the diffeomorphism type.

Furthermore one can show that up to homotopy the map $f \colon X_\delta^n \to B$ is unique, and thus we obtain from Theorems 5 and 6:

**Theorem 7.** *Given* $\alpha_i$ *($1 \leq 2i \leq n$), the set of diffeomorphism types of complete intersections with Pontrjagin classes given by* $\alpha_i$ *injects into* $\Omega_{2n}(B, \xi) \times \mathbb{Z}$.

Here $B = \mathbb{C}P^\infty \times BO\langle n + 1 \rangle$, $\xi = \xi_1 \times E$ with $\xi_1$ the bundle over $\mathbb{C}P^n$ pulling back to the normal bundle of the complete intersections under the map inducing an isomorphism on second cohomology (this bundle is determined by the $\alpha_i$'s). The map into the first factor is given by the bordism class of the (unique) normal $n$-smoothing. The map to $\mathbb{Z}$ is given by the Euler characteristic.

This reduces the classification of complete intersections to a bordism invariant. Unfortunately the corresponding bordism groups are not explicitly known. But perhaps one can decide if the complete intersections are bordant without computing the bordism group. Under conditions on the total degree this is actually possible, a program which was carried out in the *Diplomarbeit* of Claudia Traving [T]. This leads to the following result. If $d = d_1 \cdot \cdots \cdot d_r$ is the total degree of a complete intersection $X_{d_1, \cdots, d_r}^n$ of complex dimension $n$, then we assume that for all primes $p$ with $p(p - 1) \leq n + 1$, the total degree $d$ is divisible by $p^{[(2n+1)/(2p-1)]+1}$.

**Theorem 8.** *Two complete intersections* $X_\delta^n$ *and* $X_{\delta'}^n$ *of complex dimension* $n > 2$ *fulfilling the assumption above for the total degree are diffeomorphic if and only if the total degrees, the Pontrjagin classes and the Euler characteristics agree.*

It is an open and very interesting problem to determine if this theorem holds without assumptions on the total degree.

## 5. COMMENTS

All three methods have their specific strength and weakness. The embedding method, if applicable, is very straightforward and not only classifies manifolds but also identifies them with explicit models. The disadvantage is that except for highly connected manifolds there are very few cases when this method works.

The advantage of the surgery program over both other methods is that — in principle — it answers a very natural and general problem: The classification of diffeomorphism classes of manifolds simply homotopy equivalent to a given manifold. But the actual computation is very hard. In particular, the last step, passing from the homotopy smoothings to the diffeomorphism types, is a difficult problem. Of similar type is the question of when two manifolds are (simply) homotopy equivalent.

The third method can sometimes avoid the latter difficulty since the homotopy type is replaced by the $(n/2)$-type, $n$ the dimension of the manifold. And also the problem of the action of $Aut_s(X)$ on $\mathcal{S}(X)$ can become easier or disappear. The latter case can be demonstrated by classification of certain 1-connected 7-dimensional homogeneous spaces containing the Aloff-Wallach spaces. In joint work with Stephan Stolz [KS2] we give a diffeomorphism classification based on Theorem 5. In contrast to the even-dimensional case of complete intersections, here one has to analyze the obstructions in the monoid $l_8$. The homotopy type of these homogeneous spaces (and some generalizations) was determined later, by Kruggel [Kr] and Milgram [M] independently. Here it is not difficult to determine $\mathcal{S}(X)$, but to my knowledge nobody has computed the action of $Aut(X)$ on $\mathcal{S}(X)$. Concerning the Witten spaces classified in [KS1] most of these results were obtained shorty after our proof by Ib Madsen using a different approach. Recently I heard that Christine Escher [E] found a new way to classify these Witten spaces, confirming the results from [KS1], and some generalized spaces.

## REFERENCES

[A]     J. F. Adams. *On the groups $J(X)$ – IV*, Topology 5, 21–71 (1966).

[B1]    W. Browder. *Surgery on simply-connected manifolds*, Springer-Verlag, 1972.

[B2]    W. Browder. *The Kervaire invariant of framed manifolds and its generalization*, Ann. of Math. 90, 157–186 (1969).

[E]     C. Escher. *A diffeomorphism classification of generalized Einstein-Witten manifolds*, preprint (1999)

[HK1]   I. Hambleton, M. Kreck. *Smooth structures on algebraic surfaces with cyclic fundamental group*, Invent. Math. 91, 53–59 (1988).

[HK2]   I. Hambleton, M. Kreck. *Cancellation, elliptic surfaces and the topology of certain four-manifolds*, J. reine u. angew. Math. 444, 79–100 (1993).

[HT]    I. Hambleton, L. Taylor. *A guide to the calculation of the surgery obstruction groups for finite groups*, this volume.

[H]     M. Hirsch. *Differential topology*, Springer-Verlag, 1976.

[Ke]    M. Kervaire. *Le théorème de Barden-Mazur-Stallings*, Comm. Math. Helv. 40, 31–42 (1965).

[KM]    M. Kervaire, J. Milnor. *Groups of homotopy spheres I*, Ann. of Math. 77, 504–531 (1963).

[KS]    R. C. Kirby, L. C. Siebenmann. *Foundational Essays on Topological Manifolds, Smoothings and Triangulations*, Ann. of Math. Studies. vol. 88, Princeton University Press, 1977.

[K]     M. Kreck. *Surgery and Duality*, Ann. of Math. 149, no. 3, 707–754 (1999).

[KS1]   M. Kreck, S. Stolz. *A diffeomorphism classification of 7-dimensional homogeneous Einstein manifolds with $SU(3) \times SU(2) \times U(1)$-symmetry*, Ann. of Math. 127, 373–388 (1988).

[KS2]   M. Kreck, S. Stolz. *Some nondiffeomorphic homeomorphic homogeneous 7-manifolds with positive sectional curvature*, J. Diff. Geom. 33, 465–486 (1991).

[Kr]    B. Kruggel. *Kreck-Stolz invariants, normal invariants and the homotopy classification of generalised Wallach spaces*, Quart. J. Math. Oxford Ser. (2) 49, no. 196, 469–485 (1998).

[LW1]   A. Libgober, J. Wood. *Differentiable structures on complete intersections. I*, Topology 21, 469–482 (1982).

[LW2]   A. Libgober, J. Wood. *On the topological structure of even dimensional complete intersections*, Trans. Amer. Math. Soc. 267, 637–660 (1981).

[MTW]   I. Madsen, Ch. Thomas and C. T C. Wall. *The topological spherical space form problem II*, Topology 15, 375–382 (1978).

[M]     R. J. Milgram. *The classification of Aloff-Wallach manifolds and their generalizations*, this volume.

[Mi1]   J. Milnor. *Lectures on the h-cobordism theorem*, Princeton Univ. Press, 1965.

[Mi2]   J. Milnor. *Whitehead torsion*, Bull. Amer. Math. Soc. 72, 358–426 (1966).

[N]     S. P. Novikov. *Homotopy equivalent smooth manifolds I*, Izv. Akad. Nauk SSSR 28, 365–474 (1964), and Translations Amer. Math. Soc. 48, 271–396 (1965).

[S]     S. Smale. *On the structure of manifolds*, Amer. J. Math. 84, 387–399 (1962).

[St]    S. Stolz. *Hochzusammenhängende Mannigfaltigkeiten und ihre Ränder*, Lecture Notes in Math., vol. 1116, Springer-Verlag (1985).

[T]     C. Traving. *Klassifikation vollständiger Durchschnitte*, Diplomarbeit, Mainz (1985). Available from http://www.mfo.de/Staff/traving.pdf .

[W1]    C. T. C. Wall. *Classification of $(n-1)$-connected $2n$-manifolds*, Ann. of Math. 75, 163–189 (1962).

[W2]    C. T. C. Wall. *Surgery on Compact Manifolds*, London Math. Soc. Monographs, vol. 1, Academic Press, 1970.

[W3]    C. T. C. Wall. *On the classification of hermitian forms. VI: Group rings*, Ann. of Math. 103, 1–80 (1976).

Fachbereich Mathematik,
Universität Mainz
55127 Mainz
Federal Republic of Germany

and

Mathematisches Forschungsinstitut Oberwolfach
77709 Oberwolfach
Federal Republic of Germany

email: kreck@topologie.mathematik.uni-mainz.de

# Poincaré duality spaces

## John R. Klein

### INTRODUCTION

At the end of the last century, Poincaré discovered that the Betti numbers of a closed oriented triangulated topological $n$-manifold $X^n$

$$b_i(X) := \dim_{\mathbb{R}} H_i(X; \mathbb{R})$$

satisfy the relation

$$b_i(X) = b_{n-i}(X)$$

(see e.g., [Di, pp. 21–22]). In modern language, we would say that there exists a chain map $C^*(X) \to C_{n-*}(X)$ which in every degree induces an isomorphism

$$H^*(X) \cong H_{n-*}(X) .$$

The original proof used the dual cell decomposition of the triangulation of $X$. As algebraic topology developed in the course of the century, it became possible to extend the Poincaré duality theorem to non-triangulable topological manifolds, and also to homology manifolds.

In 1961, Browder [Br1] proved that a finite $H$-space satisfies Poincaré duality. This result led him to question whether or not every finite $H$-space has the homotopy type of a closed smooth (= differentiable) manifold. Abstracting further, one asks:

*Which finite complexes have the homotopy type of closed topological manifolds? of closed smooth manifolds?*

To give these questions more perspective, recall that Milnor had already shown in 1956 that there exist several distinct smooth structures on the 7-sphere [Mi1]. Furthermore, Kervaire [Ke] constructed a 10-dimensional $PL$ manifold with no smooth structure. It is therefore necessary to distinguish between the homotopy types of topological and smooth manifolds. Kervaire and Milnor [K-M] systematically studied groups of the $h$-cobordism classes of homotopy spheres, where the group structure is induced by connected sum. They showed that these groups are always finite. In dimensions $\geq 5$

the $h$-cobordism equivalence relation is just diffeomorphism, by Smale's $h$-cobordism theorem [Sm]).

Since topological manifolds satisfy Poincaré duality (with respect to suitable coefficients), the existence of a Poincaré duality isomorphism is a necessary condition for a space to have the homotopy type of a closed manifold. Such a space is called a *Poincaré duality space*, or a *Poincaré complex* for a finite $CW$ complex.

Poincaré complexes were to play a crucial role in the Browder-Novikov-Sullivan-Wall surgery theory classification of manifolds. We can view the surgery machine as a kind of *descent theory* for the forgetful functor from manifolds to Poincaré complexes:

- Given a problem involving manifolds, it is often the case that it has an analogue in the Poincaré category.
- One then tries to solve the problem in the Poincaré category, where there is more freedom. In the latter, one has techniques (e.g., homotopy theory) that weren't available to begin with.
- Supposing that there is a solution to the problem in the Poincaré category, the last step is to lift it back to a manifold solution. It is here that the surgery machine applies. Except in low (co)dimensions, the only obstruction to finding the lifting is given by the triviality of certain element of an $L$-group $L_n(\pi)$.

Thus surgery theory gives an approach for solving manifold classification problems, *modulo the solution of the corresponding problem for Poincaré complexes*.

In general, a Poincaré duality space is not homotopy equivalent to a topological manifold. Thus Poincaré duality spaces fall into more homotopy types than topological manifolds. In 1965, Gitler and Stasheff [G-S] constructed an example of a simply connected finite complex $X$ which satisfies 5-dimensional Poincaré duality, but which isn't the homotopy type of a closed topological manifold. This example has the homotopy type of a complex of the form $(S^2 \vee S^3) \cup e^5$, with respect to a suitable attaching map $S^4 \to S^2 \vee S^3$. More specifically, $X$ is the total space of a spherical fibration $S^2 \to X \to S^3$ which admits a section. By the clutching construction, such a fibration is classified by an element of $\pi_2(\mathrm{Aut}_*(S^2)) \cong \pi_4(S^2) = \mathbb{Z}/2$. We take $X$ to correspond to the generator of this group.

Returning to Browder's original question about finite $H$-spaces, it is worth remarking that at the present time there is no known example of a finite $H$-space which isn't the homotopy type of a closed smooth manifold.

*Outline.* §1 concerns homology manifolds, which are mentioned more-or-less for their historical interest. In §2 we define Poincaré complexes, following Wall. I then mention the various ways Poincaré complexes can arise.

§3 is an ode to the Spivak normal fibration. I give two proofs of its existence. The first essentially follows Spivak, and the second is due to me (probably). In §4 I outline some classification results about Poincaré complexes in low dimensions, and I also give an outline as to what happens in general dimensions in the highly connected case. In §5 I describe some results in Poincaré embedding theory and further connections to embeddings of manifolds. §6 is a (slightly impious) discussion of the Poincaré surgery programs which have been on the market for the last twenty five years or so. I've also included a short appendix on the status of the finite $H$-space problem. The bibliography has been extended to include related works not mentioned in the text.

*Acknowledgement.* I am much indebted to Andrew Ranicki for help in researching this paper. Thanks also to Teimuraz Pirashvili for help with translation from the Russian.

## 1. FORERUNNERS OF POINCARÉ DUALITY SPACES

Spaces having the homological properties of manifolds have a history which dates back to the 1930s, and are to be found in the work of Čech, Lefschetz, Alexandroff, Wilder, Pontryagin, Smith and Begle. These 'generalized $n$-manifolds' (nowadays called *homology manifolds*) were defined using the local homology structure at a point. The philosophy at the time of their introduction was that these spaces were supposedly easier to work with than smooth or combinatorial manifolds.

We recall the following very special case of the definition (for the general definition and the relevant historical background see [Di, pp 210–213]). An *(ANR) homology $n$-manifold* $X$ is a compact $ANR$ with local homology groups

$$H_*(X, X \setminus \{x\}) = H_*(\mathbb{R}^n, \mathbb{R}^n \setminus \{0\}) = \begin{cases} \mathbb{Z} & \text{if } * = n \\ 0 & \text{if } * \neq 0 \end{cases} \quad (x \in X) \ .$$

Now, if our ultimate goal is to study the homotopy properties of manifolds, this definition has an obvious disadvantage: it isn't homotopy invariant. It is easy to construct a homotopy equivalence of spaces $X \xrightarrow{\simeq} Y$ such that $X$ is a homology manifold but $Y$ is not a homology manifold. The notion of Poincaré duality space is homotopy invariant, offering a remedy for the problem by ignoring the local homology structure at each point. Any space homotopy equivalent to a Poincaré duality space is a Poincaré duality space.

## 2. The definitions

There are several different flavors of Poincaré complex in the literature [Wa3], [Wa4], [Le1], [Spi]. We shall be using Wall's definition in the finite case, without a Whitehead torsion restriction.

Suppose that $X$ is a connected finite $CW$ complex whose fundamental group $\pi = \pi_1(X)$ comes equipped with a homomorphism $w\colon \pi \to \{\pm 1\}$, which we shall call an *orientation character*. Let $\Lambda = \mathbb{Z}[\pi]$ denote the integral group ring. Define an involution on $\Lambda$ by the correspondence $g \mapsto \bar{g}$, where $\bar{g} = w(g)g^{-1}$ for $g \in \pi$. This involution will enable us to convert right modules to left modules and vice-versa. For a right module $M$, let $^wM$ denote the corresponding left module. For a left module $N$, we let $N^w$ denote the corresponding right module.

Let $C_*(\widetilde{X})$ denote the cellular chain complex of the universal covering space $\widetilde{X}$ of $X$. Since $\pi$ acts on $\widetilde{X}$ by means of deck transformations, it follows that $C_*(\widetilde{X})$ is a (finitely generated, free) chain complex of right $\Lambda$-modules.

For a right $\Lambda$-module $M$, we may therefore define

$$H^*(X;M) \quad := \quad H_{-*}(\mathrm{Hom}_\Lambda(C_*(\widetilde{X}), M))$$
$$H_*(X;M) \quad := \quad H_*(C_*(\widetilde{X}) \otimes_\Lambda {}^wM) \quad .$$

Given another right $\Lambda$-module $N$, and a class $[X] \in H_n(X; {}^wN)$ we also have a *cap product* homomorphism

$$H^*(X;M) \xrightarrow{\cap[X]} H_{n-*}(X; M \otimes_{\mathbb{Z}} {}^wN)$$

where the tensor product $M \otimes_{\mathbb{Z}} {}^wN$ is given the left $\Lambda$-module structure via

$$g \cdot (x \otimes y) := xg^{-1} \otimes gy \quad (g \in \pi, x \otimes y \in M \otimes {}^wN) \ .$$

With respect to these conventions, there is a canonical isomorphism of left modules $^w\Lambda \cong \Lambda \otimes_{\mathbb{Z}} {}^w\mathbb{Z}$.

**2.1. Definition.** The space $X$ is called a *Poincaré complex of formal dimension $n$* if there is a class $[X] \in H_n(X; {}^w\mathbb{Z})$ such that cap product with it induces an isomorphism

$$\cap[X]\colon H^*(X;\Lambda) \xrightarrow{\cong} H_{n-*}(X; {}^w\Lambda).$$

More generally, a disconnected space $X$ is a Poincaré complex of formal dimension $n$ if each of its connected components is.

We abbreviate the terminology and refer to $X$ as a *Poincaré n-complex*. For the rest of the paper, we shall be implicitly assuming that $X$ is connected. If the orientation character is trivial, we say that $X$ is *orientable*, and a choice of fundamental class $[X]$ in this case is called an *orientation* for $X$.

*2.2. Remarks.* (1). Wall proved that the definition is equivalent to the assertion that the cap product map

$$H^*(X; M) \xrightarrow{\cap [X]} H_{n-*}(X; {}^w M)$$

is an isomorphism for all left $\Lambda$-modules $M$. In particular, taking $M = \mathbb{Z}$, we obtain the isomorphism $\cap [X] \colon H^*(X, \mathbb{Z}) \cong H_{n-*}(X; {}^w \mathbb{Z})$ as a special case, which amounts to the statement of the classical Poincaré duality isomorphism when $w$ is the trivial orientation character.

(2). Every compact $n$-manifold $X$ satisfies this form of Poincaré duality.[1] A vector bundle $\eta$ over $S^1$ is trivializable if and only if

$$w_1(\eta) = +1 \in H^1(S^1; \mathbb{Z}_2) = \mathbb{Z}_2 = \{\pm 1\}.$$

The homomorphism $w \colon \pi \to \{\pm 1\}$ is defined by mapping a loop $\ell \colon S^1 \to X$ to $+1$ if the pullback of the tangent bundle of $X$ along $\ell$ is trivializable and $-1$ otherwise.

(3). In Wall's treatment of surgery theory [Wa4], the above definition of Poincaré complex is extended to include simple homotopy information. This is done as follows: the cap product homomorphism is represented by a chain map $C^*(\widetilde{X}; \Lambda) \to C_{n-*}(\widetilde{X}; {}^w \Lambda)$ of finite degreewise free chain complexes of right $\Lambda$-modules. One requires the Whitehead torsion of this chain map to be trivial. In this instance, one says that $X$ is a *simple* Poincaré $n$-complex. It is known that every compact manifold has the structure of a simple Poincaré complex.

**2.3. Poincaré pairs.** Let $(X, A)$ be a finite $CW$ pair. Assume that $X$ is connected. We assume that $X$ comes equipped with a homomorphism $w \colon \pi_1(X) \to \{\pm 1\}$. We say that $(X, A)$ is a *Poincaré n-pair* if there is a class

$$[X] \in H_n(X, A; {}^w \mathbb{Z})$$

---

[1] The standard picture of a handle in a manifold, with its core and co-core intersecting in a point, has led Bruce Williams to the following one word proof of Poincaré duality: *BEHOLD!*

such that cap product with it induces an isomorphism

$$\cap [X] \colon H^*(X; \Lambda) \xrightarrow{\cong} H_{n-*}(X, A; {}^w\Lambda) \,.$$

Moreover, it is required that $\partial_*([X]) \in H_{n-1}(A; {}^w\mathbb{Z})$ equips $A$ with the structure of a Poincaré complex, where the orientation character on $A$ is the one induced by the orientation character on $X$. Note, however, that in many important examples, $A$ is not connected, even though $X$ is.

**2.4. Examples.** We mention some ways of building Poincaré complexes.

*Gluing.* If $(M, \partial M)$ and $(N, \partial N)$ are $n$-manifolds with boundary or, more generally, Poincaré pairs, and $h\colon \partial M \to \partial N$ is a homotopy equivalence, then the amalgamated union $M \cup_h N$ is a Poincaré $n$-complex.

A special case of this is the *connected sum* $X \# Y$ of two Poincaré complexes $X^n$ and $Y^n$. To define it, we need to cite a result of Wall: every Poincaré $n$-complex $X$ has the form $K \cup D^n$, where $K$ is a $CW$ complex and dim $K < n$; this decomposition is unique up to homotopy (see 4.9 below). Converting the attaching map $S^{n-1} \to K$ into an inclusion $S^{n-1} \subset \bar{K}$, we see that $(\bar{K}, S^{n-1})$ is a Poincaré $n$-pair. Similarly, with $Y = L \cup D^n$, we may define the connected sum $X \# Y$ to be $\bar{K} \cup_{S^{n-1}} \bar{L}$.

*Fibrations.* Suppose that $F \to E \to B$ is a fibration with $F$, $E$ and $B$ having the homotopy type of finite complexes. Quinn [Qu2] has asserted that $E$ is a Poincaré complex if and only if $F$ and $B$ are. A proof using manifold techniques can be found in a paper of Gottlieb [Got].

This result is important because it explains a wide class of the known examples of Poincaré complexes:

(1) The total space of a spherical fibration over a manifold.
(2) The quotient of a Poincaré complex by a free action of a finite group.

In a somewhat different direction, if a group $G$ acts on a Poincaré complex $M$, then the orbit space $M/G$ satisfies Poincaré duality with rational coefficients. This includes for example the case of orbifolds.

*S-duality.* Let $K$ and $C$ be based spaces, and suppose that

$$d\colon S^{n-2} \to K \wedge C$$

is an *S-duality map*, meaning that slant product with the homology class $d_*([S^{n-2}]) \in \tilde{H}_{n-2}(K \wedge C)$ induces an isomorphism in all degrees $f \colon \tilde{H}^*(K) \cong \tilde{H}_{n-*-2}(C)$.

Let $P \colon \Sigma(K \wedge C) \to \Sigma K \vee \Sigma C$ denote the *generalized Whitehead product* map, whose adjoint $K \wedge C \to \Omega \Sigma(K \vee C)$ is defined by taking the loop commutator $[i_K, i_C]$ (Samelson product), where $i_K : K \to \Omega\Sigma(K \vee C)$ and $i_C : C \to \Omega\Sigma(K \vee C)$ are adjoint to the inclusions (see [B-S, p. 192]).

The $CW$ complex

$$X := (\Sigma K \vee \Sigma C) \cup_{P \circ \Sigma d} D^n$$

is a Poincaré $n$-complex, with

$$\cap[X] = \begin{pmatrix} 0 & \pm f^* \\ f & 0 \end{pmatrix} : H^{n-*}(X) = \tilde{H}^{n-*-1}(K) \oplus \tilde{H}^{n-*-1}(C)$$

$$\cong H_*(X) = \tilde{H}_{*-1}(K) \oplus \tilde{H}_{*-1}(C) \quad (* \neq 0, n) .$$

(The proof uses [B-S, 4.6, 5.14]; see also 4.10 below). Spaces of this kind arise in high dimensional knot theory, where $X$ is the boundary of a tubular neighborhood of a Seifert surface $V^n \subset S^{n+1}$ (i.e., the double $V \cup_{\partial V} V$ of $(V, \partial V)$) of a knot $S^{n-1} \subset S^{n+1}$.

Given $X^n$ as above, we can form a Poincaré $(n+2)$-complex $Y^{n+2}$ by applying the same construction to the doubly suspended $S$-duality $\Sigma^2 d : S^n \to \Sigma K \wedge \Sigma C$. Thus iterated application of the operation

$$(K, C, d) \quad \mapsto \quad (\Sigma K, \Sigma C, \Sigma^2 d)$$

gives rise to a *periodic family* of Poincaré complexes. This type of phenomenon is related to the periodicity of the high-dimensional knot cobordism groups.

## 3. THE SPIVAK FIBRATION

A compact smooth manifold $M^n$ comes equipped with a tangent bundle $\tau_M$, whose fibres are $n$-dimensional vector spaces. Embedding $M$ in a high dimensional euclidean space $\mathbb{R}^{n+k}$, we can define the *stable normal bundle* $\nu$, which is characterized by the equation

$$\tau_M \oplus \nu_M = 0$$

in the reduced Grothendieck group of stable vector bundles over $M$. By identifying a closed tubular neighborhood of $M^n$ in $\mathbb{R}^{n+k}$ with the normal disk bundle $D(\nu)$, and collapsing its complement to a point (the *Thom-Pontryagin construction*), we obtain the *normal invariant*[2]

$$\alpha : S^{n+k} = (\mathbb{R}^{n+k})^+ \xrightarrow{\text{collapse}} \mathbb{R}^{n+k}/(\mathbb{R}^{n+k} - \text{int} D(\nu_M)) \xrightarrow[\cong]{\text{excision}} T(\nu_M),$$

---

[2]The use of this term in the literature tends to vary; here we have chosen to follow Williams [Wil].

in which $T(\nu) = D(\nu_M)/S(\nu_M)$ is the *Thom space* of $\nu$ (here, $S(\nu_M)$ denotes the normal sphere bundle of $\nu_M$). The map $\alpha$ satisfies

$$U \cap \alpha_*([S^{n+k}]) = [M]\,,$$

where $U \in H^k(D(\nu_M), S(\nu_M); \mathbb{Z}^t)$ denotes a Thom class for $\nu_M$, in which the latter cohomology group is taken with respect to the local coefficient system defined by the first Stiefel-Whitney class of $\nu_M$ (i.e., the orientation character of $M$).

The above relation between the normal invariant, the Thom class and the fundamental class is reflected in an observation made by Atiyah. If $p: D(\nu_M) \to M$ denotes the bundle projection, then the assignment $v \mapsto (v, p(v))$ defines a map of pairs

$$(D(\nu_M), S(\nu_M)) \to (D(\nu_M) \times M, S(\nu_M) \times M)$$

which induces a map of associated quotients

$$T(\nu_M) \to T(\nu_M) \wedge M_+\,,$$

where $M_+$ denotes $M$ with the addition of a disjoint basepoint. Composing this map with the normal invariant, we obtain a map

$$S^{n+k} \xrightarrow{d} T(\nu_M) \wedge M_+\,.$$

**3.1. Theorem.** (*Atiyah Duality* [At]). *The map $d$ is a Spanier-Whitehead duality map, i.e., slant product with the class $d_*([S^{n+k}]) \in \widetilde{H}_{n+k}(T(\nu_M) \wedge M_+)$ yields an isomorphism*

$$\widetilde{H}^*(T(\nu_M)) \quad \cong \quad \widetilde{H}_{n+k-*}(M_+)\,.$$

With respect to this isomorphism (or rather, taking a version of it with twisted coefficients), we see that a Thom class $U$ maps to a fundamental class $[M]$ and the map is given by cap product with $\alpha_*([S^{n+k}])$. Thus, the relation $U \cap \alpha_*([S^{n+k}]) = [M]$ is a manifestation of the statement that the Thom complex $T(\nu_M)$ is a Spanier-Whitehead dual of $M_+$.

The above discussion was intended to motivate the following:

**3.2. Definition.** Let $X$ be a Poincaré $n$-complex with orientation character $w$. By a *Spivak normal fibration* for $X$, we mean

- a $(k-1)$-spherical fibration $p: E \to X$, and
- a map

$$S^{n+k} \xrightarrow{\alpha} T(p)\,,$$

where $T(p) = X \cup CE$ denotes the mapping cone of $p$.

Moreover, we require that

$$U \cap \alpha_*([S^{n+k}]) = [X],$$

where $U \in H^k(p; \mathbb{Z}^w)$ is a Thom class for the spherical fibration $p$ (here we are taking the cohomology group of the pair $(X \cup_p E \times I, E \times 0)$ defined by the mapping cylinder of $p$ and the coefficients are given by the local system on $X$ defined by the orientation character $w$).

The map $\alpha \colon S^{n+k} \to T(p)$ is called a *normal invariant*.

**3.3. Theorem.** (*Spivak*). *Every Poincaré $n$-complex $X$ admits a Spivak normal fibration with fibre $S^{k-1}$, provided that $k \gg n$. Moreover, it is unique in the following sense: given two Spivak fibrations $(E_0, p_0, \alpha_0)$ and $(E_1, p_1, \alpha_1)$ with respect to the same integer $k$, then there exists a stable fibre homotopy equivalence*

$$h \colon E_1 \xrightarrow{\simeq} E_2$$

*such that the induced map $T(h) \colon T(p_0) \to T(p_1)$ composed with $\alpha_0$ is homotopic to $\alpha_1$.*

Actually, Spivak only proves this in the 1-connected case, but a little care shows how to extend to result to the non-simply connected case.

Let me now give Spivak's construction. As $X$ is a finite complex, we can identify it up to homotopy with a closed regular neighborhood $N$ of a finite polyhedron in euclidean space $\mathbb{R}^{n+k}$. Let $p \colon E \to X$ be the result of converting the composite

$$\partial N \to N \simeq X$$

into a fibration. One now argues that the homotopy fibre of $p$ is homotopy equivalent to a $(k-1)$-sphere. To see this, we combine $n$-dimensional Poincaré duality for $X$ together with the $(n+k)$-dimensional Poincaré duality for $(N, \partial N)$ (the latter having trivial orientation character) to conclude that

$$
\begin{aligned}
H^*(X; \Lambda) &\cong H_{n-*}(X; {}^w\Lambda) \\
&\cong H_{n-*}(N; {}^w\Lambda) \\
&\cong H^{k+*}(N, \partial N; ({}^w\Lambda)^e) \\
&\cong H^{k+*}(p; ({}^w\Lambda)^e),
\end{aligned}
$$

where $({}^w\Lambda)^e$ denotes the effect of converting ${}^w\Lambda$ to a right module by means of the trivial orientation character $e(g) := 1$.

Now, it is straightforward to check that this isomorphism is induced by cup product with a class $U \in H^k(p; \mathbb{Z}^w)$, so it follows that the fibration $p \colon E \to X$ satisfies the Thom isomorphism with respect to twisted coefficients. However, by the following, such fibrations are spherical fibrations.

**3.4. Lemma.** (*Spivak* [Spi, 4.4], *Browder* [Br4, I.4.3]). *Suppose that* $p\colon E \to B$ *is a fibration of connected spaces whose fibre* $F$ *is 1-connected. Then* $F \simeq S^{k-1}$, $k \geq 2$, *if and only if the generalized Thom isomorphism holds, i.e., there exists a class* $U \in H^k(p; \mathbb{Z}^w)$ *(with respect to some choice of orientation character* $w\colon \pi_1(B) \to \{\pm 1\}$*) such that cup product induces an isomorphism*

$$U\cup\colon H^*(B; \Lambda) \to H^{*+k}(p; (^w\Lambda)^e)$$

(The original proof of this lemma involves an intricate argument with spectral sequences. For an alternative, non-computational proof see Klein [Kl1].)

To complete the proof of the existence of the normal fibration, we need to construct a normal invariant $\alpha\colon S^{n+k} \to T(p)$. By definition, $T(p)$ is homotopy equivalent to $N/\partial N$, and the latter comes equipped with a degree one map

$$S^{n+k} \to N/\partial N$$

given by collapsing the exterior of $N$ to a point. This defines $\alpha$.

Observe that when $X$ is a smooth manifold then the Spivak fibration $E \to X$ admits a reduction to a $k$-plane bundle with structure group $O(k)$, i.e., the stable normal bundle of $X$. Similar remarks apply to PL and topological manifolds. This observation gives the first order obstruction to a finding a closed (TOP, PL or DIFF) manifold which is homotopy equivalent to a given Poincaré complex: *the normal fibration should admit a (TOP, PL or DIFF) reduction.*

We illustrate the utility of this by citing a result from surgery theory.

**3.5. Theorem.** (*Browder, cf.* [Ra4, p. 210]). *If* $X$ *is a 1-connected Poincaré complex of dimension* $\geq 5$, *then* $X$ *is homotopy equivalent to a closed topological manifold if and only if the normal fibration for* $X$ *admits a TOP-reduction.*

As a corollary, we see that every finite 1-connected $H$-space of dimension $\geq 5$ is homotopy equivalent to a topological manifold: the Spivak fibration in this case is trivializable (cf. Browder and Spanier [Br-Sp]), so we may take the trivial reduction.

**3.6. An alternative approach.** The above construction of the Spivak normal fibration required us to identify the Poincaré complex $X$ with a regular neighborhood of a finite polyhedron in $\mathbb{R}^n$. From an aesthetic point of view, it is desirable to have a construction which altogether avoids the theory of regular neighborhoods. The following, which was discovered by the author, achieves this. To simplify the exposition, we shall only consider the case when $\pi_1(X)$ is trivial, and refer the reader to [K15] for the general case.

Let $G$ be a topological group (which to avoid pathology, we assume is a $CW$ complex). Consider based $G$-spaces built up inductively from a point by attaching free $G$-cells $D^j \times G$ along their boundaries $S^{j-1} \times G$. Such $G$-spaces are the based $G$-$CW$ complexes which are free away from the basepoint. We shall call such $G$-spaces *cofibrant*.

Given a cofibrant $G$-space $Y$, define the *equivariant cohomology* of $Y$ by

$$\tilde{H}_G^*(Y) := \tilde{H}^*(Y/G; \mathbb{Z})$$

where the groups on the right are given by taking reduced singular cohomology.

Similarly, we have the *equivariant homology* of $Y$

$$\tilde{H}_*^G(Y) := \tilde{H}_*(Y/G; \mathbb{Z}).$$

Given two cofibrant $G$-spaces $Y$ and $Z$, we can form their smash product $Y \wedge Z$. Give this the diagonal $G$-action, and let $Y \wedge_G Z$ denote the resulting orbit space.

**3.7. Definition.** Assume that $\pi_0(G)$ is trivial. A based map $d \colon S^m \to Y \wedge_G Z$ is said to be an *equivariant duality map* if the correspondence $x \mapsto x/d_*([S^m])$ defines an isomorphism

$$\tilde{H}_G^*(Y) \xrightarrow{\cong} \tilde{H}_{m-*}^G(Z).$$

*3.8. Remarks.* (1). Another way of saying this is that the evident composite

$$S^m \to Y \wedge_G Z \to (Y/G) \wedge (Z/G)$$

is an $S$-duality map.

(2). Our definition is a dual variation of one given by Vogell [Vo], and the set-up is similar to Ranicki [Ra1, §3] who defines an analogue for discrete groups. If $G$ is not connected, then the definition is slightly more technical in that we have to take cohomology with $\Lambda = \mathbb{Z}[\pi_0(G)]$-coefficients.

Now, using a cell-by-cell induction (basically, Spanier's exercises [Spa, pp. 462–463] made equivariant), one verifies that every finite cofibrant $G$-space $Y$ (i.e., which is built up from a point by a finite number of $G$-cells) has the property that there exists a finite $G$-space $Z$ and an equivariant duality map $S^m \to Y \wedge_G Z$ for some choice of $m \gg 0$.

It is well-known that any connected based $CW$ complex $X$ comes e-quipped with a homotopy equivalence $BG \xrightarrow{\simeq} X$, where $G$ is a suitable topological group model for the loop space of $X$ (e.g., take $G$ to be the geometric realization of the underlying simplicial set of the Kan loop group of the total singular complex of $X$). Here, $BG$ denotes the classifying space of $X$. Let $EG$ be the total space of a universal bundle over $X$. Then $EG$ is a free contractible $G$-space. Let $EG_+$ be the effect of adjoining a basepoint to $EG$. Since $BG$ is homotopy finite, it follows that $EG_+$ is the equivariant type of a finite cofibrant $G$-space. Hence, there exists an equivariant duality map

$$S^m \xrightarrow{d} EG_+ \wedge_G Z := Z_{hG}$$

for suitably large $m$, where $Z_{hG} := (EG \times_G Z)/(EG \times_G *)$ is the reduced Borel construction of $G$ acting on $Z$ (note in fact that $Z_{hG}$ is homotopy equivalent to $Z/G$ since $Z$ is assumed to be cofibrant).

In what follows, we assume that $m \gg n =: \dim X$.

**3.9. Claim.** *If $BG$ has the structure of an $n$-dimensional Poincaré complex, then $Z$ is unequivariantly homotopy equivalent to a sphere of dimension $m-n-1$.*

*Proof.* Combining Poincaré duality with equivariant duality, we obtain an isomorphism

$$\widetilde{H}_{m-n+*}(Z_{hG}) \cong \widetilde{H}^{n-*}(BG_+) \cong \widetilde{H}_*(BG_+).$$

One checks that this isomorphism is induced by cap product with a suitable class $U \in \widetilde{H}^{m-n}(Z_{hG})$. Now observe that up to a suspension, $Z_{hG}$ is the mapping cone of the evident map

$$EG \times_G Z \to BG$$

and it follows that $Z_{hG}$ amounts to the Thom complex for this map converted into a fibration. It follows that the Thom isomorphism is satisfied, and we conclude by 3.4 above that its fibre $Z$ has the homotopy type of an $(m-n-1)$-sphere.

To complete our alternative construction of the Spivak fibration, we need to specify a normal invariant $\alpha$. This is given by the duality map $d \colon S^m \to Z_{hG}$.

## 4. The Classification of Poincaré Complexes

We outline the classification theory of Poincaré complexes in two instances: (i) low dimensions, and (ii) the highly connected case. In (i), we shall see that the main invariants are of Postnikov and tangential type, and ones derived from them. In (ii), the Hopf invariant is the main tool.

**4.1. Dimension 2.** Every orientable Poincaré 2-complex is homotopy equivalent to a closed surface (see Eckmann-Linnell [E-L] and Eckmann-Müller [E-M]). Surprisingly, this is a somewhat recent result.

**4.2. Dimension 3.** Clearly, Poincaré duality implies that a 1-connected Poincaré 3-complex $X$ is necessarily homotopy equivalent to $S^3$.

Wall [Wa3] studied Poincaré 3-complexes $X$ in terms of the fundamental group $\pi = \pi_1(X)$, the number of ends $e$ of $\pi$ and the second homotopy group $G = \pi_2(X)$. The condition that $e = 0$ is the same as requiring $\pi$ to be finite. It follows that the universal cover of $X$ is homotopy equivalent to $S^3$, so $G$ is trivial in this instance.

It turns out in this case that $\pi$ is a group *period* 4, meaning that $\mathbb{Z}$ admits a periodic projective resolution of $\mathbb{Z}[\pi]$ modules of period length 4. Wall showed that the first $k$-invariant of $X$ is a generator $g$ of $H^4(\pi; \mathbb{Z})$ (the latter which is a group of order $|\pi|$). The assignment $X \mapsto (\pi_1(X), g)$ was proved to induce a bijection between the set of homotopy types of Poincaré complexes and the set of pairs $(\pi, g)$ with $\pi$ finite of period 4 and $g \in H^4(\pi; \mathbb{Z})$ a generator, modulo the equivalence relation given by identifying $(\pi, g)$ with $(\pi', g')$ if there exists an isomorphism $\pi \to \pi'$ whose induced map on cohomology maps $g'$ to $g$.

In the case when $e \neq 0$, then $\pi$ is infinite and $\widetilde{X}$ is non-compact. If $e = 1$, homological algebra shows that $\widetilde{X}$ is contractible in this case, so $X$ is a $K(\pi, 1)$.

If $e = 2$, the Wall shows that $X$ is homotopy equivalent to one of $\mathbb{RP}^3 \# \mathbb{RP}^3$, $S^1 \times \mathbb{RP}^2$ or to one of the two possible $S^2$-bundles over $S^1$. This summarizes the classification results of Wall for groups for $\pi$ in which $e \leq 2$.

In 1977, Hendriks [He] showed that the homotopy type of a connected Poincaré 3-complex $X$ is completely determined by three invariants:

- the fundamental group $\pi = \pi_1(X)$,
- the orientation character $w \in \mathrm{Hom}(\pi, \mathbb{Z}/2)$, and
- the element $\tau := u_*([X]) \in H_3(B\pi; {}^w\mathbb{Z})$ given by taking the image of the fundamental class with respect to the homomorphism $H_3(X; {}^w\mathbb{Z}) \to H_3(B\pi_1(X); {}^w\mathbb{Z})$ induced by the classifying map $u : X \to B\pi$ for the universal cover of $X$.

Call such data a *Hendriks triple*.

Shortly thereafter, Turaev [Tu] characterized those Hendriks triples $(\pi, w, \tau)$ which are realized by Poincaré complexes, thereby completing the classification. For a ring $\Lambda$, let ho-$\mathbf{mod}_\Lambda$ be the category of fractions associated to the category of right $\Lambda$-modules given by formally inverting the class of morphisms $0 \to P$, where $P$ varies over the finitely generated projective modules. Call a homomorphism $M \to N$ of right $\Lambda$-modules a *P-isomorphism* if it induces an isomorphism in ho-$\mathbf{mod}_\Lambda$.

Set $\Lambda = \mathbb{Z}[\pi]$, where $\pi$ is a finitely presented group which comes equipped with an orientation character $w: \pi \to \{\pm 1\}$. Let $I \subset \Lambda$ denote the augmentation ideal, given by taking the kernel of the ring map $\Lambda \to \mathbb{Z}$ defined on group elements by $g \mapsto 1$. In particular, $I$ is right $\Lambda$-module.

Choose a free right $\Lambda$-resolution

$$\cdots \xrightarrow{d_3} C_2 \xrightarrow{d_2} C_1 \to I \to 0$$

of $I$, with $C_1$ and $C_2$ finitely generated. Let $C^* := \hom_\Lambda(C_i, \Lambda)$ denote the corresponding complex of dual (left) modules. Let $J$ be the right module given by taking the cokernel of the map

$$(C_1^*)^w \xrightarrow{(d_2^*)^w} (C_2^*)^w.$$

Then Turaev shows that there is an isomorphism of abelian groups

$$A: \hom_{\text{ho-}\mathbf{mod}_\Lambda}(J, I) \xrightarrow{\cong} H_3(\pi; {}^w\mathbb{Z}).$$

**4.3. Theorem.** (*Turaev*). *A Hendriks triple* $x := (\pi, w, \tau)$ *is realized by a Poincaré 3-complex if in only if* $\tau = A(t)$ *for some P-isomorphism* $t: J \to I$.

**4.4. Dimension 4.** Milnor [Mi2] proved that the intersection form

$$H_2(X^4) \otimes H_2(X^4) \to \mathbb{Z}$$

(or equivalently, the cup product pairing on 2-dimensional cohomology) determines the homotopy type of a simply connected Poincaré 4-complex, and that every unimodular symmetric bilinear form over $\mathbb{Z}$ is realizable. We should perhaps also mention here the much deeper theorem of Freedman, which says that the *homeomorphism* type of a closed topological 4-manifold is determined by its intersection form and its Kirby-Siebenmann invariant (the latter is a $\mathbb{Z}/2$-valued obstruction to triangulation).

We may therefore move on to the non-simply connected case. It is well-known that any group is realizable as the fundamental group of a closed 4-manifold, and hence of a Poincaré 4-complex. Given a Poincaré 4-complex $X$ with fundamental group $\pi$, the obvious invariants which come to mind are $G := \pi_2(X)$ and the intersection form on the universal cover, which can be rewritten as $\phi: G \times G \to \mathbb{Z}$ (since $\pi_2(X) = H_2(\tilde{X})$); the group $\pi$ acts via isometries on the latter.

Wall [Wa3] studied oriented Poincaré 4-complexes $X^4$ whose fundamental group is a cyclic group of prime order $p \neq 2$. Wall showed under these assumptions that the homotopy type of $X$ is determined by $G$ and the intersection form $G \times G \to \mathbb{Z}$. However, when $\pi$ is the group of order 2, this intersection form is too weak to detect the homotopy type of $X$ (see [H-K, 4.5]).

Hambleton and Kreck [H-K] extended Wall's work to the case when $\pi$ is a finite group with periodic cohomology of order 4. To a given oriented $X^4$, they associate a 4-tuple

$$(\pi, G, \phi, k)$$

where $\pi = \pi_1(M)$, $G = \pi_2(M)$, $\phi \colon G \times G \to \mathbb{Z}$ denotes the intersection form and $k \in H^3(\pi; G)$ denotes the first Postnikov invariant of $X$. Such a system is called the *quadratic 2-type* of $X$. Moreover generally, one can consider all such 4-tuples, and define *isometry* $(\pi, G, \phi, k) \to (\pi', G', \phi', k')$ consist of isomorphisms $\pi \cong \pi'$ and $G \cong G'$ which map $\phi$ to $\phi'$ and $k$ to $k'$.

**4.5. Theorem.** (*Hambleton-Kreck*). *Let $X^4$ be a closed oriented Poincaré complex with $\pi = \pi_1(X)$ a finite group having 4-periodic cohomology. Then the homotopy type of $X$ is detected by the isometry class of its quadratic 2-type.*

Notice that the result fails to identify the possible quadratic 2-types which occur for Poincaré complexes. Bauer [Bauer] extended this to finite groups $\pi$ whose Sylow subgroups are 4-periodic. Teichner [Te] extended it to the non-orientable case where a certain additional secondary obstruction appears. Teichner also realizes the obstruction by exhibiting a non-orientable Poincaré 4-complex having the same quadratic 2-type as $\mathbb{RP}^4 \# \mathbb{CP}^2$, but the two spaces have different homotopy types. Thus Teichner's secondary obstruction may be non-trivial. Other examples in the non-orientable case were constructed by Ho, Kojima, and Raymond [H-K-R].

Another approach to classification in dimension 4 is to be found in the works of Hillman (see e.g., [Hill]).

We should also mention here the work of Baues [Baues] which a provides a (rather unwieldy but) complete set of algebraic invariants for all 4-dimensional $CW$ complexes.

**4.6. Dimension 5.** The main results in this dimension assume that the fundamental group is trivial. Madsen and Milgram [M-M, 2.8] determined all Poincaré 5-complexes with 4-skeleton homotopy equivalent to $S^2 \vee S^3$. They show that such a space is homotopy equivalent to one of the following:

(1) $S^2 \times S^3$,

(2) $S(\eta \oplus \epsilon^2)$ = the total space of the spherical fibration that is given
    by taking the fibrewise join of the Hopf fibration $S^3 \xrightarrow{\eta} S^2$ with the
    trivial fibration $\epsilon^2 \colon S^2 \times S^1 \to S^2$, or

(3) the space given by attaching a 5-cell to $S^2 \vee S^3$ by means of the map
    $S^4 \to S^2 \vee S^3$ given by $[\iota_2, \iota_3] + \eta^2 \iota_2$, where $[\iota_2, \iota_3] \colon S^4 \to S^2 \vee S^3$
    denotes the attaching map for the top cell of the cartesian product
    $S^2 \times S^3$ (= the Whitehead product), $\eta^2 \colon S^4 \to S^2$ denotes the
    composite $\Sigma \eta \colon S^4 \to S^3$ followed by $\eta$, and $\iota_2 \colon S^2 \to S^2 \vee S^3$
    denotes the inclusion.

The last of these cases is the Gitler-Stasheff example mentioned in the
introduction, and hence fails to have the homotopy type of a closed smooth
5-manifold. This can be seen by showing that the Thom space of the
associated Spivak normal bundle fails to be the Thom space of a smooth
vector bundle.

Stöcker has completely classified 1-connected Poincaré 5-complexes up
to oriented homotopy type. To a given oriented $X^5$, we may associate the
system of invariants

$$I(X) \quad := \quad (G, b, w_2, e)$$

where

- $G = H_2(X)$,
- $b \colon T(G) \times T(G) \to \mathbb{Q}/\mathbb{Z}$ is the linking form for the torsion subgroup
  $T(G) \subset G$,
- $w_2 \in \operatorname{Hom}(G, \mathbb{Z}/2)$ is the second Stiefel-Whitney class for the Spi-
  vak fibration of $X$ (which makes sense since $\operatorname{Hom}(\pi_2(BSG), \mathbb{Z}/2) = \mathbb{Z}/2$, where the space $BSG$ classifies oriented stable spherical fibrations), and
- $e \in H^3(X; \mathbb{Z}/2) \cong G \otimes \mathbb{Z}/2$ denotes the obstruction linearizing the
  Spivak-fibration over the 3-skeleton of $X$ (we are using here that
  the map $BSO \to BSG$ is 2-connected, so a linearization always
  exists over the 2-skeleton).

We remark that the first three of these invariants was used by Barden [Bar]
to classify 1-connected smooth 5-manifolds.

More generally, one can consider tuples $(G, b, w_2, e)$ in which $G$ is a
finitely generated abelian group, $b \colon T(G) \times T(G) \to \mathbb{Q}/\mathbb{Z}$ is a nonsingular
skew symmetric form, $w_2 \colon G \to \mathbb{Z}/2$ is a homomorphism and $e \in G \otimes \mathbb{Z}/2$
is an element. The data are required to satisfy $w_2(x) = b(x, x)$ for all
$x \in T(G)$ and $(w_2 \otimes \operatorname{id})(e) = 0$. It is straightforward to define isomorphism
and direct sums of these data, so we may define $J$ to be the semi-group of
isomorphism classes of such tuples.

**4.7. Theorem.** (*Stöcker* [Sto]). *The assignment $X^5 \mapsto I(X^5)$ defines an*

*isomorphism between $J$ and the semigroup of oriented homotopy types of
1-connected Poincaré 5-complexes, where addition in the latter is defined
by connected sum.*

Using a slightly different version of this, it is possible to write down
a complete list of oriented homotopy types of 1-connected Poincaré 5-
complexes in terms of 'atomic' ones and the connected sum operation (see
[loc. cit., 10.1]).

**4.8. The highly connected case.** In "Poincaré Complexes: I", Wall
announces that the classification of 'highly connected' Poincaré complexes
will appear in the forthcoming part II. Unfortunately, part II never did appear. We shall recall some of the homotopy theory which would presumably
enter into a hypothetical classification in the metastable range.

To begin with, it is well-known that a closed $n$-manifold can be given
the structure of a finite $n$-dimensional $CW$ complex with one $n$-cell. The
analogue of this for Poincaré complexes was proved by Wall [Wa3, 2.4],
[Wa4, 2.9] and is called the *disk theorem*:

**4.9. Theorem.** (*Wall*). *Let $X$ be a finite Poincaré $n$-complex. Then $X$
is homotopy equivalent to a $CW$ complex of the form $L \cup_\alpha D^n$. If $n \neq 3$
then $L$ can be chosen as a complex with $\dim L \leq n-1$ (when $n = 3$, $L$ can
be chosen as finitely dominated by a 2-complex). Moreover, the pair $(L, \alpha)$
is unique up to homotopy and orientation.*

Suppose that $X$ is a $n$-dimensional $CW$ complex of the form $(\Sigma K) \cup_\alpha
D^n$, with $K$ connected. We want to determine which attaching maps
$\alpha \colon S^{n-1} \to \Sigma K$ give $X$ the structure of a Poincaré complex. To this
end, we recall the James-Hopf invariant

$$\pi_{n-1}(\Sigma K) \xrightarrow{H} \pi_{n-1}(\Sigma K \wedge K)$$

which is defined using the using the well-known homotopy equivalence
$J(K) \xrightarrow{\simeq} \Omega \Sigma K$, where $J(K)$ denotes the free monoid on the points of $K$.
In terms of this identification, $H$ is induced by the map $J(K) \to J(K \wedge K)$
given by mapping a word $\prod_i x_i$ to the word $\prod_{i<j} x_i \wedge x_j$.

**4.10. Theorem.** (*Boardman-Steer* [B-S, 5.14]). *Up to homotopy, the
reduced diagonal $\Delta \colon X \to X \wedge X$ factors as*

$$X \xrightarrow{\text{pinch}} S^n \xrightarrow{\Sigma H(\alpha)} \Sigma K \wedge \Sigma K \xrightarrow{\subseteq} X \wedge X,$$

*where the first map in this factorization is given by collapsing $\Sigma K \subset X$ to
a point.*

Since the slant product is induced by the reduced diagonal, we obtain,

**4.11. Corollary.** *A map* $\alpha\colon S^{n-1} \to \Sigma K$ *gives rise to a Poincaré n-complex* $X = (\Sigma K) \cup_\alpha D^n$ *if and only if its Hopf invariant*

$$H(\alpha)\colon S^{n-1} \to \Sigma K \wedge K$$

*is a Spanier-Whitehead duality.*

In particular, this result says that the complex $K$ is self-dual whenever $X$ is a Poincaré complex (compare [Wa1, 3.8]).

Suppose now that we are given $CW$ complex $X = L \cup_\alpha D^n$ which $(r-1)$-connected. If $X$ is to be a Poincaré complex, then it would follow by duality that $L$ is homotopy equivalent to a $CW$ complex of dimension $\leq n-r$, so we may as well assume this is the case to begin with. If we assume moreover that $n \leq 3r - 1$ then the Freudenthal suspension theorem implies that $L$ desuspends, so we may write $L \simeq \Sigma K$, and $X$ is then of the form $\Sigma K \cup_\alpha D^n$ up to homotopy. Hence the corollary applies in this instance. Lastly, if we assume that $n \leq 3r - 2$, then $K$ is unique up to homotopy.

The above result shows that it would be too optimistic to expect an algebraic classification of Poincaré complexes in the metastable range (indeed, the classification of self-dual $CW$ complexes in the stable range would probably have to appear in any such classification). However, if we assume that we are at the very beginning of the metastable range, i.e., $n = 2r$, then $\Sigma K$ is homotopy equivalent a wedge of $r$-spheres, say

$$\Sigma K \quad = \quad \bigvee^t S^r.$$

The Hilton decomposition [Hilt] can be used to write the homotopy class of $\alpha$ in terms of summands and basic Whitehead products, i.e,

$$\alpha \quad = \quad \sum_{j=1}^t \beta_j \iota_j \quad \oplus \quad \sum_{1 \leq i < j \leq t} \gamma_{ij} [\iota_i, \iota_j],$$

where $\iota_j\colon S^r \to \Sigma K$ denotes the (homotopy class of) the inclusion into the $j$-th summand, $\beta_j \in \pi_{n-1}(S^r)$ is an element, $[\iota_i, \iota_j] \in \pi_{n-1}(S^r \vee S^r)$ denotes the basic Whitehead product (= the attaching map $S^{2r-1} \to S^r \vee S^r$ for the top cell of $S^r \times S^r$) and $\gamma_{ij}$ is an integer. Higher order Whitehead products do not appear in this formula for dimensional reasons.

It follows that the data $(\beta_i, \gamma_{ij})$ give a complete list of invariants for $X$. If $e_j$ denotes the Kronecker dual to the cohomology class defined by $\iota_j$, then the cohomology ring for $X$ is given by

$$e_i \cup e_j \quad := \quad \begin{cases} \gamma_{ij} & \text{if } i < j, \\ (-1)^r \gamma_{ij} & \text{if } j < i, \\ H(\beta_i) \in \pi_{n-1}(S^{n-1}) = \mathbb{Z} & \text{if } i = j. \end{cases}$$

Therefore, the obstruction to $X$ satisfying Poincaré duality is given by the demanding that matrix $(e_i \cup e_j)$ be invertible.

For the classification (of manifolds) in the odd dimensional case $n = 2r + 1$, see [Wa3].

## 5. POINCARÉ EMBEDDINGS

The notion of Poincaré embedding is a homotopy-theoretic imperson-ation of what one obtains from an embedding of actual manifolds. If a manifold $X$ is decomposed as a union

$$X = K \cup_A C$$

where $K, C \subset X$ are codimension zero submanifolds with common bound-ary $A := K \cap C$, then $X$ stratifies into two pieces, with $A$ as the codimension one stratum and $\text{int}(K \amalg C)$ as the codimension zero stratum. By replac-ing the above amalgamation with its homotopy invariant analogue, i.e., the homotopy colimit of $K \leftarrow A \rightarrow C$, we may recover $X$ up to homotopy equivalence.

A Poincaré embedding amounts to essentially these data, except that we do not decree the spaces to be smooth manifolds: the manifold condition is weakened to the constraint that Poincaré duality is satisfied.

Specifically, suppose that we are given a connected based finite $CW$ complex $K^k$ of dimension $k$, a Poincaré $n$-complex $X^n$ and a map $f \colon K \rightarrow X$. The definition of Poincaré embedding which we give is essentially due to Levitt [Le1].

**5.1. Definition.** We say that $f$ *Poincaré embeds* if there exists a commu-tative diagram of based spaces

$$\begin{array}{ccc} A & \longrightarrow & C \\ {\scriptstyle i}\downarrow & & \downarrow \\ K & \xrightarrow{\ f\ } & X \end{array}$$

such that

- the diagram is a homotopy pushout, i.e., the evident map from the double mapping cylinder $K \times 0 \cup A \times [0, 1] \cup C \times 1$ to $X$ is a homotopy equivalence.
- The image of $[X]$ under $H_n(X; {}^w\mathbb{Z}) \rightarrow H_n(i; f^{*w}\mathbb{Z})$ induced by the boundary map in Mayer-Vietoris sequence of the diagram gives $(\bar{K}, A)$ the structure of an $n$-dimensional Poincaré pair, where $\bar{K} := K \cup_{A \times 0} A \times [0, 1]$ denotes the mapping cylinder of $i$. Similarly, $[X]$ induces a Poincaré pair structure on $(\bar{C}, A)$.
- The map $i$ is $(n-k-1)$-connected.

The space $C$ is called the *complement*.

The above definition applies when $X$ has no boundary. If $(X, \partial X)$ is a Poincaré $n$-pair, then the definition is analogous, except: (i) we require the map $\partial X \to X$ to factor as $\partial X \to C \to X$, and (ii) the Poincaré boundary for $C$ is given by $\partial X \amalg A$.

The first condition of the definition says that $X$ is homotopy theoretically a union of $K$ with its complement. The second condition says that the 'stratification' of $X$ is 'Poincaré'. The last condition is essentially technical. In the smooth category, it would be an automatic consequence of transversality (a closed regular neighborhood $N$ a $k$-dimensional subcomplex of an $n$-manifold has the property that $\partial N \subset N$ is $(n-k-1)$-connected), so the condition that $i$ be $(n-k-1)$-connected is imposed to repair the lack of transversality in the Poincaré case. However, note when $k \leq n-3$ that $i$ is 2-connected if and only if $i$ is $(n-k-1)$-connected, by duality and the relative Hurewicz theorem.

We will assume throughout that we are in codimension $\geq 3$, i.e., $k \leq n-3$.

*5.2. Remark.* Suppose additionally that $K^k$ has the structure of a Poincaré $k$-complex. Then application of 3.4 above shows that the homotopy fibre of $i$ is homotopy equivalent to an $(n-k-1)$-sphere. Hence the map $i$ in the definition may be replaced by spherical fibration. This recovers the notion of Poincaré embedding given by Wall [Wa2, p. 113].

The following result, which has a 'folk' co-authorship, says that the descent problem for finding locally flat PL-manifold embeddings can always be solved in codimension $\geq 3$. Moreover, the smooth version can always be solved in the metastable range.

**5.3. Theorem.** (*Browder-Casson-Sullivan-Wall* [Wa2, 11.3.1]).
(*1*). *Suppose that $K^k$ and $X^n$ are PL manifolds and that $k \leq n-3$. Then $f$ is homotopic to a locally flat PL embedding if and only if $f$ Poincaré embeds.*

(*2*). *If $K^k, X^n$ are smooth manifolds with $k \leq n-3$, then $f: K^k \to X^n$ is homotopic to a smooth embedding if and only if $f$ Poincaré embeds and, additionally, one of the following holds: (i) $2n \geq 3(k+1)$, or (ii) the $(n-k-1)$-spherical fibration $A \to K$ (cf. 5.2) admits a vector bundle reduction of rank $n-k$.*

Thus, the problem of finding an embedding of PL-manifolds in codimension $\geq 3$ has been reduced to a problem in homotopy theory. *When can this homotopy problem be solved?*

A map $M^m \to N^n$ of manifolds with $n \geq 2m+1$ is always homotopic to an embedding, by transversality. It is natural to ask whether a similar result holds in the Poincaré case. Fix a map $f: K^k \to X^n$, where $K^k$ is a $k$-dimensional $CW$ complex, $X^n$ is a Poincaré complex (possibly with boundary) and $k \leq n-3$. According to Levitt [Le1], $f$ Poincaré embeds when $n \geq 2k + 2$ . One would expect that the result holds in one codimension less, in analogy with manifolds, but this isn't known in general. However, Hodgson [Ho1] asserts that $f$ will Poincaré embed when $n \geq 2k + 1$, with the additional assumptions that $K$ is a Poincaré complex and $X$ is 1-connected. Both Hodgson and Levitt used manifold engulfing techniques to arrive at these results.

Recently, the author [Kl2] proved a general result about Poincaré embeddings which implies the Levitt and Hodgson theorems as special cases:

**5.4. Theorem.** *Let $f: K^k \to X^n$ be an $r$-connected map with $k \leq n-3$. Then $f$ Poincaré embeds whenever*

$$r \quad \geq \quad 2k - n + 2 .$$

*Moreover, the Poincaré embedding is 'unique up to concordance' if strict inequality holds.*

(Two Poincaré embedding diagrams for $f$ are called *concordant* if they are isomorphic in the homotopy category of such diagrams.)

In contrast with the engulfing methods of Levitt and Hodgson, the author proves this result using purely homotopy theoretic techniques (a main ingredient of the proof is the Blakers-Massey theorem for cubical diagrams of spaces, as to be found in [Good]).

An old question about Poincaré complexes is whether or not the diagonal $X \to X \times X$ Poincaré embeds. As an application of the above, we have

**5.5. Corollary.** *Let $X^n$ be a 2-connected Poincaré $n$-complex.[3] Then the diagonal $X \to X \times X$ Poincaré embeds. Moreover, any two Poincaré embeddings of the diagonal are concordance whenever $X$ is 3-connected.*

It would be interesting to know whether or not the corollary holds without the connectivity hypothesis. Clearly, the diagonal of a manifold Poincaré embeds, by the tubular neighborhood theorem, so the existence of a diagonal Poincaré embedding for a Poincaré complex is an obstruction to finding a smoothing.

---

[3]Additional Note: The existence part of the corollary holds when $X$ is 1-connected (see [Kl4]).

**5.6. Example.** Let $X$ be a finite $H$-space with multiplication $\mu \colon X \times X \to X$. Write $X = X_0 \cup D^n$ using the disk theorem, and let $\alpha \colon D^n \to X$ be the characteristic map for the top cell of $X$. Consider the commutative diagram

$$
\begin{array}{ccc}
X \times S^{n-1} & \xrightarrow{\;\mathrm{id} \times \alpha\;} & X \times X_0 \\
{\scriptstyle \cap}\big\downarrow & & \big\downarrow{\scriptstyle s} \\
X \times D^n & \xrightarrow{\;\;\;d\;\;\;} & X \times X
\end{array}
$$

where the map $s$ is given by $(x, y) \mapsto (x, \mu(x, y))$, and the map $d$ is given by $(x, v) \mapsto (x, \mu(x, \alpha(v)))$. Then the diagram is a homotopy pushout and, moreover, the restriction of $d$ to $X \times * \subset X \times D^n$ coincides with the diagonal. Hence, the diagram amounts to a Poincaré embedding of the diagonal.

**5.7. Poincaré embeddings and unstable normal invariants.** Another type of question which naturally arises concerns the relationship between the Spivak normal fibration and Poincaré embeddings in the sphere. Suppose that $K^k$ is a Poincaré complex equipped with a choice of spherical fibration $p \colon S(p) \to K$ with fibre $S^{j-1}$. One can ask whether $K^k$ Poincaré embeds in the sphere $S^{k+j}$ with normal data $p$. That is, when does there exist a space $W$ and an inclusion $S(p) \subset W$ such that $K \cup_{S(p)} W$ is homotopy equivalent to $S^{k+j}$? Obviously, if $p$ isn't a Spivak fibration then there aren't any such Poincaré embeddings. So the first obstruction is given by the existence of a normal invariant $S^{k+j} \to T(p)$.

More generally, let $K^k$ be a $k$-dimensional $CW$ complex which is equipped with a map $g \colon A \to K$. Let $\bar{K}$ be the mapping cylinder of $g$ and assume that $(\bar{K}, A)$ is an oriented Poincaré $n$-pair. We want to know when there exists a Poincaré embedding of $K$ in $S^n$ with normal data $A \to K$, i.e., when does there exist an inclusion of spaces $A \subset W$ such that $K \cup_A W$ is homotopy equivalent to $S^n$? This problem specializes to the previous one by taking $g$ to be a spherical fibration.

Now, if the problem could be solved, then a choice of homotopy equivalence $S^n \xrightarrow{\simeq} K \cup_A W$ gives rise to a 'collapse' map

$$
S^n \xrightarrow{\simeq} K \cup_A W \simeq \bar{K} \cup_A W \to \bar{K} \cup_A * = T(g)
$$

where $T(g)$ denotes the mapping cone of $g \colon A \to K$. By correctly choosing our orientation for $(\bar{K}, A)$, we may assume that this map is of degree one. This prompts the following more general notion of normal invariant.

**5.8. Definition.** Given $g \colon A \to K$ as above together with an orientation for $(\bar{K}, A)$, we call the homotopy class of any degree one map $S^n \to T(g)$ a *normal invariant*.

The following result says that there is a bijective correspondence between normal invariants and concordance classes of Poincaré embeddings in the sphere with given normal data in the metastable range. It was first proven by Williams [Wi1], using manifold methods. A homotopy theoretic proof has been recently given by Richter [Ri1].

**5.9. Theorem.** *Suppose that* $3(k+1) \leq 2n$ *and* $n \geq 6$. *Then* $K^k$ *Poincaré embeds in* $S^n$ *with normal data* $g \colon A \to K$ *if and only if there exists a normal invariant* $S^n \to T(g)$. *Moreover, any two such Poincaré embeddings of* $K$ *which induce the same normal invariants are isotopic provided that* $3(k+1) < 2n$.

Richter [Ri2] has found some interesting applications of this result. For example, he has shown how it implies that the isotopy class of a knot $S^n \subset S^{n+2}$ is determined by its complement $X$, whenever $\pi_*(X) = \pi_*(S^1)$ for $* \leq 1/3(n+2)$; this extends a theorem of Farber by one dimension. For an extension of Theorem 5.9 to the case when the ambient space an arbitrary Poincaré complex, see [Kl4].

## 6. POINCARÉ SURGERY

Controversy seems to be one of the highlights of this subject, so to avoid potential crossfire I'll begin this section with a quote from Chris Stark's mathematical review [Stk] of the book *Geometry on Poincaré spaces*, by Hausmann and Vogel [H-V]:

> *The considerable body of work on these matters is usually referred to as "Poincaré surgery" although other fundamental issues such as transversality are involved. These efforts involve several points of view and a number of mathematicians—the authors of the present notes identify three main streams of prior scholarship in their introduction and include a useful bibliography. Because of technical difficulties and unfinished research programs, Poincaré surgery has not become the useful tool proponents of the subject once hoped to deliver.*

For the sake of simplicity, I shall only discuss the results found in [H-V], which is now the standard reference for Poincaré surgery. We begin by explaining the fundamental problem of Poincaré surgery. To keep the exposition simple, we only consider the oriented case.

**6.1. Surgery.** Quinn [Qu2] defines a *normal space* to be a $CW$ complex $X$ equipped with an (oriented) $(k-1)$-spherical fibration $p_X \colon E \to X$ and a degree one map $\alpha_X \colon S^{n+k} \to T(p_X)$, where $T(p_X)$ denotes the mapping cone = Thom space of $p_X$ (here the integer $k$ is allowed to vary). We define the *formal dimension* of $X$ to be $n$. Similarly, we have the notion of *normal pair* $(X, A)$.

A *normal map* of normal spaces from $X$ to $Y$ consists of a map $f: X \rightarrow Y$ and an oriented fibre equivalence of fibrations $b: p_X \xrightarrow{\cong} p_Y$ covering $f$ such that the composite

$$S^{n+k} \xrightarrow{\alpha_X} T(p_X) \xrightarrow{T(b)} T(p_Y)$$

coincides with $\alpha_Y$. Note that the mapping cylinder of $f$ has the structure of a normal pair whose boundary is $X \amalg Y$. Similarly, there is an evident notion of *normal cobordism* for normal maps.

The obvious example of a normal space is given by a Poincaré complex equipped with Spivak fibration. The central problem of Poincaré surgery is to decide when a given normal map $f: X \rightarrow Y$ of Poincaré complexes is normally cobordant to a homotopy equivalence. Analogously, in the language of normal pairs, one wants to know when a normal pair $(X, A)$, with $A$ Poincaré, is normally cobordant to a Poincaré pair.

The algebraic theory of surgery of Ranicki [Ra1-2], [Ra3] associates to a normal map of Poincaré complexes $f: X \rightarrow Y$ a surgery obstruction $\sigma(f) \in L_n(\pi_1(Y))$ which coincides with the classical one if the given normal map comes from a manifold surgery problem. The principal result of Poincaré surgery says that this is the only obstruction to finding such a normal cobordism, i.e., that the manifold and Poincaré surgery obstructions are the same. According to Hausmann and Vogel, there are to date three basic approaches to Poincaré surgery obstruction theory.

The first is to use thickening theory to replace a Poincaré complex with manifold with boundary, so that we can avail ourselves of manifold techniques, such as engulfing. This is the embodied in approach of several authors, including Levitt [Le2], Hodgson [Ho6] and Lannes-Latour-Morlet [L-L-M]. One philosophical disadvantage of this approach is that, in the words of Browder, "a problem in homotopy theory should have a homotopy theoretical solution" [Qu1].

The second approach, undertaken by Jones [Jo1], also uses sophisticated manifold theory. The idea here is to equip Poincaré complexes with the structure of a *patch space,* which a space having an 'atlas' of manifolds whose transition maps are homotopy equivalences, and having suitable transversality properties.

Lastly, we have the direct homotopy theoretic assault, which was first outlined by Browder and which was undertaken by Quinn [Qu1-3]. If a map $\beta: S^j \rightarrow X^n$ is an element on which one wants to do surgery, then the homotopy cofiber $X \cup_\beta D^{j+1}$ has the homotopy type of an elementary cobordism, i.e., the trace of the would-be surgery. Moreover, as Quinn observes, if the surgery can be done then there is a cofibration sequence $X' \to X \cup_\beta D^{j+1} \to S^{n-j}$ where $X'$ is the 'other end' of the cobordism. The composite map $X \subset X \cup_\beta D^{j+1} \to S^{n-j}$ is a geometric representative for a cohomology class which is Poincaré dual to the homology class defined by $\beta$. Quinn's idea [Qu3] is to find homotopy theoretic criteria (involving Poincaré duality) to decide when a map $X \cup_\beta D^{j+1} \to S^{n-j}$ extends to the left as a cofibration sequence, thus yielding $X'$.

Hausmann and Vogel point out that these three approaches are imbued with a great deal of technical difficulty and none of them were completely overcome. We pigeonhole the book of Hausmann and Vogel by placing it within the first of these schools.

**6.2. Poincaré bordism.** Under this title belong the fundamental exact sequences of Poincaré bordism found by Levitt [Le2], Jones [Jo1] and Quinn [Qu2]. Given a normal space $X$, we can let $\Omega_n^P(X)$ denote the bordism group of normal maps $(f, b): Y \to X$ with $X$ a normal space of formal dimension $n$ and $Y$ a Poincaré $n$-complex, and where cobordisms are understood in the Poincaré sense. Similarly, we can define $\Omega_n^N(X)$ to be the bordism group of normal maps $(f, b): Y \to X$. Then there is an exact sequence

$$ \cdots \to L_n(\pi_1(X)) \to \Omega_n^P(X) \xrightarrow{\text{incl}} \Omega_n^N(X) \to L_{n-1}(\pi_1(X)) \to \cdots $$

and moreover, an isomorphism $\Omega_n^N(X) \cong H_n(X; MSG)$, where the latter denotes the homology of $X$ with coefficients in the Thom spectrum $MSG$ whose $n$-th space is the Thom space of the oriented spherical fibration with fibre $S^{n-1}$ over the classifying space $BSG_n$.

**6.3. Transversality.** Let $A$ be a finite $CW$ complex and suppose that $(D, S)$ is a connected $CW$ pair such that $A$ includes in $D$ as a deformation retract. We also assume that the homotopy fibre of $S \subset D$ is $(k-1)$-spherical. Given an inclusion $S \subset C$, let $Y$ denote the union $D \cup_S C$. Roughly, we are thinking of the $Y$ as containing a 'neighborhood thickening' $D$ of $A$ in such a way that the 'link' $S$ of $A$ in $Y$ is a spherical fibration (up to homotopy).

Let $X$ be a Poincaré $n$-complex and let $f: X \to Y$ be a map. We say that $f$ is *Poincaré transverse* to $A$ when $(f^{-1}(D), f^{-1}(S))$ and $(f^{-1}(C), f^{-1}(S))$ have the structure of Poincaré $n$-pairs, and moreover, we require that the homotopy fibre of the map

$$ f^{-1}(S) \to f^{-1}(D) $$

is also $(k-1)$-spherical.

Hence, if $f$ is Poincaré transverse to $A$, we obtain a stratification of $X$ as a union of $f^{-1}(D)$ with $f^{-1}(C)$ along a common Poincaré boundary $f^{-1}(S)$. Moreover, it follows from the definition that $f^{-1}(A)$ has the structure of a Poincaré $(n-k)$-complex, so we infer that the inclusion $f^{-1}(A) \subset X$ Poincaré embeds (with normal data $f^{-1}(S)$).

The main issue now is to decide when a map $f \colon X \to Y$ can be 'deformed' (bordant, $h$-cobordant) so that it becomes Poincaré transverse to the given $A$. The philosophy is that although one can always deform a map in the smooth case to make it transverse, there are obstructions in the Wall $L$-groups for the Poincaré case, and the vanishing of these obstructions are both necessary and sufficient for Poincaré transversality up to bordism.

The algebraic $L$-theory codimension $k$ Poincaré transversality obstructions for $k = 1, 2$ are discussed in Ranicki [Ra3, Chap. 7]. Supposing in what follows that $k \geq 3$, Hausmann and Vogel provide a criterion for deciding when $f$ can be made (oriented) Poincaré bordant to a map which is transverse to $A$ [H-V, 7.11]. They define an invariant $t(f) \in L_{n-k}(\pi_1(A))$ whose vanishing is necessary and sufficient to finding the desired bordism. If in addition $f$ is 2-connected, then $t(f)$ is the complete obstruction to making $f$ transverse to $A$ up to homotopy equivalence (i.e., *Poincaré h-cobordism*) [loc. cit., 7.23]). Assertions of this kind can be found in the papers of Levitt [Le2],[Le4],[Le5], Jones [Jo1], and Quinn [Qu2]. For a general formulation, see [H-V, 7.11, 7.14].

**6.4. Handle decompositions.** Given a Poincaré $n$-pair $(Y, \partial Y)$, and a Poincaré embedding diagram

$$
\begin{array}{ccc}
S^{k-1} \times S^{n-k-1} & \longrightarrow & C \\
\cap \downarrow & & \downarrow \\
S^{k-1} \times D^{n-k} & \longrightarrow & \partial Y
\end{array}
$$

we can form the Poincaré $n$-pair

$$
(Z, \partial Z) \quad := \quad \left( Y \cup D^k \times D^{n-k}, C \cup D^k \times S^{n-k-1} \right),
$$

where $D^k \times D^{n-k}$ is attached to $Y$ by means of the composite $S^{k-1} \times D^{n-k} \to \partial Y \subset Y$ and $D^k \times S^{n-k-1}$ is attached to $\partial Y$ by means of the map $S^{k-1} \times S^{n-k-1} \to C$. Call this operation the effect of *attaching a k-handle* to $(Y, \partial Y)$. Note that there is an evident map $Y \to Z$.

A *handle decomposition* for a Poincaré complex $X^n$ consists of a sequence of spaces

$$
W_{-1} \to W_0 \to \cdots \to W_n
$$

(with $W_{-1} = \emptyset$) and a homotopy equivalence $W_n \overset{\simeq}{\to} X$. Moreover, each $W_j$ is the underlying space of a Poincaré $n$-pair with boundary $\partial W_j$ in such a way that $W_j$ is obtained from $W_{j-1}$ a a finite number of $j$-handle attachments. Handle decompositions are special cases of Jones' patch spaces [Jo1].

**6.5. Theorem.** ([H-V, 6.1]). *If $X$ is a Poincaré $n$-complex with $n \geq 5$, then $X$ admits a handle decomposition.*

6.6. APPENDIX: A QUICK UPDATE ON THE FINITE $H$-SPACE PROBLEM

When Browder posed his question: *Does every finite $H$-space have the homotopy type of a closed smooth manifold?*, it wasn't known that there exist 1-connected finite $H$-spaces which are not the homotopy type of compact Lie groups (except for products with $S^7$ or quotients thereof; see Hilton-Roitberg [H-R] and Stasheff [Sta, p. 22] for examples).

We remarked in §3 that every 1-connected finite $H$-space $X^n$ has the homotopy type of a closed topological $n$-manifold. Browder [Br5] has noted in fact that the manifold can be chosen as smooth and stably parallelizable if $n$ isn't of the form $4k+2$.

Using Zabrodsky mixing [Z] and surgery methods, Pedersen [Pe] was able to extend Browder's theorem to show that certain classes of finite $H$-spaces (some with non-trivial fundamental group) have the homotopy type of *smooth* manifolds. Recall that spaces $Y$ and $Z$ are said to have the same *genus* if $Y_{(p)} \simeq Z_{(p)}$ for all primes $p$, where $Y_{(p)}$ denotes the Sullivan localization of $Y$ at $p$. Among other things, Pedersen proved that when a finite $H$-space $X$ happens to be 1-connected and has the genus of a 1-connected Lie group, then $X$ has the homotopy type of a smooth, parallelizable manifold.

Weinberger [We] has settled the 'local' version of the problem: if $P$ denotes a finite set of primes, then a finite $H$-space is $P$-locally homotopy equivalent to a closed topological manifold.

Using localization techniques and surgery theory, Cappell and Weinberger [CW1] have shown that a finite $H$-space $X$ has the homotopy type of a closed topological manifold when $\pi_1(X)$ is either an odd $p$-group or infinite with at most cyclic 2-torsion. In another paper [CW2] they show that $X$ has the homotopy type of a closed smooth parallelizable manifold whenever $X_{(2)}$ contains a factor which is $S^7$ or a Lie group at the prime 2, and moreover, $\pi_1(X)$ is either trivial, an odd $p$-group or infinite with no 2-torsion.

# REFERENCES

[At]      M. Atiyah, *Thom complexes*, Proc. London. Math. Soc. **11** (1961), 291–310.

[Bar]     D. Barden, *Simply connected five-manifolds*, Ann. Math. **82** (1965), 365–385.

[Bauer]   S. Bauer, *The homotopy type of a four-manifold with finite fundamental group*, Proc. Top. Conf. Göttingen 1987, Springer LNM 1361, 1988, pp. 1–6.

[Baues]   H. J. Baues, *Combinatorial Homotopy and 4-dimensional Complexes*, Expositions in Mathematics, De Gruyter, 1991.

[B-S]     J. M. Boardman and B. Steer, *On Hopf invariants*, Comment. Math. Helv. **42** (1967), 180–221.

[Br1]     W. Browder, *Torsion in H-spaces*, Ann. Math. **74** (1961), 24–51.

[Br2]     _____, *Embedding 1-connected manifolds*, Bull. AMS **72** (1966), 225–231, erratum 736.

[Br3]     _____, *Embedding smooth manifolds*, Proc. ICM (Moscow 1966), Mir, 1968, pp. 712–719.

[Br4]     _____, *Surgery on Simply-connected Manifolds*, Ergebnisse der Mathematik und ihrer Grenzgebiete, Band 65, Springer-Verlag, 1972.

[Br5]     _____, *Homotopy type of differentiable manifolds (Arhus 1962)*, Novikov Conjectures, Index Theorems and Rigidity (Oberwolfach, 1993), Vol. 1, London Math. Soc. Lecture Note Ser., vol. 226, Cambridge Univ. Press, 1995, pp. 97–100.

[Br-Sp]   W. Browder and E. Spanier, *H-spaces and duality*, Pacific J. Math. **12** (1962), 411–414.

[C-W1]    S. Cappell and S. Weinberger, *Which H-spaces are manifolds? I.*, Topology **27** (1988), 377–386.

[C-W2]    _____, *Parallelizability of finite H-spaces*, Comment. Math. Helv. **60** (1985), 628–629.

[Di]      J. Dieudonné, *A History of Algebraic and Differential Topology*, Birkhäuser, 1988.

[E-L]     B. Eckmann and P. Linnell, *Poincaré duality groups of dimension two. II*, Comment. Math. Helv. **58** (1983), 111–114.

[E-M]     B. Eckmann and H. Müller, *Poincaré duality groups of dimension two*, Comment. Math. Helv. **55** (1980), 510–520.

[G-S]     S. Gitler and J. Stasheff, *The first exotic class of BF*, Topology **4** (1965), 257–266.

[Good]    T. Goodwillie, *Calculus II. Analytic functors*, K-theory **5** (1992), 295–332.

[Got]     D. Gottlieb, *Poincaré duality and fibrations*, Proc. AMS **76** (1979), 148–150.

[H-K]     I. Hambleton and M. Kreck, *On the classification of topological 4-manifolds with finite fundamental group*, Math. Ann. **280** (1988), 85–104.

[He]      H. Hendriks, *Obstruction theory in 3-dimensional topology: an extension theorem*, J. London Math. Soc. (2) **16** (1977), 160–164.

[Hill]    J. A. Hillman, *The Algebraic Characterization of Geometric 4-Manifolds.*, London Mathematical Society Lecture Note Series, vol. 198, Cambridge University Press, 1994.

[Hilt]    P. J. Hilton, *On the homotopy groups of the union of spheres*, Jour. London Math. Soc. **30** (1955), 154–172.

[H-R]     P. J. Hilton and J. Roitberg, *On principal $S^3$-bundles over spheres*, Ann. Math. **90** (1969), 91–107.

[Hol]     J. P. E. Hodgson, *The Whitney technique for Poincaré complex embeddings*, Proc. AMS **35** (1972), 263–268.

[Ho2]        _____, *General position in the Poincaré duality category*, Invent. Math. **24** (1974), 311–334.

[Ho3]        _____, *Subcomplexes of Poincaré complexes*, Bull. AMS **80** (1974), 1146–1150.

[Ho4]        _____, *General position in the Poincaré duality category*, Invent. Math. **24** (1974), 311–334.

[Ho5]        _____, *Trivializing spherical fibrations and embeddings of Poincaré complexes*, Math. Zeit. **158** (1978), 35–43.

[Ho6]        _____, *Surgery on Poincaré complexes*, Trans. Amer. Math. Soc. **285** (1984), 685–701.

[Jo1]        L. E. Jones, *Patch spaces: a geometric representation for Poincaré spaces*, Ann. Math. **97** (1973), 306–343.

[Jo2]        _____, *Corrections for "Patch Spaces"*, Ann. Math. **102** (1975), 183–185.

[Ke]         M. Kervaire, *A manifold which does not admit a differentiable structure*, Comment. Math. Helv. **34** (1960), 257–270.

[K-M]        M. Kervaire and J. Milnor, *Groups of homotopy spheres I.*, Ann. Math. **77** (1963), 504–537.

[Kl1]        J. R. Klein, *On two results about fibrations*, Manuscripta Math. **92** (1997), 77–86.

[Kl2]        _____, *Poincaré duality embeddings and fiberwise homotopy theory*, Topology **38** (1999), 597–620.

[Kl3]        _____, *On the homotopy embeddability of complexes in euclidean space I. the weak thickening theorem*, Math. Zeit. **213** (1993), 145–161.

[Kl4]        _____, *Embedding, compression and fiberwise homotopy theory*, Hopf server preprint (1998).

[Kl5]        _____, *Poincaré immersions*, to appear in Forum Math.

[L-L-M]      J. Lannes, L. Latour and C. Morlet, *Géométrie des complexes de Poincaré et chirurgie*, IHES preprint (1971).

[Le1]        N. Levitt, *On the structure of Poincaré duality spaces*, Topology **7** (1968), 369–388.

[Le2]        _____, *Poincaré duality cobordism*, Ann. Math. **96** (1972), 211–244.

[Le3]        _____, *Applications of Poincaré duality spaces to the topology of manifolds*, Topology **11** (1972), 205–221.

[Le4]        _____, *A general position theorem in the Poincaré duality category*, Comm. Pure Appl. Math. **25** (1972), 163–170.

[Le5]        _____, *On a Σ-spectrum related to G/Top*, Houston J. Math. **3** (1977), 481–493.

[M-M]        I. Madsen and R. J. Milgram, *The Classifying Spaces for Surgery and Cobordism of Manifolds*, Ann. of Math. Studies, vol. 92, Princeton University Press, 1979.

[May]        J. P. May, *Math. Review of "Extension of maps as fibrations and cofibrations"*, by Frank Quinn, MR 52#6706.

[Mi1]        J. Milnor, *On manifolds homeomorphic to the 7-sphere*, Ann. Math. **64** (1956), 399–405.

[Mi2]        _____, *On simply-connected 4-manifolds*, Symposium Internacional de Topología Algebraica (Mexico 1966), 1958, pp. 122–128.

[M-K-R]      K. Myung Ho, S. Kojima and F. Raymond, *Homotopy invariants of non-orientable 4-manifolds.*, Trans. AMS **333** (1992), 71–81.

[N]      S. P. Novikov, *Homotopy equivalent smooth manifolds*, Izv. Akad. Nauk SSSR, Ser. Mat. **28** (1964), 365–474; English translation, Amer. Math. Soc. Transl. (2) **48** (1965), 271–396.

[Pe]     E. K. Pedersen, *Smoothings of H-spaces*, Algebraic and Geometric Topology (Stanford 1976), Proc. Symp. Pure. Math., vol. 32(2), AMS, 1978, pp. 215–216.

[Qu1]    F. Quinn, *Surgery on Poincaré spaces*, Mimeographed Notes, NYU, May 1971.

[Qu2]    ———, *Surgery on Poincaré and normal spaces*, Bull. AMS **78** (1972), 262–267.

[Qu3]    ———, *Extensions of maps as fibrations and cofibrations*, Transactions AMS **211** (1975), 203–208.

[Ra1]    A. A Ranicki, *Algebraic theory of surgery I. Foundations*, Proc. London. Math. Soc. **40** (1980), 87–192.

[Ra2]    ———, *Algebraic theory of surgery II. Applications to topology*, Proc. London. Math. Soc. **40** (1980), 193–287.

[Ra3]    ———, *Exact Sequences in the Algebraic Theory of Surgery*, Mathematical Notes, vol. 26, Princeton University Press, 1981.

[Ra4]    ———, *Algebraic L-theory and Topological Manifolds*, Cambridge Tracts in Mathematics, vol. 102, Cambridge University Press, 1992.

[Ri1]    W. Richter, *A homotopy-theoretic proof of Williams's metastable Poincaré embedding theorem*, Duke Math. Jour **88** (1997), 435–447.

[Ri2]    ———, *High-dimensional knots with $\pi_1 \cong \mathbb{Z}$ are determined by their complements in one more dimension than Farber's range*, Proc. AMS **120** (1994), 285–294.

[Sm1]    S. Smale, *Generalized Poincaré's conjecture in dimensions greater than four*, Ann. of Math. **74** (1961), 391–406.

[Sm2]    ———, *On the structure of manifolds*, Amer. J. Math. **84** (1962), 387–399.

[Spa]    E. Spanier, *Algebraic Topology*, McGraw Hill, 1966.

[Spi]    M. Spivak, *Spaces satisfying Poincaré duality*, Topology **6** (1967), 767–102.

[Stk]    C. Stark, *Math. Review of "Geometry on Poincaré Spaces", by Hausmann and Vogel*, MR 95a:57042.

[Sta]    J. Stasheff, *H-spaces from the homotopy point of view*, Springer LNM 161, 1970.

[Sto]    R. Stöcker, *The structure of 5-dimensional Poincaré duality spaces*, Comm. Math. Helv. **57** (1982), 481–510.

[Su]     D. Sullivan, *Triangulating and smoothing homotopy equivalences and homeomorphisms, (Geometric Topology Seminar Notes, Princeton University, 1967)*, The Hauptvermutung Book, Kluwer, 1996, pp. 69–103.

[Te]     P. Teichner, *Topological 4-manifolds with finite fundamental group*, Dissertation, Mainz, 1992.

[Tu]     V. Turaev, *Three-dimensional Poincaré complexes: classification and splitting*, Soviet Math. Dokl. **23** (1981), 312–314.

[Vo]     W. Vogell, *The involution in the algebraic K-theory of spaces*, Springer LNM 1126, Proceedings, Rutgers 1983 (1985), 277–317.

[Wa1]    C. T. C. Wall, *Classification of $(n-1)$-connected $2n$-manifolds*, Ann. Math. **75** (1962), 163–189.

[Wa2]    ———, *Classification problems in differential topology. VI. Classification of $(s-1)$-connected $(2s+1)$-manifolds*, Topology **6** (1967), 273–296.

[Wa3]    ———, *Poincaré complexes: I*, Ann. Math. **86** (1970), 213–245.

[Wa4]    ———, *Surgery on Compact Manifolds*, Academic Press, 1970.

[We]     S. Weinberger, *Homologically trivial group actions. I. Simply connected man-ifolds*, Amer. J. Math. 108 (1986), 1005–1021.

[Wi1]    B. Williams, *Applications of unstable normal invariants–Part I*, Comp. Math. **38** (1979), 55–66.

[Wi2]    _____ , *Hopf invariants, localization and embeddings of Poincaré complexes*, Pacific J. Math. **84** (1979), 217–224.

[Z]      A. Zabrodsky, *Homotopy associativity and finite CW complexes*, Topology **9** (1970), 121–128.

DEPARTMENT OF MATHEMATICS
WAYNE STATE UNIVERSITY
DETROIT, MI 48202
*E-mail address*: klein@math.wayne.edu

# Poincaré duality groups

## Michael W. Davis

### §1. Introduction

A space $X$ is *aspherical* if $\pi_i(X) = 0$ for all $i > 1$. For a space of the homotopy type of a $CW$-complex this is equivalent to the condition that its universal covering space is contractible.

Given any group $\Gamma$, there is an aspherical $CW$-complex $B\Gamma$ (also denoted by $K(\Gamma, 1)$) with fundamental group $\Gamma$; moreover, $B\Gamma$ is unique up to homotopy equivalence (cf. [Hu]). $B\Gamma$ is called the *classifying space* of $\Gamma$. ($B\Gamma$ is also called an *Eilenberg-MacLane space* for $\Gamma$.) So, the theory of aspherical $CW$-complexes, up to homotopy, is identical with the theory of groups. This point of view led to the notion of the (co)homology of a group $\Gamma$: it is simply the (co)homology of the space $B\Gamma$.

Many interesting examples of aspherical spaces are manifolds. A principal feature of a manifold is that it satisfies Poincaré duality. Thus, one is led to define an *$n$-dimensional Poincaré duality group* $\Gamma$ to be a group such that $H^i(\Gamma; A) \cong H_{n-i}(\Gamma; A)$ for an arbitrary $\mathbb{Z}\Gamma$-module $A$. (There is also a version of this with twisted coefficients in the nonorientable case.) So, the fundamental group of a closed, aspherical $n$-dimensional manifold $M$ is an $n$-dimensional Poincaré duality group $\Gamma = \pi_1(M)$. The question of whether or not the converse is true was posed by Wall as Problem G2 in [W3]. As stated it is false: as we shall see in Theorem 7.15, Poincaré duality groups need not be finitely presented, while fundamental groups of closed manifolds must be. However, if we add the requirement that the Poincaré duality group be finitely presented, then the question of whether it must be the fundamental group of an aspherical closed manifold is still the main problem in this area.

* Partially supported by NSF grant DMS-9505003

## Examples of aspherical closed manifolds

*1) Low dimensional manifolds.*

- *Dimension 1*: The circle is aspherical.

- *Dimension 2*: Any surface other than $S^2$ or $\mathbb{R}P^2$ is aspherical.

- *Dimension 3*: Any irreducible closed 3-manifold with infinite fundamental group is aspherical. (This follows from Papakyriakopoulos' Sphere Theorem.)

*2) Lie groups.* Suppose $G$ is a Lie group and that $K$ is a maximal compact subgroup. Then $G/K$ is diffeomorphic to Euclidean space. If $\Gamma$ is a discrete, torsion-free subgroup of $G$, then $\Gamma$ acts freely on $G/K$ and $G/K \longrightarrow \Gamma\backslash G/K$ is a covering projection. Hence, $\Gamma\backslash G/K$ is an aspherical manifold. For example, if $G = \mathbb{R}^n$ and $\Gamma = \mathbb{Z}^n$, we get the $n$-torus. If $G = O(n,1)$, then $K = O(n) \times O(1)$, $G/K$ is hyperbolic $n$-space and $\Gamma\backslash G/K$ is a hyperbolic manifold. One can also obtain closed infranilmanifolds or closed infrasolvmanifolds in this fashion by taking $G$ to be a virtually nilpotent Lie group or a virtually solvable Lie group. (A group *virtually* has a property if a subgroup of finite index has that property.)

*3) Riemannian manifolds of nonpositive curvature.* Suppose $M^n$ is a closed Riemannian manifold with sectional curvature $\leq 0$. The Cartan-Hadamard Theorem then states that the exponential map, $\exp : T_x M^n \to M^n$, at any point $x$ in $M^n$, is a covering projection. Hence, the universal covering space of $M^n$ is diffeomorphic to $\mathbb{R}^n$ ($\cong T_x M^n$) and consequently, $M^n$ is aspherical.

During the last fifteen years we have witnessed a great increase in our fund of examples of aspherical manifolds and spaces. In many of these new examples the manifold is tessellated by cubes or some other convex polytope. Some of these examples occur in nature in contexts other than 1), 2) or 3) above, for instance, as the closure of an $(\mathbb{R}^*)^n$ -orbit in a flag manifold or as a blowup of $\mathbb{R}P^n$ along certain arrangements of subspaces. (See [DJS].) Some of these new techniques are discussed below.

*4) Reflection groups.* Associated to any Coxeter group $W$ there is a contractible simplicial complex $\Sigma$ on which $W$ acts properly and cocompactly as a group generated by reflections ([D1], [D3], [Mo]). It is easy to arrange that $\Sigma$ is a manifold (or a homology manifold), so if $\Gamma$ is a torsion-free subgroup of finite index in $W$, then $\Sigma/\Gamma$ is an aspherical closed manifold. Such examples are discussed in detail in §7.

*5) Nonpositively curved polyhedral manifolds.* Many new techniques for constructing examples are described in Gromov's paper [G1]. As Aleksandrov showed, the concept of nonpositive curvature often makes sense for a singular metric on a space $X$. One first requires that any two points in $X$ can be connected by a geodesic segment. Then $X$ is *nonpositively curved* if any small triangle (i.e., a configuration of three geodesic segments) in $X$ is "thinner" than the corresponding comparison triangle in the Euclidean plane. ("Thinner" means that the triangle satisfies the $CAT(0)$-inequality of [G1].) The generalization of the Cartan-Hadamard Theorem holds for a nonpositively curved space $X$: its universal cover is contractible. Gromov pointed out that there are many polyhedral examples of such spaces equipped with piecewise Euclidean metrics (this means that each cell is locally isometric to a convex cell in Euclidean space). Here are two of Gromov's techniques.

- *Hyperbolization:* In Section 3.4 of [G1] Gromov describes several different techniques for converting a polyhedron into a nonpositively curved space. In all of these hyperbolization techniques the global topology of the polyhedron is changed, but its local topology is preserved. So, if the input is a manifold, then the output is an aspherical manifold. (Expositions and applications of hyperbolization can be found in [CD3] and [DJ].)

- *Branched covers:* Let $M$ be a nonpositively curved Riemannian manifold and $Y$ a union of codimension-two, totally geodesic submanifolds which intersect orthogonally. Then the induced (singular) metric on a branched cover of $M$ along $Y$ will be nonpositively curved. (See Section 4.4 of [G1], as well as, [CD1].) Sometimes the metrics can be smoothed to get Riemannian examples as in [GT]. As Gromov points out (on pp.125-126 of [G1]) there is a large class of examples where $M^n$ is $n$-torus and $Y$ is a configuration of codimension-two subtori (see also Section 7 of [CD1]).

Using either the reflection group technique or hyperbolization, one can show that there are examples of aspherical closed (topological) $n$- manifolds $M^n, n \geq 4$, such that a) the universal covering space of $M^n$ is not homeomorphic to $\mathbb{R}^n$ ([D1], [DJ]) or b) $M^n$ is not homotopy equivalent to a smooth (in [DH]) or piecewise linear manifold (in [DJ]).

My thanks go to Tadeusz Januszkiewicz, Guido Mislin, Andrew Ranicki and the referee for their suggestions for improving earlier versions of this survey.

## §2. Finiteness conditions

The classifying space $B\Gamma$ of an $n$-dimensional Poincaré duality group $\Gamma$ is (homotopy equivalent to) an $n$-dimensional $CW$ complex. If $B\Gamma$ is a closed manifold, then it is homotopy equivalent to a finite $CW$ complex. We now investigate the cohomological versions for a group $\Gamma$ (not necessarily satisfying Poincaré duality) of the conditions that $B\Gamma$ is either a) finite dimensional or b) a finite $CW$-complex. A good reference for this material is Chapter VIII of [Br3].

Suppose that $R$ is a nonzero commutative ring and that $R\Gamma$ denotes the group ring of $\Gamma$. Regard $R$ as a $R\Gamma$-module with trivial $\Gamma$-action.

The *cohomological dimension of* $\Gamma$ *over* $R$, denoted $cd_R(\Gamma)$, is the projective dimension of $R$ over $R\Gamma$. In other words, $cd_R(\Gamma)$ is the smallest integer $n$ such there is a resolution of $R$ of length $n$ by projective $R\Gamma$-modules:

$$0 \to P_n \to \cdots \to P_1 \to P_0 \to R \to 0.$$

Our convention, from now on, will be that if we omit reference to $R$, then $R = \mathbb{Z}$. For example, $cd(\Gamma)$ means the cohomological dimension of $\Gamma$ over $\mathbb{Z}$.

If $\Gamma$ acts freely, properly and cellularly on an $n$-dimensional $CW$-complex $E$ and if $E$ is acyclic over $R$, then $cd_R(\Gamma) \leq n$. (Proof: consider the chain complex of cellular chains with coefficients in $R$:

$$0 \to C_n(E; R) \to \ldots C_0(E; R) \xrightarrow{\varepsilon} R \to 0,$$

where $\varepsilon$ is the augmentation.) In particular, if $B\Gamma$ is (homotopy equivalent to) an $n$-dimensional $CW$-complex, then $cd(\Gamma) \leq n$.

The *geometric dimension of* $\Gamma$, denoted $gd(\Gamma)$, is the smallest dimension of a $K(\Gamma, 1)$ complex (i.e., of any $CW$-complex homotopy equivalent to $B\Gamma$). We have just seen that $cd(\Gamma) \leq gd(\Gamma)$. Conversely, Eilenberg and Ganea proved in [EG] that $gd(\Gamma) \leq \max\{cd(\Gamma), 3\}$. Also, it follows from Stallings' Theorem [St] (that a group of cohomological dimension one is free) that $cd(\Gamma) = 1$ implies $gd(\Gamma) = 1$. The possibility that there exists a group $\Gamma$ with $cd(\Gamma) = 2$ and $gd(\Gamma) = 3$ remains open.

A group of finite cohomological dimension is automatically torsion-free. (Proof: a finite cyclic subgroup has nonzero cohomology in every even dimension.) Similarly, if $cd_R(\Gamma) < \infty$, the order of any torsion element of $\Gamma$ must be invertible in $R$.

A group $\Gamma$ is of *type F* if $B\Gamma$ is homotopy equivalent to a finite complex. $\Gamma$ is of *type $FP_R$* (respectively, of *type $FL_R$*) if there is a resolution of $R$ of finite length by finitely generated $R\Gamma$-modules:

$$0 \to P_n \to \cdots \to P_0 \to R \to 0,$$

where each $P_i$ is projective (respectively, free). (The key phrase here is "finitely generated.")

If $\Gamma$ acts freely, properly, cellularly and cocompactly on an $R$-acyclic $CW$-complex $E$, then $\Gamma$ is of type $FL_R$. (Consider the cellular chain complex again.) So, a group of type $F$ is of type $FL$. Similarly, if $B\Gamma$ is dominated by a finite complex, then $\Gamma$ is of type $FP$. Conversely, Wall [W1] proved that if a finitely presented group is of type $FL$, then it must be of type $F$.

There is no known example of a group which is of type $FP$ but not of type $FL$. In fact, it has been conjectured that for any torsion-free group $\Gamma$, the reduced projective class group, $\tilde{K}_0(\mathbb{Z}\Gamma)$, is zero, that is, that every finitely generated projective $\mathbb{Z}\Gamma$- module is stably free.

A group $\Gamma$ is finitely generated if and only if the augmentation ideal is finitely generated as a $\mathbb{Z}\Gamma$-module (Exercise 1, page 12 in [Br3]). Hence, any group of type $FP$ is finitely generated. However, it does not follow that such a group is finitely presented (i.e., that it admits a presentation with a finite number of generators and a finite number of relations). In fact, Bestvina and Brady [BB] have constructed examples of type $FL$ which are not finitely presented. These examples will be discussed further in §7.

## §3. POINCARÉ DUALITY GROUPS

If $\Gamma$ is the fundamental group of an aspherical, closed $n$-manifold, $M$, then it satisfies Poincaré duality:

$$H^i(\Gamma; A) \cong H_{n-i}(\Gamma; D \otimes A)$$

where $D$ is the orientation module and where the coefficients can be any $\mathbb{Z}\Gamma$-module $A$.

Since the universal covering space $\tilde{M}^n$ of an aspherical $M^n$ is contractible, it follows from Poincaré duality (in the noncompact case) that the cohomology with compact supports of $\tilde{M}^n$ is the same as that of $\mathbb{R}^n$, i.e.,

$$H_c^i(\tilde{M}^n) \cong \begin{cases} 0, & \text{for } i \neq n \\ \mathbb{Z}, & \text{for } i = n. \end{cases}$$

On the other hand, if $\Gamma$ acts freely, properly and cocompactly on an acyclic space $E$, then $H^i(\Gamma; \mathbb{Z}\Gamma) \cong H^i_c(E)$ (by Prop. 7.5, p. 209 in [Br3]). Hence, for $\Gamma = \pi_1(M^n)$,

$$H^i(\Gamma; \mathbb{Z}\Gamma) \cong \begin{cases} 0, & \text{for } i \neq n \\ \mathbb{Z}, & \text{for } i = n. \end{cases}$$

These considerations led Johnson and Wall [JW] and, independently Bieri [Bi] to the following two equivalent definitions.

**Definition 3.1.** ([Bi]) A group $\Gamma$ is a *Poincaré duality group of dimension $n$ over a commutative ring $R$* (in short, a *$PD^n_R$-group*) if there is an $R\Gamma$-module $D$, which is isomorphic to $R$ as an $R$-module, and a homology class $\mu \in H_n(\Gamma; D)$ (called the *fundamental class*) so that for any $R\Gamma$-module $A$, cap product with $\mu$ defines an isomorphism: $H^i(\Gamma; A) \cong H_{n-i}(\Gamma; D \otimes A)$. $D$ is called the *orientation module* (or *dualizing module*) for $\Gamma$. If $\Gamma$ acts trivially on $D$, then it is an *orientable $PD^n_R$-group*.

**Definition 3.2.** ([JW]). A group $\Gamma$ is a *Poincaré duality group of dimension $n$ over $R$* if the following two conditions hold:

(i)   $\Gamma$ is of type $FP_R$, and

(ii)  $H^i(\Gamma; R\Gamma) = \begin{cases} 0, \text{for } i \neq n \\ R, \text{for } i = n \end{cases}$

We note that if $R = \mathbb{Z}$, then (i) implies that $\Gamma$ is torsion-free and (ii) implies that $cd(\Gamma) = n$.

**Theorem 3.3.** ([BE2], [Br1], [Br3]). *Definitions 3.1 and 3.2 are equivalent.*

On pages 220 and 221 of [Br3] one can find three different proofs that the conditions in Definition 3.2 imply those in Definition 3.1. The dualizing module $D$ is $H^n(\Gamma; R\Gamma)$. Conversely, suppose $\Gamma$ satisfies the conditions of Definition 3.1. Since $D \otimes R\Gamma$ is free (by Cor. 5.7, page 69 of [Br3]), it is acyclic. Hence, $H^i(\Gamma; R\Gamma) \cong H_{n-i}(\Gamma; D \otimes R\Gamma)$ vanishes for $i \neq n$ and is isomorphic to $R$ for $i = n$. The main content of Theorem 3.3 is that Definition 3.1 forces $\Gamma$ to be of type $FP_R$. The reason for this is that the statement that $\Gamma$ is of type $FP_R$ is equivalent to the statement that $cd_R(\Gamma) < \infty$ and that for each $i$, the functor on $R\Gamma$-modules $A \longrightarrow H^i(\Gamma; A)$ commutes with direct limits (see [Br3]). By naturality of cap products and by Poincaré duality, this functor can be identified with $A \to D \otimes A \to H_{n-i}(\Gamma; D \otimes A)$ and this clearly commutes with direct limits.

In line with our convention from §2, for $R = \mathbb{Z}$, denote $PD_R^n$ by $PD^n$. As Johnson and Wall observed, if $\Gamma$ is a finitely presented $PD^n$- group, then $B\Gamma$ is a Poincaré complex in the sense of [W2]. In particular, $B\Gamma$ is finitely dominated.

The principal question in this area (as well as the most obvious one) is if every $PD^n$-group is the fundamental group of an aspherical closed manifold. As stated the answer is no. For, as we shall see in §7, the Bestvina-Brady examples can be promoted to examples of $PD^n$-groups, $n \geq 4$, which are not finitely presented. (Kirby and Siebenmann proved that any compact topological manifold is homotopy equivalent to a finite $CW$-complex; hence, its fundamental group is finitely presented.) So, the correct question is the following.

**Question 3.4.** *Is every finitely presented $PD^n$-group the fundamental group of an aspherical closed manifold?*

This is closely related to Borel's Question: are any two aspherical closed manifolds with the same fundamental group homeomorphic? (See [FJ] for a discussion of Borel's Question.) Thus, Question 3.4 asks if any finitely presented $PD^n$-group corresponds to an aspherical closed manifold and Borel's Question asks if this manifold is unique up to homeomorphism.

A space $X$ is a *homology manifold of dimension* $n$ *over* $R$ if for each point $x$ in $X, H_*(X, X - x; R) \cong H_*(\mathbb{R}^n, \mathbb{R}^n - 0; R)$, i.e., if

$$H_i(X, X - x; R) \cong \begin{cases} 0 \text{ , for } i \neq n \\ R \text{ , for } i = n. \end{cases}$$

The usual proof that manifolds satisfy Poincaré duality also works for homology manifolds. As we shall see in Example 7.4 there are aspherical polyhedra which are homology manifolds over some ring $R$ but not over $\mathbb{Z}$. The fundamental groups of these examples are Poincaré duality groups over $R$ but not over $\mathbb{Z}$. So, when $R \neq \mathbb{Z}$ the appropriate version of Question 3.4 is the following.

**Question 3.5.** *Is every torsion-free, finitely presented, $PD_R^n$-group the fundamental group of an aspherical closed $R$-homology manifold?*

In fact this question is relevant even when $R = \mathbb{Z}$. The reason is this. While it is true that every closed polyhedral homology manifold is homotopy equivalent to a closed manifold, Bryant, Ferry, Mio and Weinberger have shown in [BFMW] that there exist homology manifolds which are

compact $ANRs$ and which are not homotopy equivalent to closed manifolds. Thus, there is the intriguing possibility that some of these exotic
"near manifolds" of [BFMW] could be aspherical.

Ranicki has shown (in Chapter 17 of [R]) that, for $\Gamma$ of type $F$, Question
3.4 has an affirmative answer if and only if the "total surgery obstruction"
$s(B\Gamma) \in \mathbb{S}_n(B\Gamma)$ is 0 (where $\mathbb{S}_*$ means the relative homotopy groups of the
assembly map in algebraic L-theory). A similar remark applies to Question
3.5 using the "4-periodic total surgery obstruction" in Chapter 25 of [R].
In fact, as explained on page 275 of [R], the strongest form of the Novikov
Conjecture is equivalent (in dimensions $\geq 5$) to the conjecture that both
Question 3.4 and Borel's Question have affirmative answers. As explained
on page 298 of [R], a slightly weaker version of the Novikov Conjecture
(that the assembly map is an isomorphism) is equivalent to allowing the
possibility that there exist aspherical ANR homology manifolds which are
not homotopy equivalent to manifolds, as in [BFMW].

**Duality groups.** There are many interesting groups which satisfy Definition 3.1 except for the requirement that $D$ be isomorphic to $R$. The proof
of Theorem 3.3 also gives the following result.

**Theorem 3.6.** (Bieri-Eckmann [BE2], Brown [Br1]). *The following two
conditions are equivalent.*

(i)   *There exists an $R\Gamma$-module $D$ and a positive integer $n$ such that for
      any $R\Gamma$-module $A$ there is a natural isomorphism (i.e., induced by
      cap product with a fundamental class): $H^i(\Gamma; A) \cong H_{n-i}(\Gamma; D \otimes A)$.*

(ii)  $\Gamma$ *is of type $FP_R$ and*

$$H^i(\Gamma; R\Gamma) \cong \begin{cases} 0 \, , & \text{for } i \neq n \\ D \, , & \text{for } i = n. \end{cases}$$

If either of these conditions hold, then $\Gamma$ is a *duality group of dimension
$n$ over $R$* (in short, a $D_R^n$-*group*).

**Theorem 3.7.** (Farrell [F]). *Suppose $R$ is a field and that $\Gamma$ is a $D_R^n$-
group. If $\dim_R(D) < \infty$, then $\dim_R(D) = 1$ (and consequently $\Gamma$ is a
$PD_R^n$-group).*

It follows that if $\Gamma$ is a duality group over $\mathbb{Z}$, then either $D \cong \mathbb{Z}$ or $D$ is
of infinite rank. Conjecturally, $D$ must be free abelian.

## Examples of duality groups

1) Finitely generated free groups are duality groups of dimension 1.

2) Suppose that $M$ is a compact aspherical $n$-manifold with nonempty boundary. Let $\partial_1 M, \ldots, \partial_m M$ denote the components of of $\partial M$ and suppose that each $\partial_j M$ is aspherical and that $\pi_1(\partial_j M) \to \pi_1(M)$ is injective. Then $\pi_1(M)$ is a duality group of dimension $n - 1$. For example, any knot group (that is, the fundamental group of the complement of a nontrivial knot in $S^3$) is a duality group of dimension two.

3) Let $\Gamma$ be a torsion-free arithmetic group and $G/K$ the associated symmetric space. Then $\Gamma$ is a duality group of dimension $n - \ell$, where $n = \dim(G/K)$ and $\ell$ is the $\mathbb{Q}$-rank of $\Gamma$. (See [BS].)

4) Let $S_g$ denote the closed surface of genus $g > 1$. The group of outer automorphisms $\mathrm{Out}(\pi_1(S_g))$ is the *mapping class group* of $S_g$. It is a virtual duality group of dimension $4g - 5$ (i.e., any torsion-free subgroup of finite index in $\mathrm{Out}(\pi_1(S_g))$ is a duality group). (See [Ha].)

## §4. Subgroups, extensions and amalgamations

We have the following constructions for manifolds:

1) Any covering space of a manifold is a manifold.

2) If $F \to E \to B$ is a fiber bundle and if $F$ and $B$ are manifolds, then so is $E$.

3) Suppose $M$ is a manifold with boundary and that the boundary consists of two components $\partial_1 M$ and $\partial_2 M$ which are homeomorphic. Then the result of gluing $M$ together along $\partial_1 M$ and $\partial_2 M$ via a homeomorphism is a manifold. (In this construction one usually has one of two situations in mind: either a) $M$ is connected or b) $M$ consists of two components with $\partial_1 M$ and $\partial_2 M$ their respective boundaries.)

In this section we consider the analogous constructions for Poincaré duality groups.

**Subgroups.** The following analog of construction 1), is proved as Theorem 2 in [JW]. It follows fairly directly from Definition 3.2.

**Theorem 4.1.** ([Bi], [JW]). *Suppose that $\Gamma$ is a torsion -free group and that $\Gamma'$ is a subgroup of finite index in $\Gamma$. Then $\Gamma$ is a $PD_R^n$-group ($R$ a commutative ring) if and only if $\Gamma'$ is.*

By way of contrast, there is the following result of Strebel [Str].

**Theorem 4.2.** ([Str]). *If $\Gamma$ is a $PD^n$-group and $\Gamma'$ is a subgroup of infinite index, then $cd(\Gamma') < n$.*

**Extensions.** The next result, the analog of construction 2), is Theorem 3 of [JW].

**Theorem 4.3.** ([Bi], [JW]). *Suppose that $\Gamma$ is an extension of $\Gamma''$ by $\Gamma'$:*

$$1 \to \Gamma' \to \Gamma \to \Gamma'' \to 1$$

*If both $\Gamma'$ and $\Gamma''$ are Poincaré duality groups over $R$, then so is $\Gamma$. Conversely, if $\Gamma$ is a Poincaré duality group over $R$ and if both $\Gamma'$ and $\Gamma''$ are of type $FP_R$, then $\Gamma'$ and $\Gamma''$ are both Poincaré duality groups over $R$.*

The corresponding result for Poincaré spaces was stated in [Q] and proved in [Go].

**Theorem 4.4.** ([Go], [Q]). *Suppose that $F \to E \to B$ is a fibration and that $F$ and $B$ are dominated by finite complexes. Then $E$ is dominated by a finite complex and $E$ satisfies Poincaré duality if and only if both $F$ and $B$ do.*

**Corollary 4.5.** *Suppose that $\Gamma$ is a finitely presented, torsion-free group of type $FP$ and that $\Gamma$ acts freely, properly and cocompactly on a manifold $M$. If $M$ is dominated by a finite complex, then $\Gamma$ is a Poincaré duality group.*

*Proof.* Consider the fibration $M \to M \times_\Gamma E\Gamma \to B\Gamma$, where $E\Gamma$ denotes the universal covering space of $B\Gamma$. Since $M \times_\Gamma E\Gamma$ is homotopy equivalent to the closed manifold $M/\Gamma$, it satisfies Poincaré duality. $\square$

For example the corollary applies to the case where $M \cong S^k \times \mathbb{R}^n$. (See [CoP].)

**Amalgamations.** Suppose that $M$ is a compact manifold with boundary, with boundary components $(\partial_j M)_{j \in I}$, that $M$, as well as each boundary component is aspherical, and that for each $j \in I$ the inclusion $\partial_j M \subset M$ induces a monomorphism $\pi_1(\partial_j M) \to \pi_1(M)$. Set $\Gamma = \pi_1(M)$, let $S_j$ denote the image of $\pi_1(\partial_j M)$ in $\Gamma$, and let $S$ denote the family of subgroups $(S_j)_{j \in I}$. Then, following [BE3] and [E], the fact that $(M, \partial M)$ satisfies

Poincaré-Lefschetz duality can be reformulated in terms of group cohomology as follows.

Let $\Gamma$ be a group and $S = (S_j)_{j \in I}$ a finite family of subgroups. For any subgroup $H$ of $\Gamma$ let $\mathbb{Z}(\Gamma/H)$ denote the free abelian group on $\Gamma/H$ with $\mathbb{Z}\Gamma$-module structure induced from left multiplication. Let $\Delta = \ker(\bigoplus \mathbb{Z}(\Gamma/S_j) \xrightarrow{\epsilon} \mathbb{Z})$, where $\epsilon$ is defined by $\epsilon(\gamma S_j) = 1$ for all $j \in I$ and $\gamma \in \Gamma$. Set

$$H_i(\Gamma, S; A) = H_{i-1}(\Gamma; \Delta \otimes A)$$

$$H^i(\Gamma, S; A) = H^{i-1}(\Gamma; \operatorname{Hom}(\Delta, A)).$$

**Definition 4.6.** ([BE3], [E]). The pair $(\Gamma, S)$ is a *Poincaré duality pair of dimension n* (in short a $PD^n$-*pair*) with *orientation module D* (where $D$ is isomorphic to $\mathbb{Z}$ as an abelian group) if there are natural isomorphisms:

$$H^i(\Gamma; A) \cong H_{n-i}(\Gamma, S; D \otimes A)$$

$$H^i(\Gamma, S; A) \cong H^{n-i}(\Gamma; D \otimes A).$$

(It follows that each $S_j$ is a $PD^{n-1}$-group.)

As observed in [JW] one can then use Mayer-Vietoris sequences to prove the analogs of construction 3). For example, suppose that $(\Gamma_1, H)$ and $(\Gamma_2, H)$ are $PD^n$-pairs. (Here $S$ consists of a single subgroup $H$.) Then the amalgamated product $\Gamma_1 *_H \Gamma_2$ is a $PD^n$-group. Similarly, suppose $(\Gamma, S)$ is a $PD^n$-pair where $S$ consists of two subgroups $S_1$ and $S_2$ and that $\theta : S_1 \to S_2$ is an isomorphism. Then the $HNN$-extension $\Gamma*_\theta$ is a $PD^n$-group. If a group can be written as an amalgamated product or $HNN$-extension over a subgroup $H$, then it is said to *split over $H$*.) From these two constructions and induction one can prove the following more general statement.

**Theorem 4.7.** *Suppose that $\Gamma$ is the fundamental group of a finite graph of groups. Let $\Gamma_v$ denote the group associated to a vertex $v$ and $S_e$ the group associated to an edge $e$. For each vertex $v$, let $E(v)$ denote the set of edges incident to $v$ and let $S_v = (S_e)_{e \in E(v)}$ be the corresponding family of subgroups of $\Gamma_v$. If $(\Gamma_v, S_v)$ is a $PD^n$-pair for each vertex $v$, then $\Gamma$ is a $PD^n$-group.*

For the definition of a "graph of groups" and its "fundamental group", see [Se2] or [SW].

Kropholler and Roller in [KR1,2,3] have made an extensive study of when a $PD^n$-group can split over a subgroup $H$ which is a $PD^{n-1}$-group.

If a $PD^n$-group $\Gamma$ splits over a subgroup $H$, there is no reason that $H$ must be a $PD^{n-1}$-group. (We shall give examples where it is not in Example 7.3.) However, it follows from the Mayer-Vietoris sequence that $cd(H) = n - 1$. In particular, for $n \geq 3$, a $PD^n$-group cannot split over a trivial subgroup or an infinite cyclic subgroup. We restate this as the following lemma, which we will need in §6.

**Lemma 4.8.** *For $n \geq 3$, a $PD^n$-group is not the fundamental group of a graph of groups with all edge groups trivial or infinite cyclic.*

## §5. Dimensions one and two

Question 3.4 has been answered affirmatively in dimensions $\leq 2$.

A $PD^1$-group is infinite cyclic. (Since $H^1(\Gamma; \mathbb{Z}\Gamma) = \mathbb{Z}$, $\Gamma$ has two ends and since $\Gamma$ is torsion-free, a result of Hopf [H] implies that $\Gamma \cong \mathbb{Z}$.)

The affirmative answer to Question 3.4 in dimension two is the culmination of several papers by Eckmann and his collaborators, Bieri, Linnell and Müller, see [BE1], [BE2], [EL], [EM], [M] and especially [E].

**Theorem 5.1.** *A $PD^2$-group is isomorphic to the fundamental group of a closed surface.*

A summary of the proof can be found in [E]. In outline, it goes as follows: 1) using a theorem of [M] one shows that if a $PD^2$-group splits as an amalgamated product or $HNN$ extension over a finitely generated subgroup, then the theorem holds and then 2) using the Hattori-Stallings rank, it is proved in [EL] that any $PD^2$-group has positive first Betti number and hence, that it splits.

Combining Theorem 5.1 with Theorem 4.1 we get the following.

**Corollary 5.2.** *Suppose that a torsion-free group $\Gamma$ contains a surface group $\Gamma'$ as a subgroup of finite index. Then $\Gamma$ is a surface group.*

For a discussion of the situation in dimension three see Thomas' article [T].

## §6. Hyperbolic groups

Any finitely generated group $\Gamma$ can be given a "word metric," $d : \Gamma \times \Gamma \to \mathbb{N}$, as follows. Fix a finite set of generators $T$. Then $d(\gamma, \gamma')$ is the smallest integer $k$ such that $\gamma^{-1}\gamma'$ can be written as a word of length $k$ in $T \cup T^{-1}$. If $\Gamma$ is a discrete group of isometries of a metric space $Y$ and $Y/\Gamma$ is compact, then the word metric on $\Gamma$ is quasi-isometric to the induced metric on any $\Gamma$-orbit in $Y$.

Rips defined the notion of a "hyperbolic group" in terms of the word metric (For the definition, see [G1].) This idea was then developed into a vast and beautiful theory in Gromov's seminal paper [G1]. It is proved in [G1] that the property of being hyperbolic depends only on the quasi-isometry type of the word metric, in particular, it is independent of the choice of generating set. The idea behind the definition is this: in the large, $\Gamma$ should behave like a discrete, cocompact group of isometries of a metric space $Y$ which is simply connected and "negatively curved" in some sense (for example, a space $Y$ which satisfies the $\text{CAT}(\varepsilon)$-inequality for some $\varepsilon < 0$). In particular, the fundamental group of a closed Riemannian manifold of (strictly) negative sectional curvature is hyperbolic. Many more examples can be found in [G1].

Rips proved that given a hyperbolic group $\Gamma$ there is a contractible simplicial complex $E$ on which $\Gamma$ acts properly and cocompactly, see [G1]. In particular, if $\Gamma$ is torsion-free, then $E/\Gamma$ is a $K(\Gamma, 1)$ complex. So, torsion-free hyperbolic groups are automatically of type $F$.

Associated to any hyperbolic group $\Gamma$, there is a space $\partial\Gamma$, called the "ideal boundary" of $\Gamma$. The points in $\partial\Gamma$ are certain equivalence classes of sequences $(\gamma_i)_{i \in \mathbb{N}}$ in $\Gamma$ which go to infinity in an appropriate sense. (The definition of $\partial\Gamma$ can be found on page 98 of [G1].) If $\Gamma$ is the fundamental group of a negatively curved, closed Riemannian $n$-manifold, then $\partial\Gamma$ is homeomorphic to $S^{n-1}$ (in this case, $\partial\Gamma$ can be identified with the space of all geodesic rays in the universal covering space emanating from some base point).

In [BM] Bestvina and Mess proved that the Rips complex $E$ can be compactified to a space $\overline{E}$ by adding $\partial\Gamma$ as the space at infinity; moreover, $\partial\Gamma$ is homotopically inessential in $\overline{E}$ in a strong sense. (In technical terms, $\overline{E}$ is a Euclidean retract and $\partial\Gamma$ is a $Z$-set in $\overline{E}$.) It follows that $H_c^*(E) \cong \check{H}^{*-1}(\partial\Gamma)$, where $\check{H}^{*-1}(\partial\Gamma)$ denotes the reduced Čech cohomology of $\partial\Gamma$. Since we also have $H^*(\Gamma; \mathbb{Z}\Gamma) \cong H_c^*(E)$, this gives the following theorem.

**Theorem 6.1.** (Bestvina-Mess [BM]). *Let $\Gamma$ be a torsion-free hyperbolic group and $R$ a commutative ring. Then $H^*(\Gamma; R\Gamma) \cong \check{H}^{*-1}(\partial\Gamma; R)$.*

*Remark.* Suppose $\Gamma$ is a (not necessarily hyperbolic) group of type $F$ (so that $B\Gamma$ is a finite complex). Then $\Gamma$ *has a $Z$-set compactification* if $E\Gamma$

can be compactified to a Euclidean retract $\overline{E\Gamma}$ so that $\partial\overline{E\Gamma}(=\overline{E\Gamma}-E\Gamma)$ is a $Z$-set in $\overline{E\Gamma}$. It is quite possible that every group of type $F$ admits a $Z$-set compactification. (The Novikov Conjecture is known to hold for such groups, see [CaP].) We note that Theorem 6.1 holds for any such group. Thus, such a group $\Gamma$ is a $D^n$-group if and only if the Čech cohomology of $\partial\overline{E\Gamma}$ is concentrated in dimension $n-1$; it is a $PD^n$-group if and only if $\partial\overline{E\Gamma}$ has the same Čech cohomology as $S^{n-1}$.

**Theorem 6.2.** (Bestvina [Be]). *Suppose that a hyperbolic group $\Gamma$ is a $PD_R^n$-group. Then $\partial\Gamma$ is a homology $(n-1)$-manifold over $R$ (with the same $R$-homology as $S^{n-1}$).*

However, as shown in [DJ], even when $R = \mathbb{Z}$, for $n \geq 4$, there are examples where $\partial\Gamma$ is not homeomorphic to $S^{n-1}$; $\partial\Gamma$ need not be simply connected or locally simply connected (so, in these examples $\partial\Gamma$ is not even an $ANR$). In dimension three Theorem 6.2 has the following corollary.

**Corollary 6.3.** (Bestvina-Mess [BM]). *If $\Gamma$ is a hyperbolic $PD^3$- group, then $\partial\Gamma$ is homeomorphic to $S^2$.*

In this context, Cannon has proposed the following version of Thurston's Geometrization Conjecture.

**Conjecture 6.4.** *If a $PD^3$-group is hyperbolic (in the sense of Rips and Gromov), then it is isomorphic to the fundamental group of a closed hyperbolic 3-manifold (i.e., a 3-manifold of constant curvature $-1$).*

A proof of this would constitute a proof of a major portion of the Geometrization Conjecture. The issue is to show that the action of the group $\Gamma$ on $\partial\Gamma(= S^2)$ is conjugate to an action by conformal transformations. Cannon and his collaborators seem to have made progress on an elaborate program for proving this (see [C]).

**The group of outer automorphisms of a hyperbolic $PD^n$-group.** A proof of the following theorem is outlined on page 146 of [G1]. A different argument using work of Paulin [P] and Rips [Ri] can be found in [BF].

**Theorem 6.6.** (Gromov). *Let $\Gamma$ be a hyperbolic $PD^n$-group with $n \geq 3$. Then $Out(\Gamma)$, its group of outer automorphisms, is finite.*

*Sketch of Proof.* (See [BF].) Paulin [P] proved that if $\Gamma$ is hyperbolic and $Out(\Gamma)$ is infinite, then $\Gamma$ acts on an $\mathbb{R}$-tree with all edge stabilizers either virtually trivial or virtually infinite cyclic. A theorem of Rips [R] then implies that there is a $\Gamma$-action on a simplicial tree with the same type of edge stabilizers. Since $\Gamma$ is torsion-free this implies that $\Gamma$ splits as an amalgamated free product or $HNN$ extension over a trivial group or an infinite cyclic group. By Lemma 4.8 such a group cannot satisfy Poincaré duality if $n \geq 3$. $\square$

*Remarks.* i) The theorem is false for $n = 2$, i.e., for surface groups.

ii) The theorem is also false in the presence of 0 curvature. For example, $\mathrm{Out}(\mathbb{Z}^n) = GL(n, \mathbb{Z})$, which is infinite if $n > 1$.

iii) Suppose $M_1$ and $M_2$ are two aspherical manifolds with boundary with $\partial M_1 = \partial M_2 = T^{n-1}$. Let $M$ be the result of gluing $M_1$ to $M_2$ along $T^{n-1}$ and let $\Gamma = \pi_1(M)$. (If $n \geq 3$, then $\Gamma$ is not hyperbolic.) The homotopy class of any closed loop in $T^{n-1}$ then defines a Dehn twist about $T^{n-1}$. In this way we get a monomorphism $\mathbb{Z}^{n-1} \to \mathrm{Out}(\Gamma)$, so the outer automorphism group is infinite.

iv) When $\Gamma$ is the fundamental group of a closed hyperbolic manifold (either a real, complex or quaternionic hyperbolic manifold) of dimension $> 2$, then Theorem 6.6 follows from the Mostow Rigidity Theorem.

## §7. EXAMPLES

**Right-angled Coxeter groups.** Given a simplicial complex $L$, we shall describe a simple construction of a cubical cell complex $P_L$ so that the link of each vertex in $P_L$ is isomorphic to $L$. If $L$ satisfies a simple combinatorial condition (that it is a "flag complex"), then $P_L$ is aspherical.

Let $S$ denote the vertex set of $L$ and for each simplex $\sigma$ in $L$ let $S(\sigma)$ denote its vertex set. Define $P_L$ to be the subcomplex of the cube $[-1, 1]^S$ consisting of all faces parallel to $\mathbb{R}^{S(\sigma)}$ for some simplex $\sigma$ in $L$ (such a face is defined by equations of the form: $x_s = \varepsilon_s$, where $s \in S - S(\sigma)$ and $\varepsilon_s \in \{\pm 1\}$). There are $2^S$ vertices in $P_L$ and the link of each of them is naturally identified with $L$. Hence, if $L$ is homeomorphic to $S^{n-1}$, then $P_L$ is a closed $n$-manifold. Similarly, if $L$ is an $(n-1)$-dimensional homology manifold over $R$ with the same $R$-homology as $S^{n-1}$, then $P_L$ is a $R$-homology $n$-manifold.

For each $s$ in $S$ let $r_s$ be the linear reflection on $[-1, 1]^S$ which sends the standard basis vector $e_s$ to $-e_s$ and which fixes $e_{s'}$ for $s' \neq s$. The group generated by these reflections is isomorphic to $(\mathbb{Z}/2)^S$. The subcomplex $P_L$ is $(\mathbb{Z}/2)^S$- stable. A fundamental domain for the action on $[-1, 1]^S$ is $[0, 1]^S$; moreover, the orbit space $[-1, 1]^S/(\mathbb{Z}/2)^S$ is naturally identified with this subspace. Set $K = P_L \cap [0,1]^S$ and for each $s$ in $S$ let $K_s$ be the subset of $K$ defined by $x_s = 0$. ($K_s$ is called the *mirror* associated to $r_s$.) The cell complex $K$ is homeomorphic to the cone on $L$; the subcomplex $K_s$ is the closed star of the vertex $s$ in the barycentric subdivision of $L$. In order to describe the universal covering space $\Sigma_L$ of $P_L$, we first need to discuss Coxeter groups.

A *Coxeter matrix* $M$ on a set $S$ is a symmetric $S \times S$ matrix $(m_{st})$ with entries in $\mathbb{N} \cup \{\infty\}$ such that $m_{st} = 1$ if $s = t$ and $m_{st} \geq 2$ if $s \neq t$. Associated to $M$ there is a *Coxeter group* $W$ defined by the presentation

$$W = \langle S \mid (st)^{m_{st}} = 1, \ (s,t) \in S \times S \rangle$$

A Coxeter matrix is *right-angled* if all of its off-diagonal entries are 2 or $\infty$. Similarly, a Coxeter group is *right-angled* if its Coxeter matrix is.

One can associate to $L$ a right-angled Coxeter matrix $M_L$ and a right-angled Coxeter group $W_L$ as follows. $M_L$ is defined by

$$m_{st} = \begin{cases} 1, & \text{if } s = t \\ 2, & \text{if } \{s,t\} \text{ spans an edge in } L \\ \infty, & \text{otherwise.} \end{cases}$$

$W_L$ is the associated Coxeter group. Let $\Theta : W_L \to (\mathbb{Z}/2)^S$ be the epimorphism which sends $s$ to $r_s$ and let $\Gamma_L$ be the kernel of $\Theta$. Then $\Gamma_L$ is torsion-free. ($\Gamma_L$ is the commutator subgroup of $W_L$.)

The complex $\Sigma_L$ can be defined as $(W_L \times K)/\backsim$. The equivalence relation $\backsim$ on $W_L \times K$ is defined by: $(w,x) \backsim (w',x')$ if and only if $x = x'$ and $w^{-1}w' \in W_x$, where $W_x$ is the subgroup generated by $\{s \in S \mid x \in K_s\}$. It is easy to see that $P_L \cong ((\mathbb{Z}/2)^S \times K)/\backsim$, where the equivalence relation is defined similarly. The natural $\Theta$-equivariant map $\Sigma_L \to P_L$ is a covering projection. Moreover, $\Sigma_L$ is simply connected (by Corollary 10.2 in [D1]). Consequently, $\Sigma_L$ is the universal covering space of $P_L$ and $\pi_1(P_L) = \Gamma_L$.

We now turn to the question of when $\Sigma_L$ is contractible. A simplicial complex $L$ with vertex set $S$ is a *flag complex* if given any finite set of vertices $S'$, which are pairwise joined by edges, there is a simplex $\sigma$ in $L$ spanned by $S'$ (i.e., $S(\sigma) = S'$). For example, the derived complex (also called the "order complex") of any poset is a flag complex. In particular, the barycentric subdivision of any simplicial complex is a flag complex. Hence, the condition that $L$ is a flag complex does not restrict its topological type; it can be any polyhedron.

**Theorem 7.1.** ([D1], [D3]). *The complex $\Sigma_L$ is contractible if and only if $L$ is a flag complex.*

Gromov gave a different proof from that of [D1] for the above theorem (in Section 4 of [G1]); he showed that the natural piecewise Euclidean metric on a cubical complex is nonpositively curved if and only if the link of each vertex is a flag complex. Since $\Sigma_L$ is a cubical complex with the link of each vertex isomorphic to $L$, the theorem follows. (An exposition of this method is given in [D3].)

**Example 7.2.** (Topological reflection groups on $\mathbb{R}^n$). Suppose that $L$ is a $PL$ triangulation of the sphere $S^{n-1}$ as a flag complex. (To insure that $L$ is a flag complex we could take it to be the barycentric subdivision of an arbitrary $PL$ triangulation of $S^{n-1}$.) Then $K$ is homeomorphic to the $n$-disk (since it is homeomorphic to the cone on $L$). The $K_s$ are the dual cells to the vertices of $L$. If $L$ is the boundary complex of a convex polytope, then $K$ is the dual polytope. In fact this is the correct picture to keep in mind: $K$ closely resembles a convex polytope. In various special cases $W_L$ can be represented as a group generated by reflections on Euclidean $n$-space or hyperbolic $n$-space so that $K$ is a fundamental domain. The right-angledness hypothesis on the Coxeter group should be thought of as the requirement that the hyperplanes of reflection intersect orthogonally, i.e., that if two of the $K_s$ intersect, then their intersection is orthogonal. For example, suppose $L$ is a subdivision of the circle into $m$ edges. $L$ is a flag complex if and only if $m \geq 4$. $K$ is an $m$-gon; we should view it as a right-angled $m$-gon in the Euclidean plane (if $m = 4$) or the hyperbolic plane (if $m > 4$). $P_L$ is the orientable surface of Euler characteristic $2^{m-2}(4 - m)$; its universal cover $\Sigma_L$ is the Euclidean or hyperbolic plane as $m = 4$ or $m > 4$. For another example, suppose that $L$ is the boundary complex of an $n$-dimensional octahedron (the $n$-fold join of $S^0$ with itself). Then $K$ is an $n$-cube, the $K_s$ are its codimension-one faces, $P_L$ is an $n$-torus, $\Sigma_L$ is $\mathbb{R}^n$ and its tiling by copies of $K$ is the standard cubical tessellation of $\mathbb{R}^n$. If $L$ is a more random $PL$ triangulation of $S^{n-1}$, then we may no longer have such a nice geometric interpretation; however, the basic topological picture is the same ([D1]): $K$ is an $n$-cell, the $K_s$ are $(n - 1)$-cells and they give a cellulation of the boundary of $K$, $P_L$ is a closed $PL$ $n$-manifold and its universal cover $\Sigma_L$ is homeomorphic to $\mathbb{R}^n$. (The proof that $\Sigma_L$ is homeomorphic to $\mathbb{R}^n$ uses the fact that the triangulation L is $PL$; otherwise, it need not be, cf., Remark (5b.2) in [DJ].) We can think of $K$ as an orbifold; each point has a neighborhood which can be identified with the quotient space of $\mathbb{R}^n = \mathbb{R}^k \times \mathbb{R}^{n-k}$ by the action of $(\mathbb{Z}/2)^k \times 1$, for some $k \leq n$. Its orbifoldal fundamental group is $W_L$ and its orbifoldal universal cover is $\Sigma_L$.

**Example 7.3.** (Splittings). Suppose that $L$ is triangulation of $S^{n-1}$ as a flag complex and that $L_0, L_1$ and $L_2$ are full subcomplexes such that $L_1 \cup L_2 = L$ and $L_1 \cap L_2 = L_0$. Then $P_L$ is an aspherical $n$-manifold, $P_{L_1} \cup P_{L_2} = P_L$, and $P_{L_1} \cap P_{L_2} = P_{L_0}$. Thus, $\Gamma_L$ splits as an amalgamated product of $\Gamma_{L_1}$ and $\Gamma_{L_2}$ along $\Gamma_{L_0}$. It follows from Theorem B in [D4] that $\Gamma_{L_0}$ is a $PD^{n-1}$-group if and only if $L_0$ is a homology $(n-2)$-manifold with the same homology as $S^{n-2}$, and if this is the case, then $P_{L_0}$ is an aspherical homology $(n-1)$- manifold. On the other hand, the complexes $L_0, L_1$ and $L_2$ can be fairly arbitrary. For example, we could choose $L_0$ to be a triangulation of any piecewise linear submanifold of codimension one in $S^{n-1}$. ($L_0$ then separates the sphere into two pieces $L_1$ and $L_2$.) By Theorem A in [D4], for $1 < i < n$, $H^i(\Gamma_{L_0}; \mathbb{Z}\Gamma_{L_0})$ is an infinite sum of copies of $H^{i-1}(L_0)$. So, if $L_0$ is not a homology sphere, then $\Gamma_{L_0}$ will not be a $PD^{n-1}$-group and $(\Gamma_{L_1}, \Gamma_{L_0})$ and $(\Gamma_{L_2}, \Gamma_{L_0})$ will not be $PD^n$-pairs. Hence, there are many examples of splittings of $PD^n$-groups over subgroups which do not satisfy Poincaré duality.

By allowing $L$ to be a homology sphere we can use Theorem 7.1 to get many examples of aspherical homology manifolds $P_L$ over various rings $R$.

**Example 7.4.** (The fundamental group at infinity, [D1]). Suppose $L$ is a triangulation of a homology $(n-1)$-sphere as a flag complex. Then $P_L$ is an aspherical homology $n$-manifold and hence, $\Gamma_L$ is a $PD^n$-group. For $n \geq 4$, there are homology $(n-1)$-spheres which are not simply connected. If we choose $L$ to be such a homology sphere, then it is proved in [D1] that $\Sigma_L$ is not simply connected at infinity; its fundamental group at infinity (an invariant of $\Gamma_L$) is the inverse limit of $k$-fold free products of $\pi_1(L)$. (This answered Question F16 of [W3] in the negative.)

**Example 7.5.** (Nonintegral Poincaré duality groups, [D3], [D4], [DL]). Suppose that $R = \mathbb{Z}[\frac{1}{m}]$ and that $L$ is a triangulation of a lens space, $S^{2k-1}/(\mathbb{Z}/m)$ as a flag complex. With coefficients in $R$, $L$ has the same homology as does $S^{2k-1}$. Hence, $P_L$ is an aspherical $R$-homology manifold of dimension $2k$ and consequently $\Gamma_L$ is a $PD_R^n$-group, for $n = 2k$. Furthermore, one can show (as in Example 11.9 of [D3] or Example 5.4 of [D4]) that $H^n(\Gamma_L; \mathbb{Z}\Gamma_L) \cong H_c^n(\Sigma_L; \mathbb{Z}) \cong \mathbb{Z}$, while for $1 < i < n$, $H^i(\Gamma_L; \mathbb{Z}\Gamma_L)$ is a countably infinite sum of $H^{i-1}(L; \mathbb{Z})$ which is $m$-torsion whenever $i$ is odd and $i \geq 3$. So, $\Gamma_L$ is not a Poincaré duality group over $\mathbb{Z}$. By taking $L$ to be the suspension of a lens space we obtain a similar example for $n$ odd. Therefore, we have proved the following.

**Theorem 7.6.** *For $R = \mathbb{Z}[\frac{1}{m}]$ and for any $n \geq 4$ there are $PD_R^n$-groups which are not $PD^n$-groups.*

The construction in Example 7.5 suggests the following.

**Question 7.7.** *If $\Gamma$ is a $PD_R^n$-group for a nonzero ring $R$, then is $H^n(\Gamma; \mathbb{Z}\Gamma) = \mathbb{Z}$?*

There is also the following weaker version of this question.

**Question 7.8.** *For any $PD_R^n$ group $\Gamma$, is it true that the image of the orientation character $w_1 : \Gamma \to Aut(R)$ (defined by the action of $\Gamma$ on $H^n(\Gamma; R\Gamma))$ is contained in $\{\pm 1\}$?*

**The Bestvina-Brady example.** There is a similar construction to the one given above involving "right-angled Artin groups" (also called "graph groups"). Given a flag complex $L$ with vertex set $S$, let $Q_L$ be the subcomplex of the torus $T^S$ consisting of all subtori $T^{S(\sigma)}$, where $\sigma$ is a simplex in $L$. Then, as shown in [CD4], [D3], $Q_L$ is aspherical and its fundamental group is the right-angled Artin group defined by the presentation:

$$A_L = \langle S \mid [s, t] = 1, \text{ if } \{s, t\} \text{ spans an edge of } L \rangle.$$

Let $\rho : A_L \to \mathbb{Z}$ be the homomorphism defined by $\rho(s) = 1$, for all $s \in S$ (or $\rho$ could be any other "generic" homomorphism). Denote the kernel of $\rho$ by $H_L$.

The universal covering space $\tilde{Q}_L$ of $Q_L$ is naturally a cubical complex. One can find a $\rho$-equivariant map $f : \tilde{Q}_L \to \mathbb{R}$, such that its restriction to each cube is an affine map. Choose a real number $\lambda$ and let $\tilde{Y}$ denote the level set $f^{-1}(\lambda)$, and $Y = \tilde{Y}/H_L$. Then $Y$ is a finite $CW$-complex of the same dimension as $L$.

Bestvina and Brady prove in [BB] that (i) if $L$ is acyclic, then so is $\tilde{Y}$ and (ii) if $L$ is not simply connected, then $H_L$ is not finitely presented. So, if $L$ is any complex which is acyclic and not simply connected, then $H_L$ is a group of type $FL$ which is not finitely presented.

**Theorem 7.9.** (Bestvina-Brady [BB]). *There are groups of type $FL$ which are not finitely presented.*

*Remark.* In fact, the right angled Artin group $A_L$ is a subgroup of finite index in a right-angled Coxeter group $W$. To see this, for each $s \in S$, introduce new generators $r_s$ and $t_s$ and new relations: $(r_s)^2 = 1 = (t_s)^2$ and whenever $s \neq s'$, $(r_s r_{s'})^2 = 1$, $(r_s t_{s'})^2 = 1$, and $(t_s t_{s'})^2 = 1$ if $\{s, s'\}$ spans an edge in $L$. Let $W$ be the right-angled Coxeter group generated by the $r_s$ and $t_s$ and let $\theta : W \to (\mathbb{Z}/2)^S$ be the epimorphism which, for each $s \in S$, sends both $r_s$ and $t_s$ to the corresponding generator of $(\mathbb{Z}/2)^S$. Then $A_L$ is the kernel of $\theta$. The generators of $A_L$ can be identified with $\{r_s t_s\}_{s \in S}$.

**The reflection group trick.** Example 7.2 can be generalized as follows. Suppose that $X$ is an $n$-dimensional manifold with boundary and that $L$ is a $PL$ triangulation of its boundary. The cellation of $\partial X$ which is dual to $L$ gives X the structure of an orbifold. (For example, think of $X$ as a solid torus with $\partial X$ being cellulated by polygons, three meeting at each vertex.) As in Example 7.2 each point in $X$ has a neighborhood of the form $\mathbb{R}^k/(\mathbb{Z}/2)^k \times \mathbb{R}^{n-k}$ for some $k \leq n$. The orbifoldal fundamental group $G$ of $X$ is an extension of the right-angled Coxeter group $W_L$ defined previously. The epimorphism $G \to W_L \to (\mathbb{Z}/2)^S$ defines an orbifoldal covering space $P$ of $X$ and an action of $(\mathbb{Z}/2)^S$ on $P$ as a group generated by reflections. Since $X$ is an orbifold, $P$ is a closed $n$-manifold. Its fundamental group $\Gamma$ is the kernel of $G \to (\mathbb{Z}/2)^S$. It is not hard to see (cf., Remark 15.9 in [D1]) that if $X$ is aspherical and if $L$ is a flag complex, then $P$ is aspherical. Hence, its fundamental group $\Gamma$ is a $PD^n$-group.

**Example 7.10.** (The reflection group trick, first version). Suppose that $\pi$ is a group of type $F$. "Thicken" the finite complex $B\pi$ into a compact manifold with boundary $X$ (e.g., embed $B\pi$ in Euclidean space and take $X$ to be a regular neighborhood). Let $L$ be a $PL$ triangulation of $\partial X$ as a flag complex. Then $P$ is an aspherical closed manifold with fundamental group $\Gamma$. Since $X$ is a fundamental domain for the $(\mathbb{Z}/2)^S$-action on $P$, the orbit map $P \to X$ is a retraction. Hence, on the level of fundamental groups, there is a retraction from $\Gamma$ onto $\pi$. If $\pi$ has some property which holds for any group that retracts onto it, then $\Gamma$ will be a $PD^n$-group with the same property. In [Me], Mess used this construction to show that $PD^n$-groups need not be residually finite (answering Wall's Question F6 of [W3] in the negative).

**Theorem 7.11.** (Mess [Me]). *There are aspherical closed $n$-manifolds, $n \geq 4$, the fundamental groups of which are not residually finite.*

Next we want to give some more detail about the reflection group trick and at the same time weaken the hypotheses in two ways. First, we will not require $X$ to be a manifold with boundary. Second, we will not require

$X$ to be aspherical, but, rather, only that it has a covering space $\tilde{X}$ which is acyclic. So, suppose we are given the following data:

(i) a finite $CW$-complex $X$,

(ii) a group $\pi$ and an epimorphism $\varphi : \pi_1(X) \to \pi$ so that the induced covering space $\tilde{X} \to X$ is acyclic,

(iii) a subcomplex of $X$ and a triangulation of it by a flag complex $L$.

From this data we will construct a virtually torsion-free group $G$ and an action of it on an acyclic complex $\Omega$ with quotient space $X$.

Let $\tilde{L}$ denote the inverse image of $L$ in $\tilde{X}$ and let $\tilde{S}$ be its vertex set. Let $W_{\tilde{L}}$ be the right-angled Coxeter group defined by $\tilde{L}$. The group $\pi$ acts on $\tilde{S}$ (by deck transformations) and hence, on $W_{\tilde{L}}$ (by automorphisms). Define $G$ to be the semidirect product, $W_{\tilde{L}} \rtimes \pi$. There is a natural epimorphism $\theta : G \to W_L$. Set $\Gamma = \theta^{-1}(\Gamma_L)$. Then $\Gamma$ is torsion-free (since $\Gamma_L$ and $\pi$ are) and of finite index in $G$. Since $G$ is a semidirect product, so is $\Gamma$. In particular, the natural map $\Gamma \to \pi$ is a retraction.

For each $s$ in $\tilde{S}$ let $\tilde{X}_s$ denote the closed star of $s$ in the barycentric subdivision of $\tilde{L}$ and define $\Omega = (W_{\tilde{L}} \times \tilde{X})/ \backsim$ as in the first part of this section. The groups $W_{\tilde{L}}$ and $\pi$ act on $\Omega$. ($W_{\tilde{L}}$ acts by left multiplication on the first factor; $\pi$ acts by automorphisms on the first factor and deck transformations on the second.) These actions fit together to define a $G$-action on $\Omega$. This $G$-action is proper and cellular and $\Omega/G = X$. Let $P = \Omega/\Gamma$. As in [D1] or [DL], it can be shown that $\Omega$ is acyclic. ($\Omega$ is constructed by gluing together copies of $\tilde{X}$, one for each element of $W_{\tilde{L}}$. If we order the elements of $W_{\tilde{L}}$ compatibly with word length, then each copy of $\tilde{X}$ will be glued to the union of the previous ones along a contractible subspace.) The above discussion gives the following result.

**Proposition 7.12.** ([D4]).

(i) $\Gamma$ *is of type* $FL$.

(ii) *If* $\pi_1(X) = \pi$ *and* $\varphi$ *is the identity (so that* $X = B\pi$*), then* $\Omega$ *is contractible and hence,* $\Gamma$ *is of type* $F$.

The next proposition, which is stated in [DH], follows easily from the results of [D1].

**Proposition 7.13.**

(i) *If* $X$ *is a compact* $n$-*manifold with* $L = \partial X$*, then* $\Omega/\Gamma$ *is a closed* *n-manifold.*

(ii) *If $X$ is a compact homology n-manifold with boundary over a ring $R$ and if $L = \partial X$, then $\Omega/\Gamma$ is a closed R-homology n-manifold.*

(iii) *If $(X, L)$ is a Poincaré pair and if $L$ is a homology $(n-1)$-manifold, then $\Omega/\Gamma$ is a Poincaré space.*

**Example 7.14.** (The Bestvina-Brady example continued). This example is similar to Example 7.10. Let $\pi$ be one of the Bestvina-Brady examples associated to a finite acyclic 2-complex. So there is a finite 2-complex $Y$ and an epimorphism $\varphi : \pi_1(Y) \to \pi$ so that the induced covering space $\tilde{Y}$ is acyclic. Thicken $Y$ to $X$, a compact manifold with boundary, ($X$ can be of any dimension $n \geq 4$) and let $L$ be a triangulation of $\partial X$ as a flag complex. Since $\Omega$ is then an acyclic manifold and since $H^*(\Gamma; \mathbb{Z}\Gamma) = H_c^*(\Omega)$, we see that $\Gamma$ is a $PD^n$-group. Since $\pi$ is not finitely presented and since $\Gamma$ retracts onto $\pi$, $\Gamma$ is not finitely presented (Lemma 1.3 in [W1]). So, we have proved the following result (which answers Question F10 of [W3]).

**Theorem 7.15.** ([D4]). *In each dimension $n \geq 4$, there are $PD^n$-groups which cannot be finitely presented.*

## §8. THREE MORE QUESTIONS

Many open questions about aspherical manifolds make sense for Poincaré duality groups. Here are three such.

For any group $\Gamma$ of type $FP$ one can define its Euler characteristic $\chi(\Gamma)$, as in [Br2], [Br3]. If $\Gamma$ is a $PD^n$-group and $n$ is odd, then Poincaré duality implies that $\chi(\Gamma) = 0$. For $n$ even, we have the following.

**Question 8.1.** *If $\Gamma$ is a Poincaré duality group of dimension $2m$, then is*

$$(-1)^m \chi(\Gamma) \geq 0?$$

In the context of nonpositively curved Riemannian manifolds, the conjecture that this be so is due to Hopf. Thurston asked the question for aspherical manifolds. A discussion of this conjecture in the context of nonpositively curved polyhedral manifolds can be found in [CD2].

If $\Gamma$ is orientable and of dimension divisible by 4, then its *signature* $\sigma(\Gamma)$ can be defined as the signature of the middle dimensional cup product pairing. Its absolute value is independent of the choice of orientation class.

Suppose that $B\Gamma$ is a finite complex and that $E\Gamma$ denotes its universal covering space. As in Section 8 of [G2], one can then define $L_2$-cochains on $E\Gamma$ and the corresponding cohomology groups $\ell_2 H^k(E\Gamma)$ (the so-called "reduced" $L_2$-cohomology). When nonzero, these Hilbert spaces will generally be infinite dimensional. However, following [A], there is a well-defined *von Neumann dimension* or $\ell_2$-*Betti number* $h^k(\Gamma)$, which is a nonnegative real number and an invariant of $\ell_2 H^k(E\Gamma)$ with its unitary $\Gamma$-action. Atiyah proved in [A] that $\chi(\Gamma)$ is the alternating sum of the $\ell_2$-Betti numbers. He also showed that when $\Gamma$ is the fundamental group of a closed aspherical $4m$-dimensional manifold then its middle dimensional $L_2$-cohomology is a sum of two subspaces and $\sigma(\Gamma)$ is the difference of their von Neumann dimensions. Singer then observed that Question 8.1 would be answered affirmatively if the following question is answered affirmatively.

**Question 8.2.** *Suppose that $\Gamma$ is a $PD^{2m}$-group of type $F$. Is it true that $\ell_2 H^i(E\Gamma) = 0$ for $i \neq m$? In other words, is $h^i(\Gamma) = 0$ for $i \neq m$?*

Similarly, in dimensions divisible by 4, an affirmative answer to this question implies an affirmative answer to the following stronger version of Question 8.1.

**Question 8.3.** *If $\Gamma$ is an orientable Poincaré duality group of dimension $4m$, then is*
$$\chi(\Gamma) \geq |\sigma(\Gamma)|?$$

Further questions of this type can be found in Section 8 of [G2].

### REFERENCES

[A]   M.F. Atiyah, *Elliptic operators, discrete groups and von Neumann algebras*, Astérisque 32-33, Soc. Math. France (1976), 43–72.

[Be]  M. Bestvina, *The local homology properties of boundaries of groups*, Michigan Math. J. 43 (1996), 123–139.

[BB]  M. Bestvina and N. Brady, *Morse theory and finiteness properties of groups*, Invent. Math. 129 (1997), 445–470.

[BF]  M. Bestvina and M. Feighn, *Stable actions of groups on real-trees*, Invent. Math. 121 (1985), 287–321.

[BM]  M. Bestvina and G. Mess, *The boundary of negatively curved groups*, J. Amer. Math. Soc. 4 (1991), 469–481.

[Bi]  R. Bieri, *Gruppen mit Poincaré Dualität*, Comment. Math. Helv. 47 (1972), 373–396.

[BE1]     R. Bieri and B. Eckmann, *Groups with homological duality generalizing Poincaré duality*, Invent. Math. 20 (1973), 103–124.

[BE2]     ———, *Finiteness properties of duality groups*, Comment. Math. Helv. 49 (1974), 460–478.

[BE3]     ———, *Relative homology and Poincaré duality for groups pairs*, J. of Pure and Applied Algebra 13 (1978), 277–319.

[BS]      A. Borel and J-P. Serre, *Corners and arithmetic groups*, Comment. Math. Helv. 48 (1974), 244–297.

[Br1]     K.S. Brown, *Homological criteria for finiteness*, Comment. Math. Helv. 50 (1975), 129–135.

[Br2]     ———, *Groups of virtually finite dimension*, Homological Group Theory, (ed. C.T.C. Wall), London Math. Soc. Lecture Notes 36, Cambridge Univ. Press, Cambridge (1979), 27–70.

[Br3]     ———, *Cohomology of Groups*, Springer-Verlag, New York (1982).

[BFMW]    J. Bryant, S.C. Ferry, W. Mio, S. Weinberger, *Topology of homology manifolds*, Ann. of Math. 143 (1996), 435–467.

[C]       J.W. Cannon, *The combinatorial Riemann mapping theorem*, Acta Math. 173 (1994), 155–234.

[CD1]     R. Charney and M.W. Davis, *Singular metrics of nonpositive curvature on branched covers of Riemannian manifolds*, Amer. J. Math. 115 (1993), 929–1009.

[CD2]     ———, *The Euler characteristic of a nonpositively curved, piecewise Euclidean manifold*, Pac. J. Math. 171 (1995), 117–137.

[CD3]     ———, *Strict hyperbolization*, Topology 34 (1995), 329–350.

[CD4]     ———, *Finite $K(\pi,1)$s for Artin groups*, Prospect in Topology (ed. F. Quinn), Princeton Univ. Press, Princeton, Annals of Math. Studies 138 (1995), 110–124.

[CaP]     G. Carlsson and E. Pedersen, *Controlled algebra and the Novikov conjectures for K- and L-theory*, Topology 34 (1995), 731–758.

[CoP]     F. Connolly and S. Prassidis, *Groups which act freely on $\mathbb{R}^m \times S^{n-1}$*, Topology 28 (1989), 133–148.

[D1]      M.W. Davis, *Groups generated by reflections and aspherical manifolds not covered by Euclidean space*, Ann. of Math. 117 (1983), 293–325.

[D2]      ———, *The homology of a space on which a reflection group acts*, Duke Math. J. 55 (1987), 97–104, Erratum, 56 (1988), 221.

[D3]      ———, *Nonpositive curvature and reflection groups*, Handbook of Geometric Topology, (eds. R.J. Daverman and R.B. Sher), Elsevier, to appear.

[D4]      ———, *The cohomology of a Coxeter group with group ring coefficients*, Duke Math. J. 91 (1998), 297–314.

[DH]      M.W. Davis and J.-C. Hausmann, *Aspherical manifolds without smooth or PL structure*, Algebraic Topology, Lecture Notes in Math., vol. 1370, Springer-Verlag, New York, 1989.

[DJ]      M.W. Davis and T. Januszkiewicz, *Hyperbolization of polyhedra*, J. Differential Geometry 34 (1991), 347–388.

[DJS]     M.W. Davis, T. Januszkiewicz and R. Scott, *Nonpositive curvature of blow-ups*, Selecta Math. (N.S.) 4 (1998), 491–547.

[DD]      W. Dicks and M.J. Dunwoody, *Groups acting on graphs*, Cambridge Univ. Press, Cambridge, 1989.

[DL]      W. Dicks and I.J. Leary, *On subgroups of Coxeter groups*, Proc. London Math. Soc. Durham Symposium on Geometry and Cohomology in Group Theory, (eds. P. Kropholler, G. Niblo and R. Stöhr), London Math. Soc. Lecture Notes 252 (1998), 124–160, Cambridge Univ. Press.

[E]       B. Eckmann, *Poincaré duality groups of dimension two are surface groups*, Combinatorial Group Theory and Topology (eds. S.M. Gersten and J. Stallings), Annals of Math. Studies 111 (1987), 35–51, Princeton Univ. Press, Princeton.

[EL]      B. Eckmann and P.A. Linnell, *Poincaré duality groups of dimension two*, Comment. Math. Helv. 58 (1983), 111–114.

[EM]      B. Eckmann and H. Müller, *Poincaré duality groups of dimension two*, Comment. Math. Helv. 55 (1980), 510–520.

[EG]      S. Eilenberg and T. Ganea, *On the Lusternik-Schnirelmann category of abstract groups*, Ann. of Math. 65 (1957), 517–518.

[F]       F.T. Farrell, *Poincaré duality and groups of type (FP)*, Comment. Math. Helv. 50 (1975), 187–195.

[FJ]      F.T. Farrell and L.E. Jones, *Classical Aspherical Manifolds*, CBMS Regional Conference Series in Mathematics 75, Amer. Math. Soc., Providence, RI, 1990.

[FW]      S. Ferry and S. Weinberger, *A coarse approach to the Novikov Conjecture*, Novikov Conjectures, Index Theorems and Rigidity, Vol. 1 (eds. S. Ferry, A. Ranicki, J.Rosenberg), London Math. Soc. Lecture Notes 226, 147-163, Cambridge Univ. Press, Cambridge, 1995.

[Go]      D. Gottlieb, *Poincaré duality and fibrations*, Proc. of Amer. Math. Soc. 76 (1979), 148-150.

[G1]      M. Gromov, *Hyperbolic groups*, Essays in Group Theory (ed. S. Gersten), M.S.R.I. Publ. 8, Springer-Verlag, New York, 1987, pp. 75–264.

[G2]      ———, *Asymptotic invariants of infinite groups*, Geometric Group Theory, vol. 2 (eds. G. Niblo and M. Roller), London Math. Soc. Lecture Notes 182, Cambridge Univ. Press, Cambridge, 1993.

[GT]      M. Gromov and W. Thurston, *Pinching constants for hyperbolic manifolds*, Invent. Math. 89 (1987), 1–12.

[Ha]     J. Harer, *The virtual cohomological dimension of the mapping class group of an orientable surface*, Invent. Math. 84 (1986), 157–176.

[H]      H. Hopf, *Enden offener Räume und unendliche diskontinuierliche Gruppen*, Comment. Math. Helv. 16 (1943), 81–100.

[Hu]     W. Hurewicz, *Beiträge zur Topologie der Deformationen. IV. Asphärische Räume*, Nederl. Akad. Wetensch. Proc. 39 (1936), 215–224.

[KR1]    P.H. Kropholler and M.A. Roller, *Splittings of Poincaré duality groups I*, Math. Z. 197 (1988), 421–428.

[KR2]    _____, *Splittings of Poincaré duality groups II*, J. London Math. Soc. 38 (1988), 410–420.

[KR3]    _____, *Splittings of Poincaré duality groups III*, J. London Math. Soc. 39 (1989), 271–284.

[JW]     F.E.A. Johnson and C.T.C. Wall, *On groups satisfying Poincaré duality*, Ann. of Math. 96 (1972), 592–598.

[Me]     G. Mess, *Examples of Poincaré duality groups*, Proc. Amer. Math. Soc. 110 (1990), 1145-1146.

[Mo]     G. Moussong, *Hyperbolic Coxeter groups*, Ph.D. thesis, The Ohio State University, 1988.

[M]      H. Müller, *Decomposition theorems for group pairs*, Math. Z. 176 (1981), 223–246.

[P]      F. Paulin, *Outer automorphisms of hyperbolic groups and small actions on R-trees*, Arboreal Group Theory (ed. R.C. Alperin), MSRI publication 19, Springer-Verlag, New York, 331–344.

[Q]      F. Quinn, *Surgery on Poincaré and normal spaces*, Bull. Amer. Math. Soc. 78 (1972), 262–267.

[R]      A. Ranicki, *Algebraic L-theory and topological manifolds*, Cambridge Univ. Press, Cambridge, 1992.

[Ri]     E. Rips, *Group actions on $\mathbb{R}$-trees*, in preparation.

[SW]     P. Scott and C.T.C. Wall, *Topological methods in group theory*, Homological Group Theory (ed. C.T.C. Wall), London Math. Soc. Lecture Notes 36, 137–204, Cambridge Univ. Press, Cambridge, 1979.

[Se1]    J-P. Serre, *Cohomologie des groupes discrets*, Prospects in Mathematics, Annals of Math. Studies 70 (1971), 77–169, Princeton Univ. Press, Princeton.

[Se2]    _____, *Trees*, Springer-Verlag, New York, 1980.

[St]     J.R. Stallings, *On torsion-free groups with infinitely many ends*, Ann. of Math. 88 (1968), 312–334.

[Str]    R. Strebel, *A remark on subgroups of infinite index in Poincaré duality groups*, Comment. Math. Helv. 52 (1977), 317–324.

[T]     C.B. Thomas, *3-manifolds and PD(3)-groups*, Novikov Conjectures, Index Theorems and Rigidity, Vol. 1 (eds. S. Ferry, A. Ranicki, J. Rosenberg), London Math. Soc. Lecture Notes 226, 301-308, Cambridge Univ. Press, Cambridge, 1995.

[W1]    C.T.C. Wall, *Finiteness conditions for CW-complexes*, Ann. of Math. 81 (1965), 56-59.

[W2]    _____, *Poincaré complexes I*, Ann. of Math. 86 (1967), 213-245.

[W3]    C.T.C. Wall (ed), *List of problems*, Homological Group Theory, London Math. Soc. Lecture Notes 36, 369-394, Cambridge Univ. Press, Cambridge, 1979.

DEPARTMENT OF MATHEMATICS
OHIO STATE UNIVERSITY
COLUMBUS, OH 43210-1101

*E-mail address*: mdavis@math.ohio-state.edu

# Manifold aspects of the Novikov Conjecture

James F. Davis*

Let $L_M \in H^{4*}(M;\mathbb{Q})$ be the Hirzebruch $L$-class of an oriented manifold $M$. Let $B\pi$ (or $K(\pi,1)$) denote any aspherical space with fundamental group $\pi$. (A space is aspherical if it has a contractible universal cover.) In 1970 Novikov made the following conjecture.

**Novikov Conjecture.** *Let* $h : M' \to M$ *be an orientation-preserving homotopy equivalence between closed, oriented manifolds.*[1] *For any discrete group* $\pi$ *and any map* $f : M \to B\pi$,

$$f_* \circ h_*(L_{M'} \cap [M']) = f_*(L_M \cap [M]) \in H_*(B\pi;\mathbb{Q})$$

Many surveys have been written on the Novikov Conjecture. The goal here is to give an old-fashioned point of view, and emphasize connections with characteristic classes and the topology of manifolds. For more on the topology of manifolds and the Novikov Conjecture see [58], [47], [17]. This article ignores completely connections with $C^*$-algebras (see the articles of Mishchenko, Kasparov, and Rosenberg in [15]), applications of the Novikov conjecture (see [58],[9]), and most sadly, the beautiful work and mathematical ideas uncovered in proving the Novikov Conjecture in special cases (see [14]).

The level of exposition in this survey starts at the level of a reader of Milnor-Stasheff's book *Characteristic Classes*, but by the end demands more topological prerequisites. Here is a table of contents:

*Partially supported by the NSF. This survey is based on lectures given in Mainz, Germany in the Fall of 1993. The author wishes to thank the seminar participants as well as Paul Kirk, Chuck McGibbon, and Shmuel Weinberger for clarifying conversations.

[1] Does this refer to smooth, PL, or topological manifolds? Well, here it doesn't really matter. If the Novikov Conjecture is true for all smooth manifolds mapping to $B\pi$, then it is true for all PL and topological manifolds mapping to $B\pi$. However, the definition of $L$-classes for topological manifolds depends on topological transversality [25], which is orders of magnitude more difficult than transversality for smooth or PL-manifolds. The proper category of manifolds will be a problem of exposition throughout this survey.

# 1    Hirzebruch $L$-classes

The *signature* $\sigma(M)$ of a closed, oriented manifold $M$ of dimension $4k$ is the signature of its intersection form

$$\phi_M : H^{2k} M \times H^{2k} M \to \mathbb{Z}$$
$$(\alpha, \beta) \mapsto \langle \alpha \cup \beta, [M] \rangle$$

For a manifold whose dimension is not divisible by 4, we define $\sigma(M) = 0$. The key property of the signature is its bordism invariance: $\sigma(\partial W) = 0$ where $W$ is a compact, oriented manifold. The signature of a manifold[2] can be used to define the Hirzebruch $L$-class, whose main properties are given by the theorem below.

**Theorem 1.1.** *Associated to a linear, PL, or topological $\mathbb{R}^n$-bundle[3] $\xi$ are characteristic classes*

$$L_i(\xi) \in H^{4i}(B(\xi); \mathbb{Q}) \qquad i = 1, 2, 3, \ldots$$

*satisfying*

1. *$L_i = 0$ for a trivial bundle.*

2. *For a closed, oriented $4k$-manifold $M$,*

$$\langle L_k(\tau_m), [M] \rangle = \sigma(M)$$

3. *Properties 1. and 2. are axioms characterizing the $L$-classes.*

4. *Let $L = 1 + L_1 + L_2 + L_3 + \ldots$ be the total $L$-class. Then*

$$L(\xi \oplus \eta) = L(\xi)L(\eta).$$

---

[2] Gromov [17] says that the signature "is not just 'an invariant' but *the invariant* which can be matched in beauty and power only by the Euler characteristic."

[3] We require that these bundles have 0-sections, i.e. the structure group preserves the origin. We also assume that the base spaces have the homotopy type of a $CW$-complex.

5. Let $B = BO, BPL$, or $BTOP$ be the classifying spaces for stable linear, PL, or topological Euclidean bundles, respectively. Then

$$H^*(B; \mathbb{Q}) = \mathbb{Q}[L_1, L_2, L_3, \ldots]$$

where $L_i$ denotes the $i$-th L-class of the universal bundle.

We write $L_M = L(\tau_M)$ and call property 2. the *Hirzebruch signature formula*.

Properties 1.-4. are formal consequences of transversality and Serre's theorem on the finiteness of stable homotopy groups; this is due to Thom-Milnor [32], Kahn [22], and Rochlin-Svarc. Property 5 is not formal. One checks that $L_1, L_2, L_3, \ldots$ are algebraically independent by applying the Hirzebruch signature formula to products of complex projective spaces and shows that $H^*(BO; \mathbb{Q})$ is a polynomial ring with a generator in every fourth dimension by computing $H^*(BO(n); \mathbb{Q})$ inductively. That $BO, BPL$ and $BTOP$ have isomorphic rational cohomology is indicated by Novikov's result [38] that two homeomorphic smooth manifolds have the same rational L-class, but also depends on the result of Kervaire-Milnor [24] of the finiteness of exotic spheres and the topological transversality of Kirby and Siebenmann [25].

We indicate briefly how the properties above can be used to define the L-classes, because this provides some motivation for the Novikov Conjecture. By approximating a $CW$-complex by its finite skeleta, a finite complex by its regular neighborhood, and a compact manifold by the orientation double cover of its double, it suffices to define $L_M$ for a closed, oriented $n$-manifold. The idea is that this is determined by signatures of submanifolds with trivial normal bundle. Given a map $f : M \to S^{n-4i}$, the meaning of $\sigma(f^{-1}(*))$ is to perturb $f$ so that it is transverse to $* \in S^{n-4i}$ and take the signature of the inverse image. This is independent of the perturbation by cobordism invariance of the signature. Given such a map, one can show

$$\langle L_M \cup f^* u, [M] \rangle = \sigma(f^{-1}(*))$$

where $u \in H^{n-4i}(S^{n-4i})$ is a generator. Using Serre's result that

$$\pi^{n-4i}(M) \otimes \mathbb{Q} \to H^{n-4i}(M; \mathbb{Q})$$

is an isomorphism when $4i < (n-1)/2$, one sees that the above formula defines the $4i$-dimensional component of $L_M$ when $4i$ is small. To define the high-dimensional components of $L_M$ one uses the low-dimensional components of $L_{M \times S^m}$ for $m$ large.

It is more typical to define Pontryagin classes for linear vector bundles, then define the L-classes of linear bundles as polynomials in the Pontryagin classes, then prove the Hirzebruch signature theorem, then define the L-classes for PL and topological bundles (as above), and finally define the

Pontryagin classes as polynomials in the $L$-classes. But $L$-classes, which are more closely connected with the topology of manifolds, can be defined without mentioning Pontryagin classes. The Pontryagin classes are more closely tied with the group theory of $SO$, and arise in Chern-Weil theory and the Atiyah-Singer index theorem. They are useful for computations, and their integrality can give many subtle properties of smooth manifolds (e.g. the existence of exotic spheres).

We conclude this section with some remarks on the statement of the Novikov Conjecture. Given a map $f : M \to B\pi$, the 0-dimensional component of $f_*(L_M \cap [M])$ is just the signature of $M$, and its homotopy invariance provides some justification of the Novikov Conjecture. If one proves the Novikov Conjecture for a map $f : M \to B\pi_1 M$ inducing an isomorphism on the fundamental group, then one can deduce the Novikov Conjecture for all maps $M \to B\pi$. But the more general statement is useful, because it may be the case that one can prove it for $\pi$ but not for the fundamental group.

**Definition 1.2.** For $f : M \to B\pi$ and for $u \in H^*(B\pi; \mathbb{Q})$, define the *higher signature*

$$\sigma_u(M, f) = \langle L_M \cup f^*u, [M] \rangle \in \mathbb{Q}$$

When $u = 1 \in H^0$, the higher signature is just the signature of $M$. The higher signature can often be given a geometric interpretation. If the Poincaré dual of $f^*u$ can be represented by a submanifold with trivial normal bundle, the higher signature is the signature of that submanifold. Better yet, if $B\pi$ is a closed, oriented manifold and the Poincaré dual of $u$ in $B\pi$ can be represented by a submanifold $K$ with trivial normal bundle, the higher signature is the signature of the transverse inverse image of $K$. The Novikov Conjecture implies that all such signatures are homotopy invariant.

Henceforth, we will assume all homotopy equivalences between oriented manifolds are orientation-preserving and will often leave out mention of the homotopy equivalence. With this convention we give an equivalent formulation of the Novikov Conjecture.

**Novikov Conjecture.** *For a closed, oriented manifold $M$, for a discrete group $\pi$, for any $u \in H^*(B\pi)$, for any map $f : M \to B\pi$, the higher signature $\sigma_u(M, f)$ is an invariant of the oriented homotopy type of $M$.*

# 2  Novikov Conjecture for $\pi = \mathbb{Z}$

We wish to outline the proof (cf. [17]) of the following theorem of Novikov [37].

**Theorem 2.1.** *Let $M$ be a closed, oriented manifold with a map $f : M \to S^1$. Then $f_*(L_M \cap [M]) \in H_*(S^1; \mathbb{Q})$ is homotopy invariant.*

Since $\langle 1, f_*(L_M \cap [M]) \rangle = \sigma(M)$, the degree-zero component of $f_*(L_M \cap [M])$ is homotopy invariant. Let $u \in H^1(S^1)$ be a generator; it suffices to show $\langle u, f_*(L_M \cap [M]) \rangle$ is a homotopy invariant, where dim $M = 4k + 1$. Let $K^{4k} = f^{-1}(*)$ be the transverse inverse image of a point. (Note: any closed, oriented, codimension 1 submanifold of $M$ arises as $f^{-1}(*)$ for some map $f$.) Let $i : K \hookrightarrow M$ be the inclusion. Then the Poincaré dual of $i_*[K]$ is $f^*u$, since the Poincaré dual of an embedded submanifold is the image of the Thom class of its normal bundle. Thus we need to show the homotopy invariance of

$$
\begin{aligned}
\langle u, f_*(L_M \cap [M]) \rangle &= \langle f^*u, L_M \cap [M] \rangle \\
&= \langle L_M, f^*u \cap [M] \rangle \\
&= \langle L_M, i_*[K] \rangle \\
&= \langle i^*L_M, [K] \rangle \\
&= \langle L_K, [K] \rangle \\
&= \sigma(K).
\end{aligned}
$$

**Definition 2.2.** If $K^{4k}$ is a closed, oriented manifold which is a subspace $i : K \hookrightarrow X$ of a topological space $X$, let

$$
\phi_{K \subset X} : H^{2k}(X; \mathbb{Q}) \times H^{2k}(X; \mathbb{Q}) \to \mathbb{Q}
$$

be the symmetric bilinear form defined by

$$
\phi_{K \subset X}(a, b) = \langle a \cup b, i_*[K] \rangle = \langle i^*a \cup i^*b, [K] \rangle .
$$

If $X = K$, we write $\phi_K$.

**Remark 2.3.**    1. Note that

$$
\sigma(\phi_{K \subset X}) = \sigma(\phi_K : i^*H^{2k}(X; \mathbb{Q}) \times i^*H^{2k}(X; \mathbb{Q}) \to \mathbb{Q}) ,
$$

so that the signature is defined even when $X$ is not compact.

2. If $h : X' \to X$ is a proper, orientation-preserving homotopy equivalence between manifolds of the same dimension, and $K'$ is the transverse inverse image of a closed, oriented submanifold $K$ of $X$, then $h_*i'_*[K'] = i_*[K]$, so

$$
\sigma(\phi_{K' \subset X'}) = \sigma(\phi_{K \subset X}) .
$$

The Novikov conjecture for $\pi = \mathbb{Z}$ follows from:

**Theorem 2.4.** *Let $M^{4k+1}$ be a closed, oriented manifold with a map $f :$ $M \to S^1$. Let $K = f^{-1}(*)$ be the transverse inverse image of a point. Then the signature of $K$ is homotopy invariant. In fact, $\sigma(K) = \sigma(\phi_{K_0 \subset M_\infty})$ where $M_\infty \to M$ is the infinite cyclic cover induced by the universal cover of the circle and $K_0 = \tilde{f}^{-1}(\tilde{*}) \subset M_\infty$ is a lift of $K$.*

There are two key lemmas in the proof.

**Lemma 2.5.** *Let $K^{4k}$ be a closed, oriented manifold which is a subspace $i : K \hookrightarrow X$ of a CW-complex $X$. Suppose $X$ is filtered by subcomplexes*

$$X_0 \subset X_1 \subset X_2 \subset \cdots \subset X_n \subset \cdots \subset X = \bigcup_n X_n .$$

*Then there exists an $N$ so that for all $n \geq N$,*

$$\sigma(\phi_{K \subset X}) = \sigma(\phi_{K \subset X_n}).$$

*Proof.* Since $K$ is compact, $K \subset X_n$ for $n$ sufficiently large; so without loss of generality $i_n : K \hookrightarrow X_n$ for all $n$. Let $i_{n*}$ and $i_*$ ($i_n{}^*$ and $i^*$) be the maps induced by $i_n$ and $i$ on rational (co)-homology in dimension $2k$. The surjections $J_{n,n+1} : \operatorname{im} i_{n*} \to \operatorname{im} i_{n+1*}$ must be isomorphisms for $n$ sufficiently large (say for $n \geq N$) since $H_{2k}(K; \mathbb{Q})$ is finite dimensional. We claim

$$J_n : \operatorname{im} i_{n*} \to \operatorname{im} i_*$$

is also an isomorphism for $n \geq N$. Surjectivity is clear. To see it is injective, suppose $[\alpha] \in H_{2k}(K; \mathbb{Q})$ and $i_*[\alpha] = 0$. On the singular chain level $\alpha = \partial\beta$, where $\beta \in S_{2k+1}(X; \mathbb{Q}) = \cup_n S_{2k+1}(X_n; \mathbb{Q})$. Thus $i_{n*}[\alpha] = 0$ for $n$ sufficiently large, and hence for $n \geq N$ since the maps $J_{n,n+1}$ are isomorphisms.

Dualizing by applying $\operatorname{Hom}( \ ; \mathbb{Q})$ we see that

$$\operatorname{im} i^* \to \operatorname{im} i_n{}^*$$

is also an isomorphism for $n \geq N$, and hence

$$
\begin{aligned}
\sigma(\phi_{K \subset X_n}) &= \sigma(\operatorname{im} i_n{}^* \times \operatorname{im} i_n{}^* \to \mathbb{Q}) \\
&= \sigma(\operatorname{im} i^* \times \operatorname{im} i^* \to \mathbb{Q}) \\
&= \sigma(\phi_{K \subset X})
\end{aligned}
$$

$\square$

**Lemma 2.6.** *Let $X^{4k+1}$ be a manifold with compact boundary. (Note $X$ may be non-compact.)*

1. *Let $L = (\operatorname{im} i^* : H^{2k}(X; \mathbb{Q}) \to H^{2k}(\partial X; \mathbb{Q}))$ and*

$$L^\perp = \{a \in H^{2k}(\partial X; \mathbb{Q}) \mid \phi_{\partial X}(a, L) = 0\} .$$

   *Then $L^\perp \subset L$.*

2. *$\sigma(\phi_{\partial X \subset X}) = \sigma(\partial X)$.*

**Remark 2.7.** In the case of $X$ compact, part 2. above gives the cobordism invariance of the signature. Indeed

$$
\begin{aligned}
\phi_{\partial X \subset X}(a, b) &= \langle i^* a \cup i^* b, [\partial X] \rangle \\
&= \langle i^* a, i^* b \cap \partial_* [X] \rangle \\
&= \langle i^* a, \partial_* (b \cap [X]) \rangle \\
&= \langle \delta^* i^* a, b \cap [X] \rangle \\
&= 0
\end{aligned}
$$

*Proof of Lemma.* 1. Poincaré-Lefschetz duality gives a non-singular pairing

$$\phi_X : H_c^{2k+1}(X, \partial X; \mathbb{Q}) \times H^{2k}(X; \mathbb{Q}) \to \mathbb{Q}$$

Let $\delta^*$ be the coboundary map and $\delta_c^*$ be the coboundary with compact supports. Now $0 = \phi_{\partial X}(L^\perp, i^* H^{2k}(X; \mathbb{Q})) = \phi_X(\delta_c^* L^\perp, H^{2k}(X; \mathbb{Q}))$, so $\delta_c^* L^\perp = 0$, and hence $L^\perp \subset \ker \delta_c^* \subset \ker \delta^* = L$.

2. By 1., $\phi_{\partial X}(L^\perp, L^\perp) = 0$. Choose a basis $e_1, \ldots e_n$ for $L^\perp$, then find $f_1, \ldots f_n \in H^{2k}(\partial X; \mathbb{Q})$ so that $\phi_{\partial X}(e_i, f_j) = \delta_{i,j}$ and $\phi_{\partial X}(f_i, f_j) = 0$. The form $\phi_{\partial X}$ restricted to $H(L^\perp) = \operatorname{Span}(e_1, \ldots, e_n, f_1, \ldots, f_n)$ is non-singular and has zero signature. There is an orthogonal direct sum of vector spaces

$$H^{2k}(\partial X; \mathbb{Q}) = H(L^\perp) \oplus H(L^\perp)^\perp$$

The inclusion of $L$ in $H^{2k}(\partial X; \mathbb{Q})$ followed by the projection onto $H(L^\perp)^\perp$ induces an isometry between the form restricted to $H(L^\perp)^\perp$ and

$$\phi_{\partial X} : L/L^\perp \times L/L^\perp \to \mathbb{Q}$$

Thus

$$\sigma(\partial X) = \sigma(L/L^\perp \times L/L^\perp \to \mathbb{Q}) = \sigma(L \times L \to \mathbb{Q}) = \sigma(\phi_{\partial X \subset X}).$$

$\square$

*Proof of Theorem.* One has the pull-back diagram

$$
\begin{array}{ccc}
\widetilde{M} & \xrightarrow{\tilde{f}} & \mathbb{R} \\
\downarrow & & \downarrow \exp \\
M & \xrightarrow{f} & S^1
\end{array}
$$

Suppose $f$ is transverse to $* = 1 \in S^1 \subset \mathbb{C}$. Let $K = f^{-1}(*)$ and, abusing notation slightly, also let $K$ denote $\tilde{f}^{-1}(0) \subset M_\infty$. Let $M_+ = \tilde{f}^{-1}[0, \infty)$ and $X_n = \tilde{f}^{-1}[-n, n]$. Let $t : M_\infty \to M_\infty$ be the generator of the deck transformations corresponding to $x \mapsto x + 1$ in $\mathbb{R}$. Then

$$
\begin{aligned}
\sigma(K) &= \sigma(\phi_{K \subset M_+}) && \text{(by Lemma 2.6)} \\
&= \sigma(\phi_{K \subset t^N X_N}) && \text{(for } N \gg 0 \text{ by Lemma 2.5)} \\
&= \sigma(\phi_{t^{-N} K \subset X_N}) && (t^N \text{ is a homeomorphism)} \\
&= \sigma(\phi_{K \subset X_N}) && \text{(since } K \text{ and } t^{-N} K \text{ are bordant in } X_N) \\
&= \sigma(\phi_{K \subset M_\infty}) && \text{(for } N \gg 0 \text{ by Lemma 2.5 again)}
\end{aligned}
$$

$\square$

This beautiful proof of Novikov should be useful elsewhere in geometric topology, perhaps in the study of signatures of knots. The modern proof of the Novikov Conjecture is by computing the $L$-theory of $\mathbb{Z}[\mathbb{Z}]$ as in [49]; further techniques are indicated by Remark 4.2 in [47]. Later Farrell-Hsiang [19], [12] and Novikov [39] showed that the Novikov Conjecture for $\pi = \mathbb{Z}^n$ is true, although additional techniques are needed to prove it. Lusztig [27] gave an analytic proof of this result.

# 3   Topological rigidity

Given a homotopy equivalence

$$
h : M'^n \to M^n
$$

between two closed manifolds, one could ask if $h$ is homotopic to a homeomorphism or a diffeomorphism. A naive conjecture would be that the answer is always yes, after all $M'$ and $M$ have the same global topology (since $h$ is a homotopy equivalence) and the same local topology (since they are both locally Euclidean). At any rate, any invariant which answers the question in the negative must be subtle indeed.

Here is one idea for attacking this question. Let $K^k \subset M^n$ be a closed submanifold and perturb $h$ so that it is transverse to $K$. Then we have

$$
\begin{array}{ccc}
M' & \xrightarrow{\;\;h\;\;} & M \\
\uparrow & & \uparrow \\
h^{-1}K & \longrightarrow & K
\end{array}
$$

If $k$ is divisible by 4 and $M$ and $K$ are oriented, then the difference of signatures

$$\sigma(h, K) = \sigma(h^{-1}K) - \sigma(K) \in \mathbb{Z}$$

is an invariant of the homotopy class of $h$ which vanishes if $h$ is homotopic to a homeomorphism. If $n - k = 1$, then $\sigma(h, K)$ is always zero; this follows from the Novikov conjecture for $\pi = \mathbb{Z}$.

**Example 3.1.** This is basically taken from [32, Section 20]. Other examples are given in [47] and [17]. Let $\mathbb{R}^5 \hookrightarrow E \to S^4$ be the 5-plane bundle given by the Whitney sum of the quaternionic Hopf bundle and a trivial line bundle. Then it represents the generator of $\pi_4(BSO(5)) = \pi_3(SO(5)) \cong \mathbb{Z}$ and $p_1(E) = 2u$, where $u \in H^4(S^4)$ is a generator. Let $\mathbb{R}^5 \hookrightarrow E' \to S^4$ be a bundle with $p_1(E') = 48u$, say the pullback of $E$ over a degree 24 map $S^4 \to S^4$. Then $E'$ is fiber homotopically trivial (by using the $J$-homomorphism $J : \pi_3(SO(5)) \to \pi_8(S^5) \cong \mathbb{Z}/24$, see [20] for details). Thus there is a homotopy equivalence

$$h : S(E') \to S^4 \times S^4$$

commuting with the bundle map to $S^4$ in the domain and projecting on the second factor in the target. It is left as an exercise to show $\sigma(h, pt \times S^4) = 16$ and hence $h$ is not homotopic to a homeomorphism. It would be interesting to construct $h$ and $h^{-1}(pt \times S^4)$ explicitly (maybe in terms of algebraic varieties and the $K3$ surface).

The Novikov conjecture for a group $\pi$ implies that if $h : M' \to M$ is a homotopy equivalence of closed, oriented manifolds and if $K \subset M$ is a closed, oriented submanifold with trivial normal bundle which is Poincaré dual to $f^*\rho$ for some $f : M \to B\pi$ and some $\rho \in H^*(B\pi)$, then

$$\sigma(h, K) = 0$$

In particular if $M$ is aspherical, then $M = B\pi$ and $\sigma(h, K) = 0$ for all such submanifolds of $M$. Note that any two aspherical manifolds with isomorphic fundamental group are homotopy equivalent. In this case there is a conjecture much stronger that the Novikov conjecture.

**Borel Conjecture.** [4] *Any homotopy equivalence between closed aspherical manifolds is homotopic to a homeomorphism.*

See [1] for more on the Borel Conjecture and [9] for applications to the topology of 4-manifolds.

We conclude this section with a historical discussion of the distinction between closed manifolds being homotopy equivalent, homeomorphic, PL-homeomorphic, or diffeomorphic. It is classical that all the notions coincide for 2-manifolds. For 3-manifolds, homeomorphic is equivalent to diffeomorphic ([33], [35]). Poincaré conjectured that a closed manifold homotopy equivalent to $S^3$ is homeomorphic to $S^3$; this question is still open. In 1935, Reidemeister, Franz, and DeRham showed that there are homotopy equivalent 3-dimensional lens spaces $L(7; 1, 1)$ and $L(7; 1, 2)$ which are not simple homotopy equivalent (see [31], [7]), and hence not diffeomorphic. However, this was behavior based on the algebraic $K$-theory of the fundamental group. Hurewicz asked whether simply-connected homotopy equivalent closed manifolds are homeomorphic. Milnor [29] constructed exotic 7-spheres, smooth manifolds which are (PL-) homeomorphic but not diffeomorphic to the standard 7-sphere. Thom, Tamura, and Shimada gave examples in the spirit of Example 3.1, which together with Novikov's proof of the topological invariance of the rational $L$-classes showed that simply-connected homotopy equivalent manifolds need not be homeomorphic. Finally, Kirby and Siebenmann [25] gave examples of homeomorphic PL-manifolds which are not PL-homeomorphic. Thus all phenomena are realized. We will return frequently to the question of when homotopy equivalent manifolds are homeomorphic.

# 4   Oriented bordism

Let $\Omega_n$ be the oriented bordism group of smooth $n$-manifolds. $\Omega_*$ is a graded ring under disjoint union and cartesian product. Thom [53] combined his own foundational work in differential topology and with then

---

[4] A. Borel conjectured this in 1953, long before the Novikov conjecture. The motivation was not from geometric topology, but rather from rigidity theory for discrete subgroups of Lie groups.

The choice of category is important; the connected sum of an $n$-torus and an exotic sphere need not be diffeomorphic to the $n$-torus.

recent work of Serre [48] to show[5]

$$\Omega_* \otimes \mathbb{Q} = \mathbb{Q}[\mathbb{C}P^2, \mathbb{C}P^4, \mathbb{C}P^6, \mathbb{C}P^8, \dots]$$

A map $f : M \to X$ where $M$ is a closed, oriented $n$-manifold and $X$ is a topological space is called *a singular $n$-manifold over $X$*. Let $\Omega_n(X)$ be the oriented bordism group of singular $n$-manifolds over $X$, see [8]. Then $\Omega_*(X)$ is a graded module over $\Omega_*$. The map $H : \Omega_n X \otimes \mathbb{Q} \to H_n(X; \mathbb{Q})$ defined by $H(f : M \to X) = f_*[M]$ is onto. $\Omega_* X \otimes \mathbb{Q}$ is a free module over $\Omega_* \otimes \mathbb{Q}$ with a basis given by any set of singular manifolds $\{(f_\alpha : M_\alpha \to X) \otimes 1\}$ such that $\{(f_\alpha)_*[M_\alpha]\}$ is a basis of $H_*(M; \mathbb{Q})$. The *Conner-Floyd map* is

$$\Omega_* X \otimes_{\Omega_*} \mathbb{Q} \quad \to \quad H_*(X; \mathbb{Q})$$
$$(f : M \to X) \otimes r \quad \to \quad r f_*(L_M \cap [M])$$

Here $\mathbb{Q}$ is an $\Omega_*$-module via the signature homomorphism. The domain of the Conner-Floyd map is $\mathbb{Z}_4$-graded, and the Conner-Floyd map is a $\mathbb{Z}_4$-graded isomorphism, i.e. for $i = 0, 1, 2$, or 3

$$\Omega_{4*+i} X \otimes_{\Omega_*} \mathbb{Q} \cong \bigoplus_{n \equiv i(4)} H_n(X; \mathbb{Q}) .$$

All of these statements are easy consequences of the material in the previous footnote.

**Definition 4.1.** Let $M$ be a closed, smooth manifold. An element of the *structure set* $\mathbb{S}(M)$ *of homotopy smoothings* is represented by a homotopy equivalence $h : M' \to M$ where $M'$ is also a closed, smooth manifold. Another such homotopy equivalence $g : M'' \to M$ represents the same element of the structure set if there is an $h$-cobordism $W$ from $M'$ to $M''$ and a map $H : W \to M$ which restricts to $h$ and $g$ on the boundary.

---

[5]The modern point of view on Thom's work is:

$$\Omega_* \otimes \mathbb{Q} \cong \pi_*(\mathbf{MSO}) \otimes \mathbb{Q} \cong H_*(\mathbf{MSO}; \mathbb{Q}) \cong H_*(BSO; \mathbb{Q}).$$

Here **MSO** is the oriented bordism Thom spectrum, whose $n$-th space is the Thom space of the universal $\mathbb{R}^n$-bundle over $BSO(n)$. The isomorphism $\Omega_* \cong \pi_*(\mathbf{MSO})$ follows from transversality and is called the Pontryagin-Thom construction. From Serre's computations, the Hurewicz map for an Eilenberg-MacLane spectrum is an rational isomorphism in all dimensions. It follows that the Hurewicz map is a rational isomorphism for all spectra, and that the rational localization of any spectrum is a wedge of Eilenberg-MacLane spectrum. The Thom isomorphism theorem shows $H_*(\mathbf{MSO}) \cong H_*(BSO)$. After tracing through the above isomorphisms one obtains a non-singular pairing

$$H^*(BSO; \mathbb{Q}) \times (\Omega_* \otimes \mathbb{Q}) \to \mathbb{Q} \qquad (\alpha, M) \mapsto \alpha[M]$$

A computation of $L$-numbers of even-dimensional complex projective spaces shows that they freely generate $\Omega_* \otimes \mathbb{Q}$ as a polynomial algebra. References on bordism theory include [8], [50], and [28].

In particular, if $h : M' \to M$ and $g : M'' \to M$ are homotopy equivalences, then $[h] = [g] \in \mathbb{S}(M)$ if there is a diffeomorphism $f : M' \to M''$ so that $g \simeq h \circ f$. The converse holds true if all $h$-cobordisms are products, for example if the manifolds are simply-connected and have dimension greater than 4.

Given a map $f : M^n \to B\pi$, there is a function

$$\mathbb{S}(M) \to \Omega_n(B\pi)$$
$$(M' \to M) \mapsto [M' \to B\pi] - [M \to B\pi]$$

It is an interesting question, not unrelated to the Novikov Conjecture, to determine the image of this map, but we will not pursue this.

There are parallel theories for PL and topological manifolds, and all of our above statements are valid for these theories. There are variant bordism theories $\Omega^{PL}$, $\Omega^{TOP}$ and variant structure sets $\mathbb{S}^{PL}$ and $\mathbb{S}^{TOP}$. The bordism theories are rationally the same, but the integrality conditions comparing the theories are quite subtle.

# 5   A crash course in surgery theory

The purpose of surgery theory is the classification of manifolds up to homeomorphism, PL-homeomorphism, or diffeomorphism; perhaps a more descriptive name would be manifold theory. There are two main goals: *existence*, the determination of the homotopy types of manifolds and *uniqueness*, the classification of manifolds up to diffeomorphism (or whatever) within a homotopy type. For the uniqueness question, the technique is due to Kervaire-Milnor [24] for spheres, to Browder [5], Novikov [36], and Sullivan [51] for simply-connected manifolds, to Wall [55] for non-simply-connected manifolds of dimension $\geq 5$, and to Freedman-Quinn [16] for 4-manifolds. Surgery works best for manifolds of dimension $\geq 5$ due to the Whitney trick, but surgery theory also provides information about manifolds of dimension 3 and 4.

The key result of the uniqueness part of surgery theory is the *surgery exact sequence*[6]

$$\cdots \xrightarrow{\theta} L_{n+1}(\mathbb{Z}\pi_1 M) \xrightarrow{\partial} \mathbb{S}(M) \xrightarrow{\eta} \mathcal{N}(M) \xrightarrow{\theta} L_n(\mathbb{Z}\pi_1 M)$$

for a closed, smooth, oriented manifold $M^n$ with $n \geq 5$. The $L$-groups are abelian groups, algebraically defined in terms of the group ring. They are 4-periodic $L_n \cong L_{n+4}$. For the trivial group, $L_n(\mathbb{Z}) \cong \mathbb{Z}, 0, \mathbb{Z}_2, 0$ for $n \equiv 0, 1, 2, 3 \pmod 4$. The $L$-groups are Witt groups of quadratic forms (see

---

[6]This is given in [55, §10], although note that Wall deals with the $L^s$- and $\mathbb{S}^s$-theory, while we work with the $L^h$ and $\mathbb{S}^h$-theory, see [55, §17D].

Wall [55]), or, better yet, bordism groups of algebraic Poincaré complexes (see Ranicki [45]). The structure set $\mathbb{S}(M)$ and the normal invariant set $\mathcal{N}(M)$ are pointed sets, and the surgery exact sequence is an exact sequence of pointed sets. Furthermore, $L_{n+1}(\mathbb{Z}\pi_1 M)$ acts on $\mathbb{S}(M)$ so that two elements are in the same orbit if and only if that have the same image under $\eta$.

Elements of the normal invariant set $\mathcal{N}(M)$ are represented by degree one normal maps

$$
\begin{array}{ccc}
\nu_{M'} & \longrightarrow & \xi \\
\downarrow & & \downarrow \\
M' & \xrightarrow{g} & M
\end{array}
$$

that is, a map $g$ of closed, smooth, oriented[7] manifolds with $g_*[M'] = [M]$, together with normal data: a stable vector bundle $\xi$ over $M$ which pulls back to the stable normal bundle of $M'$; more precisely, the data includes a stable trivialization of $g^*\xi \oplus T(M')$. There is a notion of two such maps to $M$ being bordant; we say the maps are *normally bordant* or that one can do surgery to obtain one map from the other. The normal invariant set $\mathcal{N}(M)$ is the set of normal bordism classes. The map $\eta : \mathbb{S}(M) \to \mathcal{N}(M)$ sends a homotopy equivalence to itself where the bundle $\xi$ is the pullback of the stable normal bundle of the domain under the homotopy inverse. The map $\theta : \mathcal{N}(M) \to L_n(\mathbb{Z}\pi_1 M)$ is called the surgery obstruction map, and is defined for manifolds of any dimension, however when the dimension is greater than or equal to five, $\theta(g) = 0$ if and only if $g$ is normally bordant to a homotopy equivalence.

Sullivan computed the normal invariant set using homotopy theory; it is closely connected with characteristic classes, see [28]. In fact there is a Pontryagin-Thom type construction which identifies

$$
\mathcal{N}(M) \cong [M, G/O]
$$

Here $G(k)$ is the topological monoid of self-homotopy equivalences of $S^{k-1}$ and $G = \operatorname{colim} G(k)$. There is a fibration

$$
G/O \xhookrightarrow{\psi} BO \to BG
$$

$BG(k)$ classifies topological $\mathbb{R}^k$-bundles up to proper fiber homotopy equivalence and so $G/O$ classifies proper fiber homotopy equivalences between vector bundles. A map $M \to G/O$ corresponds to a proper fiber homotopy equivalence between vector bundles over $M$, and the transverse inverse image of the 0-section gives rise to the degree one normal map. If

---

[7]There are variant versions of surgery for non-orientable manifolds.

we take the classifying map $\hat{g} : M \to G/O$ of a degree one normal map $g : (M', \nu_{M'}) \to (M, \xi)$, then $\psi \circ \hat{g}$ is a classifying map for $\xi \oplus \tau_M$. Finally, the homotopy groups of $G$ are the stable homotopy groups of spheres, which are finite (Serre's result again), and so given any vector bundle $M \to BO$, then some non-zero multiple of it can be realized as a "$\xi$" in a degree one normal map.

Later it will be useful to consider the surgery obstruction map

$$\theta : \mathcal{N}(M^n) \to L_n(\mathbb{Z}\pi)$$

associated to a map $M \to B\pi$, which may not induce an isomorphism on the fundamental group. This is covariant in $\pi$. When $\pi$ is the trivial group and $g : (M', \nu_{M'}) \to (M, \xi)$ with $n \equiv 0 \pmod 4$,

$$\theta(g) = (1/8)(\sigma(M') - \sigma(M)) \in L_n(\mathbb{Z}) \cong \mathbb{Z}$$

For $n \equiv 2 \pmod 4$, the simply-connected surgery obstruction is called the Arf invariant of $g$.

# 6   Surgery and characteristic classes

This section is in some sense the core of this survey. We give a converse to the Novikov conjecture and rephrase the Novikov conjecture in terms of injectivity of an assembly homomorphism. The two key tools are the bordism invariance of surgery obstructions and the product formula for surgery obstructions. They are due to Sullivan in the case of simply-connected manifolds and the facts we need are due to Wall in the non-simply-connected case. Both of the tools were proved in greater generality and were given more conceptual interpretations by Ranicki.

The bordism invariance of surgery obstructions is that the surgery obstruction map associated to a $f : M \to B\pi$ factors [55, 13B.3]

$$\mathcal{N}(M^n) \xrightarrow{\beta} \Omega_n(G/O \times B\pi) \xrightarrow{\hat{\theta}} L_n(\mathbb{Z}\pi)$$

Here $\beta(g) = [(\hat{g}, f) : M \to G/O \times B\pi]$ where $\hat{g}$ classifies the degree one normal map. (The geometric interpretation of this result is that surgery obstruction is unchanged by allowing a bordism in both the domain and *range*.)

**Example 6.1.** The $K3$-surface $K^4$ is a smooth, simply-connected spin 4-manifold with signature 16. There is a degree one normal map

$$g : (K^4, \nu_K) \to (S^4, \xi)$$

where $\xi$ is any bundle with $\langle L_1(\xi), [S^4] \rangle = 16$. Then $\theta(g \times \mathrm{Id}_{S^4}) = 0$ since the classifying map $S^4 \times S^4 \to G/O$ is constant on the second factor and hence bounds. Thus $g \times \mathrm{Id}_{S^4}$ is normally bordant to a homotopy equivalence $h$ with $\sigma(h^{-1}(pt \times S^4)) = 16$. This is the same homotopy equivalence as in Example 3.1.

The second key tool is the product formula. This deals with the following situation. Given a degree one normal map

$$g : (M'^i, \nu_{M'}) \to (M^i, \xi)$$

and a closed, oriented manifold $N^j$, together with reference maps

$$M \to B\pi, \qquad N \to B\pi',$$

one would like a formula for the surgery obstruction

$$\theta(g \times \mathrm{Id}_N) \in L_{i+j}(\mathbb{Z}[\pi \times \pi'])$$

This has been given a nice conceptual answer by Ranicki [44]. There are "symmetric $L$-groups" $L^j(\mathbb{Z}\pi')$. There is the *Mishchenko-Ranicki symmetric signature map*

$$\sigma^* : \Omega_* B\pi' \to L^*(\mathbb{Z}\pi')$$

(sending a manifold to the bordism class of the chain level Poincaré duality map of its $\pi'$-cover), and a product pairing

$$L_i(\mathbb{Z}\pi) \otimes L^j(\mathbb{Z}\pi') \to L_{i+j}(\mathbb{Z}[\pi \times \pi'])$$

so that

$$\theta(g \times \mathrm{Id}_N) = \theta(g) \otimes \sigma^* N$$

Furthermore, for the trivial group, $L^j(\mathbb{Z}) \cong \mathbb{Z}, \mathbb{Z}_2, 0, 0$ for $j \equiv 0, 1, 2, 3$ (mod 4). For $j \equiv 0$ (mod 4), $\sigma^*(N) \in \mathbb{Z}$ is the signature $\sigma(N)$ and for $j \equiv 1$ (mod 4), $\sigma^*(N) \in \mathbb{Z}_2$ is called the De Rham invariant. We only need the following theorem, which follows from the above, but also from earlier work of Wall [55, 17H].

**Theorem 6.2.** *1. If $\pi = 1$, $\theta(g \times Id_N) \in L_{i+j}(\mathbb{Z}\pi') \otimes \mathbb{Q}$ depends only on the difference $\sigma(M') - \sigma(M)$ and the bordism class $[N \to B\pi'] \in \Omega_j(B\pi') \otimes \mathbb{Q}$.*

*2. If $\pi' = 1$, then*

$$\theta(g \times Id_N) = \theta(g) \cdot \sigma(N) \in L_{i+j}(\mathbb{Z}\pi) \otimes \mathbb{Q}$$

In the above theorem we are sticking with our usual convention that the signature is zero for manifolds whose dimensions are not divisible by 4.

**Example 6.3.** Let
$$g : (K, \nu_K) \to (S^4, \xi)$$
be a degree one normal map where $K$ is the $K3$-surface as in Example 6.1. Another way to see that $\theta(g \times \mathrm{Id}_{S^4}) = 0$ is by the product formula.

As a formal consequence of the bordism invariance of surgery obstructions, the product formula, and the Conner-Floyd isomorphism, it follows that the Novikov conjecture is equivalent to the injectivity of a rational assembly map. This is due to Wall [55, 17H] and Kaminker-Miller [23].

**Theorem 6.4.** *For any group $\pi$ and for any $n \in \mathbb{Z}_4$, there is a map*

$$A_n : \bigoplus_{i \equiv n(4)} H_i(\pi; \mathbb{Q}) \to L_n(\mathbb{Z}\pi) \otimes \mathbb{Q}$$

*so that*

1. *For a degree one normal map $g : (M', \nu_{M'}) \to (M, \xi)$ and a map $f \colon M \to B\pi$,*

$$\theta(g) = A_*((f \circ g)_*(L_{M'} \cap [M']) - f_*(L_M \cap [M])) \in L_*(\mathbb{Z}\pi) \otimes \mathbb{Q}$$

2. *If $A_n$ is injective then the Novikov conjecture is true for all closed, oriented manifolds mapping to $B\pi$ whose dimension is congruent to $n$ modulo 4. If the Novikov conjecture is true for all closed, oriented manifolds with fundamental group isomorphic to $\pi$ and whose dimension is congruent to $n$ modulo 4, then $A_n$ is injective.*

*Proof.* Bordism invariance, the Conner-Floyd isomorphism, and the product formula show that the surgery obstruction factors through a $\mathbb{Z}_4$-graded map

$$\widetilde{A}_n : \bigoplus_{i \equiv n(4)} H_i(G/O \times B\pi; \mathbb{Q}) \to L_n(\mathbb{Z}\pi) \otimes \mathbb{Q}$$

with

$$\widetilde{A}_n((\hat{g}, f)_*(L_M \cap [M])) = \theta(g) \in L_n(\mathbb{Z}\pi) \otimes \mathbb{Q}$$

When $\pi$ is trivial, the map $\widetilde{A}_n$ is given by Kronecker pairing $\hat{g}_*(L_M \cap [M])$ with some class $\ell \in H^*(G/O; \mathbb{Q})$. In the fibration

$$G/O \xrightarrow{\psi} BO \to BG$$

$\psi^*$ gives a rational isomorphism in cohomology, and it is not difficult to show that

$$\ell = \psi^*(\frac{1}{8}(\overline{L} - 1))$$

where $\overline{L}$ is the multiplicative inverse of the Hirzebruch $L$-class.[8] The key equation is that if $g : (M', \nu_{M'}) \to (M, \xi)$ is a degree one normal map with classifying map $\hat{g}$, then

$$8\hat{g}^*\ell \cup L_M = \overline{L}(\xi) - L_M$$

The homology of $G/O \times B\pi$ is rationally generated by cross products $\hat{g}_*(L_M \cap [M]) \times f_*(L_N \cap [N])$ where $\hat{g} : M \to G/O$ and $f : N \to B\pi$; these correspond to surgery problems of the form $M' \times N \to M \times N$ The product formula for surgery obstructions shows that

$$\widetilde{A}_*(\hat{g}_*(L_M \cap [M]) \times f_*(L_N \cap [N])) = 0$$

whenever $\langle \ell, \hat{g}_*(L_M \cap [M]) \rangle = 0$. It follows that $\widetilde{A}_n$ factors through the surjection given by the slant product

$$\ell \backslash : \bigoplus_{i \equiv n(4)} H_i(G/O \times B\pi; \mathbb{Q}) \to \bigoplus_{i \equiv n(4)} H_i(B\pi; \mathbb{Q}),$$

giving the rational assembly map

$$A_n : \bigoplus_{i \equiv n(4)} H_i(B\pi; \mathbb{Q}) \to L_i(\mathbb{Z}\pi) \otimes \mathbb{Q}$$

Tracing through the definition of $A_*$ gives the characteristic class formula in part 2. of the theorem. (I have suppressed a good deal of manipulation of cup and cap products here, partly because I believe the reader may be able to find a more efficient way than the author.)

If $A_n$ is injective, the Novikov conjecture immediately follows for $n$-manifolds equipped with a map to $B\pi$, since the surgery obstruction of a homotopy equivalence is zero.

Suppose $A_n$ is not injective; then there exists an non-zero element $a \in H_{4*+n}(B\pi; \mathbb{Q})$ so that $A_n(a) = 0 \in L_n(\mathbb{Z}\pi) \otimes \mathbb{Q}$. There is a $b \in H_{4*+n}(G/O \times B\pi; \mathbb{Q})$ so that $a = \ell \backslash b$. Next note for some $i$, there is an element $c = [(\hat{g}, f) : M \to G/O \times B\pi] \in \Omega_{4i+n}(G/O \times B\pi)$, so that $(\hat{g}, f)_*(L_M \cap [M]) = kb$ where $k \neq 0$. (Note the $i$ here. To get $c \in \Omega_{4*+n}$ one simply uses the Conner-Floyd isomorphism, but one might have to multiply the various components of $c$ by products of $\mathbb{C}P^2$ to guarantee that $c$ is homogeneous). By multiplying $c$ by a non-zero multiple, find a new $c = [(\hat{g}, f) : M \to G/O \times B\pi] \in \Omega_{4i+n}(G/O \times B\pi)$ so that $\theta(c) = 0 \in L_{4i+n}(\mathbb{Z}\pi)$. We may assume that $4i + n \mathbin{'}> 4$. By (very) low

---

[8]This class $\ell$ has a lift to $H^*(G/O; \mathbb{Z}_{(2)})$ which is quite important for the characteristic class formula for the surgery obstruction of a normal map of closed manifolds [28], [34], [52].

dimensional surgeries [55, Chapter 1], we may assume $M$ has fundamental group $\pi$. Since $\theta(c) = 0$, one may do surgery to obtain a homotopy equivalence $h : M' \to M$, where the difference of the higher signatures in $H_*(B\pi; \mathbb{Q})$ is a multiple of $a$ and hence non-zero.                    $\square$

Our next result is folklore (although seldom stated correctly) and should be considered as a converse to and a motivation for the Novikov conjecture. It generalizes Kahn's result [21] that the only possible linear combinations of $L$-classes (equivalently rational Pontrjagin classes) which can be a homotopy invariant of simply-connected manifolds is the top $L$-class of a manifold whose dimension is divisible by 4. The non-simply connected case requires a different proof; in particular one must leave the realm of smooth manifolds.

**Theorem 6.5.** *Let $M$ be a closed, oriented, smooth manifold of dimension $n > 4$, together with a map $f : M \to B\pi$ to the classifying space of a discrete group, inducing an isomorphism on fundamental group. Given any cohomology classes*

$$\mathcal{L} = \mathcal{L}_1 + \mathcal{L}_2 + \mathcal{L}_3 + \cdots \in H^{4*}(M; \mathbb{Q}), \qquad \mathcal{L}_i \in H^{4i}(M; \mathbb{Q}),$$

*so that $f_*(\mathcal{L} \cap [M]) = 0 \in H_{n-4*}(B\pi; \mathbb{Q})$, there is a non-zero integer $R$, so that for any multiple $r$ of $R$, there is a homotopy equivalence*

$$h : M' \to M$$

*of closed, smooth manifolds so that*

$$h^*(L_M + r\mathcal{L}) = L_{M'}$$

*Proof.* As motivation suppose that there is a map $\hat{g} : M \to G/O$ so that $\hat{g}^*\ell = (1/8)\mathcal{L}\overline{L_M}$, where $\ell \in H^*(G/O; \mathbb{Q})$ is as above and $\overline{L_M}$ is the multiplicative inverse of the total Hirzebruch $L$-class of $M$. Then if $g : N \to M$ is the corresponding surgery problem, a short computation shows that

$$g^*(L_M + \mathcal{L}) = L_N$$

and, by using Theorem 6.4, Part 2, that $\theta(\hat{g}) = 0 \in L_n(\mathbb{Z}\pi) \otimes \mathbb{Q}$. So the idea is to clear denominators and replace $\hat{g}$ by a multiple. Unfortunately, this is nonsense, since the surgery obstruction map is not a homomorphism of abelian groups.

To proceed we need two things. First that $G/TOP$ is an infinite loop space[9] and

$$[M, G/TOP] \xrightarrow{\theta} L_n(\mathbb{Z}\pi)$$

$$\downarrow \qquad\qquad \downarrow$$

$$\bigoplus_{i \equiv n(4)} H_i(\pi; \mathbb{Q}) \xrightarrow{A_*} L_n(\mathbb{Z}\pi) \otimes \mathbb{Q}$$

is a commutative diagram of abelian groups[10] where the left vertical map is given by $\hat{g} \mapsto f_*(\hat{g}^*\ell \cap [M])$. (See [52].)

The second thing we need is a lemma of Weinberger's [57]. Let $j : G/O \to G/TOP$ be the natural map.

**Lemma 6.6.** *For any $n$-dimensional CW-complex $M$, there is a non-zero integer $t = t(n)$ so that for any $[f] \in [M, G/TOP]$, $t[f] \in \text{im } j_* : [M, G/O] \to [M, G/TOP]$.*

Now suppose we are given $\mathcal{L} \in H^{4*}(M; \mathbb{Q})$ as in the statement of the theorem. The cohomology class $\ell$ gives a localization of $H$-spaces at 0

$$\ell : G/TOP \to \prod_{i>0} K(\mathbb{Q}, 4i)$$

and hence a localization of abelian groups

$$[M, G/TOP] \to \prod_{i>0} H^{4i}(M; \mathbb{Q})$$

In particular, there is an non-zero integer $R_1$, so that for any multiple $r_1$ of $R_1$, there is a map $\hat{g} : M \to G/TOP$ so that

$$\hat{g}^*\ell = r_1 \frac{1}{8}\overline{\mathcal{L}L_M}$$

Then $\theta(\hat{g}) = 0 \in L_n(\mathbb{Z}\pi)_{(0)}$, so there is a non-zero $R_2$ so that $R_2\theta(\hat{g}) = 0 \in L_n(\mathbb{Z}\pi)$. Finally, by Weinberger's Lemma, there is a non-zero $R_3$, so that $R_3[R_2\hat{g}]$ factors through $G/O$. Then $R = R_1R_2R_3$ works. Surgery theory giving the homotopy equivalence. $\qquad\square$

Weinberger's Lemma follows from the following; applied where $s : Y \to Z$ is the map $G/TOP \to B(TOP/O)$. Note that $B(TOP/O)$ has an infinite loop space structure coming from Whitney sum.

---

[9]$G/O$, $G/PL$, and $G/TOP$ are all infinite loop spaces with the $H$-space structure corresponding to Whitney sum. However, the surgery obstruction map is not a homomorphism. Instead, we use the infinite loop space structure on $G/TOP$ induced by periodicity $\Omega^4(\mathbb{Z} \times G/TOP) \simeq \mathbb{Z} \times G/TOP$. This will be discussed further in the next section.

[10]We can avoid references to topological surgery by using the weaker fact that the surgery obstruction map $\theta : [M, G/PL] \to L_n(\mathbb{Z}\pi)$ is a homomorphism where $G/PL$ is given the $H$-space structure provided by the Characteristic Variety Theorem [51].

**Lemma 6.7.** *Let $s : Y \to Z$ be a map of simply-connected spaces of finite type, where $Y$ is an $H$-space and $Z$ is an infinite loop space with all homotopy groups finite. Then for any $k$ there is a non-zero integer $t$ so that for any map $f : X \to Y$ whose domain is a $k$-dimensional CW-complex, then $s_*(t[f]) = 0$, where*

$$s_* : [X, Y] \to [X, Z]$$

*is the induced map on based homotopy.*

*Proof.* It suffices to prove the above when $f$ is the inclusion $i_k : Y^k \hookrightarrow Y$ of the $k$-skeleton of $Y$; in other words, we must show $s_*[i_k] \in [Y^k, Z]$ has finite order. Let $Y = Y_n$, $n = 1, 2, 3, \ldots$. Let $\alpha_n : Y_n \to Y_{n+1}$ be a cellular map so that $[\alpha_n] = n![\mathrm{Id}]$. Then $\alpha_n$ induced $\cdot n!$ on homotopy groups, since the co-$H$-group structure equals the $H$-group structure. Let $\mathrm{hocolim}_n Y_n$ denote the infinite mapping telescope of the maps $\alpha_n$. Then $Y \to \mathrm{hocolim}_n Y_n$ induces a localization at $0$ on homotopy groups.

Consider the following commutative diagram of abelian groups.

$$
\begin{array}{ccccc}
[\mathrm{hocolim}_n Y_n^{k+2}, Z] & \xrightarrow{\Phi} & \lim_n[Y_n^{k+2}, Z] & \xrightarrow{\mathrm{pr}_1} & [Y_1^{k+2}, Z] \\
\downarrow{\scriptstyle A} & & \downarrow{\scriptstyle B} & & \downarrow{\scriptstyle C} \\
[\mathrm{hocolim}_n Y_n^{k}, Z] & \xrightarrow{\Phi} & \lim_n[Y_n^{k}, Z] & \xrightarrow{\mathrm{pr}_1} & [Y_1^{k}, Z]
\end{array}
$$

To prove the lemma it suffices to show that some multiple of $s_*[i_k] \in [Y_1^k, Z]$ is in the image of $C \circ \mathrm{pr}_1 \circ \Phi$ and that $A$ is the zero map. First note $C s_*[i_{k+2}] = s_*[i_k]$. Now

$$\mathrm{im}([Y, Z] \to [Y^{k+2}, Z])$$

is a finite set by obstruction theory, so

$$\{s_*(n![i_{k+2}])\}_{n=1,2,3,\ldots}$$

sits in a finite set, hence there exists an $N$ so that $s_*(N![i_{k+2}])$ pulls back arbitrarily far in the inverse sequence. Hence by compactness of the inverse limit of finite sets, there is an $[a] \in \lim_n[Y_n^{k+2}, Z]$ so that $\mathrm{pr}_1[a] = s_*(N![i_{k+2}])$. By Milnor's $\lim^1$ result [30], $\Phi$ is onto, so that $[a] = \Phi[b]$ for some $[b]$.

We next claim that $A$ is the zero map. Indeed the homology groups of $\mathrm{hocolim}_n Y_n^{k+2}$ are rational vector spaces in dimensions less than $k+2$. By obstruction theory any map $\mathrm{hocolim}_n Y_n^{k+2} \to Z$ is zero when restricted to the $(k+1)$-st skeleton, and thus when restricted to $\mathrm{hocolim}_n Y_n^k$. Thus we have shown that $s_*(N![i_k])$ is zero by tracing around the outside of the diagram. Let $t = N!$.                                                                 $\square$

**Remark 6.8.** For a closed, oriented, topological $n$-manifold $M$, define

$$\Gamma^{TOP}(M) : \mathbb{S}^{TOP}(M) \to H^{4*}(M; \mathbb{Q})$$
$$[h : M' \to M] \mapsto \mathcal{L} \qquad \text{where } h^*(L_M + \mathcal{L}) = L_{M'}$$

If $M$ is smooth, define an analogous map $\Gamma^{DIFF}(M)$ from the smooth structure set. The above discussion shows that the image of $\Gamma^{TOP}(M)$ is a finitely generated, free abelian group whose intersection with the kernel of the map $H^{4*}(M; \mathbb{Q}) \to H_{n-4*}(B\pi_1 M; \mathbb{Q})$ is a lattice (i.e a finitely generated subgroup of full rank). Furthermore if the Novikov Conjecture is valid for $\pi_1 M$, the image of $\Gamma^{TOP}(M)$ is a precisely a lattice in the above kernel.

Similar things are true for smooth manifolds up to finite index. If $t = t(n)$ is the integer from Lemma 6.7, then

$$\frac{1}{t}\text{im } \Gamma^{TOP}(M) \subset \text{im } \Gamma^{DIFF}(M) \subset \text{im } \Gamma^{TOP}(M)$$

However, a recent computation of Weinberger's [56] shows that the image of $\Gamma^{DIFF}(M)$ is not a group when $M$ is a high-dimensional torus. It follows that $\mathbb{S}^{DIFF}(M)$ cannot be given a group structure compatible with that of $\mathbb{S}^{TOP}(M)$ and that $G/O$ cannot be given an $H$-space structure so that the surgery obstruction map is a homomorphism.

# 7  Assembly maps

The notion of an assembly map is central to modern surgery theory, and to most current attacks on the Novikov Conjecture. Assembly maps are useful for both conceptual and computation reasons. We discuss assembly maps to state some of the many generalizations of the Novikov conjecture. Assembly can be viewed as gluing together surgery problems, as a passage from local to global information, as the process of forgetting control in controlled topology, as taking the index of an elliptic operator, or as a map defined via homological algebra. There are parallel theories of assembly maps in algebraic $K$-theory and in the $K$-theory of $C^*$-algebras, but the term assembly map originated in surgery with the basic theory due to Quinn and Ranicki.

With so many different points of view on the assembly map, it is a bit difficult to pin down the concept, and it is perhaps best for the neophyte to view it as a black box and concentrate on its key properties. We refer the reader to the papers [42], [43], [46], [59], [11] for further details.

The classifying space for topological surgery problems is $G/TOP$. The generalized Poincaré conjecture and the surgery exact sequence show that

$\pi_n(G/TOP) = L_n(\mathbb{Z})$ for $n > 0$. There is a homotopy equivalence

$$\Omega^4(\mathbb{Z} \times G/TOP) \simeq \mathbb{Z} \times G/TOP$$

(Perhaps the 4-fold periodicity is halfway between real and complex Bott periodicity.) Let **L** denote the corresponding spectrum. Oriented manifolds are oriented with respect to the generalized homology theory defined by **L**. When localized at 2, **L** is a wedge of Eilenberg-MacLane spectra, and after 2 is inverted, **L** is homotopy equivalent to inverting 2 in the spectrum resulting from real Bott periodicity. The assembly map

$$A_n : H_n(X; \mathbf{L}) \to L_n(\mathbb{Z}\pi_1 X)$$

is defined for all integers $n$, is 4-fold periodic, and natural in $X$. By naturality, the assembly map for a space $X$ factors through $B\pi_1 X$.

The surgery obstruction map can interpreted in terms of the assembly map. Let **G/TOP** be the connective cover of **L**, i.e. there is a map **G/TOP** → **L** which is an isomorphism on $\pi_i$ for $i > 0$, but $\pi_i(\mathbf{G/TOP}) = 0$ for $i \leq 0$. Here **G/TOP** is an $\Omega$-spectrum whose 0-th space is $G/TOP$. The composite

$$H_n(X; \mathbf{G/TOP}) \to H_n(X; \mathbf{L}) \xrightarrow{A} L_n(\mathbb{Z}\pi_1 X)$$

is also called the assembly map. When $X = B\pi$, this assembly map tensored with $\mathrm{Id}_\mathbb{Q}$ is the same as the assembly map from the last section. In particular the Novikov conjecture is equivalent to the rational injectivity of either of the assembly maps when $X = B\pi$.

The surgery obstruction map for a closed, oriented $n$-manifold $M$

$$\theta : [M, G/TOP] \to L_n(\mathbb{Z}\pi_1 M)$$

factors as the composite of Poincaré duality and the assembly map

$$[M, G/TOP] = H^0(M; \mathbf{G/TOP}) \simeq H_n(M; \mathbf{G/TOP}) \xrightarrow{A} L_n(\mathbb{Z}\pi_1 M)$$

There is a surgery obstruction map and a structure set for manifolds with boundary (the $L$-groups remain the same however). The basic idea is that all maps are assumed to be homeomorphisms on the boundary throughout. The surgery exact sequence extends to a half-infinite sequence

$$\cdots \to \mathbb{S}^{TOP}(M \times I, \partial) \to [(M \times I)/\partial, G/TOP] \to L_{n+1}(\mathbb{Z}\pi_1 M)$$
$$\to \mathbb{S}^{TOP}(M) \to [M, G/TOP] \to L_n(\mathbb{Z}\pi_1 M)$$

Assembly maps are induced by maps of spectra; we denote the fiber of

$$X_+ \wedge \mathbf{G/TOP} \xrightarrow{A} \mathbf{L}(\mathbb{Z}\pi_1 X)$$

by $\mathbb{S}^{\mathbf{TOP}}(\mathbf{X})$. When $X$ is a manifold, the corresponding long exact sequence in homotopy can be identified with the surgery exact sequence. In particular, the structure set $\mathbb{S}^{TOP}(M) = \pi_n \mathbb{S}^{\mathbf{TOP}}(\mathbf{M})$ is naturally an abelian group, a fact which is not geometrically clear. Thus computing assembly maps is tantamount to classifying manifolds up to $h$-cobordism.

There is a parallel theory in algebraic $K$-theory. For a ring $A$, there are abelian groups $K_n(A)$ defined for all integers $n$; they are related by the fundamental theorem of $K$-theory which gives a split exact sequence

$$0 \to K_n(A) \to K_n(A[t]) \oplus K_n(A[t^{-1}]) \to K_n(A[t, t^{-1}]) \to K_{n-1}A \to 0$$

There is an $\Omega$-spectrum $\mathbf{K(A)}$ whose homotopy groups are $K_*(A)$. Abbreviate $\mathbf{K(\mathbb{Z})}$ by $\mathbf{K}$; its homotopy groups are zero in negative dimensions, $\mathbb{Z}$ in dimension 0, and $\mathbb{Z}/2$ in dimension 1. There is the $K$-theory assembly map [26], [11]

$$A_n : H_n(X; \mathbf{K}) \to K_n(\mathbb{Z}\pi_1 X).$$

Computing with the Atiyah-Hirzebruch spectral sequence and the test case where $X$ is the circle shows that $A_n$ is injective for $i < 2$, and the cokernels of $A_n$ are $K_n(\mathbb{Z}\pi_1 X), \tilde{K}_0(\mathbb{Z}\pi_1 X), Wh(\pi_1 X)$ for $n < 0, n = 0, n = 1$. The analogue of the Novikov conjecture in $K$-theory has been proven!

**Theorem 7.1 (Bökstedt-Hsiang-Madsen [4]).** *Suppose $\pi$ is any group such that $H_n(B\pi)$ is finitely generated for all $n$. Then $A_* \otimes Id_\mathbb{Q}$ is injective.*

Hopefully this result will shed light on the Novikov conjecture in $L$-theory (which has more direct geometric consequences), but so far this has been elusive.

# 8   Isomorphism conjectures

A strong version of Borel's conjecture is:

**Conjecture 8.1.** *Let $h : M' \to M$ be a homotopy equivalence between compact aspherical manifolds so that $h(M' - \partial M') \subset h(M - \partial M)$ and $h|_{\partial M'} : \partial M' \to \partial M$ is a homeomorphism. Then $h$ is homotopic rel $\partial$ to a homeomorphism.*

Applying this to $h$-cobordisms implies that $Wh(\pi_1 M) = 0$, and by crossing with tori, that $\tilde{K}_0(\mathbb{Z}\pi_1 M) = 0$ and $K_{-i}(\mathbb{Z}\pi_1 M) = 0$ for $i > 0$. Similarly, the structure groups $\mathbb{S}^{TOP}(M \times I^i, \partial) = 0$, so the assembly maps are isomorphisms. The following conjecture is motivated by these considerations.

**Borel-Novikov Isomorphism Conjecture.** [11] *For $\pi$ torsion free, the assembly maps*

$$A_* : H_*(B\pi; \mathbf{K}) \to K_*(\mathbb{Z}\pi)$$
$$A_* : H_*(B\pi; \mathbf{L}) \to L_*(\mathbb{Z}\pi)$$

*are isomorphisms.*

This implies the geometric Borel conjecture for manifolds of dimension greater than 4, and for torsion-free groups the vanishing of the Whitehead group, that all finitely generated projective modules are stably free, and the Novikov Conjecture.

What can be said for more general groups, in other words how does one compute the $L$-groups and the surgery obstruction groups? Well, for finite groups, the assembly map has been largely computed, starting with the work of Wall [55] and ending with the work of Hambleton, Milgram, Taylor, and Williams [18]. The techniques here are a mix of number theory, quadratic forms, and topology. The assembly maps are not injective, and not even rational surjective, in $K$-theory, because the Whitehead groups may be infinite, and in $L$-theory, because the multisignature (or $\rho$-invariant) show that the $L$-groups of the group ring of a finite group may be infinite.

There are also analyses of the $K$- and $L$-theory of products $\mathbb{Z} \times G$ and amalgamated free products $A *_B C$ (see [2], [49], [41], [54], [6]). While these give evidence for the Borel-Novikov Isomorphism Conjecture for torsion-free groups, for infinite groups with torsion the "nil" phenomena showed that the non-homological behavior of $L$-groups could not all be blamed on the finite subgroups. In particular there are groups $\pi$ where the assembly map is not an isomorphism, but where the assembly map is an isomorphism for all finite subgroups. To account for this, Farrell and Jones laid the blame on the following class of subgroups.

**Definition 8.2.** *A group $H$ is* virtually cyclic *if it has a cyclic subgroup of finite index.*

For example, a finite group $G$ is virtually cyclic and so is $\mathbb{Z} \times G$. Farrell-Jones have made a conjecture [13] which computes $K_*(\mathbb{Z}\pi)$ and $L_*(\mathbb{Z}\pi)$ in terms of the assembly maps for virtually cyclic subgroups and homological information concerning the group $\pi$ and the lattice of virtually cyclic subgroups of $\pi$. We give the rather complicated statement of the conjecture

---

[11] Neither Borel nor Novikov made this conjecture, but rather made weaker conjectures whose statements did not involve assembly maps.

It is a fairly bold conjecture; there exist a lot of torsion-free groups. A more conservative conjecture would be to conjecture this when $B\pi$ is a finite complex and perhaps only that the assembly map is a split injection. One might also wish to restrict the conjecture in $K$-theory to $* < 2$, where there is a geometric interpretation of the results.

below, but for now we note that the Farrell-Jones isomorphism conjecture implies:

1. The Novikov conjecture for a general group $\pi$.

2. The Borel-Novikov isomorphism conjecture for a torsion-free group $\pi$.

3. For any group $\pi$ and for any $N \in \mathbb{Z} \cup \{\infty\}$, if for all virtually cyclic subgroups $H$ of $\pi$, the assembly map

$$A_* : H_*(BH; \mathbf{K}) \to K_*(\mathbb{Z}H)$$

   is an isomorphism for $* < N$ and a surjection for $* = N$, then the assembly map for $\pi$ is an isomorphism for $* < N$, and similarly for $L$-theory.

We proceed to the statement of the isomorphism conjecture, as formulated in [11]. We work in $K$-theory, although there is an analogous conjecture in $L$-theory.[12] One can show that an inner automorphism of $\pi$ induces the identity on $K_*(\mathbb{Z}\pi)$, but not necessarily on the associated spectrum. (One needs to worry about such details to make sure that constructions don't depend on the choice of the base point of the fundamental group.) To account for this, one uses the *orbit category* $\mathrm{Or}(\pi)$, whose objects are left $\pi$-sets $\{\pi/H\}_{H \subset \pi}$ and whose morphisms are $\pi$-maps. For a family of subgroups $\mathcal{F}$ of $\pi$ (e.g. the trivial family $\mathcal{F} = 1$ or the family $\mathcal{F} = \mathcal{VC}$ of virtually cyclic subgroups of $\pi$), one defines the *restricted orbit category* $\mathrm{Or}(\pi, \mathcal{F})$ to be the full subcategory of $\mathrm{Or}(\pi)$ with objects $\{\pi/H\}_{H \in \mathcal{F}}$. A functor

$$\mathbf{K} : \mathrm{Or}(\pi) \to \mathrm{SPECTRA}$$

is constructed in [11], with $\pi_*(\mathbf{K}(\pi/H)) = K_*(\mathbb{Z}H)$. The (classical) assembly map is then given by applying homotopy groups to the map

$$\mathbf{A} : \mathop{\mathrm{hocolim}}_{\mathrm{Or}(\pi,1)} \mathbf{K} \to \mathop{\mathrm{hocolim}}_{\mathrm{Or}(\pi)} \mathbf{K}$$

induced on homotopy colimits by the inclusion of the restricted orbit category in the full orbit category.

**Farrell-Jones Isomorphism Conjecture.** *For any group $\pi$,*

$$\mathop{\mathrm{hocolim}}_{\mathrm{Or}(\pi,\mathcal{VC})} \mathbf{K} \to \mathop{\mathrm{hocolim}}_{\mathrm{Or}(\pi)} \mathbf{K}$$

*induces an isomorphism on homotopy groups.*

---

[12] In $L$-theory it is necessary to work with a variant theory, $L = L^{\langle -\infty \rangle}$.

This gives a theoretical computation of the $K$-groups, the (classical) assembly map, and in $L$-theory, the classification of manifolds with fundamental group $\pi$. For proofs of the conjecture in special cases see [13]. For applications in some special cases see [40] and [10]. The Farrell-Jones Isomorphism Conjecture is parallel to the $C^*$-algebra conjecture of Baum and Connes [3], with the family of virtually cyclic subgroups replaced by the family of finite subgroups.

# References

[1] A. Adem and J.F. Davis. Topics in transformation groups. To appear in the *Handbook of Geometric Topology*.

[2] H. Bass. *Algebraic K-theory*. Benjamin, 1968.

[3] P. Baum, A. Connes, and N. Higson. Classifying space for proper actions and $K$-theory of group $C^*$-algebras. In $C^*$-*algebras: 1943–1993 (San Antonio, TX, 1993)*, volume 167 of *Contemp. Math.*, pages 240–291. Amer. Math. Soc., 1994.

[4] M. Bökstedt, W. C. Hsiang, and I. Madsen. The cyclotomic trace and algebraic $K$-theory of spaces. *Invent. Math.*, 111:465–539, 1993.

[5] W. Browder. *Surgery on simply connected manifolds*. Springer, 1972.

[6] S. E. Cappell. A splitting theorem for manifolds. *Invent. Math.*, 33:69–170, 1976.

[7] M. M. Cohen. *A course in simple homotopy theory*. Springer, 1973.

[8] P. E. Conner and E. E. Floyd. *Differentiable periodic maps*. Springer, 1964.

[9] J. F. Davis. The Borel/Novikov conjectures and stable diffeomorphisms of 4-manifolds. Preprint.

[10] J. F. Davis and W. Lück. Computations of $K$- and $L$-groups of group rings based on isomorphism conjectures. In preparation.

[11] J. F. Davis and W. Lück. Spaces over a category and assembly maps in isomorphism conjectures in $K$-and $L$-theory. *K-Theory*, 15:201–252, 1998.

[12] F. T. Farrell and W.-C. Hsiang. Manifolds with $\pi_1 = G \times_\alpha T$. *Amer. J. Math.*, 95:813–848, 1973.

[13] F. T. Farrell and L. E. Jones. Isomorphism conjectures in algebraic $K$-theory. *J. Amer. Math. Soc.*, 6:249–297, 1993.

[14] S. C. Ferry, A. Ranicki, and J. Rosenberg. A history and survey of the Novikov conjecture. In *Novikov conjectures, index theorems and rigidity, Vol. 1 (Oberwolfach, 1993)*, pages 7–66. Cambridge Univ. Press, 1995.

[15] S. C. Ferry, A. Ranicki, and J. Rosenberg, editors. *Novikov conjectures, index theorems and rigidity. Vols. 1, 2.* Cambridge University Press, 1995.

[16] M. H. Freedman and F. Quinn. *Topology of 4-manifolds.* Princeton University Press, 1990.

[17] M. Gromov. Positive curvature, macroscopic dimension, spectral gaps and higher signatures. In *Functional analysis on the eve of the 21st century, Vol. II (New Brunswick, NJ, 1993)*, pages 1–213. Birkhäuser Boston, Boston, MA, 1996.

[18] I. Hambleton, R. J. Milgram, L. Taylor, and B. Williams. Surgery with finite fundamental group. *Proc. London Math. Soc.*, 56:349–379, 1988.

[19] W.-C. Hsiang. A splitting theorem and the Künneth formula in algebraic $K$-theory. In *Algebraic K-Theory and its Geometric Applications (Conf., Hull, 1969)*, volume 108 of *Lecture Notes in Math.*, pages 72–77. Springer, 1969.

[20] I. M. James and J. H. C. Whitehead. The homotopy theory of sphere bundles over spheres. I. *Proc. London Math. Soc.*, 4:196–218, 1954.

[21] P. J. Kahn. Characteristic numbers and oriented homotopy type. *Topology*, 3:81–95, 1965.

[22] P. J. Kahn. A note on topological Pontrjagin classes and the Hirzebruch index formula. *Illinois J. Math.*, 16:243–256, 1972.

[23] J. Kaminker and J. G. Miller. A comment on the Novikov conjecture. *Proc. Amer. Math. Soc.*, 83:656–658, 1981.

[24] M. Kervaire and J. Milnor. Groups of homotopy spheres I. *Ann. of Math.*, 77:504–537, 1963.

[25] R. C. Kirby and L. Siebenmann. *Foundational essays on topological manifolds, smoothings and triangulations.* Princeton University Press, 1977.

[26] J. L. Loday. $K$-théorie algébrique et représentations de groupes. *Ann. Sci. École Norm. Sup. (4)*, 9:309–377, 1976.

[27] G. Lusztig. Novikov's higher signature and families of elliptic operators. *J. Differential Geometry*, pages 229–256, 1972.

[28] I. Madsen and J. Milgram. *The classifying spaces for surgery and cobordism of manifolds*. Princeton University Press, 1979.

[29] J. Milnor. On manifolds homeomorphic to the 7-sphere. *Ann. of Math.*, 64:399–405, 1956.

[30] J. Milnor. On axiomatic homology theory. *Pacific J. Math.*, 12:337–341, 1962.

[31] J. Milnor. Whitehead torsion. *Bull. Amer. Math. Soc.*, 72:358–426, 1966.

[32] J. Milnor and J. D. Stasheff. *Characteristic classes*. Princeton University Press, 1974.

[33] E. E. Moise. Affine structures in 3-manifolds. V. The triangulation theorem and Hauptvermutung. *Ann. of Math.*, 56:96–114, 1952.

[34] J. W. Morgan and D. P. Sullivan. The transversality characteristic class and linking cycles in surgery theory. *Ann. of Math.*, 99:463–544, 1974.

[35] J. Munkres. Obstructions to the smoothing of piecewise-differentiable homeomorphisms. *Ann. of Math.*, 72:521–554, 1960.

[36] S. P. Novikov. Diffeomorphisms of simply-connnected manifolds. *Dokl. Akad. Nauk SSSR*, 143:1046–1049, 1962. English translation, Soviet Math. Doklady, 3:540–543, 1962.

[37] S. P. Novikov. Rational Pontrjagin classes. Homeomorphism and homotopy type of closed manifolds. I. *Izv. Akad. Nauk SSSR Ser. Mat.*, 29:1373–1388, 1965. English translation, Amer. Math. Soc. Transl. (2), 66:214–230, 1968.

[38] S. P. Novikov. On manifolds with free abelian fundamental group and their application. *Izv. Akad. Nauk SSSR Ser. Mat.*, 30:207–246, 1966. English translation, Amer. Math. Soc. Transl. (2), 71:1–42, 1968.

[39] S. P. Novikov. Algebraic construction and properties of Hermitian analogs of $K$-theory over rings with involution from the viewpoint of Hamiltonian formalism. Applications to differential topology and the theory of characteristic classes. I. II. *Math. USSR-Izv.*, 4:257–292; ibid. 4:479–505, 1970.

[40] K. Pearson. Algebraic $K$-theory of two-dimensional crystallographic groups. *K-Theory*, 14:265–280, 1998.

[41] D. Quillen. Higher algebraic $K$-theory, I. In *Proceedings of Battelle Seattle Algebraic K-theory Conference 1972, vol. I*, volume 341 of *Lecture Notes in Math.*, pages 85–147. Springer, 1973.

[42] F. Quinn. A geometric formulation of surgery. In *Topology of manifolds; Proceedings of 1969 Georgia topology conference*, pages 500–511. Markham Press, 1970.

[43] A. Ranicki. The total surgery obstruction. In *Algebraic topology, Aarhus 1978 (Proc. Sympos., Univ. Aarhus, Aarhus, 1978)*, volume 763 of *Lecture Notes in Math.*, pages 275–316. Springer, 1979.

[44] A. Ranicki. The algebraic theory of surgery I, II. *Proc. London Math. Soc.*, 40:87–287, 1980.

[45] A. Ranicki. *Exact sequences in the algebraic theory of surgery*. Princeton University Press, 1981.

[46] A. Ranicki. *Algebraic L-theory and topological manifolds*. Cambridge University Press, 1992.

[47] A. Ranicki. On the Novikov conjecture. In *Proceedings of the conference "Novikov conjectures, index theorems and rigidity" volume I, Oberwolfach 1993*, pages 272–337. Cambridge University Press, 1995.

[48] J.-P. Serre. Groupes d'homotopie et classes de groupes abéliens. *Ann. of Math.*, 58:258–294, 1953.

[49] J. Shaneson. Wall's surgery obstruction group for $G \times Z$. *Ann. of Math.*, 90:296–334, 1969.

[50] R. E. Stong. *Notes on cobordism theory*. Princeton University Press, 1968.

[51] D. P. Sullivan. Triangulating and smoothing homotopy equivalences and homeomorphisms. Geometric topology seminar notes. In *The Hauptvermutung book*, pages 69–103. Kluwer Acad. Publ., 1996.

[52] L. Taylor and B. Williams. Surgery spaces: formulae and structure. In *Algebraic topology, Waterloo, 1978 (Proc. Conf., Univ. Waterloo, Waterloo, Ont., 1978)*, volume 741 of *Lecture Notes in Math.*, pages 170–195. Springer, 1979.

[53] R. Thom. Quelques propriétés globales des variétés differentiables. *Comment. Math. Helv.*, 28:17–86, 1954.

[54] F. Waldhausen. Algebraic $K$-theory of generalized free products I,II. *Ann. of Math.*, 108:135–256, 1978.

[55] C. T. C. Wall. *Surgery on compact manifolds.* Academic Press, 1970.

[56] S. Weinberger. On smooth surgery II. Preprint.

[57] S. Weinberger. On smooth surgery. *Comm. Pure Appl. Math.*, 43:695–696, 1990.

[58] S. Weinberger. Aspects of the Novikov conjecture. In *Geometric and topological invariants of elliptic operators (Brunswick, ME, 1988)*, volume 105 of *Contemp. Math.*, pages 281–297. Amer. Math. Soc., 1990.

[59] M. Weiss and B. Williams. Assembly. In *Proceedings of the conference "Novikov conjectures, index theorems and rigidity" volume II, Oberwolfach 1993*, pages 332–352. Cambridge University Press, 1995.

Department of Mathematics
Indiana University
Bloomington, IN 47405

*E-mail address :* jfdavis@ucs.indiana.edu

# A guide to the calculation of the surgery obstruction groups for finite groups

Ian Hambleton and Laurence R. Taylor

We describe the main steps in the calculation of surgery obstruction groups for finite groups. Some new results are given and extensive tables are included in the appendix.

The surgery exact sequence of C. T. C. Wall [68] describes a method for classifying manifolds of dimension $\geq 5$ within a given (simple) homotopy type, in terms of normal bundle information and a 4–periodic sequence of obstruction groups, depending only on the fundamental group and the orientation character. These obstruction groups $L_n^s(\mathbf{Z}G, w)$ are defined by considering stable isomorphism classes of quadratic forms on finitely generated free modules over $\mathbf{Z}G$ ($n$ even), together with their unitary automorphisms ($n$ odd).

Carrying out the surgery program in any particular case requires a calculation of the surgery obstruction groups, the normal invariants, and the maps in the surgery exact sequence. For fundamental group $G = 1$, the surgery groups were calculated by Kervaire–Milnor as part of their study of homotopy spheres:

$$L_n^s(\mathbf{Z}) = 8\mathbf{Z},\ 0,\ \mathbf{Z}/2,\ 0 \quad \text{for } n = 0, 1, 2, 3 \,(\mathrm{mod}\,4) \,,$$

where the non–zero groups are detected by the signature or Arf invariant, and the notation $8\mathbf{Z}$ means that the signature can take on any value $\equiv 0 \,(\mathrm{mod}\,8)$. The Hirzebruch signature theorem can be used to understand the signature invariant, and a complete analysis of the normal data was carried out by Milgram [45], Madsen–Milgram [43] and Morgan–Sullivan [48].

The theory of non–simply connected surgery has been used to investigate three important problems in topology:

The authors wish to thank the Max Planck Institut für Mathematik in Bonn for its hospitality and support. This research was also partially supported by NSERC and the NSF.

(i)   the spherical space form problem, or the classification of free finite group actions on spheres

(ii)  the Borel and Novikov conjectures, or the study of closed aspherical manifolds and assembly maps

(iii) transformation groups, or the study of Lie group actions on manifolds.

In the first problem, surgery is applied to manifolds with finite fundamental group and the surgery obstruction groups can be investigated by methods closely related to number theory and the representation theory of finite groups. In the second problem, the fundamental groups are infinite and torsion–free, and the methods available for studying the surgery obstruction groups are largely geometrical. The case of the $n$–torus was particularly important for its applications to the theory of topological manifolds. The third problem includes both finite group actions and actions by connected Lie groups. The presence of fixed point sets introduces many interesting new features.

In this paper we consider only $L_*(\mathbf{Z}G)$ for finite groups $G$. The Novikov conjectures and other topics connected with infinite fundamental groups are outside the scope of this article.

Before giving some notation, definitions and a detailed statement of results, it may be useful to list some general properties of the surgery obstruction groups for finite groups.

(1)  The groups $L_*(\mathbf{Z}G)$ are finitely generated abelian groups, the odd–dimensional groups $L_{2k+1}(\mathbf{Z}G)$ are finite, and in every dimension the torsion subgroup of $L_*(\mathbf{Z}G)$ is 2–primary.

There is a generalization of the ordinary simply–connected signature, called the multi–signature [68, 13A], [40].

(2)  The multi–signature is a homomorphism $\sigma_G\colon L_{2k}(\mathbf{Z}G) \to R_{\mathbf{C}}^{(-)^k}(G)$ where $R_{\mathbf{C}}(G)$ denotes the ring of complex characters of $G$. The multi–signature has finite 2–groups for its kernel and cokernel.

Complex conjugation acts as an involution on $R_{\mathbf{C}}(G)$, decomposing it as a sum of $\mathbf{Z}$'s from the real–valued (type I) characters, and a sum of free $\mathbf{Z}[\mathbf{Z}/2]$ modules generated by irreducible type II characters $\chi \neq \bar{\chi}$. The $(-1)^k$–eigenspaces of the complex conjugation action are denoted $R_{\mathbf{C}}^{(-)^k}(G)$.

The theory of Dress induction [24, 25] greatly simplifies the calculation of $L$–groups. A group $G$ is called $p$–hyperelementary if $G = C \rtimes P$ where $P$ is a $p$–Sylow subgroup and $C$ is a cyclic group of order prime to $p$. Then $G$ is determined by $C$, $P$ and the structure homomorphism $t\colon P \to$

*Aut(C)*. Further, $G$ is *p–elementary* if it is *p*–hyperelementary and $t$ is trivial (equivalently $G = C \times P$).

(3) $L_*(\mathbf{Z}G)$ can be calculated from knowledge of the *L*–groups of hyperelementary subgroups of $G$, together with the maps induced by subgroup inclusions.

Moreover, one can calculate $L_*(\mathbf{Z}G) \otimes \mathbf{Z}_{(2)}$, $R_{\mathbf{C}}(G) \otimes \mathbf{Z}_{(2)}$ and $\sigma_G \otimes 1$ from the 2–hyperelementary subgroups and the maps between them. Since (1) and (2) imply that

$$
\begin{array}{ccc}
L_*(\mathbf{Z}G) & \xrightarrow{\;\sigma_G\;} & R_{\mathbf{C}}(G) \\
\downarrow & & \downarrow \\
L_*(\mathbf{Z}G) \otimes \mathbf{Z}_{(2)} & \xrightarrow{\;\sigma_G \otimes 1\;} & R_{\mathbf{C}}(G) \otimes \mathbf{Z}_{(2)}
\end{array}
$$

is a pull–back, Dress's work computes $L_*(\mathbf{Z}G)$ in terms of representation theory and the *L*–theory of 2–hyperelementary groups. For this reason, most of the calculational work has been devoted to the 2–hyperelementary case.

These general properties are fine until one needs more precise information for computing surgery obstructions. An early result of Bak and Wall (worked out as an example in Theorem 10.1) is that for $G$ of *odd order*

$$ L_n^s(\mathbf{Z}G) \; = \; \Sigma \oplus 8\mathbf{Z}, \; 0, \; \Sigma \oplus \mathbf{Z}/2, \; 0 \qquad \text{for } n = 0, 1, 2, 3 \, (\mathrm{mod} \, 4) \; . $$

The terms $\Sigma \; = \; \oplus 4(\chi \pm \bar{\chi})$ comes from the multisignatures at type II characters, and the term $\mathbf{Z}$ is the summand of $R_{\mathbf{C}}(G)$ generated by the trivial character. The term $\mathbf{Z}/2$ is detected by the ordinary Arf invariant.
    Another nice case is $G \; = \; C \times P$, where $C$ is a cyclic 2–group and $P$ has odd order (this includes arbitrary cyclic groups as well as the *p*–hyperelementary groups $G$ for $p$ odd). Assuming $C$ is non–trivial, we have:

$$ L_n^s(\mathbf{Z}G) \; = \; \Sigma \oplus 8\mathbf{Z} \oplus 8\mathbf{Z}, \; 0, \; \Sigma \oplus \mathbf{Z}/2, \; \mathbf{Z}/2 \qquad \text{for } n = 0, 1, 2, 3 \, (\mathrm{mod} \, 4) \; . $$

The signature group again has two sources, the term $\Sigma = \oplus 4(\chi \pm \bar{\chi})$ from the type II characters and the two $\mathbf{Z}$'s coming from the type I characters (just the trivial character and the linear character which sends a generator to $-1$). The $\mathbf{Z}/2$ in dimension 2 is the ordinary Arf invariant and the $\mathbf{Z}/2$ in dimension 3 is a "codimension one" Arf invariant. The special case $G = \mathbf{Z}/2^r$ is worked out in Example 11.1.
    Many geometric results have been obtained just from the vanishing of the odd–dimensional *L*–groups of odd order groups, but unfortunately

$L_n^s(\mathbf{Z}G)$ is usually not zero, and the torsion subgroup can be complicated (for example, even when $G$ is a group of odd order times an abelian 2-group).

Nor does it help to relax the Whitehead torsion requirements, and allow surgery just up to homotopy equivalence. For example, the group $L_{2k}^h(\mathbf{Z}[\mathbf{Z}/2^r])$ has torsion subgroup $([2(2^{r-2}+2)/3] - [r/2] - \epsilon)\mathbf{Z}/2$, where $\epsilon = 1$ if $k$ is even and 0 if $k$ is odd [12, Thm.A]. The notation $[x]$ means the greatest integer in $x$. The source of this torsion is $D(\mathbf{Z}G) \subseteq \tilde{K}_0(\mathbf{Z}G)$, a part of the projective class group that is often amenable to calculation [50].

The torsion subgroup of $L_n(\mathbf{Z}G)$ can also involve the ideal class groups of the algebraic number fields in the centre of the rational group algebra $\mathbf{Q}G$, and the computation of ideal class groups is a well–known and difficult problem in number theory. Another major complication is that computing the surgery obstruction groups often requires information about the Whitehead groups $Wh(\mathbf{Z}G)$, the algebraic home for the theory of Whitehead torsion.

Here the problem is that the torsion subgroup $SK_1(\mathbf{Z}G)$ of $Wh(\mathbf{Z}G)$ is highly non–trivial [49]. In particular, both the first optimistic claims for the Whitehead groups of abelian groups (tentatively quoted by Milnor in [47]) and Wall's conjecture [75, p.64,5.1.3] about the Tate cohomology of Whitehead groups, turned out to be incorrect.

In spite of these complications, the $L$–groups can be effectively computed in many cases of interest. The approach presented here (following the procedure established by Wall in [67]–[75]) will be to try and reduce the compution of $L_*(\mathbf{Z}G)$ to specific and independent questions in number theory and representation theory. From the statement of results in Section 2, we hope that the reader can get an overview of present knowledge, and useful references for further investigation. In the rest of the paper, we describe the main steps in the calculation and work out some relatively easy examples.

## TABLE OF CONTENTS

## 1. $L$–groups, decorations and geometric anti–structures

We begin with some algebraic definitions. An *antistructure* is a triple $(R, \alpha, u)$, where $R$ is a ring with unity, $u$ is a unit in $R$ and $\alpha: R \to R$ is an anti–automorphism such that $\alpha(u) = u^{-1}$ and $\alpha^2(r) = uru^{-1}$ for all $r \in R$. Such rings have $L$–groups, denoted $L_n(R, \alpha, u)$, and in [67]–[75] Wall developed effective techniques for computing them, especially for the case when $R = \mathbf{Z}G$, and $G$ a finite group. The main idea is to compare quadratic and hermitian forms over $\mathbf{Z}G$ to those over local and global fields using the "arithmetic" pull–back square

$$\begin{array}{ccc} \mathbf{Z}G & \longrightarrow & \mathbf{Q}G \\ \downarrow & & \downarrow \\ \widehat{\mathbf{Z}}G & \longrightarrow & \widehat{\mathbf{Q}}G \end{array}$$

of rings with antistructure to obtain Mayer–Vietoris sequences in $L$–theory. Here $\widehat{\mathbf{Z}}$, $\widehat{\mathbf{Q}}$ denote the completions of $\mathbf{Z}$, $\mathbf{Q}$ with respect to the primes in $\mathbf{Z}$. The $L$–groups themselves have a variety of definitions. Wall gives both a geometric and an algebraic definition in [68]. Ranicki found the "formation" version for the odd dimensional $L$–groups [56]. Ranicki later developed an approach based on chain complexes [60] which is very useful for applications. For finite groups it is the algebraic definitions which are most useful. These definitions lead to Mayer–Vietoris sequences for the arithmetic square above as well as calculations of certain carefully chosen $L$–groups involving various completed group rings.

There is a class of antistructures which suffice for applications of $L$–groups to the topology of manifolds, and which have other good properties.

We say $(\alpha, u)$ is a *geometric antistructure* on a group ring $\mathbf{Z}G$ provide that $\alpha$ is given by $\alpha(g) = w(g)\theta(g^{-1})$, where $\theta$ is an automorphism of $G$, $w\colon G \to \{\pm 1\}$ is a homomorphism and $u = \pm b$ for some $b \in G$ [33, p.110]. A geometric antistructure is *standard* if $\theta$ is trivial and $b = e$ and *oriented* if $w$ is trivial. Clearly $(\alpha, u)$ determines $\theta$, $w$ and $b$ uniquely. Conversely, given any automorphism $\theta$, any homomorphism $w\colon G \to \{\pm 1\}$ and any $b \in G$ the pairs $(\alpha, \pm b)$ are antistructures provided $\theta^2(g) = bgb^{-1}$ for all $g \in G$, $w \circ \theta = w$, $\theta(b) = b$ and $w(b) = 1$. In particular, a geometric antistructure induces an antistructure on the group ring $AG$ for any ring with unity $A$ so they fit well with the arithmetic square. In general $L_n(R, \alpha, u) = L_{n+2}(R, \alpha, -u)$, so we will usually only consider the case $u = b$. Another useful observation is that geometric antistructures induce involutions on $Wh(\mathbf{Z}G)$.

Traditionally the $L$–groups with standard antistructure are denoted $L_*(\mathbf{Z}G)$ if the antistructure is oriented and $L_*(\mathbf{Z}G, w)$ if it is not. One of the main theorems of surgery [68] states that these algebraically defined groups are naturally isomorphic to the geometrically defined surgery obstruction groups. Wall discovered the more general geometric antistructures while studying codimension one submanifolds (they give an algebraic description of the Browder-Livesay groups $LN$, see [68, 12C]).

The surgery obstruction groups come with $K$–theory *decorations* depending on the goal of the surgery process. For surgery up to homotopy equivalence (resp. simple homotopy equivalence) on compact manifolds, the relevant $L$–groups are $L_*^h(\mathbf{Z}G)$ (resp. $L_*^s(\mathbf{Z}G)$). For surgery on non–compact manifolds up to proper homotopy equivalence, the appropriate groups are $L_*^p(\mathbf{Z}G)$: see [55] and the references therein. Cappell [11] introduced "intermediate" $L$–groups for any involution invariant subgroup of $\tilde{K}_0(\mathbf{Z}G)$ or $Wh(\mathbf{Z}G)$ for use in his work on Mayer–Vietoris sequences for amalgamated free products and HNN extensions. Each of these has the form $L_n^{\tilde{X}}(\mathbf{Z}G)$, denoting an algebraic $L$–group ([56]) with decorations in an involution–invariant subgroup $\tilde{X} \subseteq \tilde{K}_1(\mathbf{Z}G)$ or $\tilde{X} \subseteq \tilde{K}_0(\mathbf{Z}G)$. In the first case, $\tilde{X}$ is always the inverse image of an involution–invariant subgroup in $Wh(\mathbf{Z}G)$, so we often refer instead to the decoration subgroup in $Wh(\mathbf{Z}G)$. For general antistructures $(\alpha, u)$ $L_*^{\tilde{X}}(\mathbf{Z}G, \alpha, u)$ is not defined unless $u \in \tilde{X}$, so again geometric antistructures provide the right setting.

Intermediate $L$–groups appear in the arithmetic Mayer–Vietoris sequence as well: in particular, $L_n'(\mathbf{Z}G)$ based on

$$\tilde{X} = SK_1(\mathbf{Z}G) \subseteq Wh(\mathbf{Z}G),$$

where $SK_1(\mathbf{Z}G) = \ker\bigl(K_1(\mathbf{Z}G) \to K_1(\mathbf{Q}G)\bigr)$, is especially important. It turns out that these $L'$-groups are more accessible to computation than either $L^s$ or $L^h$, and so they have a central role in this subject.

The $L$-groups are related by many interlocking exact sequences [60], involving change of $K$-theory and change of rings. Such sequences often have both an algebraic and an geometric interpretation, making them useful for topological applications.

REMARK: The $L$-groups mentioned so far only involve the $K$ groups $K_0$ and $K_1$, and it is natural to wonder about decorations in other $K_i$. In fact, there are geometrically interesting $L$-theories for both higher and lower $K_i$, and these are related to the ones studied here via change of $K$-theory sequences. See [77] for the higher $L$-theories and [61] for the lower ones. Carter, [16], [17], calculates $K_{-1}(\mathbf{Z}G)$ and shows $K_{-i}(\mathbf{Z}G) = 0$ for finite groups if $i \geq 2$. This should bring the calculation of the lower $L$-groups within range (see [44, §4] for lower $L$-groups of odd order cyclic groups), but we do not know of any comprehensive treatment. Calculation of the higher $L$-groups seems to much more difficult. For these reasons, this survey omits consideration of both higher and lower $L$-groups.

The exact sequences describing the change of $K$-theory decoration are often called "Ranicki–Rothenberg" sequences [62], [56]. Some important examples are

$$\ldots \to L^h_{n+1}(\mathbf{Z}G) \to H^{n+1}(Wh(\mathbf{Z}G)) \to L^s_n(\mathbf{Z}G) \to L^h_n(\mathbf{Z}G)$$
$$\to H^n(Wh(\mathbf{Z}G)) \to \ldots$$

and

$$\ldots \to L^p_{n+1}(\mathbf{Z}G) \to H^n(\tilde{K}_0(\mathbf{Z}G)) \to L^h_n(\mathbf{Z}G) \to L^p_n(\mathbf{Z}G)$$
$$\to H^{n-1}(\tilde{K}_0(\mathbf{Z}G)) \to \ldots$$

although the step between $L^s$ and $L^h$ can also be usefully divided into $L^s \to L'$, with relative group $H^*(SK_1(\mathbf{Z}G))$, and $L' \to L^h$, with relative group $H^*(Wh'(\mathbf{Z}G))$. Here $Wh'(\mathbf{Z}G) = Wh(\mathbf{Z}G)/SK_1(\mathbf{Z}G)$, and we use the convention that $H^*(X)$ denotes the Tate cohomology $H^*(\mathbf{Z}/2; X)$ for any $\mathbf{Z}/2$-module $X$: $H^*(\mathbf{Z}/2; X) = \{a \in X \mid a = (-1)^*\bar{a}\}/\{a + (-1)^*\bar{a}\}$. There are versions of these sequences for any geometric antistructure.

The exact sequences involving change of rings are particularly important for computing $L$-groups. The most important example is

$$\ldots \to L'_{n+1}(\mathbf{Z}G \to \hat{\mathbf{Z}}_2 G) \to L'_n(\mathbf{Z}G) \to L'_n(\hat{\mathbf{Z}}_2 G) \to L'_n(\mathbf{Z}G \to \hat{\mathbf{Z}}_2 G) \to \ldots$$

This sequence remains exact for any geometric antistructure.

## 2. Statement of Results

What does it mean to *compute* L–groups? Given a finite group $G$ and a geometric antistructure $(\alpha, u)$, the rational group algebra $\mathbf{Q}G$ becomes an algebra with involution. Under the Wedderburn decomposition, $\mathbf{Q}G$ splits canonically into simple algebras with involution and it is reasonable to assume that the classical invariants for such algebras (type, reduced norms, Schur indices, ideal class groups of centre fields, etc.) can be worked out for the given group $G$.

GOAL:    *Find an algorithm to compute $L_*(\mathbf{Z}G, \alpha, u)$ for geometric anti-structures, in terms of the character theory of $G$ and the classical invariants of $\mathbf{Q}G$.*

Here is a brief summary of the state of progress towards this goal. Properties (1) and (2) in the introduction hold for L–groups of finite groups with geometric antistructures and any involution–invariant torsion decoration in $Wh(\mathbf{Z}G)$ or $\tilde{K}_0(\mathbf{Z}G)$ [25], [74], [75], although the description of the multi–signature becomes more complicated. Dress's results (3) certainly hold for the standard geometric antistructures (oriented or not) with decorations $p$, $s$, $\prime$, $h$ and many others, but the general case has not been worked out. In the case of the standard antistructures, we will apply Dress induction to obtain calculations for odd order groups in §10, and $p$–hyperelementary groups, $p$ odd, in §12.

One obvious difficulty in extending Dress induction is that a given geometric antistructure on $G$ may not restrict to a geometric antistructure on enough subgroups to simplify the calculation. In any case, not much work has been done on the maps induced by subgroup inclusion (even for the standard oriented antistructure), so we will consider only 2–hyperelementary groups from now on.

(4) *For 2–hyperelementary groups $G$ with arbitrary geometric antistructure $(\alpha, u)$, the groups $L_*^p(\mathbf{Z}G, \alpha, u)$ can be computed in terms of character theory of $G$ and the number theory associated to $\mathbf{Q}G$. The torsion subgroup has exponent 2 for the standard oriented antistructure, and exponent 4 in general.* [30]

Since the goal has been achieved for the $L^p$–groups, we turn to the groups $L_*'(\mathbf{Z}G, \alpha, u)$ with $Wh(\mathbf{Z}G)$ decoration lying in the subgroup $SK_1(\mathbf{Z}G)$. The approach is to study $L_*'(\mathbf{Z}G, \alpha, u)$ using the exact sequence comparing it with $L_*'(\widehat{\mathbf{Z}}_2G, \alpha, u)$. The relative term is under control:

(5) *For 2–hyperelementary groups $G$ and any geometric antistructure, the relative groups $L_*'(\mathbf{Z}G \to \widehat{\mathbf{Z}}_2G, \alpha, u)$ can be computed in terms of char-*

acter theory of $G$ and number theory associated to $\mathbf{Q}G$. The torsion subgroup has exponent 2 (see [75] or Tables 14.12–14.15, and Theorem 7.1 for the exponent of the torsion subgroup).

The remaining obstacles are the groups $L'_*(\widehat{\mathbf{Z}}_2G, \alpha, u)$, and the maps

$$\psi_n \colon L'_*(\widehat{\mathbf{Z}}_2G, \alpha, u) \to L'_*(\mathbf{Z}G \to \widehat{\mathbf{Z}}_2G, \alpha, u) \ .$$

The groups $L'_*(\widehat{\mathbf{Z}}_2G, \alpha, u)$ can be compared to $L^h_*(\widehat{\mathbf{Z}}_2G, \alpha, u)$ using the change of $K$-theory sequence

$$\ldots \to H^{n+1}(Wh'(\widehat{\mathbf{Z}}_2G)) \to L'_n(\widehat{\mathbf{Z}}_2G, \alpha, u) \to L^h_n(\widehat{\mathbf{Z}}_2G, \alpha, u)$$
$$\to H^n(Wh'(\widehat{\mathbf{Z}}_2G)) \to \ldots$$

Every third term is easy to compute:

(6) *The groups $L^h_*(\widehat{\mathbf{Z}}_2G, \alpha, u)$ for geometric antistructures are determined by the centre of $\mathbf{F}_2G$, and the kernel of the discriminant*

$$L^h_n(\widehat{\mathbf{Z}}_2G, \alpha, u) \to H^n(Wh'(\widehat{\mathbf{Z}}_2G))$$

*is computable from the characters and types ([71] and [29, 1.16], see also Remark 8.5).*

For the Tate cohomology terms we have:

(7) *$Wh'(\widehat{\mathbf{Z}}_2G)$ is computable by restriction to the 2-elementary subgroups of $G$. There is an algorithm to calculate $Wh'(\widehat{\mathbf{Z}}_2G)$ for 2-elementary groups ([49, Thm. 6.7, 12.3]). If $w$ is trivial or if $G$ is a 2-group, the involution induced by the geometric antistructure has been computed fairly explicitly (see [49, p.163] and [50, p.61]).*

It follows that $L'_*(\widehat{\mathbf{Z}}_2G, \alpha, u)$ is algorithmically computable up to extensions in the oriented case or in the case that $G$ is a finite 2-group.

Most difficult of all is to describe the $\psi_*$ maps. Some examples can be worked out, especially for the standard antistructure.

**Example:** *If $G$ is an abelian group then $L'_*(\mathbf{Z}G, w)$ is computable in terms of the characters of $G$ (see [75], and Example 11.1 for cyclic 2-groups done in detail).*

**Example:** *If $G = G_1 \times G_2$ is a direct product where $G_1$ has odd order, then $L'_*(\mathbf{Z}G, w)$ is computable in terms of $L'_*(\mathbf{Z}G_2, w)$ and the character theory of $\mathbf{Q}G$ ([39] and Proposition 12.1).*

**Example:** *Computations (modulo some extension problems) are available for $L'_*(\mathbf{Z}G, w)$ in certain families of 2–hyperelementary groups $G$, including*

    (i) *groups of 2–power order, (see [75, §5] for partial results and Example 9.2 for a reduction of the maps $\psi_*$ to $K$–theory),*

    (ii) *groups with periodic cohomology, [42], [39]*

    (iii) *groups $G = C \rtimes P$ where $\ker(P \to Aut(C))$ is abelian and $P$ is a 2–group (see [75], and Proposition 13.4 for $G$ dihedral done as an example).*

Finally, what can one say about the $L$–groups of primary geometric interest? To study $L^s_*(\mathbf{Z}G)$ or $L^h_*(\mathbf{Z}G)$ via $L'_*(\mathbf{Z}G)$ we need the groups $H^n(SK_1(\mathbf{Z}G))$ or $H^n(Wh'(\mathbf{Z}G))$. Observe that we only need $Wh(\mathbf{Z}G) \otimes \mathbf{Z}_{(2)}$ for computing Tate cohomology, and this can result in significant simplification.

    (8) *Extensive computations are available for the groups $SK_1(\mathbf{Z}G)$, but it is not easy to determine the action of the involution induced by the antistructure (see [49] as a general reference, and [46] for a nice application to $L^s$–groups).*

There is a short exact sequence

$$0 \to Cl_1(\mathbf{Z}G) \otimes \mathbf{Z}_{(2)} \to SK_1(\mathbf{Z}G) \otimes \mathbf{Z}_{(2)} \to SK_1(\widehat{\mathbf{Z}}_2 G) \to 0$$

and Bak [3] (or [49, 5.12]) shows the standard oriented involution is trivial on $Cl_1(\mathbf{Z}G) \otimes \mathbf{Z}_{(2)}$. Oliver [49, 8.6] shows that the standard oriented involution on $SK_1(\widehat{\mathbf{Z}}_2 G)$ is multiplication by $-1$, at least for 2–groups $G$.

    (9) *The groups $Wh'(\mathbf{Z}G)$ are free abelian with rank depending on the characters of $G$. For the standard oriented antistructure, the induced involution on $Wh'(\mathbf{Z}G)$ is the identity, and $H^1(Wh'(\mathbf{Z}G)) = 0$ [72] (see [50, 4.8] for the answer when $w \not\equiv 1$).*

The study of $L^h_*(\mathbf{Z}G)$ via $L^p_*(\mathbf{Z}G)$ looks promising, but we need knowledge of $H^n(\tilde{K}_0(\mathbf{Z}G))$. In general this is not easy to compute, however:

**Example:** *The groups $L^h_*(\mathbf{Z}G, w)$ can be computed (up to extensions) in terms of the character theory of $G$, for $G$ a finite 2–group (See [31], [50], [12] for pieces of the solution, but there is no complete account in the literature).*

REMARK: In order to keep this paper reasonably short, we have omitted any discussion of hermitian $K$–theory and form parameters. This approach is fully developed in [9], [10], and [1]–[6].

## 3. Round decorations

There are two detours to be made along the way towards systematic computations of the surgery obstruction groups. The first is to use the "round" algebraic $L$-theory $L_n^X(R, \alpha, u)$ for a ring $R$ with antistructure, based on involution–invariant subgroups $X$ of $K_i(R)$, instead of the geometrically useful groups.. These groups were introduced by Wall in his sequence of calculational papers, [**67, 69, 71, 73, 74, 75**] and formalized in [**32**, §2]. These round groups have several algebraic advantages. They respect products, and are invariant under quadratic Morita equivalence of rings [**33**]. These are related to the usual $L$-groups by exact sequences which can be analyzed and largely determined in the cases considered here. The definitions also make sense for higher and lower $L$-theory, but all of our actual computations are for the $L$-groups based on subgroups of $K_0$ or $K_1$. To our knowledge, no calculational work has been done on higher $L$-theory for finite groups and very little has been done for lower $L$-theory.

A particular example of quadratic Morita equivalence is the notion of *scaling*. If $(\alpha, u)$ is an antistructure on $R$ and $v \in R$ is a unit, $(\alpha_v, u_v)$ is also an antistructure where $\alpha_v(x) = v\alpha(x)v^{-1}$ and $u_v = v\alpha(v^{-1})u$: furthermore $L_n^X(R, \alpha, u) \cong L_n^X(R, \alpha_v, u_v)$ [**33**, p.74] for any subgroup $X \subset K_i(R)$ invariant under the involution induced by $\alpha$.

The second step is to choose the "right" $K$-theory decoration. For any ring $R$, let $O_i(R) = 0 \subseteq K_i(R)$ and let $X_i(R) = \ker\big(K_i(R) \to K_i(R \otimes \mathbf{Q})\big)$. Note that $L_n^{O_i}(R) = L_n^{K_{i+1}}(R)$, so there is a Ranicki–Rothenberg sequence relating $L^{O_i} \to L^{O_{i-1}}$ with relative term $H^*(K_i(R))$. For group rings $AG$ and $i = 1$, we define

$$Y_1(AG) = X_1(AG) \oplus \{\pm G^{ab}\}$$

where $G^{ab}$ denotes the set of elements in the abelianization of $G$. If $i \leq 0$ then we define $Y_i(AG) = X_i(AG)$. For higher $L$-theory ($i \geq 2$) it seems that the right definition of $Y_i(AG)$ would be the image of the assembly map in $K$-theory.

Fortunately, the passage between round and geometric $L$-theory is very uniform so the round results suffice. Given a geometric antistructure and any invariant subgroup $\widetilde{U} \subset \widetilde{K}_i(\mathbf{Z}G)$, let $U \subset K_i(\mathbf{Z}G)$ denote the inverse image. There is a natural map $\tau_U \colon L_*^U(\mathbf{Z}G, \alpha, u) \to L_*^{\widetilde{U}}(\mathbf{Z}G, \alpha, u)$ from the round to the geometric theory. For subgroups of $\widetilde{K}_0$, $\tau$ is an isomorphism, and for $U = Y_1$, the following sequence is exact (see [**32**, 3.2]).

$$0 \to L_{2k}^{Y_1}(\mathbf{Z}G, \alpha, u) \xrightarrow{\tau_{Y_1}} L_{2k}'(\mathbf{Z}G, \alpha, u) \to \mathbf{Z}/2 \to L_{2k-1}^{Y_1}(\mathbf{Z}G, \alpha, u)$$
$$\xrightarrow{\tau_{Y_1}} L_{2k-1}'(\mathbf{Z}G, \alpha, u) \to 0.$$

The map into $\mathbf{Z}/2$ is given by the rank (mod 2) of the underlying free module.

THEOREM 3.1: *For $(\alpha, u)$ any geometric antistructure, $L_{2k}^{Y_1}(\mathbf{Z}G, \alpha, u) \cong L'_{2k}(\mathbf{Z}G, \alpha, u)$ and $L'_{2k-1}(\mathbf{Z}G, \alpha, u)$ is obtained from $L_{2k-1}^{Y_1}(\mathbf{Z}G, \alpha, u)$ by dividing out a single $\mathbf{Z}/2$ summand.*

The intermediate projective $L$–groups we can compute are the $L_n^{X_0}(\mathbf{Z}G) = L_n^{X_0}(\mathbf{Z}G)$ (denoted $L_n^{X_0}(\mathbf{Z}G)$ in [27, §3]) based on the subgroup $X_0(\mathbf{Z}G)$. These are related to the usual projective $L$–groups, $L^p = L^{K_0}$, by the exact sequence [32, 3.2], [27, 3.8])

$$0 \to L_{2k}^{X_0}(\mathbf{Z}G) \to L_{2k}^p(\mathbf{Z}G) \to \mathbf{Z}/2 \to L_{2k-1}^{X_0}(\mathbf{Z}G) \to L_{2k-1}^p(\mathbf{Z}G) \to 0 \ .$$

The map into $\mathbf{Z}/2$ is given by the rank (mod 2) of the underlying projective module.

THEOREM 3.2: *For $(\alpha, u)$ any geometric antistructure, $L_{2k}^{X_0}(\mathbf{Z}G, \alpha, u) \cong L_{2k}^p(\mathbf{Z}G, \alpha, u)$ and $L_{2k-1}^p(\mathbf{Z}G, \alpha, u)$ is obtained from $L_{2k-1}^{X_0}(\mathbf{Z}G, \alpha, u)$ by dividing out a single $\mathbf{Z}/2$ summand.*

We now follow the procedure outlined in the first two sections to compute $L_n^{Y_i}$, $i = 0$, 1 and then use Theorems 3.2 and 3.1 to compute $L_*^p$ or $L'_*$.

## 4. The main exact sequence

Exact sequences for computing $L$–groups come from the arithmetic square [74], [60, §6], where the basic form is

$$\ldots \to L_n^{X_i}(R) \to L_n^{X_i}(\hat{R}) \oplus L_n^{O_i}(S) \to L_n^{O_i}(\hat{S}) \to L_{n-1}^{X_i}(R) \to \ldots$$

where $R$ is a ring with antistructure, $S = R \otimes \mathbf{Q}$, $\hat{R} = R \otimes \hat{\mathbf{Z}}$ and $\hat{S} = \hat{R} \otimes \mathbf{Q}$. This assumes that excision holds in algebraic $L$–theory, which is known for $i \leq 1$.

Most of the difficulties involved in computing $L_n^{Y_i}(\mathbf{Z}G, \alpha, u)$ for a geometric antistructure concern the group $L_n^{Y_i}(\hat{\mathbf{Z}}_2 G, \alpha, u)$. We therefore reorganize the calculation by considering the exact sequence
(4.1)
$$\ldots \to L_{n+1}^{Y_i}(\mathbf{Z}G \to \hat{\mathbf{Z}}_2 G) \to L_n^{Y_i}(\mathbf{Z}G) \to L_n^{Y_i}(\hat{\mathbf{Z}}_2 G) \xrightarrow{\psi_n} L_n^{Y_i}(\mathbf{Z}G \to \hat{\mathbf{Z}}_2 G) \to \ldots$$

valid for any antistructure. On the other hand, we have isomorphisms of relative groups

$$L_n^{Y_i}(\mathbf{Z}G \to \hat{\mathbf{Z}}_2 G) \cong L_n^{X_i}(\mathbf{Z}G \to \hat{\mathbf{Z}}_2 G)$$

so we are free to use the $L^{X_i}$ relative groups for computation. By excision

$$(4.2) \qquad L_n^{X_i}(\mathbf{Z}G \to \widehat{\mathbf{Z}}_2 G) \cong L_n^{O_i}(\widehat{R}_{odd} \oplus S \to \hat{S}),$$

where $\widehat{R}_{odd}$ is the product of the $\ell$-adic completions of $R$ at all odd primes $\ell$.

We now introduce the groups

$$(4.3) \qquad CL_n^{O_i}(S) = L_n^{O_i}(S \to S_A)$$

where $S_A = \hat{S} \oplus (S \otimes \mathbf{R})$ is the adelic completion of S. Let $T = S \otimes \mathbf{R}$. Then by the arithmetic sequence and (4.2) we have a long exact sequence
(4.4)
$$\ldots CL_{n+1}^{O_i}(S) \to L_{n+1}^{X_i}(\mathbf{Z}G \to \widehat{\mathbf{Z}}_2 G) \to L_n^{O_i}(\widehat{R}_{odd} \oplus T) \xrightarrow{\gamma_n} CL_n^{O_i}(S) \to \ldots$$

valid for any geometric antistructure. This is the main exact sequence for computing the relative groups, and then (4.1) is used to compute the absolute groups. It is a major ingredient in Wall's program that the groups $CL_n^{O_i}(S)$ are actually computable [**73**], although not finitely generated. In fact, they are elementary abelian 2–groups depending only on the idèle class groups of the centre of $S$ (see Tables 14.9–14.11). This is a form of the Hasse principle for quadratic forms, and follows from the work of Kneser on Galois cohomology.

## 5. Dress induction and idempotents

The calculation of $L$–groups of finite groups can be reduced to calculating a limit of $L$–groups for hyperelementary subgroups. This process is known as Dress induction. More generally, Dress assumes that some Green ring, say $\mathcal{G}$, acts on a Mackey functor $\mathcal{M}$. Let $\mathcal{H}$ denote a family of subgroups of $G$ which is closed under conjugation and subgroups. Write

$$\delta_{\mathcal{G}}^{\mathcal{H}} : \bigoplus_{H \in \mathcal{H}} \mathcal{G}(H) \to \mathcal{G}(G)$$

for the sum of the induction maps.

THEOREM 5.1: *If there exists* $y \in \bigoplus_{H \in \mathcal{H}} \mathcal{G}(H)$ *such that* $\delta_{\mathcal{G}}^{\mathcal{H}}(y) = 1 \in \mathcal{G}(G)$, *then both Amitsur complexes for* $\mathcal{M}$ *are contractible.*

REMARK: One writes the conclusion as $\mathcal{M}(G) = \varprojlim_{\mathcal{H}} \mathcal{M}(H) = \varinjlim_{\mathcal{H}} \mathcal{M}(H)$ where the first limit made up of restrictions and the second of inductions.

We also say $\mathcal{M}$ is *computable* from the family $\mathcal{H}$. The result above follows from [**24**, Prop.1.2, p.305] and the remark just above [**25**, Prop.1.3, p.190].

Dress also proves a local result which says the following about $\mathcal{M}(G)$. Fix a prime $p$, let $\mathcal{H}_p$ denote the family of $p$–hyperelementary subgroups and let $\mathcal{M}(G)_{(p)}$ denote the $p$– localization of $\mathcal{M}(G)$. Then

$$(5.2) \qquad \mathcal{M}(G)_{(p)} = \varprojlim_{\mathcal{H}_p} \mathcal{M}(H)_{(p)} = \varinjlim_{\mathcal{H}_p} \mathcal{M}(H)_{(p)} \ .$$

By [**74**] the 2–localization map is an injection on $L$–groups of finite groups. To apply these results to computation of $L$–groups, Dress defined a suitable Green ring (see also [**75**]). For any commutative ring $R$, let $y(G, \theta, R)$ be the Grothendieck group of finitely–generated, projective left $R$ modules with an $R$ bilinear form $\lambda \colon M \times M \to R$ which is

   *equivariant* in that $\lambda(m_1, gm_2) = \lambda(\theta(g^{-1})m_1, m_2)$,
   *symmetric* in that $\lambda(m_2, m_1) = \lambda(b^{-1}m_1, m_2)$, and
   *non–singular* in that the adjoint of $\lambda$ is an isomorphism.

Define $GU(G, \theta, R)$ and $GW(G, \theta, R)$ by equating two forms with isomorphic Lagrangians or modding out forms with a Lagrangian. Thomas [**65**] produces the formulae needed to check that $GW(G, \theta, R)$ acts on the groups $L_n^p(RG, \alpha, b)$ where $\alpha$ is the antistructure associated to any geometric antistructure $\theta$, $b$, $w$ where $\theta$ and $b$ are fixed but $w$ is allowed to vary subject only to $w \circ \theta = w$. Dress proves generation results for the case $\theta = 1_G$ which yield

THEOREM 5.3: *For the standard geometric antistructures, the $L^p$–groups of finite groups are computable from the family of 2–hyperelementary and $p$–elementary subgroups, $p$ odd. The torsion subgroups of the $L^p$–groups and the groups $L^p \otimes \mathbf{Z}_{(2)}$ are computable from the family of 2–hyperelementary subgroups.*

REMARK 5.4: One can also show that the round groups $L^{o_i}$, $L^{x_i}$ and $L^{y_i}$ localized at 2 are 2–hyperelementary computable. This can be done either by refining the groups $GW$ or by studying the Ranicki–Rothenberg sequences.

Dress proves these results by studying the Burnside ring and its $p$–localizations. He also constructed idempotents in the $p$–local Burnside ring and in [**27**, §6] and [**49**, §11] these idempotents are combined with induction theory to do calculations. We discuss the $p$–local case on a finite group $G$. Dress constructs one idempotent $e_E(G)$ in the $p$–local Burnside ring for each conjugacy class of cyclic subgroups of order prime to $p$ in $G$

and shows that they are orthogonal. One can then split any $p$–local Mackey functor, $F$, using these idempotents into pieces $e_E(G) \cdot F(G)$ plus a piece left over since $1_G \neq \sum_E e_E(G)$ in the $p$–local Burnside ring. If $F(G)$ is generated by the images under induction from the $p$–hyperelementary subgroups, then the leftover piece vanishes. The additional result we want is Oliver's identification of the pieces [**49**, 11.5, p.256]. Let $F$ be a $p$–local Mackey functor on $G$ which is $p$–hyperelementary generated. In general Oliver describes $e_E(G) \cdot F(G)$ as a limit over subgroups $H$ of the form $E \lhd H \twoheadrightarrow P$ where $P$ is a $p$–group. He then makes the observation that the limit takes place inside $N_G(E)$ so

$$e_E(G) \cdot F(G) = e_E\big(N_G(E)\big) \cdot F\big(N_G(E)\big) \ .$$

If one could compute conjugations, induction and restriction maps for index $p$–inclusions of $p$–hyperelementary, then one could work out the limit in general. This makes 2–hyperelementary groups especially important for $L$–theory.

The $L$–theory case has an additional feature: for general geometric antistructures, the $L$–groups are not Mackey functors. The theory of "twisting diagrams", [**33**] or [**59**], can be used to overcome this difficulty.

Given an extension

$$G \lhd \widehat{G} \overset{\phi}{\twoheadrightarrow} \mathbf{Z}/2,$$

one can use $\phi$ to pull–back a non–trivial line bundle over a surgery problem with fundamental group $\widehat{G}$. This gives rise to a transfer map

$$tr \colon L_n(\mathbf{Z}\widehat{G}, w\phi) \to L_{n+1}(\mathbf{Z}G \to \mathbf{Z}\widehat{G}, w) \ .$$

Selecting a generator $t \in \widehat{G} - G$ gives rise to an automorphism $\theta$ of $G$ given by conjugation by $t$. We assume that $w$ extends over $\widehat{G}$ with $w(t) = 1$. Setting $b = t^2 \in G$ gives a geometric antistructure on $G$ and there is a long exact sequence

$$\cdots \to L_{n+2}(\mathbf{Z}G \to \mathbf{Z}\widehat{G}, w) \to L_n(\mathbf{Z}G, \alpha, b) \to L_n(\mathbf{Z}\widehat{G}, w\phi)$$
$$\overset{tr}{\longrightarrow} L_{n+1}(\mathbf{Z}G \to \mathbf{Z}\widehat{G}, w) \to \cdots$$

where $\alpha$ is the antistructure associated to $\theta$, $b$ and $s\phi$. Conversely, any geometric antistructure on $G$ arises from such a procedure. The group $L_n(\mathbf{Z}\widehat{G}, \lambda w)$ is a Mackey functor and the relative group $L_{n+1}(\mathbf{Z}G \to \mathbf{Z}\widehat{G}, w)$ sits in a long exact sequence where the other two terms are Mackey functors. This allows one to produce decompositions of $L_n(\mathbf{Z}G, \alpha, b)$ mimicking the Mackey functor case even though $L_n(\mathbf{Z}G, \alpha, b)$ has no obvious Mackey functor structure.

The most important application of these decomposition techniques applies to the 2–hyperelementary case $G = C \rtimes P$, because here there is a further identification of the $e_C(G) \cdot F(G)$ with a twisted group ring of $P$. In principle, this reduces the calculation to 2–groups where numerous simplifications occur. Wall [**75**, §4] was the first to explore this decomposition. He produced the idempotent decomposition by hand but had to restrict to groups with abelian 2–Sylow group. Hambleton–Madsen do the general case using the Burnside ring idempotents, [**27**, §6].

If $C_m$ is cyclic of odd–order $m$, let $\zeta_m$ denote a primitive $m$th root of unity. Any $d \,|\, m$ determines a unique cyclic subgroup of odd order of $G$ and we denote the corresponding summand of our functor by $F(G)(d)$. The map $t \colon P \to Aut(C_m)$ can be regarded as a map $t \colon P \to Gal\big(\mathbf{Q}(\zeta_m)/\mathbf{Q}\big)$ and we let $\mathbf{Z}[\zeta_m]^t P$ denote the corresponding *twisted group ring*. Any geometric antistructure induces an antistructure on $\mathbf{Z}[\zeta_m]^t P$ which we continue to denote by $(\alpha, u)$. The main exact sequence can be applied again.

THEOREM 5.5:([**27**, 6.13, 7.2]) *For $i = 0, 1$ and $G = C_m \rtimes P$ with any geometric antistructure, there is a natural splitting*

$$L_n^{Y_i}(\mathbf{Z}G, \alpha, u)_{(2)} = \sum{}^{\oplus}\big\{L_n^{Y_i}(\mathbf{Z}G, \alpha, u)(d) : d \,|\, m\big\}$$

(i) $L_n^{Y_i}(\mathbf{Z}G, \alpha, u)(1) \cong L_n^{X_i}(\mathbf{Z}P, \alpha, u)_{(2)}$ *via the restriction map,*
(ii) *for $d \neq 1$, $L_n^{Y_i}(\mathbf{Z}G, \alpha, u)(d) \cong L_n^{X_i}(\mathbf{Z}G, \alpha, u)(d)$,*
(iii) $L_n^{Y_i}(\mathbf{Z}G, \alpha, u)(d)$ *maps isomorphically under restriction to*
    $L_n^{Y_i}(\mathbf{Z}[C_d \rtimes P], \alpha, u)(d)$,
(iv) *for each $d \,|\, m$ there is a long exact sequence*

$$\cdots \to CL_{n+1}^{O_i}(S(d)) \to L_n^{X_i}(\mathbf{Z}G, \alpha, u)(d) \to \prod_{\ell \nmid d} L_n^{X_i}(\widehat{R}_\ell(d)) \oplus L_n^{O_i}(T(d))$$

$$\to CL_n^{O_i}(S(d)) \to \cdots$$

*where $R(d)$ is the twisted group ring $\mathbf{Z}[\zeta_d]^t P$ with antistructure $(\alpha, u)$, and $S(d) = R(d) \otimes \mathbf{Q}$, $\widehat{R}_\ell(d) = R(d) \otimes \widehat{\mathbf{Z}}_\ell$, $T(d) = R(d) \otimes \mathbf{R}$.*

There are similar splittings and calculations for the relative $L$–groups and the 2–adic $L$–groups.

For certain purposes, it is useful to be able to detect surgery obstructions. Dress's results for the standard antistructures say that we can detect by transfer to the collection of 2–hyperelementary subgroups. In some cases we can also reduce from hyperelementary groups to a smaller collection, the *basic* groups. More explicitly [**34**, 3.A.6], a 2–hyperelementary group $G = C \rtimes P$ is basic provided

$$P_1 = \ker\big(t\colon P \to Aut(C)\big)$$

is cyclic, dihedral, semi-dihedral or quaternion.

In [**34**, 1.A.4] we introduced the category $RG$–Morita, for any commutative ring $R$. The category $\mathbf{Q}G$–Morita is defined as follows. The objects are subgroups, $H \subset G$, and the morphisms from $H_1$ to $H_2$ are generated by the $H_2$–$H_1$ bisets $X$, modulo some relations spelled out in [**34**, p.249–250]. From [**34**, 1.A.12(i), p.251], $R_{\mathbf{Q}}(G)$ is a functor on $\mathbf{Q}G$–Morita defined by sending a rational representation $V$ of $H_1$ to $\mathbf{Q}[X] \otimes_{\mathbf{Q}H_1} V$. Note if $V$ is a permutation module on the $H_1$–set $Y$, then

$$\mathbf{Q}[X] \otimes_{\mathbf{Q}H_1} \mathbf{Q}[Y] = \mathbf{Q}[X \times_{H_1} Y] \ .$$

We proved in [**34**, 1.A.9, p.251] that the morphisms in $\mathbf{Q}G$–Morita are generated by generalized inductions and restrictions corresponding to homomorphisms $f\colon G_1 \to G_2$ which are either injections (subgroups) or surjections (quotient groups). Let $\mathcal{M}$ be a functor on $\mathbf{Q}G$–Morita.

THEOREM 5.6:([**34**], 1.A.11, p.251) *The sum of the generalized restriction maps,*

$$\mathcal{M}(G) \to \bigoplus_{B \in \mathcal{B}_G} \mathcal{M}(B)$$

*is a split injection where $\mathcal{B}_G$ denotes the set of basic subquotients of $G$. The sum of the generalized induction maps is a split surjection.*

This result has an analogue for quadratic Morita theory, and it applies to detect the relative groups $L_*^{X_i}(\mathbf{Z}G \to \hat{\mathbf{Z}}_2 G)(d)$ for $i = 0, 1$ since these are functors out of $(\mathbf{Q}G, -)$–Morita. To detect $L_*^p(\mathbf{Z}G, w)$ we need the $w$–basic subquotients of $G$, defined in [**34**]. Combining [**34**,1.B.7] with [**34**,6.2] gives:

THEOREM 5.7:([**30**, Thm.A]) *Let $G$ be a 2-hyperelementary group. Then the sum of all the (generalized) restriction maps*

$$L_n^p(\mathbf{Z}G, w) \longrightarrow L_n^p(\mathbf{Z}[\bar{G}], w) \oplus \sum \left\{ L_n^p(\mathbf{Z}[H/N], w) : \begin{array}{l} H/N \ \text{ a } w\text{–basic} \\ \text{subquotient of } G \end{array} \right\}$$

*is a natural (split) injection, where $\bar{G} = G/[P_1, P_1]$.*

We remark that [30, Thm.B] lists specific invariants which detect all oriented $L^p$ surgery obstructions, and [30, 5.21] shows that the torsion subgroup of the $L^p$-groups has exponent 4 in general (exponent 2 in the standard oriented case).

## 6. Central simple algebras with involution

We collect here some terminology and results about the building blocks for our calculations. These are the $L$–groups $L_n^{O_i}(R, \alpha, u)$ where $(R, \alpha, u)$ denotes an antistructure over an algebra. When the cases $i = 0, 1$ are being considered separately, we will use Wall's notation $L^s = L^{O_1}$ and $L^K = L^{O_0}$.

First we summarize some of the standard facts about quadratic forms on simple algebras with centre field continuous, local (of characteristic 0), and finite. For our purposes, the main references are [69] and [73]. Since we are only interested in the applications to surgery theory, we will restrict ourselves to the simple algebras which arise from the rational group rings of finite groups. This assumption will simplify the arguments at various points. More precisely, if $D$ denotes such a skew field with centre E, and $A \subseteq E$ the ring of integers, then $E$ is an abelian extension of $\mathbf{Q}$. We fix an odd integer $d$ such that $\hat{D}_\ell$ is split, and $E_\ell$ is an unramified extension of $\hat{\mathbf{Q}}_\ell$ for all finite primes $\ell$ with $\ell \nmid 2d$. We also assume that $D$ has "uniformly distributed invariants": the Schur indices of $D$ at all primes $\ell \in E$ over a fixed rational prime are equal, and the Hasse invariants are Galois conjugate. This holds for the algebras arising from group rings by the Benard-Schacher Theorem [78, Th. 6.1].

In addition to listing the values of the groups, we mention explicit invariants (such as signature and discriminant) used to detect them. From these facts we can compute the $CL_n^{O_i}$ and prepare for the computation of the maps $\gamma_i$. The $L^s$ to $L^K$ Rothenberg sequences are tabulated in Tables 14.1–14.8.

If $(D, \alpha, u)$ denotes an antistructure on a division algebra with centre $E$ (and $A \subseteq E$ the ring of integers), then we distinguish as usual types $U$, $Sp$ and $O$ (see [75, §1.2]). We further subdivide into types $OK$ if $D = E$, type $OD$ if $D \neq E$ and similarly for type $Sp$. If an involution-invariant factor is the product of two simple rings interchanged by the involution, this is type $GL$. Such factors make no contribution to $L$-theory. When the anti-structure is understood, we will say "D has type ..." for short. Recall that $L_n^K(D, \alpha, -u) = L_{n+2}^K(D, \alpha, u)$ and types $O$ and $Sp$ are interchanged, so we usually just list type $O$.

**(6.1) Continuous Fields:** For continuous fields ($E = \mathbf{R}$ or $\mathbf{C}$) the signature gives an explicit isomorphism of $L_0^K(\mathbf{C}, c, 1)$, $L_2^K(\mathbf{C}, c, 1)$, $L_0^K(\mathbf{R}, 1, 1)$ and $L_0^K(\mathbf{H}, c, 1)$ onto $2\mathbf{Z}$ (the types are $U$, $U$, $O$ and $Sp$); in all these cases except for $(\mathbf{H}, c, 1)$ the discriminant map $L_0^K(E) \to H^0(E^\times)$ is onto. Indeed the groups $H^n(K_1(\mathbf{H})) = 0$ so $L_n^S(\mathbf{H}, c, 1) = L_n^K(\mathbf{H}, c, 1)$. The discriminant also gives an isomorphism for $L_1^K(\mathbf{R}, 1, 1) = \mathbf{Z}/2$ and $L_1^K(\mathbf{C}, 1, 1) = \mathbf{Z}/2$. The other $L^K$-groups are zero. In the final calculation we wish to keep track of the divisibility of the signatures. The notation $2\mathbf{Z}$ stands for an infinite cyclic group of signatures taking on any even value.

**(6.2) Local Fields:** Over local fields (of characteristic 0), in type $U$: $L_{2n}^K(D) \cong H^0(E^\times) = \mathbf{Z}/2$ via the discriminant and $L_{2n+1}^K(D) = 0$. In type $OD$, $L_0^K(D) \cong H^0(E^\times)$ and the others are zero. In type $OK$, $L_1^K(E) \cong H^1(E^\times) = \mathbf{Z}/2$ by the discriminant and $L_0^K(E)$ is an extension of $H^0(E^\times)$ by $\mathbf{Z}/2$, while $L_2^K(E) = L_3^K(E) = 0$. The natural map $L_1^S(E) \to L_1^K(E)$ is zero.

**(6.3) Finite Fields:** For finite fields in type $U$, $L_n^S = L_n^K = 0$, and in type $O$ characteristic 2, $L_n^S = L_n^K = \mathbf{Z}/2$ for each $n$. For type $O$ odd characteristic, the discriminant gives isomorphisms $L_0^K \cong \mathbf{Z}/2$, $L_1^K = \mathbf{Z}/2$ and $L_2^K = L_3^K = 0$. The groups $L_n^S = 0$ for $n = 0, 3$ and $L_1^S = L_2^S = \mathbf{Z}/2$. The map $L_1^S \to L_1^K$ is zero.

# 7. Computation of the relative group, $L_n^{X_i}(\mathbf{Z}G \to \widehat{\mathbf{Z}}_2 G)$

We can now compute the map

$$\gamma_n^{O_i}(d): L_n^{O_i}(\widehat{R}_{odd}(d) \oplus T(d)) \longrightarrow CL_n^{O_i}(S(d))$$

from (4.4) for each involution-invariant factor of $S(d)$. The main result about the relative groups is:

THEOREM 7.1: *For 2–hyperelementary groups $G$ and any geometric antistructure, there is an isomorphism*

$$L_n^{X_i}(\mathbf{Z}G \to \widehat{\mathbf{Z}}_2 G)(d) \cong \operatorname{cok} \gamma_n^{O_i}(d) \oplus \ker \gamma_{n-1}^{O_i}(d)$$

*and each of the summands decomposes according to the types of the simple components of $\mathbf{Q}G$.*

The proof is just to look at the tables of the Rothenberg sequence for the relative groups. The places where we could have an extension problem in the $L^s$ tables are hit surjectively from the Tate cohomology term, hence have exponent two.

Each simple component of $\mathbf{Q}G$ is a matrix algebra over a skew field, and by Morita equivalence it suffices to study $\gamma_n$ for an antistructure $(D, \alpha, u)$ on a skew field $D$. Its centre $E$ is an abelian extension of $\mathbf{Q}$ with ring of integers $A \subseteq E$. We fix an odd integer $d$ such that $\hat{D}_\ell$ is *split*, and $E_\ell$ is an *unramified* extension of $\hat{\mathbf{Q}}_\ell$ for all finite primes $\ell$ with $\ell \nmid 2d$. Tables 14.12-14.14 (for $i = 1$) and Tables 14.16-14.22 (for $i = 0$) list the domain, range, kernel and cokernel of each summand of $\gamma_n^{O_i}(d)$.

In order to use the tables, it is necessary to determine the types $(O, Sp, U$ or $GL)$ and centre fields for all the rational representations of $G = \mathbf{Z}/d \rtimes P$, following the method given in [**29**, p.148], or [**33**, Appendix I]. Recall that for a simple factor of type $Sp$, the groups $\operatorname{cok}\gamma_n$ and $\ker\gamma_n$ are equal to $\operatorname{cok}\gamma_{n+2}$ and $\ker\gamma_{n+2}$ respectively. For the $d$–component we need to consider only those representations which are faithful on $\mathbf{Z}/d$. Here is a list of the possible types, subdivided according to the behaviour at infinite primes.

**Type $O$:**

| | |
|---|---|
| $OK(\mathbf{R})$ | if $D = E$ and $E$ has a real embedding, |
| $OK(\mathbf{C})$ | if $D = E$ and $E$ has no real embedding. |

| | |
|---|---|
| $OD(\mathbf{H})$ | if $D \neq E$ and $D$ is nonsplit at infinite primes, |
| $OD(\mathbf{R})$ | if $D \neq E$ is split at infinite primes and $E$ has a real embedding, |
| $OD(\mathbf{C})$ | if $D \neq E$ is split at infinite primes and $E$ has no real embedding. |

**Type $U$:**

| | |
|---|---|
| $U(\mathbf{C})$ | if $D_\infty$ has type $U$, |
| $U(\mathbf{GL})$ | if $D_\infty$ has type $\mathbf{GL}$. |

We remark that in type $U(\mathbf{C})$ the centre field of $D_\infty$ at each infinite place is the complex numbers with complex conjugation as the induced involution. Type $U(\mathbf{GL})$ algebras are isomorphic to matrix rings over $\mathbf{C} \times \mathbf{C}$ or $\mathbf{R} \times \mathbf{R}$, at each infinite place, with the induced involution interchanging the two factors of $\mathbf{C}$ or $\mathbf{R}$.

In the remainder of this section we give a brief discussion of the computation in the case $i = 0$, including the definition of the maps $\Phi$, $\Phi'$, the group $\Gamma(E)$, and related notation (see [**30**] or [**39**] for more details). First we consider type $U$ where $H^0(C(E)) = \mathbf{Z}/2$ lies in the sequence

$$0 \to H^0(E^\times) \to H^0(E_A^\times) \to H^0(C(E)) \to 0.$$

At finite primes $L_n^K(\hat{A}_\ell) = L_n^K(\hat{A}_\ell/Rad) = 0$, since the right-hand side is the sum of $L^K$-groups of finite fields. At the infinite places we have the signature group $L_{2n}^K(D_\infty)$. This is non-zero when $D_\infty$ remains type $U$ (a change to type **GL** is possible) and the fixed field $E_0 \subseteq E$ of the involution is real. In type $U(\mathbf{C})$, each factor $2\mathbf{Z}$ maps surjectively to $H^0(C(E)) = \mathbf{Z}/2$ so $\mathrm{cok}\, \gamma_{2n} = 0$ and $\ker \gamma_{2n} = \Sigma$, where $\Sigma$ is a subgroup of index 2 in a direct sum of factors $2\mathbf{Z}$, one for each complex place.

Next we consider type $O$. It is convenient to introduce the "discriminant part" $\tilde{\gamma}_n$ of $\gamma_n$ for a factor $(D, \alpha, u) = (E, 1, 1)$ of type $OK$ to fit into the following commutative diagram:

(7.2)

$$
\begin{array}{ccc}
\displaystyle\prod_{\ell \nmid 2d} L_n^K(\hat{A}_\ell) \times L_n^K(E_\infty) & \xrightarrow{\gamma_n} & CL_n^K(E) \\
\downarrow & & \downarrow \\
H^n(\hat{A}_{2d'}^\times) \times H^n(E_\infty^\times) & \xrightarrow{\tilde{\gamma}_n} & H^n(C(E))
\end{array}
$$

where $\hat{A}_{2d'}^\times = \prod_{\ell \nmid 2d} \hat{A}_\ell^\times$. Below we will also use the notation $\hat{A}_{2d}^\times = \prod_{\ell | 2d} \hat{A}_\ell^\times$.
Since $\tilde{\gamma}_n$ has the same kernel and cokernel as the map

$$
H^n(\hat{A}_{2d'}^\times) \times H^n(E_\infty^\times) \times H^n(E^\times) \longrightarrow H^n(E_A^\times) ,
$$

we are led to consider the following commutative diagram (for $n = 0$):

$$
\begin{array}{ccccccccc}
0 & \to & \ker \tilde{\gamma}_0 & \to & H^0(\hat{A}_{2d'}^\times) \times H^0(E_\infty^\times) \times H^0(E^\times) & \to & H^0(\hat{E}_A^\times) & \to & \mathrm{cok}\, \tilde{\gamma}_0 & \to & 0 \\
& & \downarrow & & \downarrow & & \| & & \downarrow \\
0 & \to & E^{(2)}/E_2^\times & \to & H^0(\hat{A}^\times) \times H^0(E_\infty^\times) \times H^0(E^\times) & \to & H^0(\hat{E}_A^\times) & \to & H^0(\Gamma(E)) & \to & 0 \\
& & \downarrow & & \downarrow \\
(7.3) & & H^0(\hat{A}_{2d}^\times) & = & H^0(\hat{A}_{2d}^\times)
\end{array}
$$

Let $E^{(2)}$ denotes the elements of $E$ with even valuation at all finite primes and $\Gamma(E)$ is the ideal class groups defined by

$$
1 \to E^\times/A^\times \to \hat{E}^\times/\hat{A}^\times \to \Gamma(E) \to 1 .
$$

To obtain the middle sequence, add $H^0(\hat{A}_{2d}^\times)$ to the domain of $\tilde{\gamma}_0$, then the map to $H^0(E_A^\times)$ has the same kernel and cokernel as $H^0(\hat{E}^\times) \to H^0(I(E))$ where $I(E) = \hat{E}^\times/\hat{A}^\times$ is the ideal group of $E$.

From (7.3) we obtain the following exact sequence

$$
(7.4) \quad 0 \to \ker \tilde{\gamma}_0 \to E^{(2)}/E^{\times 2} \xrightarrow{\Phi/} H^0(\hat{A}_{2d}^\times) \to \mathrm{cok}\, \tilde{\gamma}_0 \to H^0(\Gamma(E)) \to 0
$$

for the computation of $\tilde{\gamma}_0$ in type OK. In type $OD(\mathbf{H})$ when $(D, \alpha, u)$ is non-split at all infinite primes, the term $H^0(E_\infty^\times)$ is missing from the top row of 7.3. This produces instead:

(7.5)
$$0 \to \ker \tilde{\gamma}_0 \to E^{(2)}/E^{\times 2} \xrightarrow{\Phi'} H^0(\hat{A}_{2d}^\times) \oplus H^0(E_\infty^\times) \to \operatorname{cok} \tilde{\gamma}_0 \to H^0(\Gamma(E)) \to 0$$

For the map $\tilde{\gamma}_1$ in type OK a similar but easier analysis gives $\ker \tilde{\gamma}_1 = 0$ and an exact sequence

(7.6)            $$0 \to H^1(A^\times) \to H^1(\hat{A}_{2d}^\times) \to \operatorname{cok} \tilde{\gamma}_1 \to 0.$$

In type OD, nonsplit at infinite primes, $H^1(E_\infty^\times)$ is added to the middle term.

The maps $\Phi$ and $\Phi'$ occur in number theory, and the 2–ranks of their kernel and cokernels are determined by classical invariants of $E$ (see [**75**, 4.6]). A similar discussion can be carried out for the maps $\gamma^{o_1}(d)$, and it turns out that the same maps $\Phi$, $\Phi'$ appear in the calculation.

PROPOSITION 7.7:([**30**, 2.18])
  (i)   The 2-rank of $\ker \Phi_E$ (resp. $\ker \Phi'_E$) is $\gamma^*(E, 2d)$ (resp. $\gamma(E, 2d)$ ).
  (ii)  The 2-rank of $\operatorname{cok} \Phi_E$ (resp $\operatorname{cok} \Phi'_E$) is $g_{2d}(E) + r_2 + \gamma^*(E, 2d)) - \gamma(E)$
        (resp. $g_{2d}(E) + r_1 + \gamma(E, 2d)) - \gamma(E)$).
Here $\gamma(E, m)$ $(\gamma^*(E, m))$ denotes the 2–rank of the (strict) class group of $A[\frac{1}{m}]$, $g_m(E)$ is the number of primes in $E$ which divide $m$ and $r_1$ $(r_2)$ is the number of real (complex) places of $E$.

EXAMPLE 7.8: If $\Gamma(E)$ has odd order then the exact sequence

$$0 \to H^0(A^\times) \to E^{(2)}/E^{\times 2} \to H^1(\Gamma(E)) \to 0$$

gives an isomorphism $H^0(A^\times) \cong E^{(2)}/E^{\times 2}$. Then the map $\Phi$ is just the reduction map $H^0(A^\times) \to H^0(\hat{A}_{2d}^\times)$. For example, if $E \subseteq \mathbf{Q}(\zeta_{2^k})$ and $d = 1$, then $\ker \Phi = 0$ and $\operatorname{cok} \Phi = (\mathbf{Z}/2)^{r_2+1}$.  ∎

## 8. The 2–adic calculation, $L_n^{Y_i}(\widehat{\mathbf{Z}}_2 G)$

We want to state the main result of [**29**, 1.16] which computes the map

(8.1)            $$\Psi_n : L_n^K(\widehat{\mathbf{Z}}_2 G, \alpha, u) \to L_n^K(\widehat{\mathbf{Q}}_2 G, \alpha, u)$$

for $G = C \rtimes P$ a 2-hyperelementary group with a geometric antistructure $(\alpha, u)$ with $K = O_0$. This map appears in the calculation of the $\psi$ maps in the next section as well as in the Ranicki–Rothenberg sequence for computing $L_n^{Y_i}(\widehat{\mathbf{Z}}_2 G, \alpha, u)$. Since $\Psi_n$ splits as in (5.5), it is enough to give the answer for the $d$-component for each $d \mid |C|$ which we know is determined by the top component for the various $G = C_d \rtimes P$. If $T \in C$ denotes a generator, then $\theta$ induces an automorphism of $C$ given by $\theta(T) = T^\vartheta$ (there is a misprint in the formula in [29, p.148,line-9]). The domain of $\Psi_n$ is easy to compute:

THEOREM 8.2:

$$L_n^{Y_0}(\widehat{\mathbf{Z}}_2 G)(d) = L_n^K(\widehat{\mathbf{Z}}_2 G)(d) = g_2(d) \cdot (\mathbf{Z}/2)$$

where $g_2(d) = g_2\big(\mathbf{Q}(\zeta_d)^P\big)$ denotes the number of primes $\ell$ dividing 2 in the field $\mathbf{Q}(\zeta_d)^P$, where $P$ acts as Galois automorphisms via the action map $t$.

Recall that if $P_1 = \ker(t : P \to (C_d)^\times)$, then any irreducible complex character of $G$ which is faithful on $C_d$ is induced up from $\chi \otimes \xi$ on $\mathbf{Z}/d \times P_1$ where $\chi$ is a linear character of $\mathbf{Z}/d$ and $\xi$ is an irreducible character of $P_1$. These are the representations in the semi–simple algebra $S(d)$. They are divided as usual into the types $O$, $Sp$ and $U$, and we say that the *order* of a linear character $\xi$ is the order of its image $\xi(P_1)$. Let $S(d, \xi)$ denote the simple factor of $S(d)$ associated to an involution-invariant character $(\chi \otimes \xi)^*$, induced up from $\chi \otimes \xi$.

THEOREM 8.3: ([29, 1.16]) *Let $(\alpha, u)$ be a geometric antistructure. If $d > 1$ and if there is no element $g_0 \in P$ satisfying $t(g_0) = -\vartheta^{-1}$, then it follows that $L_n^K(\widehat{R}_2(d), \alpha, u) = 0$. Otherwise if $d > 1$ pick such a $g_0$ (or when $d = 1$ set $g_0 = e$), and let $m = n$ if $w(g_0) = 1$ (resp. $m = n + 2$ if $w(g_0) = -1$). For each irreducible complex character $\xi$ of $P_1$ the composite*

$$L_n^K(\widehat{R}_2(d), \alpha, u) \xrightarrow{\Psi_n(d)} L_n^K(\widehat{S}_2(d), \alpha, u) \xrightarrow{proj.} L_n^K(\widehat{S}_2(d, \xi), \alpha, u)$$

*is injective or zero and is detected by the discriminant. It is injective if and only if the character $\xi$ is*
(a) *linear type $O$ (and $m \equiv 0$ or $1 \pmod 4$)*
(b) *linear type $Sp$ (and $m \equiv 2$ or $3 \pmod 4$)*
(c) *linear type $U$ (and $m$ even), order $2^\ell$ and $\xi(b_0^{2^{\ell-1}}) = -1$.*
*Here the types refer to the antistructure $(\widehat{\mathbf{Q}}_2[P_1], \alpha_0, b_0)$, with $\alpha_0(a) = g_0 \alpha(a) g_0^{-1}$ and $b_0 = g_0 \alpha(g_0^{-1}) bw(g_0) \in \pm P_1$.*

REMARK: A type I linear character $\xi$ has type $O$ (resp. $Sp$) if $\xi(b_0) = 1$ (resp. $\xi(b_0) = -1$). If $P_1$ has a linear character $\xi$ of type 8.3(c), then (by projecting onto the $\mathbf{Z}/2$ quotient of $\xi(P_1)$) it also has linear characters of type $O$ and $Sp$.

COROLLARY 8.4:   For $X = O_1(\widehat{R}_2)$ or $X = X_1(\widehat{R}_2)$, the discriminant map

$$d_n^{K/X} : L_n^K(\widehat{R}_2(d), \alpha, b) \to H^n(K_1(\widehat{R}_2(d))/X)$$

is injective or zero. For $X = X_1(\widehat{R}_2)$, it is injective if $m \equiv 0, 1 \,(\mathrm{mod}\, 4)$, or if $m \equiv 2, 3 \,(\mathrm{mod}\, 4)$ and there exists a linear character $\xi$ with type $Sp$.

REMARK 8.5:   The first statement in Corollary 8.4 holds for any decoration subgroup $X(\widehat{R}_2)$ that decomposes completely over the primes $\ell \mid d$ in $\mathbf{Q}[\zeta_d]^p$, but $O_1$ and $X_1$ are the usual examples. If $m \equiv 2, 3 \,(\mathrm{mod}\, 4)$ and under certain assumptions, this condition is also known to be necessary for the discriminant maps $d_n^K$ to be non–zero (see [**39**, 3.14]). In either case ($X = O_1$ or $X_1$), when $d_n = 0$ the kernel $\ker d_n \cong g_2 \cdot \mathbf{Z}/2$ can be lifted to a direct summand of $L_n^s(\widehat{\mathbf{Z}}_2 G)$ or $L_n^{X_1}(\widehat{\mathbf{Z}}_2 G)$ isomorphic to $(\mathbf{Z}/4)^\kappa$ if $n$ is odd, and to $(\mathbf{Z}/2)^\kappa$ if $n$ is even. A basis for these summands is represented by flips or rank 2 quadratic forms with Arf invariant one at primes $\ell \mid 2$ in $\mathbf{Q}[\zeta_d]^p$.

It is not hard to check that these flips and Arf invariant one planes in $L_n^{X_1}(\widehat{\mathbf{Z}}_2 G)$ map to zero under

$$\Psi_n^{X_1}(d) : L_n^{X_1}(\widehat{\mathbf{Z}}_2 G) \to L_n^s(\widehat{\mathbf{Q}}_2 G) \, ,$$

so the computation of $\Psi_n^{X_1}(d)$ is reduced to the map

$$H^{n+1}(K_1(\widehat{R}_2(d))/X_1) \to H^{n+1}(K_1(\widehat{S}_2(d))) \, ,$$

which only involves $K$–theory.   ∎

When the discriminant map into $H^n(K_1(\widehat{S}_2(d, \xi)))$ is injective, its image in $H^n(\hat{A}_\ell^\times)$ for $\ell \mid 2$ is either $\langle 1 - 4\beta \rangle$ if $n = 0, 2$, or $\langle -1 \rangle$ if $n = 1, 3$. The element $\beta \in \hat{A}_\ell^\times$ is a unit whose residue class has non–zero trace in $\mathbf{F}_2$. This description allows us to identify the image of the discriminant in $H^n(K_1(\widehat{R}_2(d)))$ once the Tate cohomology group has been calculated, and thus compute $L_*^{X_1}(\widehat{\mathbf{Z}}_2 G)$.

EXAMPLE 8.6:   Consider the simplest case, where $G = 1$. Then $K_1(\widehat{\mathbf{Z}}_2) = \widehat{\mathbf{Z}}_2^\times$ is generated by the units $\langle 5, -1 \rangle$. Therefore $H^n(K_1(\widehat{\mathbf{Z}}_2)) = \mathbf{Z}/2 \oplus \mathbf{Z}/2$ ($n$ even) or $\mathbf{Z}/2$ ($n$ odd), and $L_n^K(\widehat{\mathbf{Z}}_2) = \mathbf{Z}/2$ in each dimension. In particular, the element $1 - 4\beta = 5$ is the discriminant of the generator in $L_0^K(\widehat{\mathbf{Z}}_2)$.

By the results above, we get

$$L_n^s(\widehat{\mathbf{Z}}_2) = 0, \ \mathbf{Z}/2 \oplus \mathbf{Z}/2, \ \mathbf{Z}/2 \oplus \mathbf{Z}/2, \ \mathbf{Z}/4 \quad \text{for } n \equiv 0, 1, 2, 3 \, (\text{mod } 4) \ .$$

The $\mathbf{Z}/4$ in $L_3$ is generated by the flip automorphism $\tau(e) = f$, $\tau(f) = -e$ of the hyperbolic plane over $(\widehat{\mathbf{Z}}_2, 1, -1)$. Since $X_1(\widehat{\mathbf{Z}}_2) = 0$, we also have computed $L_*^{X_1}(\widehat{\mathbf{Z}}_2) = L_*^s(\widehat{\mathbf{Z}}_2)$. ∎

## 9. The maps $\psi_n^{Y_i}: L_n^{Y_i}(\widehat{\mathbf{Z}}_2 G) \to L_n^{X_i}(\mathbf{Z}G \to \widehat{\mathbf{Z}}_2 G)$

The final step in computing the main exact sequence is to determine the maps $\psi_n$ for 2–hyperelementary groups. First we consider the case $i = 0$ needed for computing the $L^p$–groups:

$$\psi_n(d) : L_n^K(\widehat{\mathbf{Z}}_2 G)(d) \longrightarrow L_n^{X_0}(\mathbf{Z}G \to \widehat{\mathbf{Z}}_2 G)(d) \ .$$

Here $\psi_n$ factors through

$$\bar{\psi}_n: L_n^K(\widehat{\mathbf{Z}}_2 G) \xrightarrow{\Psi_n} L_n^K(\widehat{\mathbf{Q}}_2 G) \to CL_n^K(\mathbf{Q}G) \to \text{cok}\,\gamma_n^K$$

and after taking the $d$–component these can be studied one simple component of $\mathbf{Q}G$ at a time. The maps $\Psi_n(d)$ are given in Theorem 8.3, and the other maps in the composite are contained in Tables 14.16–14.22.

Computing $\bar{\psi}_n(d)$ also computes the kernel and cokernel of $\psi_n(d)$ since $\ker \psi_n(d) = \ker \bar{\psi}_n(d)$, and $\text{cok}\,\psi_n(d) \cong \text{cok}\,\bar{\psi}_n(d) \oplus \ker \gamma_{n-1}(d)$.

EXAMPLE 9.1: We continue with the example $G = 1$ from the last section. The group $\text{cok}\,\gamma_0 = \mathbf{Z}/2$ generated by the class $\langle 5 \rangle \in H^0(K_1(\widehat{\mathbf{Z}}_2))$ (see Example 7.8) and otherwise is zero. It follows that the map $\bar{\psi}_0$ is an isomorphism, and we get the values

$$L_n^{X_0}(\mathbf{Z}, 1, 1) = 8\mathbf{Z}, \ \mathbf{Z}/2, \ \mathbf{Z}/2, \ \mathbf{Z}/2 \quad \text{for } n \equiv 0, 1, 2, 3 \, (\text{mod } 4) \ .$$

To obtain the geometric $L^p$–groups, we cancel the terms $\mathbf{Z}/2$ in odd dimensions. Note that since $\tilde{K}_0(\mathbf{Z}) = 0 = Wh(\mathbf{Z})$, the other geometric $L$–groups are isomorphic to $L_*^p(\mathbf{Z})$. ∎

The calculation of the maps $\psi_n: L_n^{Y_1}(\widehat{\mathbf{Z}}_2 G) \to L_n^{Y_1}(\mathbf{Z}G \to \widehat{\mathbf{Z}}_2 G)$ needed to determine the $L'$–groups is more involved. Notice, however, that by Theorem 5.6 it is enough in principle to do the calculation for basic subquotients of $G$ and then compute some generalized restriction maps. Because of the difficulties involved in computing restriction maps on $L_*^{Y_1}(\widehat{\mathbf{Z}}_2G)$ this approach remains more a theoretical simplification than a practical one.

We can again define $\bar{\psi}_n$ as the composite

$$\bar{\psi}_n: L_n^{X_1}(\widehat{\mathbf{Z}}_2G)\xrightarrow{\Psi_n}L_n^s(\widehat{\mathbf{Q}}_2G) \to CL_n^s(\mathbf{Q}G) \to \operatorname{cok}\gamma_n^s$$

but this time we only have a commutative diagram

$$
\begin{array}{ccccc}
H^{n+1}(K_1(\widehat{\mathbf{Z}}_2G)/X_1) & \to & L_n^{X_1}(\widehat{\mathbf{Z}}_2G) & \xrightarrow{\bar{\psi}_n} & \operatorname{cok}\gamma_n^s \\
\downarrow & & \downarrow & & \downarrow \\
H^{n+1}(K_1(\widehat{\mathbf{Z}}_2G)/Y_1) & \to & L_n^{Y_1}(\widehat{\mathbf{Z}}_2G) & \xrightarrow{\psi_n} & L_n^{X_i}(\mathbf{Z}G\to\widehat{\mathbf{Z}}_2G)
\end{array}
$$

where the right–hand vertical arrow is a (split) injection. Note that the quotient group $K_1(\widehat{\mathbf{Z}}_2G)/Y_1 = Wh'(\widehat{\mathbf{Z}}_2G)$, which has been studied intensively in [49], [50]. Applying the idempotent splitting partly solves the problem: on the $d$–component for $d > 1$ the groups $L_n^{Y_1}(\widehat{\mathbf{Z}}_2G)(d) \cong L_n^{X_1}(\widehat{\mathbf{Z}}_2G)(d)$, and in Remark 8.5 we pointed out that the calculation of $\bar{\psi}_n(d)$ is now reduced to $K$–theory.

For $d = 1$ we may assume that $G$ is a finite 2–group. Then $L_n^K(\widehat{\mathbf{Z}}_2G) = \mathbf{Z}/2$ and the map $\psi_n$ also factors through $\operatorname{cok}\gamma_n^s \cong \operatorname{cok}\gamma_n^{Y_1}$. (both assertions follow because there is just one prime $\ell \mid 2$ in the centre fields of $\mathbf{Q}G$). We may assume that $\bar{\psi}_n$ is known from Theorem 8.3.

To proceed, we first compute $L_n^{Y_1}(\widehat{\mathbf{Z}}_2G)$ via the discriminant map $d_n^{K/Y_1}: L_n^{Y_1}(\widehat{\mathbf{Z}}_2G) \to H^n(Wh'(\widehat{\mathbf{Z}}_2G))$, either directly (starting with the map $d_n^{K/X_1}$ which is known), or using the long exact sequence

$$\ldots \to H^{n+1}(\{\pm1\}\oplus G^{ab}) \to L_n^{X_1}(\widehat{\mathbf{Z}}_2G) \to L_n^{Y_1}(\widehat{\mathbf{Z}}_2G) \to H^n(\{\pm1\}\oplus G^{ab}) \to \ldots$$

It is quite likely that the "twisting diagram" method introduced in [26] and [33] would be useful here.

Next, we must compute $\psi_n$. One remark that may be helpful is that the the image of $H^{n+1}(\{\pm1\}\oplus G^{ab})$ in $L_n^{X_1}(\widehat{\mathbf{Z}}_2G)$ is mapped by $\Psi_n$ into integral units, hence mapped to zero under $\bar{\psi}_n$. Hence, if $L_n^{X_1}(\widehat{\mathbf{Z}}_2G) \to L_n^{Y_1}(\widehat{\mathbf{Z}}_2G)$ happens to be surjective, we are done.

An alternate approach is to apply Remark 8.5 (valid for $Y_1(\widehat{\mathbf{Z}}_2G)$ if $G$ is a finite 2–group) to the $L^{Y_1}$ to $L^K$ Rothenberg sequence for $\widehat{\mathbf{Z}}_2G$. As before, this reduces the computation of $\psi_n$ to $K$–theory calculation. We can compute the composite

$$H^{n+1}(K_1(\widehat{\mathbf{Z}}_2G)/Y_1) \to H^{n+1}(K_1(\widehat{\mathbf{Q}}_2G)/Y_1) \to CL_n^s(\mathbf{Q}G)$$

using the algorithm from [49], [50] for computing $Wh'(\widehat{\mathbf{Z}}_2G)$. For a general geometric antistructure, this can involve a lot of book–keeping. For the standard oriented antistructure, things are not so difficult.

EXAMPLE 9.2: For the standard oriented antistructure, we can always split off $L_n(\mathbf{Z})$ from $L_n(\mathbf{Z}G)$ or $L_n(\widehat{\mathbf{Z}}_2G)$ by using the inclusion and projection $1 \to G \to 1$.

If $G$ is a finite 2–group, then

$$L_n^K(\widehat{\mathbf{Z}}_2G) \cong L_n^K(\widehat{\mathbf{Z}}_2G/Rad) \cong L_n^K(\mathbf{F}_2)$$

so the image of $L_n^{Y_1}(\widehat{\mathbf{Z}}_2G) \to CL_n^s(\mathbf{Q}G)$ is just the image of the composite

$$H^{n+1}(K_1(\widehat{\mathbf{Z}}_2G)/Y_1) \to H^{n+1}(K_1(\widehat{\mathbf{Q}}_2G)/Y_1) \to L_n^{Y_1}(\widehat{\mathbf{Q}}_2G) \to CL_n^s(\mathbf{Q}G) \ .$$

This directly reduces the calculation of $\psi_n$ to a $K_1$–calculation. ∎

## 10. Groups of odd order

We prove a well–known vanishing result, as an example of the techniques developed so far.

THEOREM 10.1:   Let $G$ be a finite group of odd order. Then in the standard oriented antistructure, $L_{2k+1}^?(\mathbf{Z}G) = 0$ for $? = s, \prime, h$ and $p$.

Proof: For groups of odd order, the 2–hyperelementary subgroups are cyclic, so it is enough to let $G = C_m$ denote a cyclic group of odd order $m$. We have a decomposition into components $L_*^{Y_i}(\mathbf{Z}G)(d)$ indexed by the divisors $d \mid m$, and there are two distinct cases according as $d = 1$ or $d \neq 1$.

Let us start with $i = 0$ or $L^p$–groups. By Theorem 5.5, when $d = 1$ we are computing $L_*^p(\mathbf{Z})$ which was done in Example 9.1. For $d > 1$, all the summands in $S(d)$ have type $U(\mathbf{C})$, so by Table 14.21 we have

$$L_n^{X_0}(\mathbf{Z}G \to \widehat{\mathbf{Z}}_2G) = 0, \ \Sigma, \ 0, \ \Sigma \quad \text{for } n \equiv 0, 1, 2, 3 \,(\text{mod}\, 4) \ .$$

Next $L_n^K(\widehat{R}_2(d)) = 0$, for all $n$, since $\widehat{R}_2(d) = \widehat{\mathbf{Z}}_2 \otimes \mathbf{Z}[\zeta_d]$ reduces modulo the radical to a product of finite fields with type $U$ antistructure. Therefore

$$L_n^p(\mathbf{Z}G)(d) = \Sigma, \ 0, \ \Sigma, \ 0 \quad \text{for } n \equiv 0, 1, 2, 3 \,(\text{mod}\, 4)$$

and in particular the $L^p$–groups vanish in odd dimensions.

Next we consider the $d > 1$ components in the main exact sequence for $L_*^{X_1}(\mathbf{Z}G)(d)$. Since type $U$ factors of $\widehat{R}_{odd}$ or $\mathbf{Q}G$ make no contribution to the relative $L^\prime$–groups, we have

$$L_n^{X_1}(\mathbf{Z}G \to \widehat{\mathbf{Z}}_2G) = 0, \ \Sigma, \ 0, \ \Sigma \quad \text{for } n \equiv 0, 1, 2, 3 \,(\text{mod}\, 4) \ .$$

as before. Now consider the 2–adic contribution. Here

$$H^{n+1}(K_1(\widehat{R}_2(d))/X_1) \cong L_n^{X_1}(\widehat{R}_2(d))$$

and $X_1(\widehat{R}_2(d)) = 0$ since the ring is abelian, so $K_1$ is just the group of units. Let $A = \widehat{\mathbf{Z}}_2[\zeta_d]$ and consider the sequences

$$1 \to (1+2A)^\times \to A^\times \to (A/2A)^\times \to 1$$

and

$$1 \to (1+2^{r+1}A)^\times \to (1+2^r A)^\times \xrightarrow{\varphi} A/2A \to 1$$

for $r \geq 1$, where $\varphi(1+2^r a) = a\,(\mathrm{mod}\,2)$. Since $(A/2A)^\times$ has odd order and $A/2A = \mathbf{F}_2[\zeta_d]$ has non-trivial involution, both are cohomologically trivial as $\mathbf{Z}/2$–modules. Therefore $H^*(A^\times) = 0$ and

$$L_n'(\mathbf{Z}G)(d) = \Sigma,\ 0,\ \Sigma,\ 0 \quad \text{for } n \equiv 0,1,2,3\,(\mathrm{mod}\,4)$$

so once again the $L$–group vanish in odd dimensions. For $G$ cyclic, we have $SK_1(\mathbf{Z}G) = 0$ and so $L' = L^s$. Also, $H^1(Wh'(\mathbf{Z}G)) = 0$ in the standard oriented antistucture, so $L'_{2k+1}(\mathbf{Z}G)$ surjects onto $L^h_{2k+1}(\mathbf{Z}G)$. ∎

REMARK 10.2: We do not want to leave the impression that all the $L$–groups of odd order groups $G$ are easy to compute. For $G$ cyclic of odd order, the groups $L^h_{2k}(\mathbf{Z}G)$ have torsion subgroup $H^0(\tilde{K}_0(\mathbf{Z}G))$ and this can be highly non–trivial. ∎

## 11. Groups of 2–power order

In [33] the groups $L^p$-groups for $\mathbf{Z}G$ were completely determined, for $G$ a finite 2-group with any geometric anti-structure. For $L'_*(\mathbf{Z}G)$ with the standard oriented antistructure, there is in principle an algorithm for carrying out the computation. We have already discussed the steps in computing the main exact sequence (see Example 9.2) and mentioned that results of Oliver give an algorithm for computing $K'_1(\widehat{\mathbf{Z}}_2 G) = K_1(\widehat{\mathbf{Z}}_2 G)/X_1$, together with the action of the antistructure, by using the integral logarithm [49, Thm. 6.6]. Thus we can regard the $L'$-groups for 2-groups as computable up to extensions, although the method can be difficult to carry out in practice.

EXAMPLE 11.1: Let's compute $L'_*(\mathbf{Z}G)$ for $G$ a cyclic 2–group of order $2^k \geq 2$ in the standard oriented antistructure (done in [75, 3.3]). Since

$SK_1(\mathbf{Z}G) = 0$, this also gives us $L_*^s(\mathbf{Z}G)$. Note that $L_*^p(\mathbf{Z}G)$ is tabulated in [**33**], and $L_*^h(\mathbf{Z}G)$ was reduced to the computation of $H^0(D(\mathbf{Z}G))$ in [**31**] or [**2**]. The final step, the computation of $H^0(D(\mathbf{Z}G))$ was carried out independently in [**50**] and [**12**].

We begin as usual with the relative groups, this time from Table 14.23 and Table 14.15. The types are $U(\mathbf{C})$ and $OK(\mathbf{R})$, where the latter are the two quotient representations arising from the projection $G \to \mathbf{Z}/2$. We get

$$L_n^{X_1}(\mathbf{Z}G \to \widehat{\mathbf{Z}}_2 G) = 0, \ \Sigma \oplus (8\mathbf{Z})^2 \oplus (\mathbf{Z}/2)^2, \ 0, \ \Sigma \quad \text{for } n = 0, 1, 2, 3 \,(\bmod\ 4)\,.$$

Here $\Sigma = \oplus 4\mathbf{Z}$ is the is the signature group from the type $U(\mathbf{C})$ representations.

Next we compute $L_n^{Y_1}(\widehat{\mathbf{Z}}_2 G)$ by comparing it to $L_n^K(\widehat{\mathbf{Z}}_2 G)$. Since the antistructure is oriented, we can split off $L_n^{Y_1}(\widehat{\mathbf{Z}}_2) = L_n^{X_1}(\widehat{\mathbf{Z}}_2)$ computed in Example 8.6, and obtain

$$L_n^{Y_1}(\widehat{\mathbf{Z}}_2 G) = L_n^{X_1}(\widehat{\mathbf{Z}}_2) \oplus H^{n+1}\big((1 + I)^\times / G\big)$$

where $I = I(\widehat{\mathbf{Z}}_2 G)$ is the augmentation ideal of $\widehat{\mathbf{Z}}_2 G$. It is not hard to see that

$$H^n\big((1 + I)^\times / G\big) = \mathbf{Z}/2, \ 0 \quad \text{for } n = 0, 1 \,(\bmod\ 2),$$

and a generator for the non-trivial element in $H^0$ is given by $\langle 3 - g - g^{-1} \rangle$ where $g \in G$ is a generator. Since this element has projection $\langle 5 \rangle$ at the non-trivial type $OK(\mathbf{R})$ representation (where $g \mapsto -1$), the map $\bar{\psi}_1$ has image $\mathbf{Z}/2$ in this summand of the relative group. This is an example of the "book-keeping" process mentioned in Example 9.2. Putting the summand from the trivial group back in, we get the well-known answer

$$L_n'(\mathbf{Z}G) = \Sigma \oplus 8\mathbf{Z} \oplus 8\mathbf{Z}, \ 0, \ \Sigma \oplus \mathbf{Z}/2, \ \mathbf{Z}/2 \quad \text{for } n = 0, 1, 2, 3 \,(\bmod\ 4)\,.$$

∎

## 12. Products with odd order groups

Here we correct an error in the statement of [30, 5.1] where $G = \sigma \times \rho$ with $\sigma$ an abelian 2–group and $\rho$ odd order. More generally, for $G = G_1 \times G_2$ where $G_1$ has odd order, we can reduce the calculation of $L_*(\mathbf{Z}G, w)$ to knowledge of $L_*(\mathbf{Z}G_2, w)$ and the character theory of $G$.

PROPOSITION 12.1: Let $G = G_1 \times G_2$, where $G_1$ has odd order. Then for $i = 0, 1$

$$L_n^{\chi_i}(\mathbf{Z}G, w) = L_n^{\chi_i}(\mathbf{Z}G_2, w) \oplus L_n^{\chi_i}(\mathbf{Z}G_2 \to \mathbf{Z}G, w)$$

where $w: G \to \{\pm 1\}$ is an orientation character. For $n = 2k$, the second summand is free abelian and detected by signatures at the type $U(\mathbf{C})$ representations of $G$ which are non–trivial on $G_1$. For $i = 0$ and $n = 2k + 1$, the second summand is a direct sum of $\mathbf{Z}/2$'s, one for each type $U(\mathbf{GL})$ representation of $G$ which is non–trivial on $\rho$.

REMARK 12.2: In the important special case when $G_2$ is an abelian 2–group, note that type $U(\mathbf{C})$ representations of $G$ exist only when $w \equiv 1$, and type $U(\mathbf{GL})$ representations of $G$ exist only when $w \not\equiv 1$. In both cases, the second summand is computed by transfer to cyclic subquotients of order $2^r q$, $q > 1$ odd, with $r \geq 2$. If $G_2$ is a cyclic 2–group, then $L^s = L'$ by [49, Ex.3,p.14].

Proof: The given direct sum decomposition follows from the existence of a retraction of the inclusion $G_2 \to G$ compatible with $w$. It also follows that

$$L_{n+1}^{\chi_i}(\mathbf{Z}G \to \widehat{\mathbf{Z}}_2 G, w) \cong L_{n+1}^{\chi_i}(\mathbf{Z}G_2 \to \widehat{\mathbf{Z}}_2 G_2, w) \oplus L_n^{\chi_i}(\mathbf{Z}G_2 \to \mathbf{Z}G, w)$$

since the map induced by inclusion $L_n^{\chi_i}(\widehat{\mathbf{Z}}_2 G_2, w) \to L_n^{\chi_i}(\widehat{\mathbf{Z}}_2 G, w)$ is an isomorphism ([39, 3.4] for $i = 1$).

The computation of the relative groups for $\mathbf{Z} \to \widehat{\mathbf{Z}}_2$ can be read off from Table 14.22: for each centre field $E$ of a type $U(\mathbf{GL})$ representation, the contribution is $H^0(C(E)) \cong \mathbf{Z}/2$ if $i \equiv 1 \pmod 2$.

The detection of $L_n^p(\mathbf{Z}G \to \widehat{\mathbf{Z}}_2 G, w)$ by cyclic subquotients is proved in [34, 1.B.7, 3.A.6, 3.B.2]. ∎

COROLLARY 12.3: Let $G = C_{2^r q}$, for $q$ odd and $r \geq 2$. If $q = 1$ assume that $r \geq 3$. Then the group

$$L_{2k-1}^p(\mathbf{Z}G, w)(q) = \bigoplus_{i=2}^{r} CL_{2k}^K(E_i) \cong (\mathbf{Z}/2)^{r-1}$$

when $w \not\equiv 1$, where the summand $CL_{2k}^K(E_i) = H^0(C(E_i))$, $2 \leq i \leq r$, corresponds to the rational representation with centre field $E_i = \mathbf{Q}(\zeta_{2^i q})$.

## 13. Dihedral groups

Wall [75], Laitinen and Madsen [42], [39] did extensive computations for the $L'$–groups of the groups $G$ with periodic cohomology, because of the importance of these computations for the spherical space form problem.

As a final, and much easier example, we will consider the dihedral groups $G = C_d \rtimes \mathbf{Z}/2$ with $d$ odd. These are the simplest kind of 2–hyperelementary groups which are not 2–elementary. Here the action map is injective, as a generator of the $\mathbf{Z}/2$ quotient group acts by inversion on $C_d$. We take the standard oriented antistructure. Note that $L' = L^s$ for dihedral groups [49, p.15].

It is enough to do the $d > 1$ component and apply Theorem 5.5. We see that $S(d)$ contains a single type $OK(\mathbf{R})$ representation with centre field $E = \mathbf{Q}(\zeta_d + \zeta_d^{-1})$ and ring of integers $A$.

For the remainder of this section, let $g_p(E)$ denote the number of primes in the field $E$ lying over the rational prime $p$. For any integer $r$, let $g_r(E) = \sum \{g_p(E) : p \mid r\}$. Let $r_E$ be the number or real places in $E$ and recall $r_E = \phi(d)/2$.

For the $L^p$ calculation, we have relative groups

$$L_n^{X_0}(\mathbf{Z}G \to \widehat{\mathbf{Z}}_2 G)(d) = \operatorname{cok} \gamma_n^K \oplus \ker \gamma_{n-1}^K$$

which can be read off from Table 14.16. The groups $L_n^K(\widehat{\mathbf{Z}}_2 G) = g_2(E) \cdot (\mathbf{Z}/2)$ in each dimension and $\Psi_n(d)$ is injective for $n \equiv 0, 1 \,(\mathrm{mod}\, 4)$ but zero for $n \equiv 2, 3 \,(\mathrm{mod}\, 4)$.

Since the image of $\Psi_1(d)$ hits the classes $\langle -1 \rangle$ at primes lying over 2, it follows that $\bar{\psi}_1(d)$ is injective with cokernel $H^1(\hat{A}_d^\times)/H^1(A^\times)$, an elementary abelian 2–group of rank $g_d(E) - 1$.

Similarly, the image of $\Psi_0(d)$ hits the classes $g_2(E)\langle 1 - 4\beta \rangle$ in $H^0(\hat{A}_2^\times)$, so we must compute the kernel and cokernel of the map

$$(13.1) \qquad \bar{\Phi} \colon E^{(2)}/E^{\times 2} \to H^0(\hat{A}_{2d}^\times)/g_2(E)(1 - 4\beta) \ .$$

It is not hard to see that $\ker \bar{\Phi} = \ker \Phi \oplus \ker \bar{\psi}_0(d)$ and $\operatorname{cok} \bar{\Phi} = \operatorname{cok} \bar{\psi}_0(d)$ (see [30, p.566]). In the short exact sequences

$$0 \to \operatorname{cok} \bar{\psi}_{n+1}(d) \oplus \ker \gamma_n(d) \to L_n^{X_0}(\mathbf{Z}G)(d) \to \ker \bar{\psi}_n(d) \to 0,$$

the only potential extension problem occurs for $n = 0$. Let $\lambda_E = g_d(E) + \gamma^*(E, d)$. Then a similar argument to that in [30, p. 551], together with [30, 5.19], shows that

PROPOSITION 13.2: *Let $G = C_d \rtimes \mathbf{Z}/2$ be a dihedral group, with $d > 1$ odd. Then*

$$L_0^p(\mathbf{Z}G)(d) = \Sigma \oplus (\mathbf{Z}/2)^{\lambda_E - 1}$$
$$L_1^p(\mathbf{Z}G)(d) = 0$$
$$L_2^p(\mathbf{Z}G)(d) = g_2(E) \cdot \mathbf{Z}/2$$
$$L_3^p(\mathbf{Z}G)(d) = g_2(E) \cdot \mathbf{Z}/2 \oplus (\mathbf{Z}/2)^{\lambda_E}$$

REMARK 13.3: The signature divisibility is given by

$$\Sigma = 8\mathbf{Z} \oplus (4\mathbf{Z})^{r_E - \bar{r}_E - 1} \oplus (2\mathbf{Z})^{\bar{r}_E}$$

where $\bar{r}_E$ is the 2–rank of the image of $(\Theta | \ker \bar{\Phi})$ as in [30, p.550] and $r_E = \phi(d)/2$ is the number of real places in $E$. The formula in [30, 5.17(ii)] is incorrect. It should read

$$(8\mathbf{Z})^{r(S)} \oplus (4\mathbf{Z})^{r_1(S) - r_O(S) - r(S)} \oplus (2\mathbf{Z})^{r_O(S)}$$

where $r(S)$ denotes the number of type $OK(\mathbf{R})$ factors in $S(d)$, and $r_1(S)$, $r_O(S)$ are as defined in [30]. ∎

For the $L'$ calculation, we have relative groups

$$L_n^{\chi_1}(\mathbf{Z}G \to \hat{\mathbf{Z}}_2 G)(d) = \operatorname{cok} \gamma_n^s \oplus \ker \gamma_{n-1}^s$$

which can be read off from Table 14.12. The groups $L_n^{\chi_1}(\hat{\mathbf{Z}}_2 G)(d)$ are computed from the Rothenberg sequence using the same method as in Section 9. We have

$$L_0^s(\hat{\mathbf{Z}}_2 G)(d) = \begin{cases} 0 & n \equiv 0 \,(\mathrm{mod}\ 4), \\ H^0(\hat{A}_2^\times) & n \equiv 1 \,(\mathrm{mod}\ 4), \\ H^1(\hat{A}_2^\times) \oplus g_2(E) \cdot \mathbf{Z}/2 & n \equiv 2 \,(\mathrm{mod}\ 4), \\ \frac{H^0(\hat{A}_2^\times)}{g_2(E)\langle 1 - 4\beta, -1\rangle} \oplus g_2(E) \cdot \mathbf{Z}/4 & n \equiv 3 \,(\mathrm{mod}\ 4), \end{cases}$$

where $H^0(\hat{A}_2^\times)$ has 2–rank $r_E + g_2(E)$. The analogous number theoretic map to (13.1) is

$$\tilde{\Phi}_E : E^{(2)}/E^{\times 2} \to H^0(\hat{A}_d^\times) .$$

and the 2–ranks of its kernel and cokernel can again be given in terms of classical invariants (see [75, p.56]). We then have the torsion subgroup of $L_1^{\chi_1}(\mathbf{Z}G)$ isomorphic to

$$\ker \bar{\psi}_1 \oplus \operatorname{cok} \Phi_E \oplus (g_d(E) - 1) \cdot \mathbf{Z}/2 \cong \ker \tilde{\Phi}_E \oplus (g_d(E) - 1) \cdot \mathbf{Z}/2 .$$

However, the exact sequence

$$0 \to \ker \Phi_E \to \ker \tilde{\Phi}_E \to H^0(\hat{A}_2^\times) \to \text{cok } \gamma_1 \to \text{cok } \tilde{\Phi}_E \to 0$$

allows us to compute the 2–rank of $\ker \tilde{\Phi}_E$ in terms of the 2–rank of $\text{cok } \tilde{\Phi}_E$ and previously defined quantities. Recall that $\gamma_E$ denotes the 2–rank of $H^0(\Gamma(E))$. It is also useful to define the quantity

$$\phi_E = \nu_E + \gamma_E$$

where $\text{cok } \tilde{\Phi}_E = (\mathbf{Z}/2)^{\nu_E}$. Putting the information together gives:

PROPOSITION 13.4: Let $G = C_d \rtimes \mathbf{Z}/2$ be a dihedral group, with $d > 1$ odd. Then

$$L_0'(\mathbf{Z}G)(d) = \Sigma \oplus (\mathbf{Z}/2)^{\phi_E}$$
$$L_1'(\mathbf{Z}G)(d) = (\mathbf{Z}/2)^{r_E + \phi_E - 1}$$
$$L_2'(\mathbf{Z}G)(d) = g_2(E) \cdot \mathbf{Z}/2$$
$$L_3'(\mathbf{Z}G)(d) = L_3^s(\hat{A}_2)(d) = g_2(E) \cdot \mathbf{Z}/4 \oplus (\mathbf{Z}/2)^{r_E - g_2(E)}$$

REMARK 13.5: The signature divisibility this time is given by

$$\Sigma = 8\mathbf{Z} \oplus (4\mathbf{Z})^{r_E - 1} .$$

These are the same divisibilities as in the relative group. ∎

## 14. Appendix: Useful Tables

### 14.A $L$–groups of fields and skew fields

We give the $L^s$ to $L^K$ change of $K$-theory sequences for the antistructures $(D, \alpha, u)$ where $D$ is a (skew) field with center $E$, and $E$ is either finite, continuous ($\mathbf{R}$ or $\mathbf{C}$) or local (a finite extension field of $\hat{\mathbf{Q}}_p$).

From the tables below, one can read off invariants determining the $L$-groups in most cases (e.g., discriminant, signature, and Pfaffian). The remaining cases are labelled $c$, $\kappa$, and $\tau$ for the Arf invariant, Hasse-Witt invariant or flip respectively ($\tau$ in $L^s$ is represented by the automorphism $\begin{pmatrix} 0 & 1 \\ u & 0 \end{pmatrix}$ of the hyperbolic plane). Note that $L^s$-groups are all zero for finite fields or local fields in type $U$, and that for a division algebra $D$ with centre $E$ the group $L_1^K(D, \alpha, u) = 0$ unless $(D, \alpha, u) = (E, 1, 1)$ and $L_1^K(E, 1, 1) = \mathbf{Z}/2$ detected by the discriminant.

### Table 14.1: Finite fields, odd characteristic, Type O

|         | $L_n^s(E, 1, 1)$ | $L_n^K(E, 1, 1)$ | $H^n(E^\times)$ |
|---------|:----------------:|:----------------:|:---------------:|
| $n = 3$ | 0                | 0                | $\mathbf{Z}/2$  |
| $n = 2$ | $\mathbf{Z}/2$   | 0                | $\mathbf{Z}/2$  |
| $n = 1$ | $\mathbf{Z}/2$   | $\mathbf{Z}/2$   | $\mathbf{Z}/2$  |
| $n = 0$ | 0                | $\mathbf{Z}/2$   | $\mathbf{Z}/2$  |

For finite fields in type $U$, both $L_n^s(E, 1, 1) = L_n^K(E, 1, 1) = 0$. In characteristic 2, $L_n^s(E, 1, 1) = L_n^K(E, 1, 1) = \mathbf{Z}/2$ in each dimension (detected by $c$ in even dimensions, and $\tau$ in odd dimensions)

**Table 14.2: Local fields, Type OK**

|  | $L_n^s(E,1,1)$ | $L_n^K(E,1,1)$ | $H^n(E^\times)$ |
|---|---|---|---|
| $n = 3$ | $0$ | $0$ | $\mathbf{Z}/2$ |
| $n = 2$ | $\mathbf{Z}/2$ | $0$ | $H^0(E^\times)$ |
| $n = 1$ | $H^0(E^\times)$ | $\mathbf{Z}/2$ | $\mathbf{Z}/2$ |
| $n = 0$ | $\mathbf{Z}/2 \langle \kappa \rangle$ | $\mathbf{Z}/2\tilde{\times}H^0(E^\times)$ | $H^0(E^\times)$ |

The extension $\mathbf{Z}/2\tilde{\times}H^0(E^\times)$ appearing in this table is split if and only if $-1 \in E^{\times 2}$.

**Table 14.3: Local fields, Type OD**

|  | $L_n^s(D,\alpha,1)$ | $L_n^K(D,\alpha,1)$ | $H^n(E^\times)$ |
|---|---|---|---|
| $n = 3$ | $0$ | $0$ | $\mathbf{Z}/2$ |
| $n = 2$ | $\mathbf{Z}/2$ | $0$ | $H^0(E^\times)$ |
| $n = 1$ | $H^0(E^\times)$ | $0$ | $\mathbf{Z}/2$ |
| $n = 0$ | $\mathbf{Z}/2$ | $H^0(E^\times)$ | $H^0(E^\times)$ |

In type $OD$ we can always scale the antistructure so that it has $u = +1$. For local fields in type $U$, we have two–fold periodicity $L_n^K(E,1,1) = L_{n+2}^K(E,1,1)$.

**Table 14.4: Local fields, Type U**

|  | $L_n^s(E,1,1)$ | $L_n^K(E,1,1)$ | $H^n(E^\times)$ |
|---|---|---|---|
| $n = 1,3$ | $0$ | $0$ | $0$ |
| $n = 0,2$ | $0$ | $\mathbf{Z}/2$ | $\mathbf{Z}/2$ |

**Table 14.5: Continuous Fields, $E = \mathbf{R}$, Type O**

|         | $L_n^s(E,1,1)$ | $L_n^\kappa(E,1,1)$ | $H^n(E^\times)$ |
|---------|:-:|:-:|:-:|
| $n = 3$ | 0 | 0 | $\mathbf{Z}/2$ |
| $n = 2$ | $\mathbf{Z}/2$ | 0 | $\mathbf{Z}/2$ |
| $n = 1$ | $\mathbf{Z}/2$ | $\mathbf{Z}/2$ | $\mathbf{Z}/2$ |
| $n = 0$ | $4\mathbf{Z}$ | $2\mathbf{Z}$ | $\mathbf{Z}/2$ |

**Table 14.6: Continuous Fields, $E = \mathbf{C}$, Type O**

|         | $L_n^s(E,1,1)$ | $L_n^\kappa(E,1,1)$ | $H^n(E^\times)$ |
|---------|:-:|:-:|:-:|
| $n = 3$ | 0 | 0 | $\mathbf{Z}/2$ |
| $n = 2$ | $\mathbf{Z}/2$ | 0 | 0 |
| $n = 1$ | 0 | $\mathbf{Z}/2$ | $\mathbf{Z}/2$ |
| $n = 0$ | 0 | 0 | 0 |

**Table 14.7: Continuous Fields, $E = \mathbf{C}$, Type $U$**

|           | $L_n^s(E,c,1)$ | $L_n^\kappa(E,c,1)$ | $H^n(E^\times)$ |
|-----------|:-:|:-:|:-:|
| $n = 1,3$ | 0 | 0 | 0 |
| $n = 0,2$ | $4\mathbf{Z}$ | $2\mathbf{Z}$ | $\mathbf{Z}/2$ |

**Table 14.8: Continuous Fields, $D = \mathbf{H}$, Type $O$**

| | $L_n^s(D, c', 1)$ | $L_n^K(D, c', 1)$ | $H^n(E^\times)$ |
|---|---|---|---|
| $n = 3$ | 0 | 0 | 0 |
| $n = 2$ | $2\mathbf{Z}$ | $2\mathbf{Z}$ | 0 |
| $n = 1$ | 0 | 0 | 0 |
| $n = 0$ | 0 | 0 | 0 |

Here $c'$ denotes the type $O$ involution on the quaternions $\mathbf{H}$. Explicitly, it is given by $c'(i) = i$, $c'(j) = j$ and $c'(k) = -k$. For the usual (type $Sp$) involution $c(i) = -i$, $c(j) = -j$, we have $L_n(D, c, 1) = L_{n+2}(D, c', 1)$.

## 14.B The Hasse principle

We will need the groups $CL_n^{O_i}(D, \alpha, u)$ describing the kernel and cokernel of the Hasse principle $L_n^{O_i}(D, \alpha, u) \to L_n^{O_i}(D_A, \alpha, u)$. We will tabulate the associated change of $K$–theory sequences

$$\ldots \to CL_n^s(D, \alpha, u) \to CL_n^K(D, \alpha, u) \to H^n(C(E)) \xrightarrow{\delta} CL_{n-1}^s(D, \alpha, u) \to$$

where $C(D) \cong C(E) = E_A^\times/E^\times$ is the idèle class group of the center field $E$ in $D$. The map $\delta$ is the coboundary map in the long exact sequence. There are short exact sequences ($n = 0, 1$):

$$0 \to H^n(E^\times) \to H^n(E_A^\times) \to H^n(C(E)) \to 0$$

and the maps are induced by the inclusions of fields.

**Table 14.9: Type $OK$**

| | $CL_n^s(E)$ | $CL_n^K(E)$ | $H^n(C(E))$ |
|---|---|---|---|
| $n = 3$ | 0 | 0 | $H^1(C(E))$ |
| $n = 2$ | $H^1(C(E))$ | 0 | $H^0(C(E))$ |
| $n = 1$ | $H^0(C(E))$ | $H^1(C(E))$ | $H^1(C(E))$ |
| $n = 0$ | $\mathbf{Z}/2$ | $\mathbf{Z}/2 \tilde\times H^0(C(E))$ | $H^0(C(E))$ |

The extension $0 \to \mathbf{Z}/2 \to CL_0^K(D) \to H^0(C(E)) \to 0$ appearing in this table is split if and only if $-1 \in E^{\times 2}$.

**Table 14.10: Type $OD$**

|         | $CL_n^s(E)$ | $CL_n^K(E)$ | $H^n(C(E))$ |
|---------|-------------|-------------|-------------|
| $n = 3$ | $0$ | $0$ | $H^1(C(E))$ |
| $n = 2$ | $H^1(C(E))$ | $0$ | $H^0(C(E))$ |
| $n = 1$ | $H^0(C(E))$ | $\ker\{\delta\colon H^1(C(E)) \to \mathbf{Z}/2\}$ | $H^1(C(E))$ |
| $n = 0$ | $\mathbf{Z}/2$ | $H^0(C(E))$ | $H^0(C(E))$ |

Some of the details of these results are clarified in [20].

**Table 14.11: Type $U$**

|           | $CL_n^s(E)$ | $CL_n^K(E)$ | $H^n(C(E))$ |
|-----------|-------------|-------------|-------------|
| $n = 1, 3$ | $0$ | $0$ | $0$ |
| $n = 0, 2$ | $0$ | $\mathbf{Z}/2$ | $\mathbf{Z}/2$ |

## 14.C The relative groups $L_n^{X_1}(\mathbf{Z}G \to \widehat{\mathbf{Z}}_2 G)$

We now suppose that $G$ is a 2–hyperelementary group and give tables for calculating the relative groups $L_n^{X_1}(\mathbf{Z}G \to \widehat{\mathbf{Z}}_2 G)$ and $L_n^{X_0}(\mathbf{Z}G \to \widehat{\mathbf{Z}}_2 G)$. Recall that by excision these split up according to the way $\mathbf{Q}G$ splits into simple algebras with involution. Then if $G = C \rtimes P$ where $C = \mathbf{Z}/d$ we can compute the $d$–component in terms of the map $\gamma_n$ defined earlier. In particular, a summand of $\gamma_n(d)$ is determined by a single algebra $(D, \alpha, u)$ with centre field $E$ and ring of integers $A \subset E$. Restricted to this summand it is the natural map

$$\gamma_n(d)\colon \prod_{\ell \nmid 2d} L_n^s(\widehat{A}_\ell) \times L_n^s(E_\infty) \longrightarrow CL_n^s(D)$$

and in the domain the terms

$$L_n^s(\widehat{A}_\ell) = L_n^s(\widehat{A}_\ell/Rad)$$

are just $L$–groups of finite fields. Thus all the terms in the domain and range are given in the previous tables for fields. The maps $\gamma_n(d)$ are also easy to relate to number theory. In particular, note that mapping a term $H^n(\widehat{A}_{2d'}^\times)$ or $H^n(E_\infty^\times)$ to $H^n(C(E))$ is the map induced by the inclusion into $H^n(E_A^\times)$ followed by the projection $H^n(E_A^\times) \to H^n(C(E))$. The symbol $\Sigma$ in the tables denotes a free abelian group of signatures at infinite primes.

### Table 14.12: Type $OK(\mathbf{R})$ or Type $OD(\mathbf{R})$

| | $\prod_{\ell \nmid 2d} L_n^s(\widehat{A}_\ell) \times L_n^s(E_\infty)$ | $CL_n^s(D)$ | $\ker \gamma_n^s(d)$ | $\mathrm{cok}\, \gamma_n^s(d)$ |
|---|---|---|---|---|
| $n = 3$ | $0$ | $0$ | $0$ | $0$ |
| $n = 2$ | $H^1(\widehat{A}_{2d'}^\times) \times H^1(E_\infty^\times)$ | $H^1(C(E))$ | $0$ | $H^1(\widehat{A}_{2d}^\times)/H^1(A^\times)$ |
| $n = 1$ | $H^0(\widehat{A}_{2d'}^\times) \times H^0(E_\infty^\times)$ | $H^0(C(E))$ | $\ker\, \Phi$ | $\mathrm{cok}\, \Phi \oplus H^0(\Gamma(E))$ |
| $n = 0$ | $0 \times \oplus 4\mathbf{Z}$ | $\mathbf{Z}/2$ | $\Sigma$ | $0$ |

### Table 14.13: Type $OK(\mathbf{C})$ or Type $OD(\mathbf{C})$

| | $\prod_{\ell \nmid 2d} L_n^s(\widehat{A}_\ell) \times L_n^s(E_\infty)$ | $CL_n^s(D)$ | $\ker \gamma_n^s(d)$ | $\mathrm{cok}\, \gamma_n^s(d)$ |
|---|---|---|---|---|
| $n = 3$ | $0$ | $0$ | $0$ | $0$ |
| $n = 2$ | $H^1(\widehat{A}_{2d'}^\times) \times H^1(E_\infty^\times)$ | $H^1(C(E))$ | $0$ | $H^1(\widehat{A}_{2d}^\times)/H^1(A^\times)$ |
| $n = 1$ | $H^0(\widehat{A}_{2d'}^\times) \times H^0(E_\infty^\times)$ | $H^0(C(E))$ | $\ker\, \Phi$ | $\mathrm{cok}\, \Phi \oplus H^0(\Gamma(E))$ |
| $n = 0$ | $0 \times 0$ | $\mathbf{Z}/2$ | $0$ | $\mathbf{Z}/2$ |

**Table 14.14: Type $OD(\mathbf{H})$**

|  | $\prod\limits_{\ell \nmid 2d} L_n^s(\widehat{A}_\ell) \times L_n^s(D_\infty)$ | $CL_n^s(D)$ | $\ker \gamma_n^s(d)$ | $\operatorname{cok} \gamma_n^s(d)(d)$ |
|---|---|---|---|---|
| $n = 3$ | $0$ | $0$ | $0$ | $0$ |
| $n = 2$ | $H^1(\hat{A}_{2d'}^\times) \times \oplus 2\mathbf{Z}$ | $H^1(C(E))$ | $\Sigma$ | $H^1(\hat{A}_{2d}^\times)/H^1(A^\times)$ |
| $n = 1$ | $H^0(\hat{A}_{2d'}^\times) \times 0$ | $H^0(C(E))$ | $\ker \Phi'$ | $\operatorname{cok} \Phi' \oplus H^0(\Gamma(E))$ |
| $n = 0$ | $0 \times 0$ | $\mathbf{Z}/2$ | $0$ | $\mathbf{Z}/2$ |

**Table 14.15: Type $U$**

|  | $\prod\limits_{\ell \nmid 2d} L_n^s(\widehat{A}_\ell) \times L_n^s(E_\infty)$ | $CL_n^s(E)$ | $\ker \gamma_n^s(d)$ | $\operatorname{cok} \gamma_n^s(d)(d)$ |
|---|---|---|---|---|
| $n = 1, 3$ | $0$ | $0$ | $0$ | $0$ |
| $n = 0, 2$ | $0 \times \oplus 4\mathbf{Z}$ | $0$ | $\oplus 4\mathbf{Z}$ | $0$ |

Since the $L^s$-groups are all zero in type $GL$, this completes the $L^s$-tables.

## 14.D The relative groups $L_n^{X_0}(\mathbf{Z}G \to \widehat{\mathbf{Z}}_2 G)$

We now give the relative group tables for the $L^p$-groups. Some additional notation is defined as it appears.

**Table 14.16: Type $OK(\mathbf{R})$**

|  | $\prod\limits_{\ell \nmid 2d} L_n^K(\widehat{A}_\ell) \times L_n^K(E_\infty)$ | $CL_n^K(E)$ | $\ker \gamma_n^K$ | $\operatorname{cok} \gamma_i^K$ |
|---|---|---|---|---|
| $n=3$ | $0$ | $0$ | $0$ | $0$ |
| $n=2$ | $0$ | $0$ | $0$ | $0$ |
| $n=1$ | $H^1(\hat{A}_{2d'}^\times) \times H^1(E_\infty)$ | $H^1(C(E))$ | $0$ | $H^1(\hat{A}_{2d}^\times)/H^1(A^\times)$ |
| $n=0$ | $H^0(\hat{A}_{2d'}^\times) \times \oplus 2\mathbf{Z}$ | $\mathbf{Z}/2 \,\tilde\times\, H^0(C(E))$ | $\Sigma \oplus \ker \Phi'$ | $\operatorname{cok} \Phi \oplus H^0(\Gamma(E))$ |

**Table 14.17: Type $OD(\mathbf{R})$**

| | $\prod_{\ell \nmid 2d} L_n^K(\widehat{A}_\ell) \times L_n^K(E_\infty)$ | $CL_n^K(D)$ | ker $\gamma_n^K$ | cok $\gamma_i^K$ |
|---|---|---|---|---|
| $n=3$ | 0 | 0 | 0 | 0 |
| $n=2$ | 0 | 0 | 0 | 0 |
| $n=1$ | $H^1(\widehat{A}_{2d'}^\times) \times H^1(E_\infty)$ | ker $\delta$ | 0 | ker $\Delta'$ |
| $n=0$ | $H^0(\widehat{A}_{2d'}^\times) \times \oplus\, 2\mathbf{Z}$ | $H^0(C(E))$ | $\Sigma \oplus$ ker $\Phi'$ | cok $\Phi \oplus H^0(\Gamma(E))$ |

Here the map $\Delta'$ is the map

$$\Delta': \frac{H^1(\widehat{A}_{2d}^\times) \oplus H^1(E_\infty^\times)}{H^1(A^\times)} \longrightarrow \{\pm 1\}$$

defined by $\Delta'(\langle -1 \rangle_\ell) = -1$ if and only if $\widehat{D}_\ell$ is nonsplit.

**Table 14.18: Type $OK(\mathbf{C})$**

| | $\prod_{\ell \nmid 2d} L_n^K(\widehat{A}_\ell) \times L_n^K(E_\infty)$ | $CL_n^K(E)$ | ker $\gamma_n^K$ | cok $\gamma_i^K$ |
|---|---|---|---|---|
| $n=3$ | 0 | 0 | 0 | 0 |
| $n=2$ | 0 | 0 | 0 | 0 |
| $n=1$ | $H^1(\widehat{A}_{2d'}^\times) \times H^1(E_\infty)$ | $H^1(C(E))$ | 0 | $H^1(\widehat{A}_{2d}^\times)/H^1(A^\times)$ |
| $n=0$ | $H^0(\widehat{A}_{2d'}^\times) \times 0$ | $\mathbf{Z}/2\,\tilde{\times}\, H^0(C(E))$ | ker $\Phi$ | $\mathbf{Z}/2\,\tilde{\times}(\text{cok } \Phi \oplus H^0(\Gamma(E)))$ |

**Table 14.19: Type $OD(\mathbf{C})$**

|  | $\prod_{\ell \nmid 2d} L_n^K(\widehat{A}_\ell) \times L_n^K(E_\infty)$ | $CL_n^K(D)$ | $\ker \gamma_n^K$ | $\operatorname{cok} \gamma_i^K$ |
|---|---|---|---|---|
| $n = 3$ | $0$ | $0$ | $0$ | $0$ |
| $n = 2$ | $0$ | $0$ | $0$ | $0$ |
| $n = 1$ | $H^1(\hat{A}_{2d'}^\times) \times H^1(E_\infty)$ | $\ker \delta$ | $0$ | $\ker \Delta$ |
| $n = 0$ | $H^0(\hat{A}_{2d'}^\times) \times 0$ | $H^0(C(E))$ | $\ker \Phi$ | $\operatorname{cok} \Phi \oplus H^0(\Gamma(E))$ |

The map $\Delta$ has the same definition as $\Delta'$ but $H^1(E_\infty^\times)$ is missing from the domain.

**Table 14.20: Type $OD(\mathbf{H})$**

|  | $\prod_{\ell \nmid 2d} L_n^K(\widehat{A}_\ell) \times L_n^K(D_\infty)$ | $CL_n^K(D)$ | $\ker \gamma_n^K$ | $\operatorname{cok} \gamma_i^K$ |
|---|---|---|---|---|
| $n = 3$ | $0$ | $0$ | $0$ | $0$ |
| $n = 2$ | $0 \times \oplus 2\mathbf{Z}$ | $0$ | $\Sigma$ | $0$ |
| $n = 1$ | $H^1(\hat{A}_{2d'}^\times) \times 0$ | $\ker \delta$ | $0$ | $\ker \Delta'$ |
| $n = 0$ | $H^0(\hat{A}_{2d'}^\times) \times 0$ | $H^0(C(E))$ | $\ker \Phi'$ | $\operatorname{cok} \Phi' \oplus H^0(\Gamma(E))$ |

**Table 14.21: Type $U(\mathbf{C})$**

|  | $\prod_{\ell \nmid 2d} L_n^K(\widehat{A}_\ell) \times L_n^K(D_\infty)$ | $CL_n^K(D)$ | $\ker \gamma_n^K$ | $\operatorname{cok} \gamma_i^K$ |
|---|---|---|---|---|
| $n = 1, 3$ | $0$ | $0$ | $0$ | $0$ |
| $n = 0, 2$ | $0 \times \oplus 2\mathbf{Z}$ | $H^0(C(E))$ | $\Sigma$ | $0$ |

**Table 14.22: Type $U(\mathbf{GL})$**

|  | $\prod_{\ell \nmid 2d} L_n^K(\widehat{A}_\ell) \times L_n^K(D_\infty)$ | $CL_n^K(D)$ | $\ker \gamma_n^K$ | $\operatorname{cok} \gamma_i^K$ |
|---|---|---|---|---|
| $n = 1, 3$ | 0 | 0 | 0 | 0 |
| $n = 0, 2$ | $0 \times 0$ | $H^0(C(E))$ | 0 | $\mathbf{Z}/2$ |

## 14.E Finite 2–groups

Here complete calculations already appear in [**33**, §3, p.80]. To compare our results with the tables there note that $\Gamma(E)$ and $\Gamma^*(E)$ have odd order (Weber's Theorem) for all the centre fields appearing in $\mathbf{Q}G$ and $g_2(E) = 1$. Hence $\Phi$ and $\Phi'$ are injective with $\operatorname{cok} \Phi_E$ of 2-rank $1 + r_2$ (resp. $\operatorname{cok} \Phi'_E$ of 2-rank $1 + r_1$). As above, the degree $[E, \mathbf{Q}] = r_1 + 2r_2$, where $r_1$ denotes the number of real places of $E$ and $r_2$ the number of complex places.

In [**33**] the basic antistructures on the simple components of $\mathbf{Q}G$ are labelled $\Gamma_N$, $F_N$, $R_N$, $H_N$, $UI$ and $UII$. These have type $OK(\mathbf{C})$, $OK(\mathbf{C})$, $OK(\mathbf{R})$, $OD(\mathbf{H})$, $U(\mathbf{C})$ and $U(\mathbf{GL})$ respectively in our notation. In our tables, the distinction between $\Gamma_N$ and $F_N$ is whether $-1 \in E^{\times 2}$. Let $\zeta_N$ denote a primitive $2^N th$ root of unity. The centres $E$ for the type $O$ factors are $\mathbf{Q}(\zeta_{N+1})$, $\mathbf{Q}(\zeta_{N+2} - \zeta_{N+2}^{-1})$, $\mathbf{Q}(\zeta_{N+2} + \zeta_{N+2}^{-1})$, $\mathbf{Q}(\zeta_N + \zeta_N^{-1})$ so that $(r_1, r_2)$ equals $(0, 2^{N-1})$, $(0, 2^{N-1})$, $(2^N, 0)$, $(2^{N-2}, 0)$ respectively. Therefore using Tables 14.12–14.14 and 14.16–14.20 we can list the contribution of the type $O$ components to $L_n^{\chi_1}(R \to \widehat{R}_2)$ or $L_n^{\chi_0}(R \to \widehat{R}_2)$. The contributions from type $U$ components are already easily read off from Tables 14.15, 14.21 and 14.22. Note that only the rank and not the divisibilities in the signature groups are given in the tables.

**Table 14.23:** $L_n^{X_1}(R \to \widehat{R}_2)$ **in Type** $O$

| $O$ | $\Gamma_N$ | $F_N$ | $R_N$ | $H_N$ |
|---|---|---|---|---|
| $n = 3$ | 0 | 0 | 0 | $\mathbf{Z}^{r_1}$ |
| $n = 2$ | 0 | 0 | 0 | 0 |
| $n = 1$ | $(\mathbf{Z}/2)^{r_2+1}$ | $(\mathbf{Z}/2)^{r_2+1}$ | $\mathbf{Z}^{r_1} \oplus \mathbf{Z}/2$ | $(\mathbf{Z}/2)^{r_1+1}$ |
| $n = 0$ | $\mathbf{Z}/2$ | $\mathbf{Z}/2$ | 0 | $\mathbf{Z}/2$ |

**Table 14.24:** $L_n^{X_0}(R \to \widehat{R}_2)$ **in Type** $O$

| $O$ | $\Gamma_N$ | $F_N$ | $R_N$ | $H_N$ |
|---|---|---|---|---|
| $n = 3$ | 0 | 0 | 0 | $\mathbf{Z}^{r_1}$ |
| $n = 2$ | 0 | 0 | 0 | 0 |
| $n = 1$ | 0 | 0 | $\mathbf{Z}^{r_1}$ | $(\mathbf{Z}/2)^{r_1-1}$ |
| $n = 0$ | $(\mathbf{Z}/2)^{r_2+2}$ | $\mathbf{Z}/4 \oplus (\mathbf{Z}/2)^{r_2}$ | $\mathbf{Z}/2$ | $(\mathbf{Z}/2)^{r_1+1}$ |

The divisibilities for $L^p$ are determined in [**30, 2.8**] to be $\Sigma = 8\mathbf{Z}\oplus(4\mathbf{Z})^{r_1-1}$ in type $R_N$ (for 2–power cyclotomic extensions $E$, the quantity $r_E = 0$), and $\Sigma = \oplus\, 2\mathbf{Z}$ in type $H_N$. Those for $L'$ are the same, by the Rothenberg sequence tables.

**Table 14.25:** $H^n(K_1(R \to \widehat{R}_2))$ **in Type** $O$

| $O$ | $\Gamma_N$ | $F_N$ | $R_N$ | $H_N$ |
|---|---|---|---|---|
| $n = 3$ | 0 | 0 | 0 | $(\mathbf{Z}/2)^{r_1}$ |
| $n = 2$ | $(\mathbf{Z}/2)^{r_2+1}$ | $(\mathbf{Z}/2)^{r_2+1}$ | $\mathbf{Z}/2$ | $(\mathbf{Z}/2)^{r_1+1}$ |
| $n = 1$ | 0 | 0 | 0 | $(\mathbf{Z}/2)^{r_1}$ |
| $n = 0$ | $(\mathbf{Z}/2)^{r_2+1}$ | $(\mathbf{Z}/2)^{r_2+1}$ | $\mathbf{Z}/2$ | $(\mathbf{Z}/2)^{r_1+1}$ |

## References.

[ 1 ] A. Bak, *Odd dimension surgery groups of odd torsion groups vanish*, Topology **14** (1975), no. 4, 367–374.

[ 2 ] _____, *The computation of surgery groups of finite groups with abelian 2-hyperelementary subgroups*, "Algebraic $K$-theory (Proc. Conf., Northwestern Univ., Evanston, Ill., 1976)", Lecture Notes in Math., vol. 551, Springer, Berlin, 1976, pp. 384–409.

[ 3 ] _____, *The involution on Whitehead torsion*, General Topology and Appl. **7** (1977), 201–206.

[ 4 ] _____, *The computation of even dimension surgery groups of odd torsion groups*, Comm. Algebra **6** (1978), no. 14, 1393–1458.

[ 5 ] _____, *Arf's theorem for trace Noetherian and other rings*, J. Pure Appl. Algebra **14** (1979), no. 1, 1–20.

[ 6 ] _____, *"$K$-Theory of Forms"*, Annals of Math. Studies, vol. 98, Princeton University Press, Princeton, 1981.

[ 7 ] A. Bak and M. Kolster, *The computation of odd-dimensional projective surgery groups of finite groups*, Topology **21** (1982), 35–63.

[ 8 ] A. Bak and W. Scharlau, *Grothendieck and Witt groups of orders and finite groups*, Invent. Math. **23** (1974), 207–240.

[ 9 ] H. Bass, *Unitary algebraic $K$-theory*, "Algebraic K-theory, III: Hermitian K-theory and geometric applications (Proc. Conf., Battelle Memorial Inst., Seattle, Wash., 1972)", Lecture Notes in Math., vol. 343, Springer-Verlag, Berlin, 1973, pp. 57–265.

[10] _____, *$L_3$ of finite abelian groups*, Ann. of Math. (2) **99** (1974), 118–153.

[11] S. E. Cappell, *Mayer-Vietoris sequences in hermitian $K$-theory*, "Algebraic K-theory, III: Hermitian K-theory and geometric applications (Proc. Conf., Battelle Memorial Inst., Seattle, Wash., 1972)", Lecture Notes in Math., vol. 343, Springer-Verlag, Berlin, 1973, pp. 478–512.

[12] S. E. Cappell and J. L. Shaneson, *Torsion in L-groups*, "Algebraic and geometric topology (New Brunswick, N.J., 1983)", Lecture Notes in Math., vol. 1126, Springer-Verlag, Berlin, 1985, pp. 22–50.

[13] G. Carlsson, *On the Witt group of a 2-adic group ring*, Quart. J. Math. Oxford Ser. (2) **31** (1980), no. 123, 283–313.

[14] G. Carlsson and R. J. Milgram, *Torsion Witt rings for orders and finite groups*, "Geometric Applications of Homotopy Theory (Proc. Conf., Evanston, Ill., 1977), I", Lecture Notes in Math., vol. 657, Springer-Verlag, Berlin, 1978, pp. 85–105.

[15] _____, *The structure of odd L-groups*, "Algebraic topology, Waterloo, 1978 (Proc. Conf., Univ. Waterloo, Waterloo, Ont., 1978)", Lecture Notes in Math., vol. 741, Springer-Verlag, Berlin, 1979, pp. 1–72.

[16] D. W. Carter, *Localizations in lower algebraic K-theory*, Comm. Algebra **8** (1980), no. 7, 603–622.

[17] _____, *Lower K-theory of finite groups*, Comm. Algebra **8** (1980), no. 20, 1927–1937.

[18] F. J.-B. J. Clauwens, *L-theory and the arf invariant*, Invent. Math. **30** (1975), no. 2, 197–206.

[19] J. F. Davis, *The surgery semicharacteristic*, Proc. London Math. Soc. (3) **47** (1983), no. 3, 411–428.

[20] _____, *Evaluation of odd-dimensional surgery obstructions with finite fundamental group*, Topology **27** (1988), no. 2, 179–204.

[21] J. F. Davis and R. J. Milgram, *Semicharacteristics, bordism, and free group actions*, Trans. Amer. Math. Soc. **312** (1989), no. 1, 55–83.

[22] J. F. Davis and A. A. Ranicki, *Semi-invariants in surgery*, K-Theory **1** (1987), no. 1, 83–109.

[23] T. t. Dieck and I. Hambleton, "Surgery theory and geometry of representations", DMV Seminar, vol. 11, Birkhäuser Verlag, Basel, 1988.

[24] A. W. M. Dress, *Contributions to the theory of induced representations*, "Algebraic K-theory, II: "Classical" algebraic K-theory and connections with arithmetic (Proc. Conf., Battelle Memorial Inst., Seattle, Wash., 1972)", Lecture Notes in Math., vol. 342, Springer-Verlag, Berlin, 1973, pp. 183–240.

[25] _____, *Induction and structure theorems for orthogonal representations of finite groups*, Ann. of Math. (2) **102** (1975), no. 2, 291–325.

[26] I. Hambleton, *Projective surgery obstructions on closed manifolds*, "Algebraic $K$-theory, Part II (Oberwolfach, 1980)", Lecture Notes in Math., vol. 967, Springer-Verlag, Berlin, 1982, pp. 101–131.

[27] I. Hambleton and I. Madsen, *Actions of finite groups on $R^{n+k}$ with fixed set $R^k$*, Canad. J. Math. **38** (1986), no. 4, 781–860.

[28] _____, *Local surgery obstructions and space forms*, Math. Z. **193** (1986), no. 2, 191–214.

[29] _____, *On the discriminants of forms with Arf invariant one*, J. Reine Angew. Math. **395** (1989), 142–166.

[30] _____, *On the computation of the projective surgery obstruction groups*, $K$-Theory **7** (1993), no. 6, 537–574.

[31] I. Hambleton and R. J. Milgram, *The surgery obstruction groups for finite 2-groups*, Invent. Math. **61** (1980), no. 1, 33–52.

[32] I. Hambleton, A. Ranicki, and L. R. Taylor, *Round L-theory*, J. Pure Appl. Algebra **47** (1987), 131–154.

[33] I. Hambleton, L. R. Taylor, and B. Williams, *An introduction to maps between surgery obstruction groups*, "Algebraic topology, Aarhus 1982 (Aarhus, 1982)", Lecture Notes in Math., vol. 1051, Springer-Verlag, Berlin, 1984, pp. 49–127.

[34] _____, *Detection theorems for $K$-theory and $L$-theory*, J. Pure Appl. Algebra **63** (1990), no. 3, 247–299.

[35] A. F. Kharshiladze, *Hermitian $K$-theory and quadratic extensions of rings*, Trudy Moskov. Mat. Obshch. **41** (1980), 3–36.

[36] _____, *Wall groups of elementary 2-groups*, Mat. Sb. (N.S.) **114** (1981), no. 1, 167–176.

[37] M. Kolster, *Computations of Witt groups of finite groups*, Math. Ann. **241** (1979), no. 2, 129–158.

[38] _____, *Even-dimensional projective surgery groups of finite groups*, "Algebraic $K$-theory, Part II (Oberwolfach, 1980)", Lecture Notes in Math., vol. 967, Springer-Verlag, Berlin, 1982, pp. 239–279.

[39] E. Laitinen and I. Madsen, *The $L$-theory of groups with periodic cohomology, I*, Tech. report, Aarhus Universitet, 1981.

[ 40 ] D. W. Lewis, *Forms over real algebras and the multisignature of a manifold*, Advances in Math. **23** (1977), no. 3, 272–284.

[ 41 ] _____, *Frobenius algebras and Hermitian forms*, J. Pure Appl. Algebra **14** (1979), no. 1, 51–58.

[ 42 ] I. Madsen, *Reidemeister torsion, surgery invariants and spherical space forms*, Proc. London Math. Soc. (3) **46** (1983), no. 2, 193–240.

[ 43 ] I. Madsen and R. J. Milgram, "The Classifying Spaces for Surgery and Cobordism of Manifolds", Annals of Mathematics Studies, vol. 92, Princeton University Press, Princeton, 1979.

[ 44 ] I. Madsen and M. Rothenberg, *On the homotopy theory of equivariant automorphism groups*, Invent. Math. **94** (1988), 623–637.

[ 45 ] R. J. Milgram, *Surgery with coefficients*, Ann. of Math. **100** (1974), 194–265.

[ 46 ] R. J. Milgram and R. Oliver, $SK_1(\hat{Z}_2[\pi])$ *and surgery*, Forum Math. **2** (1990), no. 1, 51–72.

[ 47 ] J. Milnor, *Whitehead torsion*, Bull. Amer. Math. Soc. **72** (1966), 358–426.

[ 48 ] J. Morgan and D. Sullivan, *The transversality characteristic class and linking cycles in surgery theory*, Ann. of Math. **99** (1974), 463–544.

[ 49 ] R. Oliver, "Whitehead groups of finite groups", London Mathematical Society Lecture Note Series, vol. 132, Cambridge University Press, Cambridge, 1988.

[ 50 ] R. Oliver and L. R. Taylor, "Logarithmic descriptions of Whitehead groups and class groups for $p$-groups", vol. 76, Mem. Amer. Math. Soc., no. 392, Amer. Math. Soc., Providence, 1988.

[ 51 ] W. Pardon, *Hermitian K-theory in topology: a survey of some recent results*, "Algebraic K-theory (Proc. Conf., Northwestern Univ., Evanston, Ill., 1976)", Lecture Notes in Math., vol. 551, Springer-Verlag, Berlin, 1976, pp. 303–310.

[ 52 ] _____, "Local surgery and the exact sequence of a localization for Wall groups", vol. 12, Mem. Amer. Math. Soc., no. 196, Amer. Math. Soc., Providence, 1977.

[53] _____, *Mod 2 semicharacteristics and the converse to a theorem of Milnor*, Math. Z. **171** (1980), no. 3, 247–268.

[54] _____, *The exact sequence of a localization for Witt groups. II. numerical invariants of odd-dimensional surgery obstructions*, Pacific J. Math. **102** (1982), no. 1, 123–170.

[55] E. K. Pedersen and A. A. Ranicki, *Projective surgery theory*, Topology **19** (1980), 239–254.

[56] A. A. Ranicki, *Algebraic L-theory. I. Foundations*, Proc. London Math. Soc. (3) **27** (1973), 101–125.

[57] _____, *On the algebraic L-theory of semisimple rings*, J. Algebra **50** (1978), 242–243.

[58] _____, *Localization in quadratic L-theory*, "Algebraic Topology (Proc. Conf. Univ. Waterloo, Waterloo, Ont., 1978)", Lecture Notes in Math., vol. 741, Springer-Verlag, Berlin, 1979, pp. 102–157.

[59] _____, *The L-theory of twisted quadratic extensions*, Canad. J. Math. **39** (1987), 345–364.

[60] _____, "*Exact sequences in the algebraic theory of surgery*", Mathematical Notes, vol. 26, Princeton University Press, Princeton, N.J., 1991.

[61] _____, "*Lower K- and L-theory*", London Math. Soc. Lecture Notes, vol. 178, Cambridge University Press, Cambridge, 1992.

[62] J. L. Shaneson, *Wall's surgery obstruction groups for $G \times Z$*, Ann. of Math. (2) **90** (1969), 296–334.

[63] _____, *Hermitian K-theory in topology*, "Algebraic K-theory, III: Hermitian K-theory and geometric applications (Proc. Conf., Battelle Memorial Inst., Seattle, Wash., 1972)", Lecture Notes in Math., vol. 343, Springer-Verlag, Berlin, 1973, pp. 1–40.

[64] L. R. Taylor, *Surgery groups and inner automorphisms*, "Algebraic K-theory, III: Hermitian K-theory and geometric applications (Proc. Conf. Battelle Memorial Inst., Seattle, Wash., 1972)", Lecture Notes in Math., vol. 343, Springer-Verlag, Berlin, 1973, pp. 471–477.

[65] C. B. Thomas, *Frobenius reciprocity of Hermitian forms*, J. Algebra **18** (1971), 237–244.

[66] C. T. C. Wall, *On the axiomatic foundations of the theory of Hermitian forms*, Proc. Cambridge Philos. Soc. **67** (1970), 243–250.

[67] _____, *On the classification of hermitian forms. I. Rings of algebraic integers*, Compositio Math. **22** (1970), 425–451.

[68] _____, "Surgery on compact manifolds", Academic Press, London, 1970, London Mathematical Society Monographs, No. 1.

[69] _____, *On the classification of Hermitian forms. II. Semisimple rings*, Invent. Math. **18** (1972), 119–141.

[70] _____, *Foundations of algebraic L-theory*, "Algebraic $K$-theory, III: Hermitian $K$-theory and geometric applications (Proc. Conf., Battelle Memorial Inst., Seattle, Wash., 1972)", Lecture Notes in Math., vol. 343, Springer-Verlag, Berlin, 1973, pp. 266–300.

[71] _____, *On the classification of Hermitian forms. III. Complete semilocal rings*, Invent. Math. **19** (1973), 59–71.

[72] _____, *Norms of units in group rings*, Proc. London Math. Soc. (3) **29** (1974), 593–632.

[73] _____, *On the classification of Hermitian forms. IV. Adele rings*, Invent. Math. **23** (1974), 241–260.

[74] _____, *On the classification of Hermitian forms. V. Global rings*, Invent. Math. **23** (1974), 261–288.

[75] _____, *Classification of Hermitian Forms. VI. Group rings*, Ann. of Math. (2) **103** (1976), no. 1, 1–80.

[76] M. Weiss, *Visible L-theory*, Forum Math. **4** (1992), no. 5, 465–498.

[77] M. Weiss and B. Williams, *Automorphisms of manifolds and algebraic $K$-theory, Part II*, J. Pure Appl. Algebra **69** (1989), 47–107.

[78] T. Yamada, "The Schur subgroup of the Brauer group", Lecture Notes in Math., vol. 397, Springer-Verlag, Berlin, 1974.

DEPARTMENT OF MATHEMATICS & STATISTICS
MCMASTER UNIVERSITY
HAMILTON, ONTARIO L8S 4K1, CANADA
*Email:* ian@icarus.Math.McMaster.CA

DEPARTMENT OF MATHEMATICS
UNIVERSITY OF NOTRE DAME
NOTRE DAME, IN 46556, USA
*Email:* taylor.2@nd.edu

# Surgery theory and infinite fundamental groups

## C. W. Stark*

# 1 Introduction

We begin with a sketch of early work in surgery on manifolds with infinite fundamental groups, state some of the major problems in surgery and manifold topology concerning such manifolds, and then describe surgical aspects of certain classes and constructions of geometrically interesting infinite groups. The themes and outline of this survey may soon be outdated, especially as new aspects of the Borel and Novikov conjectures come to light.

Like many readers the author would be grateful for an authoritative history of these developments, but such an account is unlikely to appear since the intensity of activity around some of these problems makes precise attributions difficult, especially after the passage of ten to thirty years. The author apologizes to any researchers whose work and favorite problems in manifold topology and related subjects may be slighted or omitted in this terse account.

The books of Wall [155] and Ranicki [130], and other papers in the present volume should be consulted for information on the $L$–groups and other features of surgery theory. A brief guide to the relevant literature of surveys and problem lists concludes this paper.

A few important terms may be unfamiliar. A path–connected topological space $X$ is *aspherical* if the homotopy groups $\pi_i(X) \cong 0$ for $i \geq 2$; if $X$ is homotopy equivalent to a CW complex then an equivalent statement is that the universal covering space of $X$ is contractible. (Many examples of aspherical manifolds appear below.) A closed manifold $X$ is *topologically rigid* if any other closed manifold homotopy equivalent to $X$ is homeomorphic to $X$.

*Partially supported by an NSA grant.

Algebra, topology, and geometry are fruitful sources of manifolds with infinite fundamental groups. Specialists in combinatorial group theory sometimes warn topologists that we tend to consider only a small sample of a large, unruly population of infinite groups and the reader should bear this in mind.

# 2  Free Abelian Groups and Homotopy Tori

Manifolds with free Abelian fundamental group assumed a distinguished place with Novikov's arguments for the topological invariance of rational Pontrjagin classes and the first versions of the Novikov conjecture on higher signatures [114, 115, 116, 63].

After Wall had established a surgery theory for non–simply connected manifolds and Poincaré complexes [153], free Abelian fundamental groups were among the first to be studied closely (in the latter half of the 1960's). These groups have the advantage of permitting an attack by induction on rank via the split surjection $\mathbf{Z}^n \twoheadrightarrow \mathbf{Z}$ with kernel $\mathbf{Z}^{n-1}$. A model for such an argument was already one of the best–known achievements of $K$–theory, the Bass–Heller–Swan theorem [3], which also showed that Whitehead groups vanish for free Abelian groups and therefore that their $L$–groups will be insensitive to $K$–theoretic decorations.

The key results describe the $L$–groups of a product $\pi \times \mathbf{Z}$ [141, 154], or more generally of a group $\Pi$ realized as an extension $1 \to \pi \to \Pi \to \mathbf{Z} \to 1$.

**Theorem 2.1 (Shaneson, Wall).** *For free Abelian groups with trivial orientation character,*

$$L_m(\mathbf{Z}^r) \cong \bigoplus_{0 \leq i \leq r} \binom{r}{i} L_{m-i}(1).$$

See [155, Theorems 12.6 and 13A.8] for this and related results, especially those involving nontrivial orientation characters.

The classification of piecewise linear structures on tori is the main application of these computations of $L$–groups, and this classification was obtained by Hsiang–Shaneson and Wall [82, 154] circa 1969; the summaries in the books of Wall and Kirby–Siebenmann [155, 94] are the standard references. The $L$–group results are accompanied by computations of the $PL$ surgery obstruction for tori [155, Proposition 13.B.8] and of the long exact sequence of surgery for tori [155, Chapter 15A].

**Theorem 2.2 (Hsiang–Shaneson, Wall).** *The piecewise linear structure set of the torus is*

$$\mathcal{S}_{PL}\left(T^n \times D^k, \, T^n \times S^{k-1}\right) \cong H^{3-k}(T^n; \, \mathbf{Z}/2),$$

*for $n + k \geq 5$, and the bijection is natural for finite covering projections.*

The naturality claim for this bijection and the consequent vanishing of every element of the *PL* structure set of the torus under transfer to suitable finite–sheeted covering spaces provide the last step in the application by Kirby and Siebenmann of Kirby's torus trick in the solution of the Hauptvermutung [93, 94]. For more information on the Hauptvermutung see [132] for historical remarks and earlier papers and see also the papers collected as appendices to [94].

The classification of *PL* structures on the torus turns out to preserve the normal invariant of the standard torus, so all these *PL* structures are parallelizable and smoothable, and the smooth structure set $\mathcal{S}_{DIFF}(T^n) \cong [T^n, PL/O]$ (see Wall [155, p. 227]).

Hsiang and Wall [83] showed that any closed topological manifold homotopy equivalent to $T^n$ is homeomorphic to it:

**Theorem 2.3 (Hsiang–Wall).** *The topological structure set*

$$\mathcal{S}_{TOP}\left(T^n \times D^k, T^n \times \partial D^k\right)$$

*has a single element for $n + k \geq 4$.*

The *TOP* version of the Farrell fibering theorem [41] and Siebenmann periodicity [94, 112] yield the result for $n + k \geq 5$, while Freedman's work on four–dimensional surgery is applicable since $\mathbf{Z}^n$ is a good group for his methods [69].

The ideas developed in the attack on free Abelian groups and homotopy tori reappear in product and transfer formulae (see Section 4), fibering and splitting theorems (Section 5), and topological rigidity results for aspherical manifolds. The explicit determination of piecewise linear and smooth structure sets for tori is rarely duplicated for other aspherical manifolds.

Exotic differentiable structures on the torus have received a limited amount of subsequent attention. In one result of note, Farrell and Jones showed that these smooth manifolds can carry expanding self–maps [51]. The models for this class of maps are the self–covering of $T^n$ induced by the linear maps $x \to ax$ on $\mathbf{R}^n$, for $a \in \mathbf{Z}$ ($> 1$).

# 3   Grand Challenge Problems

Some of the major problems of manifold topology concern manifolds with infinite fundamental groups and surgery on such manifolds. Only a few of the problems stated below are completely settled and some, such as the Novikov conjecture, are receiving intense attention.

## 3.1 Novikov's Theorem: Topological Invariance of Rational Pontrjagin Classes

The influence of this theorem is considerable, if sometimes indirect. Novikov's proof that if $f: M^n \to N^n$ is a homeomorphism of smooth manifolds then $f^*(p(N)) = p(M)$, where $p(V) \in H^*(V; \mathbf{Q})$ is the total Pontrjagin class of the smooth manifold $V$, involved an analysis of noncompact manifolds of the form $T^r \times B^s$, sometimes called "anchor rings" on the model of the solid torus in dimension three.

The argument is a precursor of splitting arguments (see Section 5) and of non–simply connected finiteness, simple homotopy, and surgery arguments. We recommend the survey [63] as well as Novikov's papers for the interested reader.

## 3.2 The Novikov conjecture

One of the most important outgrowths of Novikov's theorem is the Novikov conjecture, or more accurately the family of conjectures which grow out of Novikov's original conjecture. See [117] for the original statement, the survey volumes [64, 63], and the chapter [32] in this volume.

## 3.3 The Borel conjecture

Modern versions of the Novikov conjecture for various functors assert that a rationalized assembly map is a split injection. The much stronger assertion for a closed aspherical manifold $M^n$ that the assembly map for topological surgery is an isomorphism translates to the following claim.

**Conjecture 3.1.** (*The Borel Conjecture*)
     If $M^n$ is a closed aspherical manifold then the topological structure set $\mathcal{S}_{TOP}(M^n \times D^k, M^n \times \partial D^k)$ is a one–element set.

This is the modern, strong form of the Borel conjecture on topological rigidity for aspherical manifolds. See chapter [32] in this volume and the survey volumes [64, 63]. The brief sketch in Kirby's problem list is a good starting point for readers who have not encountered the problem before [92, Problem 5.29].

## 3.4 Structure of open manifolds

Versions of the following question have been extremely influential, beginning with [13]; Siebenmann's thesis [142] has been particularly useful in later work. See Section 5 for more information.

**Problem 3.2.** Given a connected, open (that is, noncompact) manifold $V^n$, is this space the interior of a compact manifold with boundary?

Note that this conclusion forces a number of finiteness properties upon $V^n$; the question is whether algebraic finiteness properties are equivalent to the geometric finiteness operation of compactifying $V^n$ to a manifold with boundary.

The classification problem for open manifolds is available in a variety of flavors to suit one's taste and applications. If open manifolds $V^n$ and $W^n$ are homotopy equivalent, we then ask if they are simple homotopy equivalent in a more stringent sense (allowing only maps and homotopies which are proper, metrically bounded, or controlled against a variable gauge), and if such homotopy equivalences or simple homotopy equivalences imply homeomorphism, which may also be obliged to satisfy a control hypothesis. The work of Chapman and Ferry [23] and Quinn's sequence of papers on ends of maps [125, 126, 127] were followed by a long stream of developments which are described in [121] elsewhere in this volume.

## 3.5 Placement problems

Surgical questions concerning manifold imbeddings $N^k \hookrightarrow M^n$ have been studied since the early 1960s in the context of knot theory and as problems in relative or ambient surgery which might serve as stepping stones toward theorems concerning $M^n$. Note that in this context one tends to do surgery on $N^k$ with normal cobordisms contained within $N^n$: the two manifolds separately and the imbedding $N^k \hookrightarrow M^n$ could all work to obstruct a sequence of surgeries.

Wall develops a variety of placement problems and their obstruction groups in [155, Chapter 12]. Splitting methods (Section 5) are the main placement problems considered in the present paper. We are concerned with codimension–one submanifolds of $M^n$ and approaches to the structure set or other features of $M^n$ which proceed by induction over a sequence of cut–and–paste constructions for this manifold.

## 3.6 Topological space–form problems

Classically a "space–form" is a Riemannian manifold of constant sectional curvature and the "space–form problem" asks for a classification of closed or complete Riemannian manifolds of constant sectional curvature.

These questions may be reduced to group theory, since we are asking for a classification of faithful representations with discrete image in the isometry group of the sphere, Euclidean space, or hyperbolic space. Excellent results are known for the first two cases [108, 160], while a characterization of the fundamental groups of closed hyperbolic manifolds is still lacking [161, Problem 9], even after the remarkable characterization of manifolds of nonpositive curvature whose rank is two or greater (we cite [1, 2] from a much larger body of work in this area).

Wall [156] stated classification problems in topology for manifolds with simple universal covering spaces (spheres and Euclidean spaces), largely modelled on his work then in progress on groups which act upon $S^n$ as groups of covering transformations.

**Problem 3.3.** Classify the cocompact, free, and properly discontinuous group actions (in $TOP$, $PL$, or $DIFF$) on $S^n$ and $\mathbf{R}^n$.

**Problem 3.4.** Which homotopy spheres and contractible manifolds appear as the universal covering spaces of closed manifolds?

The phrase "which contractible manifolds" may seem odd above, but in dimensions $n \geq 3$ there are contractible manifolds $V^n$ which are not homeomorphic to $\mathbf{R}^n$. Stallings showed in 1962 that a contractible manifold is homeomorphic to Euclidean space if and only if it is simply connected at infinity [145]; contractible manifolds which do not have this property had been known since early work of J. H. C. Whitehead [159]. In 1982 M. Davis constructed the first examples of closed manifolds whose universal covering spaces are contractible but not Euclidean [34, 35]. Problem 3.4 has received a great deal of attention in dimension three; see Kirby's problem list [92].

## 3.7  Poincaré duality groups

If $\pi$ is the fundamental group of a closed $n$–manifold with contractible universal covering space then the group cohomology $H^*(\pi)$ satisfies Poincaré duality (in a twisted sense if $M^n$ is nonorientable). The properties of such groups were abstracted by Bieri and Johnson–Wall [6, 88] in the definition of a Poincaré duality group of formal dimension $n$, or $PD(n)$–group. (A convenient source is Brown's textbook [16].) The main problem in this direction is an existential version of the Borel conjecture (see versions in [157] and [92, Problem 5.29]):

**Problem 3.5.** If $\pi$ is a $PD(n)$–group, must $\pi$ be the fundamental group of a closed aspherical manifold?

## 3.8  Mixed spaceform problems

A generalized version of the periodicity conditions seen in the spherical spaceform problems appears in the Farrell–Tate cohomology groups of the fundamental group of a manifold covered by $S^n \times \mathbf{R}^k$ [43, 14, 15]. Wall asked if every group satisfying these periodicity conditions is the fundamental group of a closed manifold covered by $S^n \times \mathbf{R}^k$ [157], and although a number of authors have worked on the question, a full characterization of such manifolds and groups has not been found [27, 60, 78, 87, 122, 147].

Less classical, but almost as interesting because of close connections to the study of discrete subgroups of Lie groups, are variations on these questions in which we ask that an infinite–sheeted regular covering space be a product of a closed manifold and a Euclidean space or that such a covering space have the homotopy type of a finite complex.

## 3.9 Nielsen problems

Let $\pi$ be a group, let $\mathrm{Aut}(\pi)$ denote its automorphism group, and let $\mathrm{Inn}(\pi)$ denote the subgroup of inner automorphisms (those homomorphisms $\alpha\colon \pi \to \pi$ with an associated element $g \in \pi$ such that $\alpha(x) = gxg^{-1}$). The inner automorphisms are a normal subgroup of the group of all automorphisms of $\pi$ and the quotient $\mathrm{Out}(\pi)$ is known as the group of outer automorphisms of $\pi$.

If $X$ is a path–connected topological space and $h\colon X \to X$ is a homeomorphism then base–point difficulties may not allow the association of an element of $\mathrm{Aut}(\pi_1(X, x))$ to $h$, but we get a homomorphism of groups $\mathrm{Homeo}(X) \to \mathrm{Out}(\pi_1(X, x))$.

The problem of lifting subgroups of the outer automorphism group of $\pi_1(M^n)$ to groups acting on $M^n$ is often called the Nielsen problem in honor of its original version for surfaces. The survey paper of Conner–Raymond [26] is a good source for results which are non–surgical and mostly related to group actions on the most interesting case of this question, aspherical manifolds. See 8.

## 3.10 Harmonic homotopy equivalences

For $i = 1, 2$ suppose that $(M_i, g_i)$ is a Riemannian manifold and suppose that $M_1$ is closed. The $L_2$–norm of the covariant derivative of a smooth map $f\colon M_1 \to M_2$ determines an energy functional $E\colon C^\infty(M_1, M_2) \to \mathbf{R}$ whose critical points (which may lie in a Sobolev space or other completion of $C^\infty$) are called *harmonic maps*.

Many aspects of harmonic maps resemble minimal surface theory and, like minimal surfaces, these maps are useful in surprising ways. For example, Corlette's method for proving superrigidity theorems is a close analysis of harmonic sections of certain bundles [31].

Variational methods and heat flow techniques establish existence results and some uniqueness results, which are particularly satisfactory if $M_2$ has negative sectional curvature.

Regularity or singularity properties of harmonic maps are a matter of ongoing work in geometric analysis and S.–T. Yau posed the following regularity problem c. 1980 [161, Problem 111]:

**Problem 3.6 (Yau).** Let $f\colon M_1 \to M_2$ be a diffeomorphism between two compact manifolds with negative curvature. If $h\colon M_1 \to M_2$ is a harmonic map which is homotopic to $f$, is $h$ a univalent map?

Farrell and Jones have produced several examples of badly behaved harmonic maps using smooth nonrigidity results [59].

## 3.11   Four–dimensional surgery

The remaining and perhaps immensely difficult challenge in completing M. Freedman's extension of topological surgery into dimension four concerns infinite fundamental groups which grow quickly (in the sense of the volume of balls in the word length metric). Current technology for handle manipulation breaks down for such groups, so advances as radical as Casson's original introduction of kinky handles appear necessary. See the last chapter of [69], Freedman's papers [21, 66, 67, 68], and the portions of the sequel to this volume on four–dimensional topology for more information.

# 4   Products, bundles, and transfers

## 4.1   Extensions, Bundles, and Singular Bundles

Bundles and orbifold bundles (also known as singular bundles or Seifert fibered spaces) often lead to extensions of fundamental groups, thanks to the long exact sequence in the homotopy groups of a fibration. Although an extension of groups yields a fibration of Eilenberg–MacLane classifying spaces corresponding to those groups, work is still under way on the realization of extensions by manifold and orbifold bundles. Conner, Raymond, and their collaborators are responsible for much of our knowledge of these constructions [25, 26, 89, 97, 98, 133].

Closed manifolds carrying effective toral actions are among the best understood aspherical manifolds known, at least if the torus action is of low codimension. Other bundles or singular bundles are usually more mysterious.

**Problem 4.1.** Must a manifold $M^n$ which is homotopy equivalent to the total space $E$ of a fiber bundle also admit a bundle structure? Is the generalization to orbifold bundles tractable?

Versions of this question influenced Quinn's development of the surgery assembly map [124] and was considered by Casson [20, 22].

## 4.2   Product and transfer formulae

After the computations of Shaneson [141] and Wall described in Section 2, a number of authors studied formulae for surgery obstructions in products and fiber bundles, along with $K$–theory formulas for the finiteness obstruction and Whitehead torsion.

The geometric splitting theorem of Shaneson [141]

$$L_n^s(\pi \times \mathbf{Z}) = L_n^s(\pi) \oplus L_{n-1}^h(\pi)$$

was obtained algebraically by Novikov [117] and Ranicki [131]. See also Milgram and Ranicki [107].

These results often compute obstructions for a problem induced in the total space of a fiber bundle by a problem in the bundle's base space. When the bundle is a product, we are often able to get vanishing results for the induced obstruction. Product formulae for the finiteness obstruction and the Whitehead torsion are perhaps the most familiar theorems in this genre [70, 96], especially in the guise of vanishing results for products with manifolds of Euler characteristic zero.

The most reliable resource on these constructions in surgery theory is the work of Lück–Ranicki [100, 101].

Ranicki has studied the stabilizing effect of taking products with tori in his work on lower $K-$ and $L-$theory [131].

# 5 Graphs of groups and splitting problems

## 5.1 Graphs of groups

See [136] for a discussion of combinatorial group theory along the lines indicated below. More algebraic treatments are found in [102, 104].

Let $\Gamma$ be a connected graph with vertex set $V(\Gamma)$ and edge set $E(\Gamma)$. We assume that the edges are oriented and we denote the vertices joined by an edge $e$ by $\partial_0(e)$, $\partial_1(e)$; note that we allow $\partial_0(e) = \partial_1(e)$. Regard $\Gamma$ as a category whose objects are the elements of $V(\Gamma) \cup E(\Gamma)$ and whose morphisms are the assignments $e \to \partial_0 e$ and $e \to \partial_1(e)$ as $e$ runs over the edge set.

Let $G \colon \Gamma \to \mathbf{GroupsInc}$ be a covariant functor from this category to the category of groups and inclusions. The resulting pushout group is known as the "fundamental group of the graph of groups with vertex groups $G(v)$ and edge groups $G(e)$."

In topology this construction arises when we paste spaces together and use the Seifert–van Kampen theorem to compute fundamental groups.

The simplest graphs lead to well–established constructions in combinatorial group theory. If $\Gamma$ has one edge $e$ and two vertices joined by $e$ then our pushout is the "free product with amalgamations" $G(\partial_0(e)) *_{G(e)} G(\partial_1(e))$, while if $\Gamma$ has one edge $e$ and one vertex $v$ then our pushout is then "HNN extension" determined by $G(v)$ and two inclusions of $G(e)$ into $G(v)$.

Terminology and technique for these constructions has been heavily influenced by Serre's demonstration [140] that this structure for a group $\pi$ is equivalent to the existence of a simplicial action of $\pi$ on a tree $T$

(connected simplicial 1–complex) which has the property that if a group element $\gamma$ carries an edge $e$ to itself then $\gamma$ fixes both vertices of $e$. Given an action $(\pi, T)$, $\pi$ is constructed from the stabilizer subgroups for edges and vertices of $T$ by the pushout construction described above.

## 5.2 Splitting manifolds and maps

This algebraic construction is often encountered in manifold topology through constructions which build up a manifold by pasting together two boundary components of a not–necessarily–connected manifold, or which break down a manifold through dissections along well–imbedded codimension–one submanifolds with trivial normal bundle.

These ideas are readily extended from manifolds to maps; the following description is given in the differentiable category but generalizes. Given manifold pairs $(M^n, N^{n-1})$ and $(V^n, W^{n-1})$ where the submanifolds are smoothly imbedded with trivial normal bundles, along with a homotopy equivalence $g$ of $M^n$ to $V^n$, we ask if $g$ is homotopic or cobordant to a homotopy equivalence $f \colon (M^n, N^{n-1}) \to (V^n, W^{n-1})$ of manifold pairs, where $f$ is transverse to $W^{n-1}$ and $N^{n-1} = f^{-1}(W^{n-1})$.

If we can deform $g$ to such a homotopy equivalence $f$ then we say that we can split the homotopy equivalence $g$ along $W^{n-1}$. This analysis of homotopy equivalent manifolds by dissection is familiar and venerable in dimension two; the class of three–manifolds which may be broken down to a union of disks by well–controlled dissections form the "Haken manifolds" for which we have proved the sharpest theorems in three–manifold topology, including the work of [149] and the settled cases of Thurston's geometrization conjecture.

In higher dimensions the basic move of cutting a homotopy equivalence of manifolds open along a locally flat, two–sided, codimension–one submanifold may be obstructed, as we explain below.

## 5.3 Early splitting arguments

Novikov's proof of the topological invariance of rational Pontrjagin classes is an induction which was recognized as a splitting argument for open manifolds.

A similar theme was taken up by Browder [10]: when is $M^n$ diffeomorphic to a product $N^{n-1} \times \mathbf{R}$? This question is imbedded within the study of manifolds which fiber over the circle [12], since a fiber bundle projection $M^n \to S^1$ induces a projection with the same fiber from an infinite cyclic covering space $V^n$ of $M^n$ to the real line, and hence a product structure on $V^n$. At the time the fibering question was taken up in the surgical dimensions a definitive fibering theorem for 3–manifolds had already been proved by Stallings [144]. Stallings' theorem served as a model in higher dimensions, in part by emphasizing the role of homotopy finiteness of an infinite cyclic covering space.

## 5.4 Algebra and fibering problems

If a group is a product with an infinite cyclic group, $\Gamma = \pi \times \mathbf{Z}$, then the group ring $\mathbf{Z}\Gamma \cong \mathbf{Z}\pi[t, t^{-1}]$, the ring of Laurent polynomials over $\mathbf{Z}\pi$. If $\Gamma$ is the fundamental group of a fibration $N^{n-1} \to M^n \to S^1$ with fiber $N^{n-1}$ then $\mathbf{Z}\Gamma$ is a twisted Laurent extension of $\mathbf{Z}\pi_1(N^n)$.

The Bass–Heller–Swan theorem [3] analyzed the $K$–theory of products with a circle by describing the $K$–theory of polynomial and Laurent polynomial rings. The crucial discovery for later developments is that the Mayer–Vietoris sequences for such extensions are cluttered by unanticipated $K$–groups generally known as Nil–groups.

These results were extended to fiber bundles over a circle by Farrell–Hsiang [46, 47] and then clarified by the Loday $K$–theory assembly map [99] and Waldhausen's results on the algebraic $K$–theory of free products with amalgamation and HNN extensions [150].

## 5.5 Compactification of open manifolds

Section 3 described the problem of compactifying an open manifold $V^n$ to a manifold with boundary. Because this is usually approached internally, by looking for submanifolds which may serve as cross–sections in a product structure for an end, the classic results on finding a boundary for an open manifold are closely related to splitting theorems.

Browder–Levine–Livesay [13] addressed this problem for the simply connected case in high dimensions. The pattern of argument here and in subsequent papers is heavily influenced by $h$–cobordism results: an end of a manifold is the union of a sequence of cobordisms, which one hopes to show may be taken to be $h$–cobordisms and eventually to be product $h$–cobordisms. Stallings' characterization of Euclidean space [145] is also of influence on later papers.

The major geometric issue in this problem is that each end of $V^n$ must be modelled on a product with a half–line, so one needs algebraic recognition criteria for this situation. Siebenmann [142] succeeds both in stating homological prerequisites to compactifiability (tameness conditions) and in the identification of an obstruction related to Wall's finiteness obstruction [151, 152]. See [86] for a detailed modern account of some of this material.

## 5.6 Fibering a manifold over a circle

Criteria for fibering a manifold over a circle in the surgical dimensions were provided by Farrell in his 1967 thesis [41], which exhibited complete obstructions in $K$-theory (see also [46, 47, 143]). These results have been extended to all dimensions, within the limitations of current four–dimensional methods [158]. The fibering obstruction lies in a Whitehead group and is a complete invariant for the problem in high dimensions, there exist four– and five–dimensional manifolds which satisfy the preliminary hypotheses for fibering and have vanishing Whitehead fibering obstruction, but do not fiber over a circle [90, 158].

See [131, Section 20] for an algebraic treatment of the obstruction to fibering a manifold over a circle.

## 5.7   Splitting theorems following Waldhausen

Waldhausen's $K$-theory splitting theorem [150] shows that the pattern of the Bass–Heller–Swan theorem persists for the basic constructions in a graph of groups: free products with amalgamations and HNN extensions both lead to Mayer–Vietoris sequences in $K$-theory involving secondary terms known as Nil groups. Roughly speaking, the algebraic version of a splitting problem over a group $\pi$ given as a free product with amalgamation $\pi = A *_C B$ amounts to a rewriting problem for a finitely generated free chain complex $C_*$ over $\mathbf{Z}\pi$. The challenge is to make a Mayer–Vietoris decomposition of $C_*$ over $\mathbf{Z}A$ and $\mathbf{Z}B$, with relations and trade-offs passing through a chain complex over $\mathbf{Z}C$; the trading process may be blocked at some stage, and the Nil groups are the repository of this obstruction. (The name "Nil" is a nod toward the filtered structure observed in the trading process, which arises from the fact that a handle in a split manifold may cross the splitting submanifold many times.)

Cappell's $L$-theory splitting theory [18, 17, 19] continues the stream of development: Mayer–Vietoris sequences are exhibited, with the modification that exotic terms known as UNil groups (for "unitary nilpotent" structures) appear as obstructions to splitting problems.

## 5.8   Nil groups

One would usually prefer to do without the Nil and UNil groups. They are prime candidates for the production or detection of counterexamples to the Borel Conjecture, but they tend to resist direct computation. The computational difficulties are in part due to the fact that these groups are attributes of a pushout diagram and are not determined by any single group in the diagram. Even if the edge and vertex groups in a graph of group are very well understood in $K$- or $L$-theory we might have almost no access to the Nil or UNil groups except through a vanishing theorem.

Sufficient conditions are known for the vanishing of Nil and UNil groups which are adequate for some striking applications of these theories to manifold classification problems and the Novikov Conjecture, but every worker in the subject would like to see more vanishing conditions and structural results.

We concentrate on the UNil groups here, with an occasional glance at the Nil groups. Cappell found that the UNil groups vanish for a free product with amalgamation $\pi = A *_C B$ or HNN extension $\pi = A*_C$, provided that the amalgamating subgroup $C$ is square–root closed in the sense that if $g \in \pi$ and $g^2 \in C$, then $g \in C$. The principal example of a graph of of groups structure which is not square root closed is the infinite dihedral group $\pi = \mathbf{Z}/2 * \mathbf{Z}/2$: the square of any element of either free summand is trivial, hence an element of the trivial amalgamating subgroup.

Cappell showed that the UNil group of $\mathbf{Z}/2 * \mathbf{Z}/2$ is nontrivial, and in fact is infinite. Most and probably all of the nonvanishing UNil groups known are closely related to this example; this observation has produced more guesswork than theorems to date, unfortunately.

The nonvanishing result is striking because the UNil groups for a splitting of $M^n$ act on the structure set of $M^n$ through the action of $L_{n+1}$, and Cappell shows that this action is faithful. The exhibition of an infinite family of nontrivial elements in the UNil group of $\mathbf{Z}/2 * \mathbf{Z}/2$ therefore shows that in appropriate dimensions, the connected sum of two real projective spaces has an infinite structure set: these spaces are spectacularly non–rigid in all three manifold categories.

Farrell showed that if a Nil group is nontrivial then it is infinitely generated [42]. He also refined Cappell's analysis of the exponents of UNil groups [44], so that we have some qualitative sense of the behavior of these infinitely generated Abelian 2–groups. Connolly has investigated the behavior of Nil–type groups as modules, in which sense they can be finitely generated [28].

Ranicki has treated the splitting problem within the algebraic theory of surgery in [129] and especially in [130, Section 23]. This view of the theory is valuable for applications because finiteness hypotheses are often unnecessary in the algebraic theory: for instance, the algebraic view of twisted Laurent polynomial extensions can be applied to the extension associated to an infinite cyclic covering space of manifold which might have no connection to a true circle fibering.

## 5.9  Further applications

Aspherical manifolds built using toral actions of low codimension can often be analyzed with a combination of splitting and Frobenius induction techniques [113].

Graphs of free Abelian groups have been analyzed using the algebraic splitting theory [146]. This is an example of a not uncommon problem with known vanishing results: a class of groups formed by taking repeated graphs of groups can be partially, but not completely described in $K$– or $L$–theory because the vanishing criterion weakens after repeated use. (Similar problems arise with Waldhausen's vanishing results: he is able to induce along with a hierarchy specifying the way a Haken 3–manifold is built up from 2–balls by a finite number of sums along boundary surfaces, which are similarly built up from 2–balls, but his vanishing criterion for Nil groups is not strong enough to carry on to the 4–manifolds which might be expected to come next.)

Right–angled Coxeter groups have been carefully analyzed from the splitting point of view [123], and we have good rigidity results in consequence.

# 6   Frobenius reciprocity for infinite groups

The solution by Wall and his co–authors of the spherical spaceform problem relies upon reduction steps using Frobenius reciprocity theorems for surgery groups. This technique is borrowed from representation theory and is described for finite groups in [148, 40, 103].

Hsiang's student Nicas [112] proved a version of such a reduction or induction theorem for infinite groups with finite quotients. If $\phi \colon \Gamma \to Q$ is an epimorphism to a finite group then we may decide nontriviality of elements in terms in the exact sequence of topological surgery for a Poincaré complex $X$ with fundamental group $\pi$ by considering the transfers of those elements to covering spaces $X(\phi^{-1}(E))$ as $E$ runs over the family of 2–hyperelementary subgroups of $Q$. (This is a brief summary of a technique which can often employ a smaller family of test covers; see [112] for details.)

The technique of hyper–elementary induction for infinite fundamental groups is employed by Farrell–Hsiang for Riemannian flat manifolds and related spaces [48, 49, 50]. It has been useful in other contexts, but never so spectacularly successful as in [48].

# 7   Groups with geometric origins
## and controlled methods

This section begins with brief descriptions of geometric sources of infinite groups and continues with results derived by controlled methods. Discrete subgroups of Lie groups and rigidity results concerning them lie at the root of the Borel Conjecture; curvature conditions on finitely generated groups are a later development largely arising with Gromov [73, 74], although curvature conditions on polyhedra had been heavily studied by Alexandrov and others.

## 7.1 Geometric sources: discrete subgroups of Lie groups

Geometry and arithmetic often construct manifolds with infinite fundamental groups via discrete subgroups of Lie groups. If $\Gamma$ is a discrete subgroup of a Lie group $G$ then the coset space $\Gamma\backslash G$ is a smooth manifold; if $H$ is a closed subgroup of $G$ then it may also happen that the double coset space $\Gamma\backslash G/H$ is a manifold (see [95] for properness criteria for the action $(\Gamma, G/H)$). Locally symmetric spaces are among the familiar manifolds which may be described as double coset spaces of Lie groups.

Analytic and dynamical methods for the study of discrete subgroups of Lie groups are most effective if we assume that the discrete subgroup $\Gamma$ is a lattice in the Lie group $G$, i.e. that for some left–invariant volume form on $G$ the volume of the quotient space $\Gamma\backslash G$ is finite. This hypothesis often leads to topological as well as volumetric finiteness results. The standard reference for basic properties is [128].

Certain discrete subgroups of linear algebraic Lie groups are defined by integrality conditions, and are known as arithmetic groups; a variation on this class leads to $S$-arithmetic groups [139]. The superrigidity theorem of Margulis [105, 106] asserts that under commonly satisfied hypotheses, if $\Gamma$ is a lattice in a linear algebraic group of $\mathbf{R}$–rank two or more then $\Gamma$ is arithmetic.

Lattices in semisimple linear groups are plentiful thanks to an existence result of Borel [7] based on arithmetic constructions. Arithmetic groups have most of the important geometrical or homological finiteness properties [138, 8, 9], thanks in part to careful compactification constructions involving Tits buildings. Similar results for more general lattices may be extracted from compactifications obtained by other methods such as Morse theory [128].

For more information see the surveys [138, 139] and books [128, 106, 162]. See also the volumes on Lie groups in the Encyclopædia of Mathematics [119, 120].

## 7.2 Geometric classes of groups: fundamental groups of nonpositively curved polyhedra

The Cartan–Hadamard theorem shows that any complete Riemannian manifold of nonpositive sectional curvature is diffeomorphic to Euclidean space via the exponential map at any basepoint. Because of this, the most familiar and geometric aspherical manifolds have long been the closed manifolds of constant, nonpositive sectional curvature: Riemannian flat and real hyperbolic manifolds.

Gromov [74] showed that many of the best features of complete Riemannian manifolds of negative sectional curvature are reproduced in the setting of geodesic length spaces which possess negative curvature as detected by methods of Alexandrov and Toponogov, based on geodesic triangle comparisons. If the Cayley graph of a finitely generated group $\pi$ is a length space of curvature bounded above by some $c < 0$ then we say that $\pi$ is a hyperbolic group (or word hyperbolic group, if we care to emphasize the use of the word length metric).

Hyperbolic groups and Gromov's account of their properties have been explicated in a number of publications, of which we cite only a few [29, 30, 71, 72]. The hyperbolization construction introduced by Gromov and placed on solid footing by M. Davis and his co–workers [36, 24] is both useful and delicate (mistakes have been made and the reader who is new to the topic should place more faith in [24] than in earlier references): in particular, thanks to hyperbolization we have a very large supply of word hyperbolic groups.

Olshanskiĭ [118] confirmed a claim made by Gromov in [74] that almost every finitely presented group is hyperbolic, in an asymptotic statistical sense. Despite this result, a class of groups which properly contains the hyperbolic groups is of great interest: fundamental groups of compact, polyhedral length spaces of nonpositive curvature are an environment in which the tools of controlled topology have fair chances of success.

Geometric constructions relevant to the applications of controlled topology in this setting include Gromov's construction of a boundary for a group which is a counterpart to the sphere at infinity in a nonpositively curved manifold and several features with origins in hyperbolic dynamics, such as a sort of geodesic flow.

The similarity between hyperbolic groups and hyperbolic manifolds can be pressed too far, and the reader should beware of this tendency. Benakli and Dranishnikov have produced examples of hyperbolic groups or fundamental groups of nonpositively curved polyhedra whose Gromov boundary is far from spherical [4, 37, 39, 38].

## 7.3    Geometric constructions:  Coxeter groups, Artin groups

Among the groups arising as fundamental groups of compact, polyhedral length spaces of nonpositive curvature we find some which are geometrically and combinatorially pleasant, if not as familiar as they first appear.

Coxeter groups are finitely presented groups generated by elements of order and satisfying relations modelled on reflections in a totally geodesic hyperplane within a symmetric space. These groups are known to admit faithful linear representations, and therefore possess finite–index subgroups which are torsion–free, by the Selberg lemma [137].

A Coxeter group admits several combinatorial descriptions, of which the clearest for topological purposes labels a space which serves as a fundamental domain with reflection walls and finite isotropy subgroup data. M. Davis used such a description of Coxeter groups in his beautiful construction of closed aspherical manifolds with non–Euclidean universal covers [34, 35, 33]. More recently, infinite Coxeter groups have been shown to be fundamental groups of nonpositively curved polyhedra [111], so they are amenable to metric as well as to combinatorial analysis.

The familiar braid groups generalize to a family of groups known as Artin groups, which may be attacked with some the descriptive tools developed for Coxeter groups. Working in this context, Bestvina and Brady [5] produce some remarkable counterexamples to some of the stronger conjectures concerning Poincaré duality groups.

## 7.4 Consequences of controlled methods

See the papers in this volume on controlled topology for details – we list a few consequences of controlled methods here.

Fundamental groups of closed manifolds of nonpositive curvature have as many rigidity properties as we are able to establish in any context, including positive solutions of the Novikov and Borel conjectures for these groups. Many hands have been involved in the Novikov conjecture for these groups, but the Borel conjecture results are largely due to [53] and other work of Farrell-Jones. The surveys [54, 55, 57, 45] may be particularly useful.

Jones' student Bizhong Hu is responsible for extending some of the $K$–theoretic arguments of Farrell and Jones to fundamental groups of polyhedra of probably nonpositive curvature [84, 85]. The Novikov conjecture has been well studied for hyperbolic groups and the nonpositively curved case has received attention there as well. See also [32] for more information.

## 7.5 Mapping class groups

Let $\Sigma_g^2$ be a closed, orientable Riemann surface of genus $g \geq 2$. The mapping class group $\mathrm{Out}(\pi_1(\Sigma_g^2))$ has been studied in algebraic geometry and geometric group theory. It can serve as a chastening example for over–ambitious topologists since it shares many of the properties of groups discussed above without quite falling into any of those classes.

$\mathrm{Out}(\pi_1(\Sigma_g^2))$ has many of the properties of arithmetic groups. In particular, this group has a torsion–free subgroup $\Gamma$ of finite index, and the quotient $\mathcal{T}_g/\Gamma$ of the Teichmüller space of $\Sigma_g^2$ by such a subgroup is an open aspherical manifold which is known to have good compactifications [79, 80, 81].

$\mathcal{T}_g/\Gamma$ is an interesting testbed for geometrical approaches to rigidity and $L$–theoretic problems since the most familiar geometric structures on Teichmüller space are geodesically incomplete, while compactifications defined in algebraic geometry are notably explicit in some ways.

# 8 Uncharted Territory

This section lists some questions which have been visited by exploring parties but for which we do not have good maps of the terrain. The most important of the grand challenge problems listed in Section 3 will not be repeated here.

## 8.1 Nielsen problems for aspherical manifolds

The most familiar settled cases are for surfaces [91, 109] or are due to Mostow rigidity [110].

Farrell and Jones have investigated outer automorphism groups where their methods apply, along with pseudo–isotopy.

The following Nielsen–type problem remains of interest, and should be extended to other contractible universal covering spaces.

**Problem 8.1 (Problem G1 of [157]).** Let $\Gamma'$ be a subgroup of finite index in $\Gamma$ and suppose given a free, proper action of $\Gamma'$ on $\mathbf{R}^n$ with compact quotient. Does the action necessarily extend to $\Gamma$?

## 8.2 Gromovian constructions

Surgical aspects of two of Gromov's geometric constructions have not been much studied.

Gromov and Thurston used branched covers to construct negatively curved manifolds which are close to hyperbolic manifolds in the Gromov–Hausdorff metric, but which admit no hyperbolic metric [76].

Gromov and Piatetski–Shapiro mixed arithmetic lattices by a splitting and regluing procedure to produce numerous examples of non–arithmetic hyperbolic manifolds [75].

A surgery in the style of Dehn surgery is part of the Farrell–Jones program for producing exotic smooth structures [52, 56, 58]. We hope that their methods may be useful in other aspherical manifolds.

## 8.3 Hyperplane arrangements

Proper, controlled, or stratified classification and rigidity theorems for hyperplane arrangements in $\mathbf{C}^n$ may be in reach of current methods. See [92, Problems 4.144-ff.] for some remarks and references.

## 8.4 Infinite groups without 2–torsion

The surgery sequence, especially away from 2, may be particularly tractable for infinite groups without 2–torsion. Conjectures have been offered in a quiet way on this for several years and most workers hope that this will eventually be seen to be a relatively pleasant class of fundamental groups.

## 8.5 Dimension three

The Poincaré–duality group counterpart to the Seifert fiber space conjecture in dimension three is unsettled as of [92, Problem 3.77(B)]. (I.e., if a PD(3) group has infinite cyclic center, must it be the fundamental group of a closed aspherical 3–manifold?)

## 8.6 Virtually standard smooth structures

If $M^n$ is a closed, aspherical, smooth manifold with residually finite fundamental group, is there a finite sheeted covering projection $N \to M$ such that all smooth or piecewise linear structures on $M$ become equivalent on pullback to $N$? (This is a counterpart of the property Kirby and Siebenmann needed for exotic tori.)

# 9 Surveys and problem lists

Kirby's revised problem list includes a bibliography of problem lists [92]. No one working in high–dimensional topology seems to have the strength to emulate Kirby's list, but the problem lists of [64, 65] and [157], among others, were consulted for this paper.

Textbooks and surveys of manifold topology and related material include surveys by Farrell [45], Farrell–Jones [54], and Ferry [61]. Ferry's forthcoming textbook [62], Rosenberg's text on algebraic $K$–theory [135], and the book of Freedman and Quinn on four–dimensional surgery [69] are more detailed.

The books of Wall [155], Browder [11] and Ranicki [129, 130, 131] are probably our most reliable book–length references on surgery. Discussions of specialized variations, such as equivariant surgery or surgery on stratified spaces are worth consulting in part for the light they throw on the garden–variety theory.

Reports on the Borel and Novikov conjectures include the Farrell–Jones CBMS lectures [54], the Novikov conjecture conference volumes [64, 65], and Roe's CBMS lectures [134].

Sources of mainly historical interest may also yield fruitful ideas; we list a few items in this category [77, 94, 132].

# References

[1] W. Ballmann, M. Brin, and P. Eberlein, *Structure of manifolds of nonpositive curvature. I*, Ann. of Math. **122** (1985), 171–203.

[2] W. Ballmann, M. Gromov, and V. Schroeder, *Manifolds of Nonpositive Curvature*, Birkhäuser, Boston, 1985.

[3] H. Bass, A. Heller, and R. Swan, *The Whitehead group of a polynomial extension*, Inst. Hautes Études Sci. Publ. Math. **22** (1964), 64–79.

[4] N. Benakli, *Groupes hyperboliques de bord la courbe de Menger ou la courbe de Sierpinski*, Ph.D. thesis, Université de Paris–Sud (Orsay), 1992.

[5] M. Bestvina and N. Brady, *Morse theory and finiteness properties of groups*, Invent. Math. **129** (1997), 445–470.

[6] R. Bieri, *Gruppen mit Poincaré Dualität*, Comment. Math. Helv. **47** (1972), 373–396.

[7] A. Borel, *Compact Clifford–Klein forms of symmetric spaces*, Topology **2** (1963), 111–122.

[8] A. Borel and J.-P. Serre, *Corners and arithmetic groups*, Comment. Math. Helv. **48** (1973), 436–491.

[9] ———, *Cohomologie d'immeubles et de groupes S–arithmétiques*, Topology **15** (1976), 211–232.

[10] W. Browder, *Structures on $M \times \mathbf{R}$*, Proc. Camb. Phil. Soc. **61** (1965), 337–345.

[11] ———, *Surgery on Simply Connected Manifolds*, Springer – Verlag, Berlin, 1972.

[12] W. Browder and J. Levine, *Fibering manifolds over a circle*, Comment. Math. Helv. **40** (1966), 153–160.

[13] W. Browder, J. Levine, and G. R. Livesay, *Finding a boundary for an open manifold*, Amer. J. Math. **87** (1965), 1017–1028.

[14] K. S. Brown, *High dimensional cohomology of discrete groups*, Proc. Nat. Acad. Sci. U.S.A. **73** (1976), 1795–1797.

[15] ———, *Groups of virtually finite dimension*, Homological Group Theory (Durham, 1977) (C. T. C. Wall, ed.), London Math. Soc. Lecture Notes, no. 36, Cambridge University Press, Cambridge, 1979, pp. 27–70.

[16] ———, *Cohomology of Groups*, Graduate Texts in Math, no. 87, Springer–Verlag, Berlin, 1982.

[17] S. E. Cappell, *Manifolds with fundamental group a generalized free product. I*, Bull. Amer. Math. Soc. **80** (1974), 1193–1198.

[18] ———, *Unitary Nilpotent groups and Hermitian K-theory. I*, Bull. Amer. Math. Soc. **80** (1974), 1117–1122.

[19] ———, *A splitting theorem for manifolds*, Invent. Math. **33** (1976), 69–170.

[20] A. J. Casson, *Fibrations over spheres*, Topology **6** (1967), 489–499.

[21] A. Casson and M. Freedman, *Atomic surgery problems*, Four-Manifold Topology (Durham, NH, 1982), Contemp. Math., no. 35, Amer. Math. Soc., Providence, 1984, pp. 181–199.

[22] A. Casson and D. Gottlieb, *Fibrations with compact fibres*, Amer. J. Math. **99** (1977), 159–189.

[23] T. A. Chapman and S. Ferry, *Approximating homotopy equivalences by homeomorphisms*, Amer. J. Math. **101** (1979), 567–582.

[24] R. M. Charney and M. W. Davis, *Strict hyperbolization*, Topology **34** (1995), 329–350.

[25] P. E. Conner and F. Raymond, *Actions of compact Lie groups on aspherical manifolds*, Topology of Manifolds (Athens, Ga., 1969) (J. C. Cantrell and Jr. C. H. Edwards, eds.), Markham, Chicago, 1970, pp. 227–264.

[26] ———, *Deforming homotopy equivalences to homeomorphisms in aspherical manifolds*, Bull. Amer. Math. Soc. **83** (1977), 36–85.

[27] F. Connolly and S. Prassidis, *Groups which act freely on $\mathbf{R}^m \times S^{n-1}$*, Topology **28** (1989), 133–148.

[28] F. X. Connolly and M. O. M. Da Silva, *The groups $N^r K_0(\mathbf{Z}\pi)$ are finitely generated $\mathbf{Z}[N^r]$ modules if $\pi$ is a finite group*, K-Theory **9** (1995), 1–11.

[29] M. Coornaert, T. Delzant, and A. Papadopoulos, *Géométrie et Théorie des Groupes: les Groupes Hyperboliques de Gromov*, Lecture Notes in Math., no. 1441, Springer–Verlag, Berlin, 1990.

[30] M. Coornaert and A. Papadopoulos, *Symbolic Dynamics and Hyperbolic Groups*, Lecture Notes in Mathematics, no. 1539, Springer–Verlag, Berlin, 1993.

[31] K. Corlette, *Archimedean superrigidity and hyperbolic geometry*, Ann. of Math. **135** (1992), 165–182.

[32] J. Davis, *Manifold aspects of the Novikov Conjecture*, this volume.

[33] M. Davis and J.-C. Hausmann, *Aspherical manifolds without smooth or PL structures*, Algebraic Topology (Arcata, 1986), Lecture Notes in Math., no. 1370, Springer–Verlag, New York, 1989, pp. 135–142.

[34] M. W. Davis, *Groups generated by reflections and aspherical manifolds not covered by Euclidean space*, Annals of Math. **117** (1983), 293–324.

[35] ———, *Coxeter groups and aspherical manifolds*, Algebraic Topology, Aarhus 1982 (I. Madsen and B. Oliver, eds.), Lecture Notes in Math., no. 1051, Springer–Verlag, Berlin, 1984, pp. 197–221.

[36] M. W. Davis and T. Januszkiewicz, *Hyperbolization of polyhedra*, J. Diff. Geom. **34** (1991), 347–388.

[37] A. N. Dranishnikov, *On boundaries of Coxeter groups*, University of Florida preprint, 1994.

[38] ———, *Boundaries of Coxeter groups and simplicial complexes with given links*, J. Pure Appl. Algebra **137** (1999), 139–151.

[39] ———, *On the virtual cohomological dimensions of Coxeter group*, Proc. Amer. Math. Soc. **125** (1997), 1885-1891.

[40] A. W. M. Dress, *Induction and structure theorems for orthogonal representations of finite groups*, Annals of Math. **102** (1975), 291–325.

[41] F. T. Farrell, *The obstruction to fibering a manifold over a circle*, Indiana Univ. Math. J. **21** (1971/1972), 315–346.

[42] ———, *The nonfiniteness of Nil*, Proc. Amer. Math. Soc. **65** (1977), 215–216.

[43] ———, *An extension of Tate cohomology to a class of infinite groups*, J. Pure Appl. Algebra **10** (1977/78), 153–161.

[44] ———, *The exponent of UNil*, Topology **18** (1979), 305–312.

[45] ———, *Surgical Methods in Rigidity*, Published for Tata Institute, Bombay by Springer (1996).

[46] F. T. Farrell and W. C. Hsiang, *A formula for $K_1 R_\alpha[T]$*, Categorical Algebra, Proc. Symp. Pure Math., no. 17, Amer. Math. Soc., Providence, 1970, pp. 192–218.

[47] _____, *Manifolds with $\pi_1 = G \times_\alpha T$*, Amer. J. Math. **95** (1973), 813–848.

[48] _____, *The topological-Euclidean space form problem*, Invent. Math. **45** (1978), 181–192.

[49] _____, *The Whitehead group of poly-(finite or cyclic) groups*, J. London Math. Soc. (2) **24** (1981), 308–324.

[50] _____, *Topological characterization of flat and almost flat Riemannian manifolds $M^n$ ($n \neq 3, 4$)*, Amer. J. Math. **105** (1983), 641–672.

[51] F. T. Farrell and L. E. Jones, *Examples of expanding endomorphisms on exotic tori*, Invent. Math. **45** (1978), 175–179.

[52] _____, *Negatively curved manifolds with exotic smooth structures*, J. American Math. Soc. **2** (1989), 899–908.

[53] _____, *A topological analogue of Mostow's rigidity theorem*, J. Amer. Math. Soc. **2** (1989), 257–370.

[54] _____, *Classical Aspherical Manifolds*, CBMS Regional Conference Series in Mathematics, no. 75, American Mathematical Society, Providence, 1990.

[55] _____, *Rigidity in geometry and topology*, Proceedings of the International Congress of Mathematicians (Kyoto, 1990), Math. Soc. Japan, Tokyo, 1991, pp. 653–663.

[56] _____, *Nonuniform hyperbolic lattices and exotic smooth structures*, J. Differential Geom. **38** (1993), 235–261.

[57] _____, *Topological rigidity for compact nonpositively curved manifolds*, Differential Geometry, Proc. Symp. Pure. Math., no. 54, part 3, Amer. Math. Soc., Providence, 1993, pp. 229–274.

[58] _____, *Complex hyperbolic manifolds and exotic smooth structures*, Invent. Math. **117** (1994), 57–74.

[59] _____, *Some non-homeomorphic harmonic homotopy equivalences*, Bull. London Math. Soc. **28** (1996), 177–182.

[60] F. T. Farrell and C. W. Stark, *Cocompact spherical-euclidean space-form groups of infinite VCD*, Bull. London Math. Soc. **25** (1993), 189–192.

[61] S. Ferry, *Controlled topology and the characterization of topological manifolds*, Notes from an NSF–CBMS Regional Conference, Knoxville, TN, 1994.

[62] _____, *Topological Manifolds and Polyhedra*, Oxford University Press, to appear.

[63] S. Ferry, A. Ranicki, and J. Rosenberg, *A history and survey of the Novikov conjecture*, Novikov Conjectures, Index Theorems and Rigidity, Volume 1 (S. C. Ferry, A. Ranicki, and J. Rosenberg, eds.), London Math. Soc. Lecture Notes, no. 226, Cambridge University Press, Cambridge, 1995, pp. 7–66.

[64] S. Ferry, A. Ranicki, and J. Rosenberg (eds.), *Novikov Conjectures, Index Theorems and Rigidity, Volume 1*, London Math. Soc. Lecture Notes, no. 226, Cambridge University Press, Cambridge, 1995.

[65] _____, *Novikov conjectures, index theorems and rigidity, volume 2*, London Math. Soc. Lecture Notes, no. 227, Cambridge University Press, Cambridge, 1995.

[66] M. H. Freedman, *A surgery sequence in dimension four: the relations with knot concordance*, Invent. Math. **68** (1982), 195–226.

[67] _____, *The disk theorem for four dimensional manifolds*, Proc. ICM (Warsaw, 1983), vol. 1,2, PWN, Warsaw, 1984, pp. 647–663.

[68] _____, *Poincaré transversality and four–dimensional surgery*, Topology **27** (1988), 171–175.

[69] M. H. Freedman and F. Quinn, *Topology of 4–Manifolds*, Princeton Mathematical Series, no. 39, Princeton University Press, Princeton, 1990.

[70] S. Gersten, *A product formula for Wall's obstruction*, Amer. J. Math. **88** (1966), 337 – 346.

[71] E. Ghys and P. de la Harpe (eds.), *Sur les groupes hyperboliques d'après Mikhael Gromov*, Progress in Mathematics, no. 83, Birkhäuser, Boston, 1990.

[72] E. Ghys, A. Haefliger, and A. Verjovsky (eds.), *Group Theory from a Geometrical Viewpoint (ICTP, Trieste, 1990)*, World Scientific, Singapore, 1991.

[73] M. Gromov, *Hyperbolic manifolds, groups and actions*, Riemann Surfaces and Related Topics: Proceedings of the 1978 Stony Brook Conference (I. Kra and B. Maskit, eds.), Annals of Math. Studies, no. 97, Princeton University Press, Princeton, 1981, pp. 183–213.

[74] _____ , *Hyperbolic groups*, Essays in Group Theory (S. M. Gersten, ed.), Math. Sci. Res. Inst. Publ., no. 8, Springer–Verlag, New York, 1987, pp. 75–263.

[75] M. Gromov and I. Piatetski-Shapiro, *Non–arithmetic groups in Lobachevsky spaces*, Inst. Hautes Études Sci. Publ. Math. **66** (1988), 93–103.

[76] M. Gromov and W. Thurston, *Pinching constants for hyperbolic manifolds*, Invent. Math. **89** (1987), 1–12.

[77] L. Guillou and A. Marin (eds.), *À la Recherche de la Topologie Perdue*, no. 62, Birkhäuser, Boston, 1986.

[78] I. Hambleton and E. K. Pedersen, *Bounded surgery and dihedral group actions on spheres*, J. Amer. Math. Soc. **4** (1991), 105–126.

[79] J. Harer, *The virtual cohomological dimension of the mapping class group of an orientable surface*, Invent. Math. **84** (1986), 157–176.

[80] W. J. Harvey, *Boundary structure for the modular group*, Riemann Surfaces and Related Topics (Stony Brook, 1978), Ann. of Math. Studies, no. 97, Princeton University Press, Princeton, 1978, pp. 245–251.

[81] _____ , *Geometric structure of surface mapping class groups*, Homological Group Theory (C. T. C. Wall, ed.), London Math. Soc. Lecture Note Series, no. 36, Cambridge University Press, Cambridge, 1979, pp. 255–269.

[82] W. C. Hsiang and J. Shaneson, *Fake tori*, Topology of Manifolds (Athens, Ga., 1969) (J. C. Cantrell and Jr. C. H. Edwards, eds.), Markham, Chicago, 1970, pp. 18–51.

[83] W.-C. Hsiang and C. T. C. Wall, *On homotopy tori II*, Bull. London Math. Soc. **1** (1969), 341–342.

[84] B. Hu, *Whitehead groups of finite polyhedra with nonpositive curvature*, J. Diff. Geom. **38** (1993), 501–517.

[85] _____ , *Relations between simplicial and manifold nonpositive curvature*, Geometric Group Theory (Proceedings of a Special Research Quarter at The Ohio State University, Spring 1992) (R. Charney, M. Davis, , and M. Shapiro, eds.), Ohio State University Math. Research Institute Publications, no. 3, de Gruyter, Berlin, 1995.

[86] B. Hughes and A. Ranicki, *Ends of complexes*, Cambridge Tracts in Mathematics, no. 123, Cambridge University Press, 1996.

[87] F. E. A. Johnson, *On groups which act freely on* $\mathbf{R}^m \times S^n$, J. London Math. Soc. (2) **32** (1985), 370–376.

[88] F. E. A. Johnson and C. T. C. Wall, *On groups satisfying Poincaré duality*, Ann. of Math. **96** (1972), 592–598.

[89] Y. Kamishima, K. B. Lee, and F. Raymond, *The Seifert construction and its applications to infranilmanifolds*, Quart. J. Math. Oxford (2) **34** (1983), 433–452.

[90] C. Kearton, *Some non–fibred 3–knots*, Bull. London Math. Soc. **15** (1983), 365–367.

[91] S. P. Kerckhoff, *The Nielsen realization problem*, Ann. of Math. **117** (1983), 235–265.

[92] R. Kirby, *Problems in low–dimensional topology*, Geometric Topology (Athens, Georgia, 1993), AMS/IP Studies in Advances Mathematics, no. 2, Part 2, Amer. Math. Soc. – International Press, Providence, 1997, pp. 35–473.

[93] R. C. Kirby and L. C. Siebenmann, *On the triangulation of manifolds and the Hauptvermutung*, Bull. Amer. Math. Soc. **75** (1969), 742–749.

[94] ———, *Foundational Essays on Topological Manifolds, Smoothings, and Triangulations*, Annals of Mathematics Studies, no. 88, Princeton University Press, Princeton, 1977.

[95] R. S. Kulkarni, *Proper actions and pseudo–Riemannian space forms*, Adv. in Math. **40** (1981), 10–51.

[96] K. W. Kwun and R. H. Szczarba, *Sum and product theorems for Whitehead torsion*, Ann. of Math. **82** (1965), 183–190.

[97] K. B. Lee and F. Raymond, *Geometric realization of group extensions by the Seifert construction*, Combinatorics and Algebra (C. Greene, ed.), Contemporary Math., no. 34, Amer. Math. Soc., Providence, 1984, pp. 353–411.

[98] ———, *The role of Seifert fiber spaces in transformation groups*, Group Actions on Manifolds (R. Schultz, ed.), Contemporary Math., no. 36, Amer. Math. Soc., Providence, 1985, pp. 367–425.

[99] J.-L. Loday, *K–théorie algébrique et représentations de groupes*, Ann. Sci. École Norm. Sup. (4) **9** (1976), 309–377.

[100] W. Lück and A. Ranicki, *The surgery transfer*, Konferenzbericht der Göttinger Topologie Tagung 1987 (T. tom Dieck, ed.), Lecture Notes in Math., no. 1361, Springer – Verlag, Berlin, 1988, pp. 167–246.

[101] W. Lück and A. Ranicki, *Surgery obstructions of fibre bundles*, J. Pure and Applied Algebra **81** (1992), 139–189.

[102] R. C. Lyndon and P. E. Schupp, *Combinatorial Group Theory*, Springer – Verlag, Berlin, 1977.

[103] I. Madsen, C. B. Thomas, and C. T. C. Wall, *The topological spherical form problem - II. Existence of free actions*, Topology **15** (1976), 375–382.

[104] W. Magnus, A. Karass, and D. Solitar, *Combinatorial Group Theory (Second Revised Edition)*, Dover, New York, 1976.

[105] G. A. Margulis, *Discrete groups of motions of manifolds of non-positive curvature*, Proc. Int. Congress. Math. (Vancouver, 1974), Amer. Math. Soc. Transl., vol. 109, Amer. Math. Soc., Providence, 1977, pp. 33–45.

[106] ――――, *Discrete Subgroups of Semisimple Lie Groups*, Springer-Verlag, Berlin, 1991.

[107] R. J. Milgram and A. A. Ranicki, *The L-theory of Laurent polynomial extensions and genus 0 function fields*, J. reine angew. Math. **406** (1990), 121–166.

[108] J. W. Milnor, *Hilbert's Problem 18: On crystallographic groups, fundamental domains, and on sphere packing*, Mathematical Developments Arising From Hilbert Problems, Proc. Symp. Pure Math., no. 28, Amer. Math. Soc., Providence, 1976, pp. 491–506.

[109] S. Morita, *Characteristic classes of surface bundles*, Invent. Math. **90** (1987), 551–577.

[110] G. D. Mostow, *Strong Rigidity of Locally Symmetric Spaces*, Annals of Math. Studies, no. 78, Princeton University Press, Princeton, 1973.

[111] G. Moussong, *Hyperbolic Coxeter groups*, Ph.D. thesis, Ohio State University, 1988.

[112] A. J. Nicas, *Induction theorems for groups of homotopy manifold structures*, Memoirs Amer. Math. Soc., no. 267, Amer. Math. Soc., Providence, 1982.

[113] A. J. Nicas and C. W. Stark, *K-theory and surgery of codimension-two torus actions on aspherical manifolds*, J. London Math. Soc. (2) **31** (1985), 173–183.

[114] S. P. Novikov, *The homotopy and topological invariance of certain rational Pontrjagin classes*, Dokl. Akad. Nauk SSSR **162** (1965), 1248–1251, English translation: Soviet Math. Dokl. **6** (1965), 854–857.

[115] _____, *Topological invariance of rational classes of Pontrjagin*, Dokl. Akad. Nauk SSSR **163** (1965), 298–300, English translation: Soviet Math. Dokl. **6** (1965), 921–923.

[116] _____, *On manifolds with free abelian fundamental group and their application*, Izv. Akad. Nauk SSSR, Ser. Mat. **30** (1966), 207–246, English translation: Amer. Math. Soc. Transl. (2), Vol. 71, 1968, pp. 1–42.

[117] _____, *Algebraic construction and properties of hermitian analogues of K-theory over rings with involution from the viewpoint of the hamiltonian formalism. applications to differential topology and the theory of characteristic classes*, Izv. Akad. Nauk SSSR, Ser. Mat. **34** (1970), 253–288, 475–500.

[118] A. Yu. Ol'shanskiĭ, *Almost every group is hyperbolic*, Internat. J. Algebra Comput. **2** (1992), 1–17.

[119] A. L. Onishchik (ed.), *Lie Groups and Lie Algebras I*, Encyclopædia of Math. Sciences, no. 20, Springer–Verlag, Berlin, 1993.

[120] A. L. Onishchik and E. B. Vinberg (eds.), *Lie Groups and Lie Algebras II*, Encyclopædia of Math. Sciences, no. 21, Springer–Verlag, Berlin, 1993.

[121] E. Pedersen, *Continuously controlled surgery theory*, this volume.

[122] S. Prassidis, *Groups with infinite virtual cohomological dimension which act freely on* $\mathbf{R}^m \times S^{n-1}$, J. Pure Appl. Algebra **78** (1992), 85–100.

[123] S. Prassidis and B. Spieler, *Rigidity of Coxeter groups*, Vanderbilt University preprint, 1996.

[124] F. Quinn, *A geometric formulation of surgery*, Topology of Manifolds (Athens, Ga., 1969) (J. C. Cantrell and Jr. C. H. Edwards, eds.), Markham, Chicago, 1970, pp. 500–511.

[125] _____, *Ends of maps I*, Ann. of Math. **110** (1979), 275–331.

[126] _____, *Ends of maps, II*, Invent. Math. **68** (1982), 353–424.

[127] _____, *Ends of maps. III: Dimensions 4 and 5*, J. Diff. Geom. **17** (1982), 503–521.

[128] M. S. Raghunathan, *Discrete Subgroups of Lie Groups*, Ergebnisse der Math., no. 68, Springer–Verlag, Berlin, 1972.

[129] A. A. Ranicki, *Exact Sequences in the Algebraic Theory of Surgery*, Princeton University Press, Princeton, 1981.

[130] ———, *Algebraic L–Theory and Topological Manifolds*, Cambridge Tracts in Mathematics, no. 102, Cambridge University Press, Cambridge, 1992.

[131] ———, *Lower K– and L–theory*, London Math. Soc. Lecture Notes, no. 178, Cambridge University Press, Cambridge, 1992.

[132] A. A. Ranicki (ed.), *The Hauptvermutung Book*, K–monographs in Mathematics, no. 1, Kluwer, 1996.

[133] F. Raymond and D. Wigner, *Construction of aspherical manifolds*, Geometric Applications of Homotopy Theory, Lecture Notes in Math., no. 657, Springer–Verlag, Berlin, 1978, pp. 408–422.

[134] J. Roe, *Index Theory, Coarse Geometry, and Topology of Manifolds*, CBMS Regional Conference Series in Math., no. 90, Amer. Math. Soc., Providence, 1996.

[135] J. Rosenberg, *Algebraic K–Theory and its Applications*, Graduate Texts in Math., no. 147, Springer–Verlag, New York, 1994.

[136] P. Scott and C. T. C. Wall, *Topological methods in group theory*, Homological Group Theory (C. T. C. Wall, ed.), London Math. Soc. Lecture Notes, no. 36, Cambridge University Press, Cambridge, 1979, pp. 137–203.

[137] A. Selberg, *On discontinuous groups in higher–dimensional symmetric spaces*, Contributions to Function Theory (Int. Colloq. on Function Theory, Bombay, 1960), Tata Institute, Bombay, 1960.

[138] J.-P. Serre, *Cohomologie des groupes discrets*, Prospects in Mathematics, Ann. of Math. Studies, no. 70, Princeton University Press, Princeton, 1971, pp. 77–169.

[139] ———, *Arithmetic groups*, Homological Group Theory (C. T. C. Wall, ed.), London Math. Soc. Lecture Note Series, no. 36, Cambridge University Press, Cambridge, 1979, pp. 105–136.

[140] ———, *Trees*, Springer–Verlag, Berlin, 1980.

[141] J. L. Shaneson, *Wall's surgery obstruction groups for* $\mathbf{Z} \times G$, Ann. of Math. **90** (1969), 296–334.

[142] L. Siebenmann, *The obstruction to finding a boundary for an open manifold of dimension greater than five*, Ph.D. thesis, Princeton University, 1965.

[143] _____, *A total Whitehead obstruction to fibering over the circle*, Comm. Math. Helv. **45** (1970), 1–48.

[144] J. Stallings, *On fibering certain 3–manifolds*, Topology of 3–manifolds and Related Topics (Jr. M. K. Fort, ed.), Prentice–Hall, Englewood Cliffs, N.J., 1962, pp. 95–100.

[145] _____, *The piecewise linear structure of Euclidean space*, Proc. Cambridge Philos. Soc. **58** (1962), 481–488.

[146] C. W. Stark, *L–theory and graphs of free Abelian groups*, J. Pure Appl. Algebra **47** (1987), 299–309.

[147] _____, *Groups acting on $S^n \times \mathbf{R}^k$: generalizations of a construction of Hambleton and Pedersen*, K–Theory **5** (1992), 333–354.

[148] C. B. Thomas and C. T. C. Wall, *The topological spherical space form problem I*, Compositio Math. **23** (1971), 101–114.

[149] F. Waldhausen, *On irreducible 3–manifolds which are sufficiently large*, Ann. of Math. **87** (1968), 56–88.

[150] _____, *Algebraic K–theory of generalized free products*, Ann. of Math. **108** (1978), 135–256.

[151] C. T. C. Wall, *Finiteness conditions for CW complexes*, Ann. Math. **81** (1965), 56–69.

[152] _____, *Finiteness conditions for CW complexes. II*, Proc. Roy. Soc. A **295** (1966), 129–139.

[153] _____, *Surgery of non–simply–connected manifolds*, Annals of Math. **84** (1966), 217–276.

[154] _____, *On homotopy tori and the annulus theorem*, Bull. London Math. Soc. **1** (1969), 95–97.

[155] _____, *Surgery on Compact Manifolds*, Academic Press, London, 1970.

[156] _____, *The topological space–form problems*, Topology of Manifolds (Proc. of The University of Georgia Topology of Manifolds Institute, 1969) (J. C. Cantrell and Jr. C. H. Edwards, eds.), Markham, Chicago, 1970, pp. 319–331.

[157] ———, *List of problems*, Homological Group Theory (Durham, 1977) (C. T. C. Wall, ed.), London Math. Soc. Lecture Notes, no. 36, Cambridge University Press, Cambridge, 1979, pp. 369–394.

[158] S. Weinberger, *On fibering four- and five-manifolds*, Israel J. Math. **59** (1987), 1–7.

[159] J. H. C. Whitehead, *A certain open manifold whose group is unity*, Quart. J. Math. **6** (1935), 268–279.

[160] J. A. Wolf, *Spaces of Constant Curvature*, fourth ed., Publish or Perish, Berkeley, 1977.

[161] S.-T. Yau, *Problem section*, Seminar on Differential Geometry (S.-T. Yau, ed.), Annals of Math. Studies, no. 102, Princeton University Press, Princeton, 1982, pp. 669–706.

[162] R. J. Zimmer, *Ergodic Theory and Semisimple Groups*, Monographs in Math., no. 81, Birkhäuser, Boston, 1984.

University of Florida
Department of Mathematics
Gainesville, FL 32611–8105
email: cws@math.ufl.edu

# Continuously controlled surgery theory

Erik Kjær Pedersen

## 0. Introduction

One of the basic questions in surgery theory is to determine whether a given homotopy equivalence of manifolds is homotopic to a homeomorphism. This can be determined by global algebraic topological invariants such as the normal invariant and the surgery obstruction (the Browder–Novikov–Sullivan–Wall theory). Another possibility is to impose extra geometric hypothesis on the homotopy equivalence. Such conditions are particularly useful when working in the topological category. Novikov's proof of the topological invariance of the rational Pontrjagin classes only used that a homeomorphism is a homotopy equivalence with contractible point inverses, as was first observed by Sullivan. Siebenmann [30] proved that every homotopy equivalence of manifolds in dimension bigger than five, with contractible point inverses is in fact homotopic to a homeomorphism by a small homotopy. Chapman and Ferry [10] generalized this to showing that $\varepsilon$-controlled homotopy equivalences are homotopic to homeomorphisms.

Controlled algebra was developed in order to guide geometric constructions maintaining control conditions, where the smallness is measured in some metric space. Such a theory was first proposed by Connell and Hollingsworth [11]. One of the aims was to prove the topological invariance of Whitehead torsion for homeomorphisms of polyhedra. In fact the first proof of topological invariance of Whitehead torsion [8] was developed without the use of controlled algebra. Such proofs have been developed later [14, 29]. Chapman developed a controlled Whitehead torsion theory using geometric methods [9]. Quinn [25, 26] developed the theory of Connell and Hollingsworth into a usable, computable tool. However there are technical difficulties in any kind of $\varepsilon$ control because of the lack of functoriality. The composite of two $\varepsilon$ maps is a $2\varepsilon$ map, so one needs to apply squeezing to regain control.

The basic idea in bounded topology and algebra is to keep control bounded, but let the metric space "go to infinity". This is obtained as follows. Assume $K \subset S^{n-1}$. We then define *the open cone*

$$O(K) = \{t \cdot x \mid t \in [0, \infty) \subset \mathbb{R}^n, \, x \in K\}.$$

---

The author was partially supported by NSF grant DMS 9104026.

The subset $t \cdot K \subset O(K)$ is a copy of $K$, but the metric has been enlarged by the factor $t$. This approach was developed in [21, 24, 15] and the controlled torsion and surgery obstructions live in the $K$ and $L$-theory of additive categories (with involution). The basic fact being used is that bounded + bounded is bounded. Similarly as $\varepsilon$ goes to 0 we may use that $0 + 0 = 0$, and that is the basis of continuously controlled algebra and topology. Again the obstruction groups live in the $K$ and $L$-theory of additive categories (with involution).

The object of this paper is to study the obstructions to deforming a homotopy equivalence of manifolds to a homeomorphism using continuously controlled algebra [1]. Suppose given a homotopy equivalence of manifolds $f : M \to N$ and a common closed subspace $K \subset M$, $K \subset N$ such that $f$ restricted to $K$ is the identity. We also assume $f : (M, K) \to (N, K)$ is a *strict map* i. e. that $f$ sends $M - K$ to $N - K$. The basic question of continuously controlled surgery is whether it is possible to find a strict homotopy of $f$ relative to $K$ to a homeomorphism. A *strict homotopy* is a homotopy through strict maps. We shall also study an existence question corresponding to the uniqueness question above.

This type of question has typically been studied using bounded surgery [15], at least when $K$ is compact, by methods as follows: Choose an embedding of $K$ in a large dimensional sphere, and define $O(K) = \{t \cdot x \in \mathbb{R}^{N+1} | t \in [0, \infty), x \in K\}$. If $K$ is a neighborhood retract, it is easy to produce a proper map from $N - K$ to $O(K)$ with the property that if we compactify $O(K)$ radially by adding a copy of $K$, the map extends continuously to $N$, using the identity on $K$. Thus when we approach $K$ in $N$, the image goes to infinity in $O(K)$. If the map $M - K \to N - K$ can be homotoped to a homeomorphism which does not move any point more than a bounded amount when measured in $O(K)$, we may obviously complete the homotopy by the identity on $K$ to obtain a strict homotopy of $f : M \to N$ relative to $K$, to a homeomorphism. This follows because the open cone construction blows up the metric near $K$, so bounded moves measured in $O(K)$ become arbitrarily small as we approach $K$ in the original manifold. The method works well in the case when $M$ and $K$ are compact, and can be related to compact surgery theory via the torus. This is because homotopy equivalences parameterized by a torus become bounded homotopy equivalences parameterized by Euclidean space when passing to the universal cover of the torus, thus giving a way to use compact surgery theory for this kind of problem. The choice of the reference map to $O(K)$ however, is a bit unnatural, even though it can be shown not to matter, and in case $M$ and $K$ are not compact, this method does not work so well.

We shall indicate how to develop a continuously controlled surgery theory, in the locally simply connected case, and in the non-simply connected case, in the special case of a group action. The algebra described here was developed in [18] in the case of finite isotropy groups at the singular set. The case developed here allowing infinite isotropy groups is new. This algebra is relevant to generalized assembly maps of the type considered in the Baum–Connes and Farrell–Jones conjectures, [2, 13] see [12]. This is discussed in §4 .

A lot of the arguments are very similar to the bounded surgery theory [15], and will not be repeated here. We shall choose to emphasize the points where there are essential differences, and try to state precise definitions.

Following the path initialized by Wall in [32] we need to develop algebra that determines when a strict map

$$(f, 1_K) : (M, K) \to (N, K)$$

is a strict homotopy equivalence relative to $K$. The problem will then be solved by establishing a continuously controlled surgery exact sequence.

## 1. THE SIMPLY CONNECTED CASE

As a warmup let us consider the simply connected case. We assume $M$ and $N$ are manifolds, and by simply connected we shall mean $N - K$ is simply connected and $N$ is locally connected and locally simply connected at $K$. Specifically for each point $x \in K$ and each neighborhood $U$ in $N$, there must exist a neighborhood $V$ so that every two points in $V - K$ can be connected in $V - K$, and every loop in $V - K$ bounds a disk in $U - K$. These conditions are satisfied if $N$ is simply connected and $K$ is of codimension at least three. Recall the definition of the continuously controlled category from [1] and [5].

**1.1. Definition.** Let $R$ be a ring and $(\overline{X}, \partial X)$ a pair of topological spaces, $X = \overline{X} - \partial X$ with $\partial X$ closed in $\overline{X}$ and $X$ dense in $\overline{X}$. We define the category $\mathcal{B}(\overline{X}, \partial X; R)$ as follows: An object $A$ is a collection $\{A_x\}_{x \in X}$ of finitely generated free $R$-modules so that $\{x | A_x \neq 0\}$ is locally finite in $X$. A morphism $\phi : A \to B$ is an $R$-module morphism $\oplus A_x \to \oplus B_y$, satisfying a continuously controlled condition: For every $z \in \partial X$ and for every neighborhood $U$ of $z$ in $\overline{X}$, there exists a neighborhood $V$ of $z$ in $\overline{X}$ such that $\phi_x^y = 0$ and $\phi_y^x = 0$ if $x \in V \cap X$ and $y \in X - U$.

The continuous control condition thus requires that non trivial components of a morphism must be "small" near $\partial X$.

In the case discussed above, we could put $(\overline{X}, \partial X) = (N, K)$. We may then triangulate $N - K$ in such a fashion that simplices become small near $K$. Specifically this means that for every $z \in K$ and every neighborhood $U$ there exists a neighborhood $V$ such that if a simplex $\sigma$ intersects $V$, it must be contained in $U$. The cellular chain complex of $N - K$ can thus be thought of as a chain complex in $\mathcal{B}(N, K; \mathbb{Z})$ by associating each $\mathbb{Z}$-module generated by a simplex to the barycenter of that simplex. Since the boundary maps in a cellular chain complex are given by geometric intersection, boundary maps will indeed be continuously controlled in the sense of definition 1.1. The following condition gives an algebraic condition for strict homotopy equivalence and thus provides the key to a continuously controlled surgery theory.

**1.2. Proposition.** *Suppose $M$ and $N$ are manifolds, and that $(M, K)$ and $(N, K)$ are simply connected and locally simply connected at $K$. Given a strict map $f : (M, K) \to (N, K)$, which is the identity on $K$. Then $f$ is a strict homotopy equivalence relative to $K$, if and only if $f_\sharp : C_\sharp(M - K) \to C_\sharp(N - K)$ is a chain homotopy equivalence in $\mathcal{B}(N, K; \mathbb{Z})$*

**Remark.** In the proposition above, we do not actually need $M$ and $N$ to be manifolds, it would suffice to have locally compact Hausdorff pairs $(M, K)$ and $(N, K)$ with a CW structure on $M - K$ and $N - K$ satisfying some extra conditions. We shall return to this.

The proof of this proposition is a straightforward handle argument. However translation to algebra depends very strongly on the pair $(N, K)$. To remedy this situation and show that it really only depends on $K$ we present the following lemma from [5]:

**1.3. Lemma.** *If $(\overline{X}, \partial X)$ is a compact metrizable pair then, denoting the cone of $\partial X$ by $C\partial X$ we have an equivalence of categories*

$$\mathcal{B}(\overline{X}, \partial X; R) \cong \mathcal{B}(C\partial X, \partial X; R)$$

*Proof.* The isomorphism is given by moving the modules $A_x$, $x \in X$ to points in $C\partial X$, the same module. If two are put at the same place we take the direct sum. On morphisms the isomorphism is induced by the identity, so we have to ensure the continuously controlled condition is not violated. We proceed as follows: Choose a metric on $\overline{X}$ so that all distances are $\leq 1$. Given $z \in X$, let $y$ be a point in $\partial X$ closest to $z$, and send $z$ to $(1 - d(z, y))y$. Clearly, as $z$ approaches the boundary it is moved very little. In the other direction send $t \cdot y$ to a point in $B(y; 1 - t)$, the closed ball with center $y$ and radius $1 - t$, which is furthest away from $\partial X$. Again moves become small as $t$ approaches 1 or equivalently as the point approaches $\partial X$. It is easy to see that we never take more than a finite direct sum, and that the local finiteness condition is preserved. $\qquad \square$

This lemma shows that in the metrizable case the algebra only depends on $K$.

We need duality in the category $\mathcal{B}(\overline{X}, \partial X; R)$ . The duality we need is that $\mathcal{B}(\overline{X}, \partial X; R)$ is an additive category with involution in the sense of Ranicki [27]. The duality is given by $(A^*)_x = (A_x)^*$. This codifies the local nature of Poincaré duality. The dual cell of a cell sitting near $x$ will also be near $x$. Given this duality the algebraic $L$-groups are defined in [27] using forms and formations. These will be the appropriate obstruction groups. Using this duality, a chain complex $C_\sharp$ in $\mathcal{B}(\overline{X}, \partial X; R)$ has a "dual" chain complex $C^\sharp$ in $\mathcal{B}(\overline{X}, \partial X; R)$.

Before continuing to develop the continuously controlled surgery theory, of which we have only touched upon the uniqueness aspects, notice that for the preceding discussion we do not really need that $M$ and $N$ are manifolds. All we needed was that $M - K$ and $N - K$ are manifolds, that allow a triangulation (or handle body decomposition) with small simplices near $K$. This is important in developing the existence aspects of a continuously controlled surgery theory which we shall proceed to do.

We need to codify a continuously controlled simply connected Poincaré duality space. We model this on a simply connected manifold $M$ with a closed subset $K$ of codimension at least three. Consider a pair $(\overline{X}, \partial X)$ as above with $\partial X$ closed in $\overline{X}$ and $X = \overline{X} - \partial X$ dense in $\overline{X}$.

**1.4. Definition.** The pair $(\overline{X}, \partial X)$ is $-1$-*connected at* $\partial X$ if for every point $z \in \partial X$ and every neighborhood $U$ we have $U \cap X$ is nonempty. The pair $(\overline{X}, \partial X)$ is $0$-*connected at* $\partial X$, if for every $z \in \partial X$ and every neighborhood $U$ of $z$ in $\overline{X}$, there is a neighborhood $V$ of $z$ in $\overline{X}$ so that any two points in $V \cap Z$ can be connected by a path in $U \cap X$. The pair is $1$-*connected at* $\partial X$ if if for every $z \in \partial X$ and every neighborhood $U$ of $z$ in $\overline{X}$, there is a neighborhood $V$ of $z$ in $\overline{X}$ so that every loop in $V \cap X$ bounds a disc in $U \cap X$.

**1.5. Definition.** A *continuously controlled* CW-*structure* on the pair $(X, \partial X)$ is a CW structure on $(X - \partial X)$ such that the cells are small at $\partial X$ i. e. such that for every $z \in \partial X$ and every neighborhood $U$ of $z$ in $\overline{X}$, there is a neighborhood $V$ of $z$ in $\overline{X}$ so that if a cell in the CW structure intersects $V$ then the cell is contained in $U$. A *continuously controlled* CW-*complex* $(X, \partial X)$ is such a pair endowed with a continuously controlled CW-structure. We shall call the CW-structure *locally finite* if the CW-structure on $X - \partial X$ is locally finite. Similarly there is an obvious notion of locally finitely dominated in the continuously controlled sense.

Obviously a manifold $M$ with a codimension at least three subcomplex $K$ is $-1$-, $0$- and $1$-connected at $K$ and can be given a CW- structure which is continuously controlled at $K$.

**1.6. Definition.** A *simply connected continuously controlled Poincaré Duality space at* $\partial X$ is a continuously controlled, locally finite CW-complex $(\overline{X}, \partial X)$, $X = \overline{X} - \partial X$, such that $X$ is simply connected, $(\overline{X}, \partial X)$ must be $-1$-, $0$- and $1$-connected at $\partial X$ and the CW-structure must be continuously controlled at $\partial X$. Given this the cellular chains of $X$, define a chain complex in the category $\mathcal{B}(\overline{X}, \partial X; \mathbb{Z})$ which we denote by $C_\sharp(X)$, with dual chain complex denoted $C^\sharp(X)$. We then further require the existence of a homology class $[X] \in H^{l.f.}(X; Z)$ such that cap product with

$[X] \cap -$, which by its geometric nature defines a map of chain complexes in $\mathcal{B}(\overline{X}, \partial X; \mathbb{Z})$, is a homotopy equivalence of chain complexes

$$C^\sharp(X; \mathbb{Z}) \to C_\sharp(X; \mathbb{Z})$$

as chain complexes in the category $\mathcal{B}(\overline{X}, \partial X; \mathbb{Z})$.

**Remark.** The locally finite homology referred to above is singular homology based on locally finite chains

Once again it is clear that a simply connected manifold with a codimension three subcomplex can be triangulated to satisfy these conditions.

Given a pair $(\overline{X}, \partial X)$ and a proper map $f : Y \to X$ we say *f is a continuously controlled homotopy equivalence at* $\partial X$ if $f$ induces a strict homotopy equivalence of pairs $(\overline{Y}, \partial X) \xrightarrow{(\overline{f},1)} (\overline{X}, \partial X)$ relative to $\partial X$. Here $\overline{Y}$ is a completion of $Y$ by $\partial X$ through the map $f$. As a set $\overline{Y}$ is the disjoint union of $Y$ and $\partial X$, and the topology is given by the open sets being the open sets of $Y$ and sets of the form $V \cap \partial X \cup f^{-1}(V)$, where $V$ is open in $\overline{X}$. The aim of continuously controlled surgery is to turn a degree one normal map (in the proper sense) into a continuously controlled homotopy equivalence. Proper surgery [31] fits into this picture by completing an open Poincaré duality space by precisely one point for each end. Since a continuously controlled Poincaré duality space is automatically a proper Poincaré duality space, we can use the theory of Spivak normal fibrations from the proper theory. Alternatively it is not very difficult to see that the inclusion of a the boundary of a regular neighborhood in Euclidean space produces a spherical fibration.

At this point we proceed exactly as in [15] to do surgery below the mid-dimension. The method introduced by Wall [32] for surgery below the mid-dimension is completely geometric: Given a surgery problem $M \to X$ we may replace the map by an inclusion, replacing $X$ by the mapping cylinder. Doing a surgery for each of the cells in $X - M$ until we reach the mid-dimension produces a surgery problem $N \to X$ where $X$ is obtained from $N$ by attaching cells above the mid-dimension, see [15] for more details on this. This makes sense in the continuously controlled setting if we triangulate the manifold so that we obtain a continuously controlled CW-complex.

At this point we get obstructions to obtain continuously controlled homotopy equivalences with values in the algebraic $L$-theory of the additive categories with involution $\mathcal{B}(\overline{X}, \partial X; \mathbb{Z})$. The proof of this once again follows [15] closely and is a translation of the arguments given in [32] avoiding homology at all points, i. e. working directly with the chain complexes in the additive category. Given a controlled simply connected Poincaré pair $(\overline{X}, \partial X)$ with $X = \overline{X} - \partial X$ with a reduction of the Spivak fibration to

$B$ CAT where CAT $=$ Top, PL or $O$, we obtain a surgery exact sequence

$$\ldots \to L_{n+1}(\mathcal{B}(\overline{X}, \partial X; \mathbb{Z})) \to$$
$$\mathcal{S}^{c.c.}(\overline{X}, \partial X) \to [X, G/\text{CAT}] \to L_n(\mathcal{B}(\overline{X}, \partial X; \mathbb{Z}))$$

deciding topological, PL, and smooth, continuously controlled structures on $X$ respectively. The continuously controlled structure set consists of CAT manifolds $M$ together with a continuously controlled homotopy equivalence to $X = \overline{X} - \partial X$. Two structures $M_1 \to X$ and $M_2 \to X$ are equivalent, if there is a continuously controlled h-cobordism (in the CAT-category) between $M_1$, $M_2$ and a map extending the given maps.

**Remark.** There are all the usual possible modifications of a surgery theory. There is an obvious notion of a simple Poincaré complex in this context allowing the $h$-cobordism to be replaced by $s$-cobordism. Notice it is standard to prove a continuously controlled $h$- or $s$-cobordism theorem along the lines of of [22]. Similarly by allowing locally finitely dominated Poincaré complexes in the continuously controlled sense we would obtain a projective version of such a surgery theory along the lines of [23]. In the simply connected case these theories coincide, but later we shall indicate how to weaken the simply connected assumption.

## 2. GERM CATEGORIES

One would obviously want to generalize this to a non simply connected situation, but before discussing that we shall consider a less obvious generalization involving germs, that turns out to be very useful for computations. Suppose $U$ is an open subset of $\partial X$. We wish to develop a surgery theory where the aim is only to obtain a homotopy equivalence in a neighborhood of $U$. By this we mean that the "homotopy inverse" is only defined in a neighborhood of $U$, and that the compositions are homotopic by small homotopies to the identity in a neighborhood of $U$. We shall proceed to describe the algebra that describes this situation. Denote the complement of $U$ in $\partial X$ by $Z$. We need a definition

**2.1. Definition.** An object $A$ in $\mathcal{B}(\overline{X}, \partial X; R)$ has support at infinity contained in $Z$ if

$$\overline{\{x | A_x \neq 0\}} \cap \partial X \subset Z$$

We denote the full subcategory of $\mathcal{B}(\overline{X}, \partial X; R)$ on objects with support at infinity contained in $Z$ by $\mathcal{B}(\overline{X}, \partial X; R)_Z$. This is a typical example of an additive category $\mathcal{U} = \mathcal{B}(\overline{X}, \partial X; R)$ which is $\mathcal{A} = \mathcal{B}(\overline{X}, \partial X; R)_Z$-filtered in the sense of Karoubi [20], see also [4]. We recall the notion of an $\mathcal{A}$-filtered additive category $\mathcal{U}$ in the following

**2.2. Definition.** We say $\mathcal{U}$ is $\mathcal{A}$-filtered if every object $U$ has a family of decompositions $\{U = E_\alpha \oplus U_\alpha\}$ (called a filtration of $U$) satisfying the following axioms: (We denote objects in $\mathcal{A}$ by $A, B, \ldots$ and in $\mathcal{U}$ by $U, V, \ldots$)

F1: For each $U$, the decompositions form a filtered poset under the partial order $E_\alpha \oplus U_\alpha \leq E_\beta \oplus U_\beta$ whenever $U_\beta \subset U_\alpha$ and $E_\alpha \subset E_\beta$.

F2: Every map $A \to U$ factors $A \to E_\alpha \to E_\alpha \oplus U_\alpha = U$ for some $\alpha$.

F3: Every map $U \to A$ factors $U = E_\alpha \oplus U_\alpha \to E_\alpha \to A$ for some $\alpha$.

F4: For each $U$, $V$ the filtration on $U \oplus V$ is equivalent to the sum of filtrations $\{U = E_\alpha \oplus U_\alpha\}$ and $\{V = F_\beta \oplus V_\beta\}$, i. e. to $\{U \oplus V = (E_\alpha \oplus F_\beta) \oplus (U_\alpha \oplus V_\beta)\}$.

This is a precise analogue in the category of small additive categories of an ideal in a ring. The quotient category $\mathcal{U}/\mathcal{A}$ is defined to have the same objects as $\mathcal{U}$, but two morphisms $\phi_1 : U \to V$ and $\phi_2 : U \to V$ are identified if the difference $\phi_1 - \phi_2$ factors through the category $\mathcal{A}$, i. e. if there exists an object $A$ in $\mathcal{A}$ and a factorization $\phi_1 - \phi_2 : U \to A \to V$. The axioms ensure that $\mathcal{U}/\mathcal{A}$ is a category.

In the case we are considering $\mathcal{U} = \mathcal{B}(\overline{X}, \partial X; R)$ and $\mathcal{A} = \mathcal{B}(\overline{X}, \partial X; R)_Z$ it is easy to see the axioms above are satisfied: as indexing set we may use open neighborhoods of $\partial X - Z$ in $X$, and decompose an object $U = \{U_x\}$ in a part where $U_x$ is replaced by 0 if $x$ belongs to the given neighborhood, and another part where $U_x$ is replaced by 0 if $x$ does not belong to the given neighborhood. We denote the quotient category $\mathcal{U}/\mathcal{A}$ by $\mathcal{B}(\overline{X}, \partial X; R)^{\partial X - Z}$. Evidently two morphisms are identified if and only if they agree in a neighborhood of $U = \partial X - Z$, and it is not difficult to see that the category $\mathcal{B}(\overline{X}, \partial X; \mathbb{Z})^{\partial X - Z}$ measures morphisms that are homotopy equivalences in a neighborhood of $U$. To us however the real strength of these categories is in conjunction with Lemma 1.3, they allow computation of the $L$-groups, see Theorem 2.4 below. To prepare for this we need the idempotent completion of an additive category [16, p. 61]. The idempotent completion $\mathcal{A}^\wedge$ of $\mathcal{A}$ has objects $(A, p)$ where $A$ is an object of $\mathcal{A}$ and $p$ is an idempotent morphism $p^2 = p$. A morphism $\phi : (A, p) \to (B, q)$ is an $\mathcal{A}$-morphism $\phi : A \to B$ such that $q\phi p = \phi$. Intuitively $(A, p)$ represents the image of $p$, and the condition says that $\phi$ only depends on the image and lands in the image. The category $\mathcal{A}$ is embedded in its idempotent completion by sending $A$ to $(A, 1)$. Given a subgroup $k \subset K_0(\mathcal{A}^\wedge)$, we can perform a partial idempotent completion $\mathcal{A}^{\wedge k}$, the full subcategory of $\mathcal{A}^\wedge$ on objects $(A, p)$ with $[(A, p)] \in k \subset K_0(\mathcal{A}^\wedge)$. The following theorem is proved in [5] based on ideas from [28].

**2.3. Theorem.** *Given an additive category with involution $\mathcal{U}$ which is $\mathcal{A}$-filtered by a $*$-invariant subcategory $\mathcal{A}$, there is a fibration of 4-periodic $L$-spectra*

$$\mathbb{L}(\mathcal{A}^{\wedge k}) \to \mathbb{L}(\mathcal{U}) \to \mathbb{L}(\mathcal{U}/\mathcal{A})$$

*where k is the inverse image of* $K_0(\mathcal{U})$ *in* $K_0(\mathcal{U}^\wedge)$

The proof of this theorem goes as follows: It follows from standard bordisms methods that there is a fibration of spectra $\mathbb{L}(\mathcal{A}) \to \mathbb{L}(\mathcal{U}) \to \mathbb{L}(\mathcal{U}, \mathcal{A})$. Next one proves that a chain complex in $\mathcal{U}$ is dominated by a chain complex in $\mathcal{A}$ if and only if the chain complex induces a contractible chain complex in $\mathcal{U}/\mathcal{A}$. This means that an attempt to prove $\mathbb{L}(\mathcal{U}, \mathcal{A})$ is homotopy equivalent to $\mathbb{L}(\mathcal{U}/\mathcal{A})$ is off by a finiteness obstruction, and adjusting with idempotent completion as in the statement above solves the problem. For the ultimate statement regarding the decorations on the $L$-groups see [17, Theorem 6.7]. It is now fairly straightforward to compute the $L$-groups of $\mathcal{B}(\overline{X}, \partial X; \mathbb{Z})$ in the case $(\overline{X}, \partial X)$ is metrizable using Lemma 1.3 as in [6]. The result is

**2.4. Theorem.** *If* $(\overline{X}, \partial X)$ *is a compact metrizable pair, then*

$$L_*(\mathcal{B}(\overline{X}, \partial X; \mathbb{Z}))$$

*is isomorphic to the Steenrod homology theory (see axioms below) of* $\partial X$ *associated to the spectrum* $\Sigma \mathbb{L}(\mathbb{Z})$.

In case $\partial X$ is a CW-complex this is just the standard generalized homology theory associated with a spectrum. In the general case it satisfies the Steenrod axioms. The Steenrod axioms for a homology theory $h$ [19] say

(i) given any sequence of compact metrizable spaces $A \subset B \to B/A$ there is a long exact sequence in homology

$$\ldots \to h_i(A) \to h_i(B) \to h_i(B/A) \to h_{i-1}(A) \ldots$$

(ii) Given a countable collection $X_i$ of compact metrizable spaces letting $\bigvee X_i \subset \prod X_i$ be given the subset topology (the strong wedge) we have an isomorphism

$$h_*\left(\bigvee X_i\right) \cong \prod h_*(X_i)$$

It is also required that $h$ is homotopy invariant.

### 3. THE EQUIVARIANT CASE

Finally we want to discuss to what extent we can avoid the simple connectedness assumption. We shall not try to deal with the most general case, even though something could be said using the germ methods mentioned above. We shall satisfy ourselves with the following situation where the local variation in the fundamental group is given by a global group action, in other words, we shall consider the situation where we have a group $\Gamma$ acting freely cellularly on $X = \overline{X} - \partial X$, and $(\overline{X}, \partial X)$ is a simply connected, continuously controlled Poincaré duality space. We need to be able to decide when an equivariant proper map from $M$ to $X$ is a continuously controlled equivariant homotopy equivalence in order to setup a surgery theory as in the simply connected case.

The action of $\Gamma$ is not assumed to be free on $\partial X$. In the case where we only have finite isotropy it is fairly easy to define a category which measures this kind of continuously controlled equivariant homotopy equivalence. This was done in [15] and [18]. Here we want to deal with the more general situation where we do not have an assumption about isotropy. This is of interest in connection with the kind of generalizations of the assembly map studied in the Baum–Connes and the Farrell–Jones conjectures [2, 13]. An extra complication is to be able to deal with duality on the category. Assume $(\overline{X}, \partial X)$ is a compact Hausdorff pair with a $\Gamma$-action (if the pair is only locally compact Hausdorff one point compactify ). Choose once and for all a large $R[\Gamma]$ module say $U = R[\Gamma \times \mathbb{N}]$. We shall think of $U$ as a universe. This will help make categories small and thus allow talking about group actions on categories. We shall call a subset $Z \subset \overline{X}$ *relatively $\Gamma$- compact* if $Z \cdot \Gamma/\Gamma \subset \overline{X}/\Gamma$ is contained in a compact subset of $\overline{X}/\Gamma$.

**3.1. Definition.** The category $\mathcal{D}(\overline{X}, \partial X; R)$ has objects A, an $R$-submodule of $U$, together with a map $f : A \longrightarrow F(X \times \Gamma)$, where $F(X \times \Gamma)$ is the set of finite subsets of $X \times \Gamma$, $X = \overline{X} - \partial X$, satisfying

(i) $A_x = \{a \in A | f(a) \subseteq \{x\}\}$ is a finitely generated free sub $R$-module for each $x \in X \times \Gamma$.
(ii) As an $R$-module $A = \bigoplus_{x \in X \times \Gamma} A_x$
(iii) $f(a + b) \subseteq f(a) \cup f(b)$
(iv) $\{x \in X \times \Gamma | A_x \neq 0\}$ is locally finite and relatively $\Gamma$-compact in $\overline{X} \times \Gamma$ with the diagonal $\Gamma$-action.

A morphism $\phi : A \longrightarrow B$ is a morphism of $R$- modules, satisfying the continuous control condition at $\partial X$ when we forget the extra factor of $\Gamma$, i. e. for every point $z$ in $\partial X$ and for every neighborhood $U$ of $z$ in $\overline{X}$ there is a neighborhood $V$ of $z$ in $\overline{X}$ such that if $x \in (X - U) \times \Gamma$ and $y \in (V - \partial X) \times \Gamma$, then $\phi_y^x$ and $\phi_x^y$ are 0.

Combining the action of $\Gamma$ on $U$ and on $\overline{X}$ we get a $\Gamma$-action on $\mathcal{D}(\overline{X}, \partial X; R)$ by conjugation. An object $A$ is fixed under this action if $A$ is invariant under the $\Gamma$-action on $U$, thus inheriting an $R[\Gamma]$-module structure which has to be free, and the reference map is equivariant. A morphism is fixed under the $\Gamma$-action if it is $\Gamma$-equivariant. We denote the fixed category by $\mathcal{D}_\Gamma(\overline{X}, \partial X; R)$. These are the categories that determine the relevant equivariant continuously controlled homotopy equivalences.

**Remark.** In case $\Gamma$ is the trivial group and $\overline{X}$ is compact we recover our old definition of the $\mathcal{B}$-categories. In case $\overline{X}$ is not compact we get a different category which corresponds to homology rather than homology with locally finite coefficients when applying $K$ or $L$-theory.

The definition above is designed to make it easy to define a duality on the category. We define the dual of an object $A$ to be the set of $R$-module homomorphisms from $A$ to $R$ that are locally finite i. e. only nontrivial on $A_x$ for finitely many $x$. The reference map to $F(X \times \Gamma)$ is given by the set of $x$ for which the homomorphism is non-zero. As usual we inherit a left $\Gamma$-action which we may turn into a right action, possibly using an orientation homomorphism in the process.

In the definition above we crossed with $\Gamma$. It is easy to see we get an equivalent category if instead we crossed with some other free $\Gamma$ space. In particular we could cross with $E\Gamma$. This is relevant for developing this kind of theory in $A$-theory where the underlying homotopy type does play a role.

Suppose $E$ is a locally compact Hausdorff $\Gamma$-space. Consider $E \subset E \times I$ included as $E \times 1$. The category $\mathcal{A} = \mathcal{D}_\Gamma(E \times I, E; R)_\emptyset$ is the full subcategory of $\mathcal{U} = \mathcal{D}_\Gamma(E \times I, E; R)$ on objects with empty support at infinity i.e. on objects $A$ such that the closure of $\{x \in E \times [0,1) | \exists g \in \Gamma : A_{(x,g)} \neq 0\}$ intersects $E = E \times 1$ trivially. It is easy to see that $\mathcal{U}$ is $\mathcal{A}$- filtered. $\mathcal{D}_\Gamma(E \times I, E; R)_\emptyset$ is equivalent to the category of finitely generated free $R[\Gamma]$-modules. We denote the quotient category by $\mathcal{D}_\Gamma(E \times I, E; R)^E$. It is a germ category. Objects are the same as in $\mathcal{D}_\Gamma(E \times I, E; R)$, but morphisms are identified if they agree close to $E$.

When studying the $L$-theory we have to deal with the variations in the upper index. This is necessary in the geometric application. Here we choose to use $L^{-\infty}$, avoiding that problem. Similarly in algebraic $K$- theory we need to use $K^{-\infty}$, the $K$- theory functor that includes the negative $K$-theory groups. It was proved in [24], see [4] for a more modern proof, that we have a fibration of spectra

$$K^{-\infty}(\mathcal{A}) \to K^{-\infty}(\mathcal{U}) \to K^{-\infty}(\mathcal{U}/\mathcal{A})$$

whenever we have an $\mathcal{A}$- filtered category $\mathcal{U}$. This leads to the following

**3.2. Theorem.** *The functors from locally compact Hausdorff spaces with $\Gamma$-action and $\Gamma$-equivariant maps, sending $E$ to*

$$K^{-\infty}(\mathcal{D}_\Gamma(E \times I, E; R)^E) \text{ and } L^{-\infty}(\mathcal{D}_\Gamma(E \times I, E; R)^E)$$

*are homotopy invariant and excisive. If $\Gamma$ acts transitively on $S$, the value on $S$ is homotopy equivalent to $\Sigma K^{-\infty}(R[H])$ and $\Sigma L^{-\infty}(R[H])$ respectively, where $H$ denotes the isotropy subgroup.*

The proof of the first two statements follows the methods in [5] closely. It is an application of the basic $\mathcal{A} \to \mathcal{U} \to \mathcal{U}/\mathcal{A}$ fibrations. In the last statement one should notice that the identification is not canonical, it depends on choosing a basepoint in $S$. To see the last statement notice that we are considering $R[\Gamma]$-modules parameterized by $S \times [0,1)$, and germs of morphisms near $S \times 1$. The control conditions imply that near $S \times 1$, a morphism can not reach from one component of $S$ to another. Hence this category is equivalent to the category of $R[H]$-modules parameterized by $[0, 1]$ with germs taken near 1, since the group action tells us what to do everywhere else. The $K$ or $L$- theory of this category is a deloop of the $K$ or $L$-theory of $R[H]$.

These methods are generalized to the $C^*$-situation in [7]. The advantage of this method is that we not only get the kind of description needed in the study of assembly maps in [12], but we also obtain the relevant spectra as fixed spectra under a $\Gamma$-action.

## 4. ASSEMBLY MAPS

Denote $K^{-\infty}$ or $L^{-\infty}$ applied to $\mathcal{D}_\Gamma(E \times I, E; R)^E$ by $F$. It is then easy to see, using the methods of [5], specifically [5, Theorem 1.28, Theorem 4.2] and their corollaries, that $F$ is a homotopy invariant excisive functor from the category of $\Gamma$-spaces and $\Gamma$-maps to spectra, and thus fits precisely into the framework developed by Davis and Lück [12] for generalized assembly maps of the type considered by Quinn, Farrell-Jones, and Baum–Connes. That $F$ is a functor on all $\Gamma$ maps without any properness assumption uses the fact that we have a $\Gamma$-compactness assumption on the support of the modules considered.

Davis and Lück describe assembly maps of this type as the induced map $F(E) \to F(*)$, where $E$ is a $\Gamma$-space. The Farrell-Jones conjecture is then the statement that $F(E) \to F(*)$ is an isomorphism when $E$ is the universal space for $\Gamma$ actions with isotropy groups virtually cyclic groups, and the Baum-Connes conjecture is a similar statement where $F$ is defined using topological $K$-theory and $E$ is the universal space for $\Gamma$-actions with finite isotropy. Consider

$$\mathcal{D}_\Gamma(E \times I, E; R)_\emptyset \to \mathcal{D}_\Gamma(E \times I, E; R) \to \mathcal{D}_\Gamma(E \times I, E; R)^E$$

Applying $K^{-\infty}$ or $L^{-\infty}$ we get a fibration of spectra (by [5, Theorem 1.28 and Theorem 4.2])

**4.1. Theorem.** *The generalized assembly map is the connecting homomorphism in the above mentioned fibration.*

*Proof.* The category $\mathcal{D}_\Gamma(E \times I, E; R)_\emptyset$ is equivalent to the category of finitely generated $R\Gamma$-modules since the support conditions make the modules finitely generated $R\Gamma$-modules and the control conditions vacuous. Consider the diagram:

$$
\begin{array}{ccc}
\mathcal{D}_\Gamma(E \times I, E; R)_\emptyset \longrightarrow \mathcal{D}_\Gamma(E \times I, E; R) \longrightarrow \mathcal{D}_\Gamma(E \times I, E; R)^E \\
\downarrow a \qquad\qquad\qquad \downarrow b \qquad\qquad\qquad \downarrow c \\
\mathcal{D}_\Gamma(* \times I, *; R)_\emptyset \longrightarrow \mathcal{D}_\Gamma(* \times I, *; R) \longrightarrow \mathcal{D}_\Gamma(* \times I, *; R)^*
\end{array}
$$

Here $a$ is an equivalence of categories, since both categories are equivalent to the category of finitely generated free $R\Gamma$-modules. The $K$- and $L$-theory of $\mathcal{D}_\Gamma(* \times I, *; R)$ is trivial since the category admits a flasque structure by shifting modules towards 1. The map $c$ is the induced map of $E \to *$. The result now follows.

$\square$

We finish by mentioning a result which is essentially contained in [5] but not explicitly stated.

**4.2. Theorem.** *Let $M$ be an $n$-dimensional topological manifold. Then the (4-periodical) structure set of $M$ is isomorphic to*

$$L^h_{n+1}(\mathcal{D}_\Gamma(\widetilde{M} \times I, \widetilde{M}; \mathbb{Z}))$$

*where $\Gamma$ is the fundamental group of $M$ acting on the universal cover $\widetilde{M}$.*

**Remark.** The surgery exact sequence for topological manifolds is 4-periodic except the periodicity breaks down in the bottom since the normal invariant is given by $[M, G/\operatorname{Top}]$, not $[M, G/\operatorname{Top} \times \mathbb{Z}]$. The result above identifies the $L$-group with a periodical structure set. See [3] for a further discussion of this phenomenon.

*Proof.* Consider

$$\mathcal{D}_\Gamma(\widetilde{M} \times I, \widetilde{M}; \mathbb{Z})_0 \to \mathcal{D}_\Gamma(\widetilde{M} \times I, \widetilde{M}; \mathbb{Z}) \to \mathcal{D}_\Gamma(\widetilde{M} \times I, \widetilde{M}; \mathbb{Z})^{\widetilde{M}}$$

which, in the following we will discuss as

$$\mathcal{A} \to \mathcal{U} \to \mathcal{U}/\mathcal{A}.$$

According to [5, Theorem 4.1] we get a fibration of $\mathbb{L}$-spectra if we choose appropriate decorations. We can, as mentioned above, always use the $-\infty$ decoration. but we need to improve on that a bit. we have

$$\mathcal{U}/\mathcal{A} = \mathcal{D}_\Gamma(\widetilde{M} \times I, \widetilde{M}; \mathbb{Z})^{\widetilde{M}} = \mathcal{B}(M \times I, M; \mathbb{Z})^M$$

and by [1] $K_i(\mathcal{B}(M \times I, M; \mathbb{Z}) = h_{i-1}(M_+; K\mathbb{Z})$ where $K\mathbb{Z}$ is the $K$-theory spectrum for the integers. Hence $K_0(\mathcal{U}/\mathcal{A}) = \mathbb{Z}$ and the boundary map (which is the $K$-theory assembly map) hits the subgroup of $K_0(\mathbb{Z}\Gamma)$ given by the free modules. This means we get a fibration of spectra

$$\mathbb{L}^h(\mathcal{D}_\Gamma(\widetilde{M} \times I, \widetilde{M}; \mathbb{Z})_0) \to$$
$$\mathbb{L}^h(\mathcal{D}_\Gamma(\widetilde{M} \times I, \widetilde{M}; \mathbb{Z})) \to \mathbb{L}^h(\mathcal{D}_\Gamma(\widetilde{M} \times I, \widetilde{M}; \mathbb{Z})^{\widetilde{M}})$$

which we identify with

$$\mathbb{L}^h(\mathbb{Z}\Gamma) \to \mathbb{L}^h(\mathcal{D}_\Gamma(\widetilde{M} \times I, \widetilde{M}; \mathbb{Z})) \to \Sigma M_+ \wedge \mathbb{L}(\mathbb{Z})$$

as in [5]. The classifying map $\Sigma M_+ \wedge \mathbb{L}(\mathbb{Z}) \to \Sigma \mathbb{L}^h(\mathbb{Z}\Gamma)$ was identified with the (suspension of) the assembly map in [5]. Hence the fibre represents the structure set, and the result follows. $\square$

REFERENCES

1. D. R. Anderson, F. Connolly, S. C. Ferry, and E. K. Pedersen, *Algebraic K-theory with continuous control at infinity*, J. Pure Appl. Algebra **94** (1994), 25–47.

2. P. Baum and A. Connes, *K-theory for discrete groups*, Operator Algebras and Applications, Vol. 1, London Math. Soc. Lecture Notes, vol. 135, Cambridge University Press, Cambridge – New York, 1988, pp. 1–20.

3. J. Bryant, S. Ferry, W. Mio, and S. Weinberger, *The topology of homology manifolds*, Ann. of Math. (2) **143** (1996), 435–467.

4. M. Cárdenas and E. K. Pedersen, *On the Karoubi filtration of a category*, K-Theory **12** (1994), 165–191.

5. G. Carlsson and E. K. Pedersen, *Controlled algebra and the Novikov conjectures for K- and L-theory*, Topology **34** (1995), 731–758.

6. _____, *Čech homology and the Novikov conjectures*, Math. Scand. **82** (1998), 5–47.

7. G. Carlsson, E. K. Pedersen, and J. Roe, *Controlled C*-algebra theory and the injectivity of the Baum-Connes map*, Preprint, 1994.

8. T. A. Chapman, *Topological invariance of Whitehead torsion*, J. Amer. Math. Soc. **96** (1974), 488–497.

9. _____, *Controlled Simple Homotopy Theory and Applications*, Lecture Notes in Mathematics, vol. 1009, Springer-Verlag, Berlin-New York, 1983.

10. T. A. Chapman and S. C. Ferry, *Approximating homotopy equivalences by homeomorphisms*, Amer. J. Math. **101** (1979), 583–607.

11. E. H. Connell and J. Hollingsworth, *Geometric groups and Whitehead torsion*, Trans. Amer. Math. Soc. **140** (1969), 161–181.

12. J. Davis and W. Lück, *Spaces over a category and assembly maps in isomorphism conjectures in K- and L-theory*, K-Theory **15** (1998), no. 3, 201–252.

13. F. T. Farrell and L. E. Jones, *Isomorphism conjectures in algebraic K-theory*, J. Amer. Math. Soc. **6** (1993), 249–297.

14. S. C. Ferry and E. K. Pedersen, *Controlled algebraic K-theory*, (1987), Submitted.

15. _____, *Epsilon surgery Theory*, Novikov Conjectures, Index Theorems and Rigidity, Vol. 2 (Oberwolfach, 1993), London Math. Soc. Lecture Notes, vol. 227, Cambridge University Press, Cambridge – New York, 1995, pp. 167–226.

16. P. Freyd, *Abelian Categories*, Harper and Row, New York, 1966.

17. I. Hambleton and E. K. Pedersen, *Topological equivalence of linear representations*, (To appear).

18. _____, *Bounded surgery and dihedral group actions on spheres*, J. Amer. Math. Soc. **4** (1991), 105–126.

19. D. S. Kahn, J. Kaminker, and C. Schochet, *Generalized homology theories on compact metric spaces*, Michigan Math. J. **24** (1977), 203–224.

20. M. Karoubi, *Foncteur derivees et K-theorie*, Lecture Notes in Mathematics, vol. 136, Springer-Verlag, Berlin-New York, 1970.

21. E. K. Pedersen, *On the $K_{-i}$ functors*, J. Algebra **90** (1984), 461–475.

22. _____, *On the bounded and thin h-cobordism theorem parameterized by $\mathbb{R}^n$*, Transformation Groups (Poznan, 1985), Lecture Notes in Mathematics, vol. 1217, Springer-Verlag, Berlin-New York, 1986, pp. 306–320.

23. E. K. Pedersen and A. A. Ranicki, *Projective surgery theory*, Topology **19** (1980), 239–254.

24. E. K. Pedersen and C. Weibel, *K-theory homology of spaces*, Algebraic Topology (Arcata, 1986), Lecture Notes in Mathematics, vol. 1370, Springer-Verlag, Berlin-New York, 1989, pp. 346–361.

25. F. Quinn, *Ends of maps, I*, Ann. of Math. (2) **110** (1979), 275–331.

26. _____, *Ends of maps, II*, Invent. Math. **68** (1982), 353–424.

27. A. A. Ranicki, *Additive L-theory*, *K*-Theory **3** (1989), 163–195.
28. _____ , *Algebraic L-theory and Topological Manifolds*, Cambridge Tracts in Math., vol. 102, Cambridge University Press, Cambridge – New York, 1992.
29. A. A. Ranicki and M. Yamasaki, *Symmetric and quadratic complexes with geometric control*, Proceedings of TGRC-KOSEF, vol. 3, 1993, available on WWW from http://www.maths.ed.ac.uk/~aar and http://math.josai.ac.jp/~yamasaki, pp. 139–152.
30. L. C. Siebenmann, *Approximating cellular maps by homeomorphisms*, Topology **11** (1972), 271–294.
31. L. Taylor, *Surgery on paracompact manifolds*, Ph.D. thesis, UC at Berkeley, 1972.
32. C. T. C. Wall, *Surgery on Compact Manifolds*, Academic Press, New York, 1970.

DEPARTMENT OF MATHEMATICAL SCIENCES
SUNY AT BINGHAMTON
BINGHAMTON, NEW YORK 13901
*E-mail address*: erik@math.binghamton.edu

# Homology manifolds

## Washington Mio

The study of the local-global geometric topology of homology manifolds has a long history. Homology manifolds were introduced in the 1930s in attempts to identify local homological properties that implied the duality theorems satisfied by manifolds [23, 56]. Bing's work on decomposition space theory opened new perspectives. He constructed important examples of 3-dimensional homology manifolds with non-manifold points, which led to the study of other structural properties of these spaces, and also established his *shrinking criterion* that can be used to determine when homology manifolds obtained as decomposition spaces of manifolds are manifolds [4]. In the 1970s, the fundamental work of Cannon and Edwards on the double suspension problem led Cannon to propose a conjecture on the nature of manifolds, and generated a program that culminated with the Edwards-Quinn characterization of higher-dimensional topological manifolds [15, 24, 21]. Starting with the work of Quinn [44, 46], a new viewpoint has emerged. Recent advances [10] use techniques of controlled topology to produce a wealth of previously unknown homology manifolds and to extend to these spaces the Browder-Novikov-Sullivan-Wall surgery classification of compact manifolds [53], suggesting a new role for these objects in geometric topology, and tying together two strands of manifold theory that have developed independently. In this article, we approach homology manifolds from this perspective. We present a summary of these developments and discuss some of what we consider to be among the pressing questions in the subject. For more detailed treatments, we refer the reader to article [10] by Bryant, Ferry, Mio and Weinberger, and the forthcoming lecture notes by Ferry [26]. The survey papers by Quinn [45] and Weinberger [54] offer overviews of these developments.

## 1. EARLY DEVELOPMENTS

Localized forms of global properties of topological spaces and continuous mappings often reveal richer structures than their global counterparts alone. The identification of these local properties and the study of their influence on the large scale structure of spaces and mappings have a history

Partially supported by the National Science Foundation.

that dates back to the beginning of this century. Wilder's work [56] reflects the extensive study of local homology conducted by many authors, a line of investigation that has its roots in the search – started by Čech [20] and Lefschetz [38] – for local homological conditions that implied the duality and separation properties known to be satisfied by triangulable manifolds.

**Definition 1.1.** A *topological n-manifold* is a separable metrizable space that is locally homeomorphic to euclidean $n$-space $\mathbb{R}^n$.

Early proofs that a closed oriented manifold $M^n$ satisfies Poincaré duality assumed the existence of a triangulation of $M$ [37, 42]. Orientability was defined as a global property of the triangulation, and the Poincaré duality isomorphism

$$\cap[M]\colon H^*(M;\mathbb{Z}) \to H_{n-*}(M;\mathbb{Z})$$

was established by analysing the pattern of intersection of simplices with "cells" of the dual block structure on $M$ obtained from the triangulation.

If $M$ is an $n$-manifold and $x \in M$, then $x$ has arbitrarily small $n$-disk neighborhoods which have $(n-1)$-dimensional spheres as boundaries. By excision, homologically this local structure can be expressed as $H_*(M, M \smallsetminus \{x\}) \cong H_*(D^n, S^{n-1})$, for every $x \in M$.

**Definition 1.2.** $X$ is a *homology n-manifold* if for every $x \in X$

$$H_i(X, X \smallsetminus \{x\}) \cong \begin{cases} \mathbb{Z}, & \text{if } i = n \\ 0, & \text{otherwise.} \end{cases}$$

The local homology groups $H_*(X, X \smallsetminus \{x\})$ of these *generalized manifolds* can be used to define and localize the notion of orientation for these spaces, and to formulate proofs (at various degrees of generality) that compact oriented generalized manifolds satisfy Poincaré and Alexander duality. For a historical account of these developments, we refer the reader to [23].

Topological manifolds are homology manifolds; however, the latter form a larger class of spaces. (As we shall see later, there are numerous homology manifolds without a single manifold point.) Spaces $X$ satisfying the Poincaré duality isomorphism with respect to a fundamental class $[X] \in H_n(X)$ are called *Poincaré spaces* of formal dimension $n$. We thus have three distinct classes of spaces related by forgetful functors:

$$\left\{\begin{matrix} \text{Topological} \\ \text{manifolds} \end{matrix}\right\} \longrightarrow \left\{\begin{matrix} \text{Homology} \\ \text{manifolds} \end{matrix}\right\} \longrightarrow \left\{\begin{matrix} \text{Poincaré} \\ \text{spaces} \end{matrix}\right\}.$$

Classical surgery theory studies topological-manifold structures on Poincaré spaces [53]. Our discussion will be focused on the differences between topological and homology manifolds, a problem that is usually treated in two stages:

(i) determine whether or not a given homology manifold $X$ is a "fine" quotient space of a topological manifold (we shall elaborate on this later), and

(ii) exhibit conditions under which a quotient space $X$ of a manifold $M$ is homeomorphic to $M$.

The latter is a central question in decomposition space theory, an area that originated with the work of Moore [40]. He proved that if $X$ is Hausdorff and $f: S^2 \to X$ is a surjection such that $S^2 \smallsetminus f^{-1}(x)$ is non-empty and connected, for every $x \in X$, then $X$ is homeomorphic to $S^2$. This result is a precursor to the characterization of the 2-sphere in terms of separation properties obtained by Bing. If $X$ is a compact, connected, locally connected metrizable space with more than one point, then $X$ is homeomorphic to $S^2$ if and only if the complement of any two points in $X$ is connected and the complement of any subspace of $X$ homeomorphic to a circle is disconnected [3].

Bing's work on decompositions of 3-manifolds defined an important chapter in decomposition theory. While focused on the geometry of decompositions of low dimensional manifolds, his work was influential in subsequent developments in higher dimensions. Given a quotient map $f: M \to X$, exploiting the interplay between the local structure of $X$ near points $x \in X$ and the local geometry of the embeddings $f^{-1}(x) \subseteq M$ of the corresponding point inverses, he constructed examples of generalized 3-manifolds with non-manifold points, which led to the first considerations of general position properties of generalized manifolds. Conversely, Bing's *shrinking criterion* uses the geometry of the point inverses of $f$ to provide conditions under which the quotient space $X$ is homeomorphic to $M$ [4]. For metric spaces, the criterion can be stated as follows.

**Theorem 1.3** (R. H. Bing). *A surjection $f: M \to X$ of compact metric spaces can be approximated by homeomorphisms if and only if for any $\epsilon > 0$, there is a homeomorphism $h: M \to M$ such that:*

(i) $d(f \circ h, f) < \epsilon$.

(ii) $\mathrm{diam}\, h(f^{-1}(x)) < \epsilon$, *for every $x \in X$.*

Applications of the shrinking criterion in low dimensions include the construction of a $\mathbb{Z}_2$-action on $S^3$ which is not topologically conjugate to a linear involution [4].

Generalized manifolds also arise in the study of dynamics on manifolds. Smith theory [50, 7] implies that fixed points of topological semifree circle actions on manifolds are generalized manifolds, giving further early evidence of the relevance of these spaces in geometric topology.

## 2. The recognition problem

How can one decide whether or not a given topological space $X$ is a manifold? A reference to the definition of manifolds simply reduces the question to a characterization of euclidean spaces, a problem of essentially the same complexity. The proposition that a characterization of higher dimensional manifolds in terms of their most accessible properties might be possible evolved from groundbreaking developments in decomposition space theory in the 1970s. We begin our discussion of the recognition problem with a list of basic characteristic properties of topological manifolds. For simplicity, we assume that $X$ is compact, unless otherwise stated.

(i) *Manifolds are finite dimensional.*

**Definition 2.1.** The (covering) dimension of a topological space $X$ is $\leq n$, if any open covering $\mathcal{U}$ of $X$ has a refinement $\mathcal{V}$ such that any subcollection of $\mathcal{V}$ containing more than $(n+1)$ distinct elements has empty intersection. The *dimension* of $X$ is $n$, if $n$ is the least integer for which dimension of $X$ is $\leq n$. If no such integer exists, $X$ is said to be infinite dimensional.

Topological $n$-manifolds, and euclidean $n$-space $\mathbb{R}^n$ in particular, are examples of $n$-dimensional spaces.

(ii) *Local contractibility.*

Every point in a manifold has a contractible neighborhood. The following weaker notion of local contractibility is, however, a more manageable property.

**Definition 2.2.** $X$ is *locally contractible* if for any $x \in X$ and any neighborhood $U$ of $x$ in $X$, there is a neighborhood $V$ of $x$ such that $V \subseteq U$ and $V$ can be deformed to a point in $U$, i.e., the inclusion $V \subseteq U$ is nullhomotopic.

Absolute neighborhood retracts (ANR) are important examples of locally contractible spaces. (Recall that $X$ is an ANR if there is an embedding of $X$ as a closed subspace of the Hilbert cube $I^\infty$ such that some neighborhood $N$ of $X$ retracts onto $X$.) Conversely, if $X$ is finite dimensional and locally contractible, then $X$ is an ANR [6]. Since any $n$-dimensional space can be properly embedded in $\mathbb{R}^{2n+1}$ [31], it follows that $X$ is a finite dimensional locally contractible space if and only if $X$ is an euclidean neighborhood retract (ENR). The definition of ENR is analogous to that of ANR with the Hilbert cube replaced by some euclidean space. Hence, conditions (i) and (ii) above can be elegantly summarized in the requirement that $X$ be an ENR.

(iii) *Local homology.*

Topological $n$-manifolds are homology $n$-manifolds. The assumption that $X$ is an ENR homology manifold encodes all separation properties satisfied by closed manifolds, since compact oriented ENR homology manifolds satisfy Poincaré and Alexander duality. (As usual, in the nonorientable case we twist homology using the orientation character.) Moreover, since the dimension of finite dimensional spaces can be detected homologically, ENR homology $n$-manifolds are $n$-dimensional spaces.

An ENR homology $n$-manifold $X$ is an $n$-dimensional locally contractible space in which points have homologically spherical "links". Thus, to this hypothesis, it is necessary to incorporate a local fundamental group condition that will guarantee that "links" of points in $X$ are homotopically spherical, as illustrated by the following classical example.

Let $H^n$ be a homology $n$-sphere (i.e., a closed manifold such that $H_*(H; \mathbb{Z}) \cong H_*(S^n; \mathbb{Z})$) with nontrivial fundamental group, and let $X = \Sigma H$ be the suspension of $H$. $X$ is a simply connected homology manifold, but arbitrarily close to the suspension points there are loops $\alpha$ that are nontrivial in the complement of the suspension points. Therefore, $X$ is not a manifold since any small punctured neighborhood of a suspension point is non-spherical. Nonetheless, an important result of Cannon establishes that the double suspension of $H$ is a topological manifold [14]. Since any bounding disk $D_\alpha^2$ for the loop $\alpha$ must intersect one of the suspension points, the presence of a nontrivial local fundamental group can be interpreted as a failure of general position, if $n \geq 4$. $D_\alpha^2$ cannot be moved away from itself by small deformations.

(iv) *The disjoint disks property.*

Manifolds satisfy general position. If $P^p$ and $Q^q$ are complexes tamely embedded in a manifold $M$, under arbitrarily small perturbations, we can assume that $P \cap Q$ is tamely embedded in $M$ and that $\dim(P \cap Q) \leq p + q - n$. In particular, if $n \geq 5$, 2-dimensional disks can be positioned away from each other by small moves.

**Definition 2.3.** $X$ has the *disjoint disks property* (DDP) if for any $\epsilon > 0$, any pair of maps $f, g \colon D^2 \to X$ can be $\epsilon$-approximated by maps with disjoint images.

The fact that the DDP is the appropriate general position hypothesis for the recognition problem became evident in Cannon's work on the double suspension problem. Later, Bryant showed that if $X^n$ is an ENR homology manifold with the DDP, $n \geq 5$, then tame embeddings of complexes into $X$ can be approximated by maps in general position. [8].

ENR homology manifolds with the DDP have the local-global algebraic topology and general position properties of topological manifolds. In 1977, motivated largely by his solution of the double suspension problem, Cannon formulated the following conjecture [14, 15].

**The characterization conjecture.** *ENR homology n-manifolds with the disjoint disks property, $n \geq 5$, are topological n-manifolds.*

**Definition 2.4.** A mapping $f: M \to X$ of ENRs is *cell-like (CE)*, if $f$ is a proper surjection and for every $x \in X$, $f^{-1}(x)$ is contractible in any of its neighborhoods. A $CE$-map $f$ is a *resolution* of $X$ if $M$ is a topological manifold.

All examples of ENR homology manifolds known at the time these developments were taking place could be obtained as cell-like quotients of topological manifolds. In addition, if $M$ is a manifold and $f$ is cell-like, then $X$ is a homology manifold [36]. The fact that suspensions of homology spheres are resolvable follows from a theorem of Kervaire that states that homology spheres bound contractible manifolds [34].

The following result, of which the double suspension theorem is a special case, is a landmark in decomposition space theory [21, 24].

**Theorem 2.5** (R. D. Edwards). *Let $X^n$ be an ENR homology manifold with the DDP, $n \geq 5$. If $f: M \to X$ is a resolution of $X$, then $f$ can be approximated by homeomorphisms.*

In light of Edwards' theorem, the completion of the manifold characterization program is reduced to the study of the following conjecture.

**The resolution conjecture.** *ENR homology manifolds of dimension $\geq 5$ are resolvable.*

Early results supporting this conjecture assumed that the homology manifolds under consideration contained many manifold points. Cannon and Bryant-Lacher showed that $X$ is resolvable if the dimension of the singular set of $X$ is in the stable range [16]. Galewski and Stern proved that polyhedral homology manifolds are resolvable, so that non-resolvable homology manifolds, if they exist, must not be polyhedral [29].

A major advance toward the solution of the resolution conjecture is due to F. Quinn. He showed that the existence of resolutions can be traced to a single locally defined integral invariant that can be interpreted as an index [44, 46].

**Theorem 2.6** (Quinn). *Let $X$ be a connected ENR homology n-manifold, $n \geq 5$. There is an invariant $I(X) \in 8\mathbb{Z} + 1$ such that:*

(a) *If $U \subseteq X$ is open, then $I(X) = I(U)$.*
(b) *$I(X \times Y) = I(X) \times I(Y)$.*
(c) *$I(X) = 1$ if and only if $X$ is resolvable.*

*Remark* . The local character of Quinn's invariant implies that if $X$ is connected and contains at least one manifold point, then $X$ is resolvable.

Thus, a non-resolvable ENR homology $n$-manifold, $n \geq 5$, cannot be a cell complex, since the interior of a top cell would consist of manifold points.

Combined, Theorems 2.5 and 2.6 yield the celebrated characterization of higher dimensional topological manifolds.

**Theorem 2.7** (Edwards-Quinn). *Let $X$ be an ENR homology $n$-manifold with the DDP, $n \geq 5$. $X$ is a topological manifold if and only if $I(X) = 1$.*

The resolution conjecture, however, remained unsolved. Are there ENR homology manifolds with $I(X) \neq 1$?

## 3. Controlled surgery

This is a brief review of results of simply-connected controlled surgery theory needed in our discussion of the resolution problem. Proofs and further details can be found in [27, 28].

In classical surgery theory one studies the existence and uniqueness of manifold structures on a given Poincaré complex $X^n$ of formal dimension $n$. Controlled surgery addresses an estimated form of this problem, when $X$ is equipped with a map to a control space $B$. For simplicity, we assume that $\partial X = \emptyset$, although even in this case bounded versions are needed in considerations of uniqueness of structures.

**Definition 3.1.** Let $p \colon X \to B$ be a map to a metric space $B$ and $\epsilon > 0$. A map $f \colon Y \to X$ is an $\epsilon$-*homotopy equivalence* over $B$, if there exist a map $g \colon X \to Y$ and homotopies $H_t$ from $g \circ f$ to $1_Y$ and $K_t$ from $f \circ g$ to $1_X$, respectively, such that the tracks of $H$ and $K$ are $\epsilon$-small in $B$, i.e., $\operatorname{diam}(p \circ f \circ H_t(y)) < \epsilon$ for every $y \in Y$, and $\operatorname{diam}(p \circ K_t(x)) < \epsilon$, for every $x \in X$. The map $f \colon Y \to X$ is a *controlled equivalence* over $B$, if it is an $\epsilon$-equivalence over $B$, for every $\epsilon > 0$.

In order to use surgery theory to produce $\epsilon$-homotopy equivalences, we need the notion of $\epsilon$-Poincaré spaces. Poincaré duality can be estimated by the diameter of cap product with a fundamental class as a chain homotopy equivalence.

**Definition 3.2.** Let $p \colon X \to B$ be a map, where $X$ is a polyhedron and $B$ is a metric space. $X$ is an $\epsilon$-*Poincaré complex of formal dimension $n$* over $B$ if there exist a subdivision of $X$ such that simplices have diameter $\ll \epsilon$ in $B$ and an $n$-cycle $y$ in the simplicial chains of $X$ so that $\cap y \colon C^{\sharp}(X) \to C_{n-\sharp}(X)$ is an $\epsilon$-chain homotopy equivalence in the sense that $\cap y$ and the chain homotopies have the property that the image of each generator $\sigma$ only involves generators whose images under $p$ are within an $\epsilon$-neighborhood of $p(\sigma)$ in $B$.

The next definition encodes the fact that the local fundamental group of $X$ is trivial from the viewpoint of the control space $B$.

**Definition 3.3.** A map $p: X \to B$ is $UV^1$ if for any $\epsilon > 0$, and any polyhedral pair $(P, Q)$ with $\dim(P) \leq 2$, and any maps $\alpha_0: Q \to X$ and $\beta: P \to B$ such that $p \circ \alpha_0 = \beta|_Q$,

$$
\begin{array}{ccc}
Q & \xrightarrow{\alpha_0} & X \\
{\scriptstyle i}\downarrow & {\scriptstyle \alpha}\nearrow & \downarrow{\scriptstyle p} \\
P & \xrightarrow[\beta]{} & B
\end{array}
$$

there is a map $\alpha: P \to X$ extending $\alpha_0$ so that $d(p \circ \alpha, \beta) < \epsilon$.

*Remark* . When both $X$ and $B$ are polyhedra and $p$ is $PL$, this is the same as requiring that $p^{-1}(b)$ be simply connected, for every $b \in B$.

**Definition 3.4.** Let $p: X \to B$ be an $\epsilon$-Poincaré complex over the metric space $B$, where $p$ is $UV^1$. An $\epsilon$-surgery problem over $p: X \to B$ is a degree-one normal map

$$
\begin{array}{ccc}
\nu_M & \xrightarrow{F} & \xi \\
\downarrow & & \downarrow \\
M & \xrightarrow{f} & X
\end{array}
$$

where $\xi$ is a bundle over $X$, $\nu_M$ denotes the stable normal bundle of $M$, and $F$ is a bundle map covering $f$.

**Theorem 3.5.** *Let $B$ be a compact metric ENR and $n \geq 5$. There exist an $\epsilon_0 > 0$ and a function $T: (0, \epsilon_0] \to (0, \infty)$ satisfying $T(t) \geq t$ and $\lim_{t \to 0} T(t) = 0$, such that for any $\epsilon$, $0 < \epsilon \leq \epsilon_0$, if $f: M \to X$ is an $\epsilon$-surgery problem with respect to the $UV^1$ map $p: X \to B$, associated to the normal bordism class of $f$, there is an obstruction $\sigma_f \in H_n(B; \mathbb{L})$ which vanishes if and only if $f$ is normally bordant to a $T(\epsilon)$-equivalence over $B$. Here, $H_n(B; \mathbb{L})$ denotes the nth generalized homology group of $B$ with coefficients in the simply-connected periodic surgery spectrum.*

Theorem 3.5 requires that $X$ be a polyhedron. Nonetheless, if $X$ is an ENR homology $n$-manifold, a normal map $f: M \to X$ has a well-defined controlled surgery obstruction over $B$. Let $U$ be a mapping cylinder neighborhood of $X$ in a large euclidean space $\mathbb{R}^N$ with projection $\pi: U \to X$ [55, 43]. For any $\epsilon > 0$, $U$ is an $\epsilon$-Poincaré complex of formal dimension $n$ over $B$ under the control map $p \circ \pi: U \to B$ [11]. Hence, the composition $M \xrightarrow{f} X \subseteq U$ can be viewed as an $\epsilon$-surgery problem $f'$ over $B$. By Theorem 3.5, for each $\epsilon$, $0 < \epsilon \leq \epsilon_0$, $f'$ has a well-defined $T(\epsilon)$-surgery obstruction $\sigma_{f'} \in H_n(B; \mathbb{L})$ over $B$, where $\lim_{t \to 0} T(t) = 0$. The *controlled surgery obstruction* of $f$ is defined by $\sigma_f = \sigma_{f'}$.

**Theorem 3.6.** *Let $p\colon X \to B$ be a $UV^1$ map, where $X$ is a compact ENR homology $n$-manifold, $n \geq 5$. The controlled surgery obstruction $\sigma_f \in H_n(B; \mathbb{L})$ of $f\colon M \to X$ is well-defined, and $\sigma_f$ vanishes if and only if, for any $\epsilon > 0$, $f$ is normally bordant to an $\epsilon$-homotopy equivalence over $B$.*

**Definition 3.7.** Let $X$ be a compact ENR homology manifold, and let $p\colon X \to B$ be a control map. An $\epsilon$-structure on $p\colon X \to B$ is an $\epsilon$-homotopy equivalence $f\colon M \to X$ over $B$, where $M$ is a closed manifold. Two structures $f_i\colon M_i \to X$, $i \in \{1,2\}$, are equivalent if there is a homeomorphism $h\colon M_1 \to M_2$ such that $f_1$ and $f_2 \circ h$ are $\epsilon$-homotopic over $B$. The collection of equivalence classes of $\epsilon$-structures is denoted by $S_\epsilon\left(\begin{smallmatrix} X \\ \downarrow \\ B \end{smallmatrix}\right)$.

Given a Poincaré space $X$ of formal dimension $n$, let $\mathcal{N}_n(X)$ denote the collection of normal bordism classes of degree-one normal maps to $X$.

**Theorem 3.8.** *Let $X$ be a compact ENR homology $n$-manifold, $n \geq 5$, and let $p\colon X \to B$ be a $UV^1$ control map, where $B$ is a compact metric ENR. There exist an $\epsilon_0 > 0$ and a function $T\colon (0, \epsilon_0] \to (0, \infty)$ that depends only on $n$ and $B$ such that $T(t) \geq t$, $\lim_{t \to 0} T(t) = 0$, and if $S_{\epsilon_0}\left(\begin{smallmatrix} X \\ \downarrow \\ B \end{smallmatrix}\right) \neq \emptyset$, there is an exact sequence*

$$\cdots \longrightarrow H_{n+1}(B; \mathbb{L}) \longrightarrow S_\epsilon\left(\begin{smallmatrix} X \\ \downarrow \\ B \end{smallmatrix}\right) \longrightarrow \mathcal{N}_n(X) \longrightarrow H_n(B; \mathbb{L}),$$

*for each $0 < \epsilon \leq \epsilon_0$, where*

$$S_\epsilon\left(\begin{smallmatrix} X \\ \downarrow \\ B \end{smallmatrix}\right) = im\left( S_\epsilon\left(\begin{smallmatrix} X \\ \downarrow \\ B \end{smallmatrix}\right) \longrightarrow S_{T(\epsilon)}\left(\begin{smallmatrix} X \\ \downarrow \\ B \end{smallmatrix}\right) \right).$$

*Moreover, $S_\epsilon\left(\begin{smallmatrix} X \\ \downarrow \\ B \end{smallmatrix}\right) \cong S_{\epsilon_0}\left(\begin{smallmatrix} X \\ \downarrow \\ B \end{smallmatrix}\right)$ if $\epsilon \leq \epsilon_0$.*

## 4. THE RESOLUTION OBSTRUCTION

In this section we discuss various geometric aspects of Quinn's work on the resolution conjecture that lead to the invariant $I(X)$, adopting a variant of his original formulation. For simplicity, we assume that $X$ is a compact oriented ENR homology $n$-manifold, $n \geq 5$.

Resolutions are fine homotopy equivalences that desingularize homology manifolds. A map $f\colon M \to X$ is a resolution if and only if $f|_{f^{-1}(U)}\colon f^{-1}(U) \to U$ is a homotopy equivalence, for every open set $U \subseteq X$ [36]. This implies that $f\colon M \to X$ is a resolution if and only if $f$ is a controlled homotopy equivalence with the identity map of $X$ as control map.

In [27], Ferry and Pedersen showed that there is a degree-one normal map $f\colon M \to X$. Our goal is to understand the obstructions to finding a controlled homotopy equivalence over $X$ within the normal bordism class of $f$. Notice that if an obstruction is encountered, we can try to eliminate it by changing the normal map to $X$. Therefore, in trying to construct resolutions, it is more natural to consider the collection $\mathcal{N}_n(X)$ of all normal bordism classes of $n$-dimensional degree-one normal maps to $X$. Recall that there is a one-to-one correspondence between $\mathcal{N}_n(X)$ and (stable) topological reductions of the Spivak normal fibration $\nu_X$ of $X$. A topological reduction of $\nu_X$ corresponds to a fiber homotopy class of lifts to $BTop$ of the map $\nu_X \colon X \to BG$ that classifies the Spivak fibration of $X$.

Any two reductions differ by the action of a unique element of $[X, G/Top]$, where $G/Top$ is the homotopy fiber of $BTop \to BG$. Hence, $[X, G/Top]$ acts freely and transitively on $\mathcal{N}_n(X)$, since $\mathcal{N}_n(X) \neq \emptyset$. When $X$ is a manifold, this action induces a canonical identification $\eta \colon \mathcal{N}_n(X) \to [X, G/Top]$ since there is a preferred element of $\mathcal{N}_n(X)$, namely, the bordism class of the identity map of $X$, which corresponds to the (stable) $Top$ reduction of $\nu_X$ given by the normal bundle of an embedding of $X$ in a large euclidean space. We refer to $\eta(f) \in [X, G/Top]$ as the *normal invariant* of $f$.

To motivate our discussion, we first consider the case where $X$ is a closed manifold, although this assumption trivializes the problem from the standpoint of existence of resolutions. Siebenmann's CE-approximation theorem states that cell-like maps of closed $n$-manifolds, $n \geq 5$, can be approximated by homeomorphisms [49]. Hence, if $X$ is a manifold, we are to consider the obstructions to finding a homeomorphism in the normal bordism class of $f$. Such homeomorphism exists if and only if the normal invariant $\eta(f)$ vanishes [53].

Sullivan's description of the homotopy type type of $G/Top$ [52] shows that, rationally, the normal invariant is detected by the difference of the rational $\mathcal{L}$-classes of $M$ and $X$, respectively. Let

$$L_X = 1 + \ell_1 + \ell_2 + \ldots \in H^{4*}(X; \mathbb{Q})$$

be the total $\mathcal{L}$-class of $X^n$. The $i$th class $\ell_i \in H^{4i}(X; \mathbb{Q})$ is determined (after stabilizing $X$ by crossing it, say, with a sphere if $4i \geq \frac{n-1}{2}$) by the signature of $4i$-dimensional submanifolds $N^{4i} \subseteq X$ with framed normal bundles. Hence, up to finite indeterminacies, the normal invariant of $f$ is detected by the difference of the signatures of these characteristic submanifolds and their transverse inverse images. Notice that when $X$ is a manifold, we can disregard 0-dimensional submanifolds, since the transverse inverse image of a point under a degree-one map can be assumed to be a point.

Carrying out this type of program for studying the existence of resolutions involves, among other things, defining (at least implicitly) characteristic classes for ENR homology manifolds. This has been done in [17], but following [10, 27] we take a controlled-surgery approach to the problem and argue that the Spivak normal fibration of an ENR homology manifold has a canonical *Top* reduction.

By Theorem 3.6, associated to a normal map $f: M \to X$ there is a controlled surgery obstruction $\sigma_f \in H_n(X; \mathbb{L}) \cong [X, G/Top \times \mathbb{Z}]$ such that $\sigma_f = 0$ if and only if, for any $\epsilon > 0$, $f$ is normally bordant to an $\epsilon$-homotopy equivalence. Under the natural (free) action of $[X, G/Top]$ on $H_n(X; \mathbb{L}) \cong [X, G/Top \times \mathbb{Z}]$, controlled surgery obstructions induce a $[X, G/Top]$-equivariant injection

$$\mathcal{N}_n(X) \longrightarrow H_n(X; \mathbb{L}).$$

Let $f: M \to X$ be a normal map. Letting $[X, G/Top]$ act on $f$, we can assume that the image of $\sigma_f$ under the projection $H_n(X; \mathbb{L}) \cong [X, G/Top \times \mathbb{Z}] \to [X, G/Top]$ vanishes, so that $\sigma_f \in [X, \mathbb{Z}] \subseteq [X, G/Top \times \mathbb{Z}]$. Hence, if $X$ is connected, $\sigma_f$ is an integer. The local index of $X$ is defined by

$$I(X) = 8\sigma_f + 1 \in 8\mathbb{Z} + 1.$$

Since the $\mathbb{Z}$-component of $\sigma_f$ is persistent under the action of $[X, G/Top]$, this is the closest we can get to a resolution. This construction yields a preferred normal bordism class of normal maps to $X$ (and therefore, a canonical *Top* reduction of $\nu_X$) and induces an identification $\eta: \mathcal{N}_n(X) \to [X, G/Top]$. Rationally, the action of $[X, G/Top]$ on $f$ can be interpreted as the analogue of changing the normal map $f$ so that the signatures of the transverse preimages $f^{-1}(N)$ and $N$ be the same for (stable) framed submanifolds $N^{4i} \subseteq X$, $i > 0$, when $X$ is a manifold. This suggests that the $\mathbb{Z}$-component of $\sigma_f$ be interpreted as a difference of signatures in dimension zero and that $I(X)$ be viewed as the 0-dimensional $\mathcal{L}$-class of $X$. This is the approach taken by Quinn in [44, 46], which explains the local nature of the invariant.

If $I(X) = 1$, there is a normal map $f: M \to X$ such that $\sigma_f = 0$. Let $\epsilon_i \to 0$ be a decreasing sequence. Theorem 3.6 implies that, for each $i > 0$, there is an $\epsilon_i$-structure $f_i: M_i \to X$, so that $\mathcal{S}_{\epsilon_i} \neq \emptyset$. Under the identification $\mathcal{N}_n(X) \cong [X, G/Top] \cong H_n(X; G/Top)$, the controlled surgery sequence of Theorem 3.8 can be expressed as

$$H_{n+1}(X; G/Top) \to H_{n+1}(X; \mathbb{L}) \to$$
$$\mathcal{S}_\epsilon \begin{pmatrix} X \\ \downarrow \\ X \end{pmatrix} \to H_n(X; G/Top) \to H_n(X; \mathbb{L}).$$

It follows from the Atiyah-Hirzebruch spectral sequence that $H_i(X; G/Top)$ $\to H_i(X; \mathbb{L})$ is injective if $i = n$, and an isomorphism if $i = n+1$. This shows that $\mathcal{S}_{\epsilon_i} \begin{pmatrix} X \\ \downarrow \\ X \end{pmatrix} = 0$, if $\epsilon_i$ is small enough. Then, viewing $f_i$ and $f_{i+1}$ as equivalent $\epsilon_i$-structures on $X$, we obtain homeomorphisms $h_i \colon M_i \to M_{i+1}$ such that $f_{i+1} \circ h_i$ and $f_i$ are $T(\epsilon_i)$-homotopic over $X$. Consider the sequence

$$f_i^* = f_i \circ h_{i-1} \circ \cdots \circ h_1 \colon M_1 \to X.$$

For each $i > 0$, $f_i^*$ is an $\epsilon_i$-equivalence over $X$ and

$$d(f_{i+1}^*, f_i^*) = d(f_{i+1} \circ h_i, f_i) < T(\epsilon_i).$$

If $\epsilon_i > 0$ is so small that $\sum T(\epsilon_i) < \infty$, the sequence $\{f_i^*\}$ converges to a resolution of $X$.

## 5. PERIODICITY IN MANIFOLD THEORY

A beautiful periodicity phenomenon emerges from the surgery classification of compact manifolds. All essential elements in the theory exhibit an almost 4-periodic behavior with respect to the dimension $n$. Siebenmann periodicity is the most geometric form of this phenomenon.

**Definition 5.1.** Let $X$ be a compact manifold. A *structure* on $X$ is a simple homotopy equivalence $f \colon M \to X$ that restricts to a homeomorphism $f \colon \partial M \to \partial X$ on the boundary, where $M$ is a topological manifold. The structures $f_i \colon M_i \to X$, $i \in \{1, 2\}$, are equivalent if there is a homeomorphism $h \colon M_1 \to M_2$ making the diagram

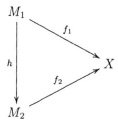

homotopy commute rel $(\partial)$. The *structure set* $\mathcal{S}(X)$ is the collection of all equivalence classes of structures on $X$.

The following theorem is proved in [35], with a correction by Nicas in [41].

**Theorem 5.2** (Siebenmann periodicity). *If $X^n$ is a compact connected manifold of dimension $\geq 5$, there is an exact sequence*

$$0 \longrightarrow \mathcal{S}(X) \overset{\wp}{\longrightarrow} \mathcal{S}(X \times D^4) \overset{\sigma}{\longrightarrow} \mathbb{Z}.$$

*Moreover, $\wp$ is an isomorphism if $\partial X \neq \emptyset$.*

The map $\sigma$ associates to a structure $f \colon W \to X \times D^4$, the signature of the transverse inverse image of $\{*\} \times D^4$. Siebenmann's construction of the map $\wp$ was indirect. In [18], Cappell and Weinberger describe a geometric realization of the periodicity map $\wp \colon \mathcal{S}(M) \to \mathcal{S}(M \times D^4)$ using the Casson-Sullivan embedding theorem and branched circle fibrations.

The structure set $\mathcal{S}(S^n)$ of the $n$-sphere, $n \geq 4$, contains a single element, by the generalized Poincaré conjecture. However, it can be shown that $\mathcal{S}(S^n \times D^4) \cong \mathbb{Z}$, so that periodicity does fail for closed manifolds. This suggests that there may be "unidentified manifolds" that yield a fully periodic theory of manifolds.

Quinn's work on the resolution problem shows that the local index that obstructs the existence of resolutions and the $\mathbb{Z}$-factor that prevents periodicity from holding for closed manifolds have the same geometric nature, a fact to our knowledge first observed by Cappell. This indicates that the non-resolvable homology manifolds in the recognition problem are the same as the missing manifolds in Siebenmann periodicity, and creates an interesting link between the classification theory of manifolds and the resolution conjecture.

## 6. Classification of ENR homology manifolds

The first examples of nonresolvable ENR homology manifolds were produced in 1992 by Bryant, Ferry, Mio and Weinberger using techniques of controlled topology [9]. In this section, we outline the construction of examples modeled on simply-connected $PL$ manifolds, where the central ideas are already present. For a more general discussion, we refer the reader to [10].

**Theorem 6.1** (BFMW). *Let $M^n$ be a simply-connected closed PL manifold, $n \geq 6$. Given $\sigma \in 8\mathbb{Z} + 1$, there exists a closed ENR homology $n$-manifold $X$ homotopy equivalent to $M$ such that $I(X) = \sigma$.*

Variants of the methods employed in the construction yield an $s$-cobordism classification of ENR homology $n$-manifolds within a fixed simple homotopy type and an identification of the (simple) types realized by closed homology manifolds of dimension $\geq 6$ in terms of Ranicki's total surgery obstruction [47]. We only state the classification theorem [10, 11], whose proof requires relative versions of the arguments to be presented.

**Definition 6.2.** Let $M^n$ be a compact manifold. A *homology manifold structure* on $M$ is a simple homotopy equivalence $f: (X, \partial X) \to (M, \partial M)$, where $X$ is an ENR homology $n$-manifold with the DDP and $f$ restricts to a homeomorphism on the boundary. The *homology structure set* $\mathcal{S}^H(M)$ of $M$ is the set of all $s$-cobordism classes of homology manifold structures on $M$.

*Remark* . We consider $s$-cobordism classes of structures since the validity of the $s$-cobordism theorem in this category is still an open problem.

Since a structure $f: X \to M$ restricts to a homeomorphism on the boundary, if $\partial M \neq \emptyset$ we have that $\partial X$ is a manifold. Adding a collar $\partial X \times I$ to $X$ gives a homology manifold $Y$ containing manifold points. Since Quinn's index is local, $I(X) = I(Y) = 1$ and $X$ is a manifold. By the manifold $s$-cobordism theorem, $\mathcal{S}^H(M) = \mathcal{S}(M)$, so that $\mathcal{S}^H(M)$ consists entirely of manifold structures if $\partial M \neq \emptyset$.

**Theorem 6.3** (BFMW). *If $M^n$ is a closed manifold, $n \geq 6$, there is an exact sequence*

$$\ldots \to H_{n+1}(M; \mathbb{L}) \to L_{n+1}(\mathbb{Z}\pi_1(M)) \to$$
$$\mathcal{S}^H(M) \to H_n(M; \mathbb{L}) \xrightarrow{A} L_n(\mathbb{Z}\pi_1(M)),$$

*where $L_i$ is the $i$th Wall surgery obstruction group of the group $\pi_1(M)$, $\mathbb{L}$ is the simply-connected periodic surgery spectrum, and $A$ denotes the assembly map.*

This classification implies that homology manifold structures produce a fully periodic manifold theory.

**Corollary 6.4.** *The Siebenmann periodicity map $\wp: \mathcal{S}^H(M) \to \mathcal{S}^H(M \times D^4)$ is an isomorphism, if $M^n$ is a compact manifold, $n \geq 6$.*

*Sketch of the proof of Theorem 6.1.* We perform a sequence of cut-paste constructions on the manifold $M$ to obtain a sequence $\{X_i\}$ of Poincaré complexes that converges (in a large euclidean space) to an ENR homology manifold $X$ with the required properties. There are two properties of the sequence that must be carefully monitored during the construction:

(i) *Controlled Poincaré duality.*

As pointed out earlier, homology manifolds satisfy a local form of Poincaré duality. Therefore, the approximating complexes are constructed so that $X_i$, $i \geq 2$, are Poincaré complexes with ever finer control over $X_{i-1}$.

(ii) *Convergence.*

We need the limit space $X$ to inherit the fine Poincaré duality and the local contractibility of the complexes $X_i$. This is achieved by connecting successive stages of the construction via maps $p_i: X_{i+1} \to X_i$ which are fine homotopy equivalences over $X_{i-1}$.

If the maps $p_i: X_{i+1} \to X_i$ are fine equivalences, they are, in particular, finely 2-connected. Control improvement theorems imply that once enough control has been obtained at the fundamental group level, arbitrarily fine control can be achieved under a small deformation [1, 25]. Hence, throughout the construction we require that all maps be $UV^1$ (see Definition 3.3) so that the construction of controlled homotopy equivalences can be reduced to homological estimates via appropriate forms of the Hurewicz theorem [43].

*Constructing $X_1$.* Gluing manifolds by a homotopy equivalence of their boundaries, we obtain Poincaré spaces. We use a controlled version of this procedure to construct $\epsilon$-Poincaré spaces. Let $C_1$ be a regular neighborhood of the 2-skeleton of a triangulation of $M$, $D_1$ be the closure of the complement of $C_1$ in $M$, and $N_1 = \partial C_1$.

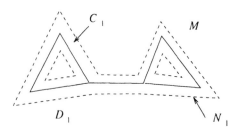

FIGURE 6.1

If the triangulation is fine enough, there is a small deformation of the inclusion $N_1 \hookrightarrow M$ to a $UV^1$ map $q: N_1 \to M$. A controlled analogue of Wall's realization theorem (Theorem 5.8 of [53]) applied to the control map $q: N_1 \to M$ gives a degree-one normal map $F_\sigma: (V, N_1, N_1') \to (N_1 \times I, N_1 \times \{0\}, N_1 \times \{1\})$ satisfying:

(a) $F_\sigma|_{N_1} = id$.
(b) $f_\sigma = F_\sigma|_{N_1'}$ is a fine homotopy equivalence over $M$.
(c) The controlled surgery obstruction of $F_\sigma$ rel $\partial$ over $M$ is $\sigma \in H_n(M; \mathbb{L})$.

Since the image of $\sigma$ under the surgery forget-control map $H_n(M; \mathbb{L}) \to L_n(e)$ is trivial, doing surgery on $V$ we can assume that $V = N_1 \times I$, and in particular that $N_1' = N_1$. Using $f_\sigma: N_1 \to N_1$ as gluing map, form the complex $X_1 = D_1 \cup_{f_\sigma} C_1$ as indicated in Figure 6.2.

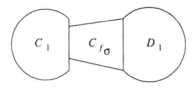

FIGURE 6.2

Here, $C_{f_\sigma}$ is the mapping cylinder of $f_\sigma$. The construction of $X_1$ allows us to extend the control map $q\colon N_1 \to M$ to a $UV^1$ homotopy equivalence $p_0\colon X_1 \to M$ such that the restrictions of $p_0$ to $C_1$ and $D_1$ are close to the respective inclusions. In a large euclidean space $\mathbb{R}^L$, gently perturb $p_0$ to an embedding. This defines a metric on $X_1$ and completes the first stage of the construction. Notice that the control on the Poincaré duality of $X_1$ over $M$ is only constrained by the magnitude of the controlled equivalence $f_\sigma\colon N_1 \to N_1$, which can be chosen to be arbitrarily fine.

*Constructing $X_2$.* Starting with a $UV^1$ homotopy equivalence $M \to X_1$, we perform a similar cut-paste construction on $M$ along the boundary $N_2$ of a regular neighborhood $C_2$ of the 2-skeleton of a much finer triangulation of $M$. As in the construction of $p_0\colon X_1 \to M$, we obtain a $UV^1$ homotopy equivalence $p_1'\colon X_2' \to X_1$. The difference in this step is that we modify $p_1'$ to a fine equivalence over $M$, with a view toward fast convergence. By construction, the controlled surgery obstruction of $p_1'$ with respect to the control map $p_0\colon X_1 \to M$ is zero. Surgery on $X_2'$ can be done as in the manifold case, by moving spheres off of the 2-dimensional spine of $C_2$ and pushing them away from the singular set under small deformations. This gives a fine $UV^1$ homotopy equivalence $p_1\colon X_2 \to X_1$ over $M$. Control on $p_1$ is only limited by the Poincaré duality of $X_1$ over $M$, since $X_2$ can be constructed to be a much finer Poincaré space than $X_1$.

Mildly perturb $p_1\colon X_2 \to X_1$ to an embedding of $X_2$ into a small regular neighborhood $V_1$ of $X_1 \subseteq \mathbb{R}^L$. By the thin $h$-cobordism theorem [43], we can assume that the region between $V_1$ and a small regular neighborhood $V_2$ of $X_2$ in $V_1$ admits a fine product structure over $M$.

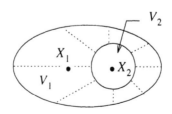

FIGURE 6.3

Iterating the construction, we obtain fine homotopy equivalences $p_i: X_{i+1} \to X_i$ over $X_{i-1}$. The control on $p_{i+1}$ depends only on the Poincaré duality of $X_i$ over $X_{i-1}$ which can be chosen to be so fine that the region between small regular neighborhoods $V_i$ and $V_{i+1}$ of $X_i$ and $X_{i+1}$, respectively, admits a controlled product structure over $X_{i-1}$. As before, the Poincaré duality of $X_{i+1}$ over $X_i$ can be assumed to be as fine as necessary in the next stage of the construction.

Let $X = \bigcap\limits_{i=1}^{\infty} V_i$ be the intersection of the nested sequence $V_i$ of regular neighborhoods of $X_i$. Concatenating the product structures on $V_i \setminus \text{int} (V_{i+1})$, $i \geq 1$, gives a deformation retraction $p: V_1 \to X$, provided that the product structures are sufficiently fine. This shows that $X$ is an ENR. The map $p$ actually defines a mapping cylinder structure on the neighborhood $V_1$ of $X$.

In order to show that $X$ is a homology manifold, we first reinterpret controlled Poincaré duality in terms of lifting properties, via a controlled analogue of Spivak's thesis [51]. Let $\rho = p|_{\partial V_1}: \partial V_1 \to X$, and let $\rho_i: \partial V_i \to X_i$ denote the restriction of the regular neighborhood projection $V_i \to X_i$ to $\partial V_i$. Proposition 4.5 of [10] implies that given $\delta > 0$, $\rho_i$ has the $\delta$-homotopy lifting property, provided that $i$ is large enough. Hence, the projection $\partial V_1 \to X_i$ obtained from the product structure connecting $\partial V_1$ to $\partial V_i$ also has the $\delta$-homotopy lifting property, for $i$ large enough. Since the homotopy equivalences $X_i \to X$ become finer as $i \to \infty$, it follows that $\rho: \partial V_1 \to X$ has the $\epsilon$-homotopy lifting property, for every $\epsilon > 0$, i.e., $\rho$ is a manifold approximate fibration over $X$. This implies that $X$ is a homology manifold [22].

The approximate homology manifolds $X_i$ were constructed to carry the resolution obstruction $\sigma$, in the sense that there is a normal map $\phi_i: M \to X_i$ with controlled surgery obstruction $\sigma \in H_n(X_{i-1}; \mathbb{L})$. Since the sequence $\{X_i\}$ converges to $X$, a change of control space argument implies that $I(X) = \sigma$. This concludes the construction. $\quad\square$

## 7. Concluding remarks

The existence of nonresolvable ENR homology manifolds raises numerous questions about the geometric topology of these spaces. In [10], we summarized several of these questions in a conjecture.

**Conjecture** (BFMW). *There exist spaces $\mathbb{R}_k^4$, $k \in \mathbb{Z}$, such that every connected DDP homology $n$-manifold $X$ with local index $I(X) = 8k + 1$, $n \geq 5$, is locally homeomorphic to $\mathbb{R}_k^4 \times \mathbb{R}^{n-4}$. ENR homology $n$-manifolds with the DDP are topologically homogeneous, the $s$-cobordism theorem holds for these spaces, and structures on closed DDP homology manifolds $X^n$ are* *classified (up to homeomorphisms) by a surgery exact sequence*

$$\cdots \to H_{n+1}(X; \mathbb{L}) \to L_{n+1}(\mathbb{Z}\pi_1(X)) \to$$
$$\mathbb{S}^H(X) \to H_n(X; \mathbb{L}) \to L_n(\mathbb{Z}\pi_1(X)).$$

*Remark* . This exact sequence has been established in [10] up to $s$-cobordisms of homology manifolds.

Recall that a topological space $X$ is homogeneous if for any pair of points $a, b \in X$, there is a homeomorphism $h \colon X \to X$ such that $h(a) = b$. The topological homogeneity of DDP homology manifolds seems to be a problem of fundamental importance. A positive solution would strongly support the contention that DDP homology manifolds form the natural class in which to develop manifold theory in higher dimensions and would also settle the long standing question *"Are homogeneous ENRs manifolds?"*, proposed by Bing and Borsuk in [5].

The validity of Edward's CE-approximation theorem in this class of spaces is a recurring theme in the study of the topology of homology manifolds. Can a cell-like map $f \colon X \to Y$ of DDP homology $n$-manifolds, $n \geq 5$, be approximated by homeomorphisms? Homogeneity and many other questions can be reduced to (variants of) this approximation problem.

Homology manifolds are also related to important rigidity questions. For example, the existence of a nonresolvable closed aspherical ENR homology $n$-manifold $X$, $n \geq 5$, would imply that either the integral Novikov conjecture or the Poincaré duality group conjecture are false for the group $\pi_1(X)$. Indeed, if the assembly map

$$\mathcal{A} \colon H_*(X; \mathbb{L}) \longrightarrow L_*(\mathbb{Z}\pi_1(X))$$

is an isomorphism, the homology-manifold structure set $\mathbb{S}^H(X)$ contains a single element. Therefore, if $M$ is a closed manifold homotopy equivalent to $X$, then $X$ is $s$-cobordant to $M$. This implies that $I(X) = 1$, contradicting the assumption that $X$ is not resolvable. Hence, $\pi_1(X)$ would be a Poincaré duality group which is not the fundamental group of any closed aspherical manifold [27].

Can a map of DDP homology manifolds be made transverse to a codimension $q$ tamely embedded homology manifold? In her thesis, Johnston established map transversality (up to $s$-cobordisms) in the case the homology submanifolds have bundle neighborhoods [32] (see also [33]). Although the existence of such neighborhoods is, in general, obstructed (since indices satisfy a product formula), it seems plausible that there exist an appropriate notion of normal structure for these subobjects that yield general map transversality. When the ambient spaces are topological manifolds, $q \geq 3$, and the homology submanifolds have dimension $\geq 5$, mapping cylinders of spherical manifold approximate fibrations appear to provide the right structures [43]. This is consistent with the fact that, for manifolds, Marin's topological transversality is equivalent to the neighborhood transversality of Rourke and Sanderson [39]. In [12], approximate fibrations are used to extend to homology manifolds the classification of tame codimension $q$ manifold neighborhoods of topological manifolds, $q \geq 3$, obtained by Rourke and Sanderson [48]. This classification is used to prove various embedding theorems in codimensions $\geq 3$.

Smith theory [50] and the work of Cappell and Weinberger on propagation of group actions [19] indicate that nonstandard homology manifolds may occur as fixed sets of semifree periodic dynamical systems on manifolds. Homology manifolds also arise as limits of sequences of riemannian manifolds in Gromov-Hausdorff space [30]. Results of Bestvina [2] show that boundaries of Poincaré duality groups are homology manifolds, further suggesting that exotic ENR homology manifolds may become natural geometric models for various phenomena.

*Acknowledgements.* I wish to thank Andrew Ranicki for many comments and suggestions, and John Bryant for numerous discussions during several years of collaboration.

REFERENCES

[1] M. Bestvina, *Characterizing k-dimensional Menger compacta*, Memoirs Amer. Math. Soc. **71 380** (1988).

[2] _____, *Local homology properties of boundaries of groups*, Michigan Math. J. **43** (1996), 123–139.

[3] R. H. Bing, *The Kline sphere characterization problem*, Bull. Amer. Math. Soc. **52** (1946), 644–653.

[4] _____, *A homeomorphism between the 3-sphere and the sum of two solid horned spheres*, Ann. of Math. **56** (1952), 354–362.

[5] R. H. Bing and K. Borsuk, *Some remarks concerning topologically homogeneous spaces*, Ann. of Math. **81** (1965), 100–111.

[6] K. Borsuk, *On some metrizations of the hyperspace of compact sets*, Fund. Math. **41** (1955), 168–201.

[7] G. E. Bredon, *Introduction to compact transformation groups*, Academic Press, 1972.

[8] J. Bryant, *General position theorems for generalized manifolds*, Proc. Amer. Math. Soc. **98** (1986), 667–670.

[9] J. Bryant, S. Ferry, W. Mio and S. Weinberger, *Topology of homology manifolds*, Bull. Amer. Math. Soc. **28** (1993), 324–328.

[10] _____, *Topology of homology manifolds*, Ann. of Math. **143** (1996), 435–467.

[11] _____, in preparation.

[12] J. Bryant and W. Mio, *Embeddings of homology manifolds in codimension $\geq$ 3*, Topology **38** (1999), 811–821.

[13] J. W. Cannon, *The recognition problem: what is a topological manifold?*, Bull. Amer. Math. **84** (1978), 832–866.

[14] _____, *Shrinking cell-like decompositions of manifolds in codimension three*, Ann. of Math. **110** (1979), 83–112.

[15] ——, *The characterization of topological manifolds of dimension* $n \geq 5$, Proc. Internat. Cong. Mathematicians, Helsinki (1980), 449–454.

[16] J. Cannon, J. Bryant and R. C. Lacher, *The structure of generalized manifolds having nonmanifold set of trivial dimension*, in Geometric Topology, ed. J. C. Cantrell, 261–300, Academic Press, 1979.

[17] S. Cappell, J. Shaneson and S. Weinberger, *Topological characteristic classes for group actions on Witt spaces*, C. R. Acad. Sci. Paris **313** (1991), 293–295.

[18] S. Cappell and S. Weinberger, *A geometric interpretation of Siebenmann's periodicity phenomenon*, Proc. 1985 Georgia Conference on Topology and Geometry, Dekker (1985), 47–52.

[19] S. Cappell and S. Weinberger, *Replacement theorems for fixed sets and normal representations for PL group actions on manifolds*, Prospects in Topology, Annals of Maths. Study 138, Princeton, 1995, 67–109.

[20] E. Čech, *Höherdimensionalen Homotopiegruppen*, Verhandl. des Intern. Math. Kongresses, Zürich 1932, Bd. 2, 203.

[21] R. J. Daverman, *Decompositions of manifolds*, Academic Press, New York, 1986.

[22] R. J. Daverman and L. Husch, *Decompositions and approximate fibrations*, Michigan Math. J. **31** (1984), 197–214.

[23] J. Dieudonné, *A history of algebraic and differential topology 1900-1960*, Birkhäuser, Boston, 1989.

[24] R. D. Edwards, *The topology of manifolds and cell-like maps*, Proc. Internat. Cong. Mathematicians, Helsinki (1980), 111–127.

[25] S. Ferry, *Mapping manifolds to polyhedra*, preprint.

[26] ——, *Controlled topology and the characterization of topological manifolds*, NSF-CBMS Lecture Notes (to appear).

[27] S. Ferry and E. K. Pedersen, *Epsilon surgery theory*, Proceedings 1993 Oberwolfach Conference on the Novikov Conjectures, Rigidity and Index Theorems Vol. 2, LMS Lecture Notes 227, Cambridge (1995), 167–226.

[28] ——, *Squeezing structures*, preprint.

[29] D. E. Galewski and R. Stern, *Classification of simplicial triangulations of topological manifolds*, Ann. of Math. **111** (1980), 1–34.

[30] K. Grove, P. Petersen and J. Wu, *Geometric finiteness theorems in controlled topology*, Invent. Math. **99** (1990), 205–213.

[31] W. Hurewicz and H. Wallman, *Dimension theory*, Princeton University Press, Princeton, NJ, 1941.

[32] H. Johnston, *Transversality for homology manifolds*, Topology, **38** (1999), 673–697.

[33] H. Johnston and A. Ranicki, *Homology manifold bordism*, preprint.

[34] M. Kervaire, *Smooth homology spheres and their fundamental groups*, Trans. Amer. Math. Soc. **114** (1969), 67–72.

[35] R. Kirby and L. C. Siebenmann, *Foundational essays on topological manifolds, smoothings, and triangulations*, Princeton University Press, Princeton, NJ, 1977.

[36] R. C. Lacher, *Cell-like mappings and their generalizations*, Bull. Amer. Math. Soc. **83** (1977), 495–552.

[37] S. Lefschetz, *Topology*, Amer. Math. Soc. Coll. Publ. 12, Providence, RI, 1930.

[38] ——, *On generalized manifolds*, Amer. J. Maths. **55** (1933), 469–504.

[39] A. Marin, *La transversalité topologique*, Ann. of Math. **106** (1977), 269–293.

[40] R. L. Moore, *Concerning uppersemicontinuous collections of compacta*, Trans. Amer. Math. Soc. **27** (1925), 416–428.

[41] A. Nicas, *Induction theorems for groups of manifold structure sets*, Mem. Amer. Math. Soc. **267**, Providence, RI, 1982.

[42] L. Pontrjagin, *Über den algebraischen Inhalt topologischer Dualitätssätze*, Math. Ann. **105** (1931), 165–205.

[43] F. Quinn, *Ends of maps I*, Ann. of Math. **110** (1979), 275–331.

[44] _____ , *Resolutions of homology manifolds, and the topological characterization of manifolds*, Invent. Math. **72** (1983), 267–284.

[45] _____ , *Local algebraic topology*, Notices Amer. Math. Soc. **33** (1986), 895–899.

[46] _____ , *An obstruction to the resolution of homology manifolds*, Michigan Math. J. **301** (1987), 285–292.

[47] A. Ranicki, *Algebraic L-theory and topological manifolds*, Cambridge University Press, 1992.

[48] C. P. Rourke and B. J. Sanderson, *On topological neighbourhoods*, Compositio Math. **22** (1970), 387–424.

[49] L. C. Siebenmann, *Approximating cellular maps by homeomorphisms*, Topology **11** (1973), 271–294.

[50] P. A. Smith, *New results and old problems in finite transformation groups*, Bull. Amer. Math. Soc. **66** (1960), 401–415.

[51] M. Spivak, *Spaces satisfying Poincaré duality*, Topology **6** (1967), 77–102.

[52] D. Sullivan, *Triangulating and smoothing homotopy equivalences and homeomorphisms*, Geometric Topology Seminar Notes, Princeton Univ., 1967. Reprinted in The Hauptvermutung Book, $K$-Monographs in Mathematics 1, Kluwer (1996), 69–103.

[53] C. T. C. Wall, *Surgery on compact manifolds*, Academic Press, 1970.

[54] S. Weinberger, *Nonlocally linear manifolds and orbifolds*, Proc. Internat. Cong. Mathematicians, Zürich (1995), 637–647.

[55] J. E. West, *Mapping Hilbert cube manifolds to ANR's: a solution to a conjecture of Borsuk*, Ann. of Math. **106** (1977), 1–18.

[56] R. L. Wilder, *Topology of manifolds*, Amer. Math. Soc., 1963.

DEPARTMENT OF MATHEMATICS
FLORIDA STATE UNIVERSITY
TALLAHASSEE, FL 32306
*E-mail address*: mio@math.fsu.edu

# A survey of applications of surgery to knot and link theory

## Jerome Levine and Kent E. Orr

### 1. Introduction

Knot and link theory studies how one manifold embeds in another. Given a manifold embedding, one can alter that embedding in a neighborhood of a point by removing this neighborhood and replacing it with an embedded disk pair. In this way traditional knot theory, the study of embeddings of spheres in spheres, impacts the general manifold embedding problem. In dimension one, the manifold embedding problem *is* knot and link theory.

This article attempts a rapid survey of the role of surgery in the development of knot and link theory. Surgery is one of the most powerful tools in dealing with the question "To what extent are manifolds (or manifold embeddings) uniquely determined by their homotopy type?" As we shall see, roughly speaking, knots and links are determined by their homotopy type (more precisely, Poincaré embedding type) in codimension $\geq 3$ and are much more complicated in codimension two. We proceed, largely, from an historical perspective, presenting most of seminal early results in the language and techniques in which they were first discovered. These results in knot theory are among the most significant early applications of surgery theory and contributed to its development. We will emphasize knotted and linked spheres, providing only a brief discussion of more general codimension two embedding questions. In particular, the theory of codimension two embedding, from the standpoint of classifying within a Poincaré embedding type, deserves a long overdue survey paper. The present paper will not fill this void in the literature. Cappell and Shaneson give an excellent introduction to this subject in [CS78].

By no means is this survey comprehensive, and we apologize in advance for the omission of many areas where considerable and important work has been done. For example, we will omit the extensive subject of equivariant knot theory. We will also not include any discussion of the techniques of Dehn surgery that have proven so valuable in the study of three manifolds

Both authors partially supported by a National Science Foundation grant.

and classical knots. Furthermore, we will not touch on the related subject of immersion theory, and barely mention singularity theory. We urge the reader to consult one of the many excellent surveys which have covered the early (before 1977) development of codimension two knot theory in more depth. The articles by Cameron Gordon [Gor77] and Kervaire-Weber [KW77] on, respectively, low-dimensional and high-dimensional knot theory are excellent. A detailed discussion of surgery and embedding theory can be found in Ranicki's books, [Ran81], [Ran98]. On the other hand, we are not aware of any previously existing comprehensive survey of recent developments in link theory.

## 2. CODIMENSION > 2

Perhaps the first use of surgery techniques in knot theory was in the work of Andre Haefliger. In 1961 Haefliger [Hae61] proved a basic theorem which showed that, for appropriately highly connected manifolds, the isotopy classification of embeddings coincided with the homotopy classification of maps, as long as one was in the *metastable* range of dimensions. More specifically Haefliger showed that if $M$ is a compact manifold, then any $q$-connected map $f : M^n \to V^m$ (where superscripts denote dimension) is homotopic to an embedding, if $m \geq 2n - q$, and any two homotopic $q$-connected maps $M^n \to V^m$ are isotopic if $m > 2n - q$. This theorem required, also, the restriction $2m \geq 3(n + 1)$ and $2m > 3(n + 1)$, respectively. In particular, any homotopy $n$-sphere embeds in $S^m$ and any two such embeddings are homotopic as long as $2m > 3(n + 1)$.

His proof proceeded by examining the singular set of a smooth map and eliminating it by handle manipulations — a generalization of Whitney's method for $n$-manifolds in $2n$-space. Meanwhile Zeeman in the $PL$-category, and Stallings in the topological category, had shown, using the technique of *engulfing*, that there were no non-trivial knots as long as $m > n + 2$ [Zee60] [Sta63].

Haefliger, in a seminal paper [Hae62], showed that when $m = \frac{3}{2}(n + 1)$, the analogous smooth result was already false. Here was the first real use of surgery to study embedding problems. In this paper Haefliger developed the technique of *ambient* surgery, i.e. surgery on embedded manifolds, and used this technique to give a classification of knotted $(4k - 1)$-spheres in $6k$-space (which was, shortly after, extended to a classification of $(2k - 1)$-spheres in $3k$-space). He first observed that the set $\Theta^{n,k}$ of $h$-cobordism classes of embedded homotopy $n$-spheres in $(n + k)$-space was an abelian group under connected sum (by results of Smale, $h$-cobordism and isotopy are synonymous if $k > 2$ and $n > 4$). He then showed that $\Theta^{4k-1,2k+1} \cong \mathbb{Z}$ by constructing an invariant in the following manner.

If $K^{4k-1} \subseteq S^{6k}$ is a smooth knot, then choose a framed properly embedded submanifold $N \subseteq D^{6k+1}$ bounded by $K$. A $2k$-cocycle of $N$ is defined by considering the linking number of any $2k$-cycle of $N$ with a translate of $N$ in $D^{6k+1}$. The square of this cocycle is the desired invariant. It turned out to be the complete obstruction to ambiently "surgering" $N$ to a disk . A similar argument showed that $\Theta^{4k+1,2k+2} \cong \mathbb{Z}_2$.

In 1964 Levine [Lev65a] used the methods of Kervaire-Milnor's ground-breaking work [KM63] on the classification of homotopy-spheres, together with Haefliger's ambient surgery techniques, to produce a *non-stable* version of the Kervaire-Milnor exact sequences for $k > 2$ and $n > 4$:

$$\cdots \to \pi_{n+1}(G_k, SO_k) \to P_{n+1} \xrightarrow{d} \Theta^{n,k} \xrightarrow{\tau} \pi_n(G_k, SO_k) \xrightarrow{\sigma} P_n \to \cdots .$$

Here $P_n$ is defined to be $\mathbb{Z}$, if $n \equiv 0 \pmod 4$, $\mathbb{Z}_2$ if $n \equiv 2 \pmod 4$, and $0$ if $n$ is odd. $G_k$ is the space of maps $S^{k-1} \to S^{k-1}$ of degree 1. The map $d$ is defined as follows. Choose a proper embedding $N^{n+1} \subseteq D^{n+k+1}$, where $N$ is some framed manifold with spherical boundary and signature or Kervaire invariant a given element $a \in P_{n+1}$. Then $d(a)$ is defined to be the knot $\partial N \subseteq S^{n+k}$ and is independent of the choice of $N$. If $K \subseteq S^{n+k}$, then $\tau([K])$ is defined from the homotopy class of the inclusion $\partial T \subseteq S^{n+k} - K \simeq S^{k-1}$, where $T$ is a tubular neighborhood of $K$. This sequence essentially reduced the classification of knots in codimension $> 2$ to the computation of some homotopy groups of spheres and the relevant $J$-homomorphisms, modulo some important group extension problems including the infamous Kervaire invariant conjecture.

Shortly after this, Haefliger [Hae66a] produced an alternative classification of knots using triad homotopy groups. He considered the group $C^{n,k}$ of $h$-cobordism classes of *embeddings* of $S^n$ in $S^{n+k}$. The relation between $C^{n,k}$ and $\Theta^{n,k}$ is embodied in an exact sequence due to Kervaire:

$$\cdots \to \Theta^{n+1} \to C^{n,k} \to \Theta^{n,k} \to \Theta^n \xrightarrow{\partial} C^{n-1,k} \to \cdots .$$

Here $\Theta^n$ denotes the group of $h$-cobordism classes of homotopy $n$-spheres, and $\partial$ is defined by associating to any $\Sigma \in \Theta^n$ its gluing map $h$, defined by the formula $\Sigma = D^n \cup_h D^n$, and then considering the embedding

$$S^{n-1} \xrightarrow{h} S^{n-1} \subseteq S^{n+k-1} .$$

Haefliger showed that $C^{n,k} \cong \pi_{n+1}(G; G_k, SO)$, where $G = \lim_{q \to \infty} G_q$ and $SO = \lim_{q \to \infty} SO_q$.

All of these results are interconnected by a "braided" collection of exact sequences (see [Hae66a]).

In [Hae66b] Haefliger applied these techniques to the classification of *links* in codimension $> 2$ and the result was another collection of exact sequences which reduced the classification of links to the classification of the knot components and more homotopy theory. For any collection of positive integers $p_1, \cdots p_r, m$, where $m > p_i + 2$, the set of $h$-cobordism classes of disjoint embeddings $S^{p_1} + \cdots + S^{p_r} \subseteq S^m$ forms an abelian group under component-wise connected sum. It contains, as a summand, the direct sum $\bigoplus_i C^{p_i, m-p_i}$, representing the *split* links. The remaining summand $L_p^m$ was shown by Haefliger to lie in an exact sequence:

$$(1) \qquad \cdots \to A_{p+1}^m \to B_{p+1}^m \to L_p^m \to A_p^m \xrightarrow{W} B_p^m \to \cdots$$

where $p$ stands for the sequence $p_1, \cdots p_r$. The terms $A_p^m$ and $B_p^m$ are made up from homotopy groups of spheres and $W$ is defined by Whitehead products.

After the development of the surgery sequence of Browder, Novikov, Sullivan and Wall [Wal70] these earlier knot and link classification results were given a more concise treatment in [Hab86]. In fact the methods of Browder and Novikov had already been extended to give a surgery-theoretic classification of embeddings of a simply-connected manifold in another simply-connected manifold. A general classification of embeddings in the meta-stable range, using the homotopy theory of the Thom space of the normal bundle was given by Levine in [Lev63]. For any closed simply-connected manifold $M^n$ and vector bundle $\xi^k$ over $M$, with $n < 2k - 3$, which is stably isomorphic to the stable normal bundle of $M$, there is a one-one correspondence (with some possible exceptions related to the Kervaire invariant problem) between the set of $h$-cobordism classes of embeddings of $M$ into $S^{n+k}$ and normal bundle $\xi$ and the set $h^{-1}(\omega)$, where $\omega \in H_{n+k}(T(\xi)) \cong \mathbb{Z}$ and $h : \pi_{n+k}(T(\xi)) \to H_{n+k}(T(\xi))$ is the Hurewicz homomorphism. Here, $T(\xi)$ is the Thom space of $\xi$. Browder, in [Bro66], gives a classification of smooth simply connected embeddings in codimension $> 2$ in terms of a homotopy-theoretic model of the complement. Here the fundamental notion of a Poincaré embedding first appeared, and was later refined by Levitt [Lev68] and Wall [Wal70].

A Poincaré embedding of manifolds $X$ in $Y$ is a spherical fibration $\xi$ over $X$, a Poincaré pair $(C, B)$, a homotopy equivalence of $B$ with the total space $S(\xi)$ of $\xi$, and of $Y$ with the union along $B$ of $C$ and the mapping cylinder of the map $S(\xi) \to X$. $C$ is a homotopy theoretic model for the complement of the embedding. A theorem of Browder (extended by Wall to the non-simply connected case) says that if if X is an $m$ manifold and $Y$ is an $n$ manifold, and $n - m \geq 3$ then a (locally flat) topological or $PL$ embedding determines a unique Poincaré embedding and a Poincaré embedding corresponds to a unique locally flat $PL$ or topological embedding (See, for instance, [Wal70].) For smooth embeddings one must first specify a linear reduction for $\xi$ as well. This extended an earlier result of Browder, Casson, Haefliger and Wall that said that any homotopy equivalence $M^n \to V^{n+q}$ of $PL$-manifolds is homotopic to an embedding if $q \geq 3$. (The more general result has been sometimes referred to as the Browder, Casson, Haefliger, Sullivan, Wall theorem.) A broad extension of this result to stratified spaces can be found in [Wei94].

## 3. Knot theory in codimension two

3.1. **Unknotting.** One of the earliest applications of surgery to codimension two knot theory was the unknotting theorem of Levine [Lev65b] which states that a smooth or piecewise-linearly embedded homotopy $n$-sphere $K \subseteq S^{n+2}$, for $n > 2$, is smoothly isotopic to the standard embedding $S^n \subseteq S^{n+2}$ if and only if the complement $S^{n+2} - K$ is homotopy equivalent to the circle. Earlier Stallings had established that topological locally flat codimension 2 knots, of dimension $> 2$, whose complements have the homotopy type of a circle, are unknotted [Sta63]. His proof used the method of engulfing. Levine's proof of this fact (in dimensions $> 4$, extended by Wall [Wal65] to $n = 3$) in the smooth or piecewise-linear category proceeded by showing that one could do ambient surgery on a Seifert surface of the knot to convert it to a disk.

These surgery techniques were later used by Levine, in [Lev70], to give a classification of *simple* odd-dimensional knots of dimension $> 1$— i.e. knots whose complements are homotopy equivalent to that of the trivial knot below the middle dimension— in terms of the *Seifert matrix* of the knot. The Seifert matrix of a knot $K^{2n-1} \subseteq S^{2n+1}$ is a representative matrix of the Seifert pairing which is defined as follows. Choose any $(n-1)$-connected *Seifert surface* for $K$, i.e. a submanifold $M^{2n} \subseteq S^{2n+1}$ whose boundary is $K$. The existence of such $M$ is equivalent to $K$ being simple. The Seifert pairing associated to $M$ is a bilinear pairing $\sigma : H_n(M) \otimes H_n(M) \to \mathbb{Z}$. If $\alpha, \beta \in H_n(M)$ choose representative cycles $z, w$, respectively and define $\sigma(\alpha, \beta) = \ell k(z', w)$, where $\ell k$ denotes linking number and $z'$ is a translate of $z$ off $M$ in the positive normal direction. Different choices of $M$ give different Seifert matrices but any two are related by a sequence of simple moves called *S-equivalence*. The classification of simple knots is then given by the S-equivalence class of its Seifert matrix.

Classification of simple even-dimensional knots was achieved, in special cases, by Kearton [Kea76] and Kojima [Koj79] and, in full generality, by Farber in [Far80]. The classification scheme here is considerably more complex than in the odd-dimensional case. For a simple knot $K^{2n} \subseteq S^{2n+2}$ let $X = S^{2n+2} - K$ and $\tilde{X}$ denote the infinite cyclic cover of $X$. Then the invariants which classify, in Farber's formulation, are: the $\mathbb{Z}[t, t^{-1}]$-modules $A = H_n(\tilde{X}), B = \pi_{n+2}^S(\tilde{X})$ (the stable homotopy group), the map $\alpha : A \otimes \mathbb{Z}_2 \to B$, defined by composition with the non-zero element of $\pi_{n+2}(S^n)$, and two pairings $l : T(A) \otimes T(A) \to \mathbb{Q}/\mathbb{Z}$ ($T(A)$ is the $\mathbb{Z}$-torsion submodule of $A$) and $\psi : B \otimes_{\mathbb{Z}} B \to \mathbb{Z}_4$ which are defined from Poincaré duality.

This result is, in fact, a consequence of a more general result of Farber's [Far84] which gives a homotopy-theoretic classification of *stable* knots, i.e. knots $K^n \subseteq S^{n+2}$ whose complements are homotopy equivalent to that of the trivial knot below dimension $(n+3)/3$. The classification is via the stable homotopy type of a Seifert surface $M$ together with a product structure $u: M \wedge M \to S^{n+1}$, representing the intersection pairing, and a map $z: \Sigma M \to \Sigma M$ ($\Sigma M$ is the suspension of $M$) representing the Seifert pairing, i.e. translation into the complement of $M$ in $S^{n+2}$ combined with Alexander duality. In a somewhat different direction, Lashof-Shaneson [LS69], used the surgery theory of Wall [Wal70] to show that the isotopy class of a knot is determined by the homotopy type of its *complementary pair* $(X, \partial X)$, where $X$ is the complement of the knot, as long as $\pi_1(X) = \mathbb{Z}$.

A specific problem which received some attention was the question of how well the complement of a knot determined the knot (we restrict ourselves to knots of dimension $> 1$). Gluck [Glu67] showed that there could be at most two knots with the same complement in dimension 2. Later Browder [Bro67] obtained this result in all dimensions $\geq 5$. Lashof and Shaneson extended this to the remaining high dimensional cases, $n = 3, 4$ [LS69]. It followed from Farber's classification that stable knots were determined by their complement, but Gordon [Gor76], Cappell-Shaneson [CS76b] and Suciu [Suc92] constructed examples of knots which were not determined by their complements. These examples all had non-abelian fundamental group and it remains a popular open conjecture that, when $\pi_1(\text{complement}) = \mathbb{Z}$, the knot is determined by its complement.

3.2. **Knot invariants.** Surgery methods were also used to describe the various algebraic invariants associated to knots. For example in [Lev66] Levine gave another proof of Seifert's result characterizing which polynomials could be the Alexander polynomial of a knot (also see [Rol75]). This generalized Seifert's result to a wider array of knot polynomials, defined for higher-dimensional knots as the Fitting invariants of the homology $\mathbb{Z}[t, t^{-1}]$-modules of the canonical infinite cyclic covering of the complement of the knot. In [Ker65a] Kervaire gave a complete and simple characterization of which groups $\pi$ could be the fundamental group of the complement of a knot of dimension $> 2$. The proof used plumbing constructions to construct the knot complement with the desired group, and then invoked the Poincaré conjecture to recognize that a given manifold was a knot complement. This last idea at least partially foreshadowed the homology surgery techniques of Cappell and Shaneson of the next decade. The conditions Kervaire obtained were:

(i) $H_1(\pi) \cong \mathbb{Z}$
(ii) $H_2(\pi) = 0$

(iii) $\pi$ is normally generated by a single element

By replacing condition (ii) by the stronger condition:

(ii') $\pi$ has a presentation with one more generator than relators

he described a large class of groups which are the fundamental group of the complement of some 2-dimensional knot (the process of *spinning* shows that any 2-knot group is a 3-knot group). Using Poincaré duality in the universal cover of the complement, several people found further properties of 2-knot groups which enabled them to produce examples of 3-knot groups which were not 2-knot groups, but the problem of characterizing 2-knot groups is still open (as is, of course, 1-knot groups). See Farber [Far75], Gutierrez [Gut72], Hausmann and Weinberger [HW85], Hillman [Hil80], Levine [Lev77b], and especially, see Hillman's book [Hil89] for an extensive study of this question. An old example of Fox showed that (ii') was not a necessary condition for 2-knot groups. In [Ker65a] Kervaire also gave a complete characterization of the lowest non-trivial homotopy group of the complement of a knot with $\pi_1$(complement) = $\mathbb{Z}$, as a $\mathbb{Z}[t, t^{-1}]$-module. In [Lev77a], Levine gives a complete characterization of the $\mathbb{Z}[t, t^{-1}]$-modules which can arise as any given homology module of the infinite cyclic covering of a knot of dimension > 2 (except for the torsion submodule of $H_1$).

### 3.3. Knot concordance.

In codimension two, the relation of $h$-cobordism (more often called concordance today) is definitely weaker than isotopy and so the group $\Theta^{n,2}$, known as the knot concordance group, measures this weaker relation. Its computation required drastically different techniques.

The application of surgery techniques in this context was begun by Kervaire. In [Ker65b] he showed that all even-dimensional knots were slice. In [Lev69b] Levine gave an algebraic determination of the odd-dimensional knot concordance group in dimensions > 1 in terms of the *algebraic cobordism* classes of Seifert matrices. Two Seifert matrices $A, B$ are cobordant if the block sum $A \oplus (-B)$ is congruent to a matrix of the form $\left( \begin{smallmatrix} 0 & X \\ Y & Z \end{smallmatrix} \right)$, where $X, Y, Z$ and the zero matrix 0 are all square. It was then shown [Lev69a], using results of Milnor [Mil69], that the knot concordance group is a sum of an infinite number of $\mathbb{Z}, \mathbb{Z}/2$ and $\mathbb{Z}/4$ summands. More detailed information on the structure of this group was obtained by Stoltzfus in [Sto77].

In summation, knot concordance is now reasonably well-understood in dimensions > 1; the (smooth and $PL$) knot concordance group is periodic of period 4 (except it is a subgroup of index 2 for 3-dimensional knots). The topological knot concordance group preserves this periodicity at dimension 3 and is otherwise the same as the smooth and $PL$ groups (see [CS73]).

### 3.4. Homology surgery.

In [CS74], Cappell and Shaneson attacked the problem of classifying codimension two embeddings within a fixed $h$-Poincaré embedding type. (See [CS78] for a precise definition of an $h$-Poincaré embedding.) Here, the key idea was to interpret codimension two embedding problems as problems in the classification of spaces up to homology type. The motivating example should illustrate this well.

By the high-dimensional Poincaré conjecture, a manifold with boundary $S^n \times S^1$ is the complement of a knot if and only if it is a homology circle and the fundamental group is normally generated by a single element (the meridian.) Thus, the classification of knot complements is the classification of homology circles, a calculation carried out in [CS74]. (In contrast, Levine's unknotting theorem tells us that only the trivial knot has the *homotopy* type of $S^1$.) Similarly, a homology cobordism between knot complements (again with extra $\pi_1$ condition) extends, by the $h$-cobordism theorem, to a concordance of knots. Hence the classification of knot concordance reduces to computing the structure group of homology $S^1 \times D^{n+1}$'s. A general discussion tying together the various surgery theoretic tools for codimension two placement, known as of 1981, can be found in [Ran81].

Cappell and Shaneson's applications of these techniques were quite rich. For example they showed that concordance classes of embeddings of a simply-connected manifold in a codimension two tubular neighborhood of itself were in one-one correspondence with the knot concordance group of the same dimension and this bijection was produced by adding local knots to the 0-section embedding. (See Matsumoto [Mat73] for related results.) This allowed for a geometric interpretation of the periodicity of knot concordance from a more natural surgery theoretic point of view [CS74], than those given via tensoring knots [KN77], or groups actions [Bre73]. In turn, as an example of how knot theory fertilizes the more general subject of manifold theory, knot theoretic ideas (in particular, branched fibrations) provided a geometric description of Siebenmann periodicity [CW87].

Cappell and Shaneson applied their homology surgery techniques to the study of singularities of codimension two $PL$-embeddings (i.e. non-locally flat embeddings) in [CS76a] and gave definitive results on the existence of such embeddings as well as an obstruction theory for removing the singularities. A codimension two $PL$ locally flat embedding has a trivial tubular neighborhood. Thus one might hope to study non-locally flat embeddings with isolated singularities via the knot types of the links of the singularities. Indeed, they showed that the classifying space of oriented codimension two thickenings has the knot concordance groups as its homotopy groups. More recently, Cappell and Shaneson studied non-isolated singularities by observing that, with appropriate perversity, the link of a singularity looks like a knot to intersection homology [CS91].

Cappell and Shaneson prove that for closed oriented odd dimensional $PL$ manifolds $M^n$ and $W^{n+2}$, with $n \geq 3$, a map $f : M \to W$ is homotopic to a (in general, non-locally flat) $PL$ embedding if and only if $f$ is the underlying map of an $h$-Poincaré embedding. This is often false in even dimensions, but still holds if $W$ is simply connected. In fact, they show the existence of even dimensional spineless manifolds, i.e., manifolds $W^{n+2}$ with the homotopy type of an $n$ manifold and such that $W$ contains no codimension two embedded submanifold within its homotopy type. See [CS78] for an extensive discussion of these and other results and the techniques used to derive them.

**3.5. Four-dimensional surgery and classical knot concordance.** For the case of classical one-dimensional knots it was clear that the classification scheme of Kervaire and Levine must fail but it took some time before it was actually proved by Casson and Gordon [CG76], in a paper that is among the deepest in the literature of knot theory. All the higher-dimensional knot concordance invariants are invariants of knotted circles as well, and knots for which these invariants vanish are often called *algebraically slice*. Casson and Gordon defined secondary slicing obstructions using signatures associated to metabelian coverings of the knot complement, and gave explicit examples of very simple one-dimensional knots that were *algebraically slice* but not (even topologically) slice. These remain among the most obscure invariants in geometric topology, and very little progress has been made in understanding them. Several papers of interest include Gilmer [Gil83], and Letsche [Let95] for traditional Casson Gordon invariants, and results of Cappell and Ruberman [CR88], Gilmer-Livingston [GL92], Ruberman [Rub83], and Smolinsky [Smo86] that investigate the use of Casson Gordon invariants to study *doubly slice* knots in the classical and higher dimensional context.

The knot slice problem seeks to classify the structure set of homology circles, and it seems natural to suppose that the Casson-Gordon invariants manifest the existence of secondary four-dimensional homology surgery invariants. Freedman's work suggests that any secondary obstructions to topological surgery obstruct building Casson handles. It is a central question to relate these ideas, and see what role Casson-Gordon invariants play in the general problem of computing homology structure groups in dimension four, and in creating Casson handles in general.

In a remarkable application of Freedman's topological surgery machine, Freedman has shown that a classical knot is slice with a slicing complement with fundamental group $\mathbb{Z}$ (called $\mathbb{Z}$-slice) if and only if the Alexander module of the knot vanishes [Fre82]. (Donaldson's work implies not all of these knots are smoothly slice, giving counterexamples to the topological ribbon slice problem! Freedman's work predicts that an analogous class of links, called *good boundary links*, are slice. However, such links have free (or nearly so) fundamental groups, and it is still an open question whether topological surgery works for such groups. In fact, Casson and Freedman showed that good boundary links are slice if and only if every four-dimensional normal map with vanishing surgery obstruction is normally cobordant to a homotopy equivalence [CF84]. The Whitehead double of any link, with pairwise vanishing linking numbers zero, is a good boundary link, and the slicing problem for the Whitehead double of the Borromean rings may be the archetypal example on which this problem's solution rests. A discussion of these and other connections between the four-dimensional topological surgery conjecture and the link slice problem can be found in [FQ90].

Among the most important open problems that surgery theory gives hope of answering is the ribbon-slice problem. It is conjectured that a knot is ribbon if and only if it is smoothly slice. In the topological category, one seeks to determine if a knot is slice if and only if it is homotopy ribbon. A knot is *homotopy ribbon* if it is slice by a locally flat, topologically embedded two disk where the inclusion of the complement of the knot to the complement of the slicing disk induces an epimorphism on fundamental group. The Casson-Gordon invariants give potential obstructions to this, as they may detect the failure of this map to induce an epimorphism of fundamental groups. A more complete theory of topological homology surgery in dimension 4 would give deeper invariants, and possibly realization techniques for solving the homotopy ribbon-slice problem. For instance, reducing the classification of classical knot concordance to the four dimensional topological surgery conjecture might reduce this problem to a surgery group computation.

## 4. LINK CONCORDANCE

**4.1. Boundary links.** Following success in the classification of knot concordance in high dimensions, attention focused on classifying links up to concordance. The knot concordance classification theorems made explicit use of the Seifert surface for the knot. The existence of this Seifert surface meant that $S^1$ split from the knot complement. A link for which the components bound pairwise disjoint Seifert surfaces is called a *boundary link*. A boundary link complement splits a wedge of circles. It is natural to suspect that concordance of boundary links might be computable using similar techniques to those used to classify knots. In fact, the trivial link gives a nice Poincaré embedding and the classification of concordance of boundary links is, roughly, the classification of homology structures on the trivial link complement.

The arguments of Kervaire used to slice even-dimensional knots were easily seen to slice even-dimensional boundary links as well. Cappell-Shaneson applied their homology surgery machinery to calculate the boundary concordance group of boundary links of dimension $> 1$ [CS80], where *boundary link concordance* is the natural notion of concordance for boundary links. More precisely, the components of the concordance, together with the Seifert surface systems for the links, are assumed to bound pairwise disjoint, oriented, and embedded manifolds. They prove that in all odd dimensions there exist infinitely many distinct concordance classes of boundary links none of which contain split links. Their argument is somewhat delicate. The homology surgery group which computes boundary concordance of boundary links detects links not *boundary* concordant to a split link. Further arguments were needed to show that these same links were not concordant to split links.

Later, Ko [Ko87] and Mio [Mio87] used Seifert matrices to give an alternative classification of boundary link concordance of boundary links and Duval [DuV86] obtained the classification using Blanchfield pairings. The complete computation of these surgery groups has not been attempted to our knowledge, and remains an interesting open problem. An isotopy classification of simple odd-dimensional (boundary) links was carried out by Liang [Lia77].

**4.2. Non-boundary links.** We have seen that the classification of boundary links, up to concordance, followed similar lines to the classification of knot concordance. But the concordance classification of non-boundary links has proven more difficult, requiring new ideas and techniques. With the work of Cappell and Shaneson, attention naturally focused on these two questions:

1. Are all links concordant to boundary links?
2. Is boundary link concordance the same as link concordance?

The first question has only recently been answered and the second remains open.

Perhaps the first suggestion of how to proceed appeared in a small concluding section of [CS80], where the authors anticipate and motivate many of the techniques which continue to dominate research on link concordance. The authors suggested that one may study general link concordance (as opposed to boundary link concordance) by considering limiting constructions which serve as a way of measuring the failure of a given link to be a boundary link. We elaborate further.

Boundary links are accessible to surgery techniques because there is a terminal boundary link complement (the trivial link) to which all boundary link complements map by a degree one map. This gives a manifold to which all boundary link complements can be compared. Similarly, a slice complement for the trivial link is terminal among all boundary link slice complements. Since the fundamental group can change dramatically under a homology equivalence (and under a concordance), no simple terminal object exists for general links. Cappell and Shaneson suggested that a limit of link groups might be used to construct such a terminal object for links. This suggestion launched a flurry of research activity.

The missing idea, needed to make Cappell and Shaneson's suggestion work, was discovered by Vogel in an unpublished manuscript, and implemented in a paper of Le Dimet [Dim88]. Vogel suggested that instead of taking the limit through link groups, one should take the limit through spaces of the homology type one seeks to classify, thus constructing a terminal object within a homology class. Homotopy theory had long studied similar limiting constructions, i.e., Bousfield's homology localization of a space [Bou75]. Bousfield's space was far too big for the study of compact manifolds. The Vogel localization was a limit through maps of finite CW complexes with contractible cofiber.

In [Dim88], Le Dimet uses Vogel's idea to classify concordance classes of disk links. A *disk link* is a collection of codimension two disks disjointly embedded in a disk so that the embedding is standard on the boundary [Dim88]. (We believe disk links first appeared in this paper. Links in the 3-disk were later referred to as *string links*.) The inclusion of the meridians (a wedge of circles) of a disk link into the complement is a map of finite complexes with contractible cofiber, and thus becomes a homotopy equivalence after localization. Restricting a homotopy inverse to the boundary of the disk link complement gives a map whose homotopy class is a concordance invariant of the disk link. We will refer to this as *Le Dimet's homotopy invariant*. Le Dimet proved that $m$ component, $n$-dimensional disks links, modulo disk link concordance, form a group $C_{n,m}$ and that, for $n \geq 2$, this group and his homotopy invariant fit into an exact sequence involving Cappell-Shaneson homology surgery groups. In particular, Le Dimet gives a long exact sequence as follows:

$$\cdots \to C_{n+1,m} \to H_{n+1} \to \Gamma_{n+3} \to C_{n,m} \to H_n \to \cdots$$

Here, $H_n = [\#S^n \times S^1, E(\vee S^1)]$ (homotopy classes of maps) is the home of Le Dimet's homotopy invariant. $E(\vee S^1)$ is the Vogel localization of a wedge of circles, and $\Gamma_{n+3}$ is a relative homology surgery group involving $\pi_1(E(\vee S^1))$ which acts on $C_{n,m}$.

Concordance classes of links are a quotient set of Le Dimet's disk link concordance group, and so the computation of this group is of fundamental importance. Unfortunately, the Vogel localization is difficult to compute and almost nothing is known about it. Understanding this space and its fundamental group remains among the most central open problems in the study of high-dimensional link concordance. For example, if this space is a $K(\pi, 1)$, then even-dimensional links are always slice.

In [CO90, CO93], Cochran and Orr gave the first examples of higher odd-dimensional links not concordant to boundary links. Although this work was partly motivated by Le Dimet's work (and other sources as well) the paper gave obstructions in terms of localized Blanchfield pairings of knots lying in branched covers of the given link. They obtained examples of 2-torsion and 4-torsion, Brunnian examples (remove one component and the link becomes trivial) as well as other interesting phenomena. They gave similar examples for links in $S^3$. (Here the interesting problem was to find links with vanishing Milnor $\bar{\mu}$-invariants that are not concordant to boundary links.) All these examples are odd-dimensional and realize non-trivial surgery group obstructions from Le Dimet's sequence. After their work several alternative approaches have provided more examples. (See Gilmer-Livingston [GL92] and Levine [Lev94]. The latter paper investigates the invariance of signatures and the Atiyah-Patodi-Singer invariant under homology cobordism.)

In [Coc87], Cochran began an investigation of homology boundary links and concordance. Homology boundary links, like boundary links, have a rank preserving homomorphism from their link groups to a free group. But unlike boundary links, this homomorphism is not required to take meridians to a generating set. (Using the Pontryagin-Thom construction one can obtain what is called a *singular* Seifert surface system for this class of links (see Smythe [Smy66]). All known examples of higher-dimensional links not concordant to boundary links are sublinks of homology boundary links.

Realizing Le Dimet's homotopy obstruction is a difficult problem about which almost nothing is known. For even-dimensional links it is the sole obstruction to slicing. For links in $S^3$, Levine showed that it (or equivalently, an invariant he defined independently – see below) vanishes if and only if the link is concordant to a sublink of a homology boundary link [Lev89]. Shortly afterwards, Levine, Mio and Orr proved the same result for links of higher odd dimension [LMO93]. An easy calculation shows Le Dimet's invariant vanishes on even dimensional sublinks of homology boundary links as well, implying these links are always slice! Thus, homology boundary links provide a geometric interpretation for the vanishing of Le Dimet's homotopy invariant.

In [CO94], Cochran and Orr classified homology boundary links. Of particular interest here was a new construction for creating a homology boundary link from a boundary link and a ribbon link with a fixed normal generating set, creating a link with prescribed properties and realizing given surgery invariants. It seems likely that this construction can be generalized, potentially providing examples for a wide class of related problems in knot and link theory.

The work of Vogel and Le Dimet was not unprecedented. In [Coc84], Cochran employed Cappell and Shaneson's suggestion of taking a limit through link groups to classify links of two spheres in $S^4$. He used the observation that link groups had the homology type of link complements through half the dimensions of the link complement for links in $S^4$. In fact,

Le Dimet's homotopy invariant followed a flurry of mathematical activity in the study of homotopy theoretic invariants of link concordance.

Prior to 1980, Milnor's $\bar{\mu}$-invariants for classical links (equivalently, Massey products) were the only known homotopy theoretic obstructions to slicing a link. They remain among the deepest and most important invariants of knot theory and play an important role in the study of topological surgery in dimension four [FQ90]. In [Sat84], Sato (and Levine, independently) introduced a concordance invariant for higher-dimensional links that generalized the $\bar{\mu}$-invariant, $\bar{\mu}_{1212}$, which detects the Whitehead link in dimension one. These invariants were greatly extended using geometric techniques by Cochran [Coc85, Coc90], and homotopy-theoretic techniques by Orr [Orr87, Orr89]. But the only invariant among these that was not later shown to vanish, or to be roughly equivalent to Milnor's invariants was a single invariant from [Orr89]. This invariant remains obscure and unrealized.

One outgrowth of this study was the formulation of a group theoretic construction called *algebraic closure* by Levine in [Lev89], a smaller version of the nilpotent completion. This work provides a combinatorial description of the fundamental group of a Vogel local space, and has proven useful both in defining new invariants, and as a tool for computing local groups. There are two variations of this construction. For a group $\pi$, $\bar{\pi}$ lives in the nilpotent completion of $\pi$ while the possibly larger group $\hat{\pi}$ is defined by a universal property. $\hat{\pi}$ is the fundamental group of the Vogel localization of any space with fundamental group $\pi$. Levine used this latter algebraic closure construction to define a new invariant for certain classical links. First of all, the $\bar{\mu}$-invariants of Milnor can be viewed as living in $\bar{F}$. (This observation allows one to prove a realizability theorem for the $\bar{\mu}$ invariants; one of the conditions for realizability is the vanishing of a class in $H_2(\bar{F})$). One can define a slight generalization of the $\bar{\mu}$ invariants which are just liftings into $\hat{F}$. Then, for (classical) links on which these invariants vanish, a new, possibly non-trivial, concordance invariant lives in $H_3(\hat{F})$. This latter invariant vanishes if and only if Le Dimet's invariant vanishes. It was then proved in [Lev89] that this invariant vanishes if and only if the link is concordant to a sublink of a homology-boundary link, and that every element of $H_3(\hat{F})$ is realized by some link. (Unfortunately we do not know if this homology group is non-zero.) This result suggested the higher-dimensional analogue in [LMO93].

4.3. **Poincaré embeddings again.** In summary, it is still unknown whether all even-dimensional links are slice and whether every higher-dimensional link is concordant to a sublink of a homology boundary link. Both of these problems would be solved by computing Le Dimet's homotopy invariant. But, more generally, we should ask what is the larger role of Vogel local spaces in surgery and embedding theory?

Implicit in Le Dimet's work is the notion that, for codimension two placement and the classification theory of manifolds within a homology type, one should consider a weakened version of Poincaré embedding, where spaces are replaced with their Vogel local counterparts. For the study of high-codimension embeddings all spaces considered are usually simply connected, and therefore already local. For this reason, earlier results did not need this operation of localization. This helps account for both the early progress in high codimension, and the long delay in dealing effectively with the codimension-two case. It is a fundamental problem to develop this theory to its conclusion, and to consider the more general theory for stratified spaces (see [Wei94].) Examples where this is used (at least implicitly) to study general embedding theory can be found in the classification results of Mio, for links with one codimension component [Mio92], and the torus knotting results of Miller [Mil94].

Another problem is to derive the surgery exact sequence for this type of classification. Normal maps with coefficients were classified in [Qui75] and [TW79] for subrings of $\mathbb{Q}$, but a general theory for the localization of an arbitrary ring does not exist at this time.

4.4. **Open problems.** The following list of problems is by no means exhaustive, representing a small subset of difficult problems we think are particularly interesting. They are either problems in surgery theory motivated by knot theory and from whose solution knot theory should benefit, or problems in knot theory that should be approachable through surgery theory.

1. Are knots determined by their complement when $\pi_1$ is $\mathbb{Z}$?
2. Give an algebraic characterization of 2-knot groups.
3. Find tools for computing the algebraic closure of a group.
4. For the free group $F$, compute the homology of $\bar{F}$ and the algebraic closure of the free group $\hat{F} \cong \pi_1(E(\vee S^1))$. Is $\hat{F} \cong \bar{F}$?
5. Are all even-dimensional links slice?
6. Is every odd dimensional link (with vanishing Milnor's $\bar{\mu}$-invariants if $n = 1$) concordant to a sublink of a homology boundary link? Are all higher dimensional links sublinks of homology boundary links?
7. Is every sublink of a homology boundary link concordant to a homology boundary link?
8. Compute the homotopy type of the Vogel localization of a wedge of circles, and Le Dimet's homotopy invariant. More generally, find tools for computing the homotopy type of Vogel local spaces.
9. Find more invariants and, ultimately, compute the homology surgery group classifying boundary link concordance. In particular, if two boundary links are concordant, are they also boundary concordant?

10. Find more invariants and, ultimately, compute the homology surgery group in LeDimet's exact sequence which classifies sublinks of homology boundary links up to concordance.

11. Derive a surgery exact sequence for homology structures with coefficients. In particular, classify normal maps.

12. Find a complete set of invariants for classical knot concordance.

13. If a classical knot is topologically slice in a homology three disk, is it topologically slice? This problem measures the possible difference between computing concordance of classical knots, and the solution of a homology surgery problem.

14. Develop a theory of homology surgery in dimension four (at least modulo Freedman's four dimensional topological surgery problem.)

15. Solve the homotopy ribbon-topological slice problem.

16. Relate Casson-Gordon invariants to a topological four-dimensional homology surgery machine, and in particular, find the relation to Casson handles.

## REFERENCES

[Bou75]   A. K. Bousfield. The localization of spaces with respect to homology. *Topology*, 14:133–150, 1975.

[Bre73]   G. Bredon. Regular $O(n)$ manifolds, suspensions of knots, and periodicity. *Bull. Am. Math. Soc.*, 79:87–91, 1973.

[Bro66]   W. Browder. Embedding 1-connected manifolds. *Bull. Am. Math. Soc.*, 72:225–31, 1966.

[Bro67]   W. Browder. Diffeomorphisms of 1-connected manifolds. *Transactions AMS*, 128:155–163, 1967.

[CF84]    A. Casson and M. H. Freedman. Atomic surgery problems. *AMS Contemporary Math.*, 35:181–200, 1984.

[CG76]    A. Casson and C. McA. Gordon. On slice knots in dimension three. In *Proc. Symp. Pure Math. XXXII*, volume 2, pages 39–53. AMS, 1976.

[CO90]    T. D. Cochran and K. E. Orr. Not all links are concordant to boundary links. *Bull. Amer. Math. Soc.*, 23(1), 1990.

[CO93]    T. D. Cochran and K. E. Orr. Not all links are concordant to boundary links. *Annals of Math*, 138:519–554, 1993.

[CO94]    T. D. Cochran and K. E. Orr. Homology boundary links and Blanchfield forms: Concordance classification and new tangle-theoretic constructions. *Topology*, 33:397–427, 1994.

[Coc84]   T. D. Cochran. Slice links in $S^4$. *Transactions AMS*, 285:389–401, 1984.

[Coc85]   T. D. Cochran. Geometric invariants of link cobordism. *Comment. Math. Helv.*, 60:291–311, 1985.

[Coc87]   T. D. Cochran. Link concordance invariants and homotopy theory. *Invent. Math.*, 90:635–645, 1987.

[Coc90]   T. D. Cochran. Derivatives of links: Milnor's concordance invariants and Massey products. *Memoirs AMS*, 84:291–311, 1990.

[CR88]    S. Cappell and D. Ruberman. Imbeddings and cobordisms of lens spaces. *Comm. Math. Helv.*, 63:75–88, 1988.

[CS73]    S. Cappell and J. L. Shaneson. On topological knots and knot cobordism. *Topology*, 12:33–40, 1973.

[CS74]     S. Cappell and J. L. Shaneson. The codimension two placement problem and homology equivalent manifolds. *Annals of Math.*, 99:277–348, 1974.

[CS76a]    S. Cappell and J. L. Shaneson. Piecewise linear embeddings and their singularities. *Annals of Math.*, 103:163–228, 1976.

[CS76b]    S. Cappell and J. L. Shaneson. There exit inequivalent knots with the same complement. *Annals of Math.*, 103:349–353, 1976.

[CS78]     S. Cappell and J. L. Shaneson. An introduction to embeddings, immersions and singularities in codimension two. In *Proc. of Symp. in Pure Math. XXXII*, volume 2, pages 129–149. AMS, 1978.

[CS80]     S. Cappell and J. L. Shaneson. Link cobordism. *Comment. Math. Helv.*, 55:20–49, 1980.

[CS91]     S. Cappell and J. L. Shaneson. Singular spaces, characteristic classes, and intersection homology. *Annals of Math.*, 134:325–374, 1991.

[CW87]     S. Cappell and S. Weinberger. A geometric interpretation of Siebenmann's periodicity phenomenon. In *Proc. 1985 Georgia Topology Conference*, pages 476–52. Dekker, 1987.

[Dim88]    J. Y. Le Dimet. Cobordisme d'enlacement de disques. *Bull. Soc. Math. France*, 116, 1988.

[DuV86]    J. DuVal. Forme de Blanchfield et cobordisme d'entrelacs bords. *Comment. Math. Helv.*, 61:617–635, 1986.

[Far84]    M. Farber. Knots and stable homotopy. *Lecture Notes*, 1060:140–150, 1884.

[Far75]    M. Farber. Linking coefficients and codimension two knots. *Soviet Math. Doklady*, 16:647–650, 1975.

[Far80]    M. Farber. An algebraic classification of some even-dimensional spherical knots I, II. *Trans. AMS*, 261:185–209, 1980.

[FQ90]     M. H. Freedman and F. Quinn. *Topology of 4-manifolds*. Princeton University Press, 1990.

[Fre82]    M. H. Freedman. The topology of four dimensional manifolds. *Jour. Diff. Geom.*, 17:357–453, 1982.

[Gil83]    P. M. Gilmer. Slice knots in $S^3$. *Quart. J. Oxford*, 34(3):305–322, 1983.

[GL92]     P. M. Gilmer and C. Livingston. The Casson-Gordon invariants and link concordance. *Topology*, 31:475–492, 1992.

[Glu67]    H. Gluck. The embeddings of 2-spheres in the 4-sphere. *Bull. AMS*, 67:586–589, 1967.

[Gor76]    C. McA. Gordon. Knots in the 4-sphere. *Comment. Math. Helv.*, 51:585–596, 1976.

[Gor77]    C. McA. Gordon. Some aspects of classical knot theory. In *Proc. Knot Theory Conference, Plan-sur-Bex, Lecture Notes in Math. 685*, pages 1–60. Springer, 1977.

[Gut72]    M. Gutierrez. Boundary links and an unlinking theorem. *Trans. AMS*, 171:491–499, 1972.

[Hab86]    N. Habegger. Knots and links in codimension greater than 2. *Topology*, 25:253–260, 1986.

[Hae61]    A. Haefliger. Plongements différentiables de variété dans variétés. *Comment. Math. Helv.*, 36:47–82, 1961.

[Hae62]    A. Haefliger. Knotted $(4k-1)$ spheres in $6k$ space. *Annals of Math.*, 75:452–466, 1962.

[Hae66a]   A. Haefliger. Differentiable embeddings of $S^n$ in $S^{n+q}$ for $q > 2$. *Annals of Math.*, 83:402–436, 1966.

[Hae66b]   A. Haefliger. Enlacements de sphère en codimension superieure a 2. *Comment Math. Helv.*, 41:51–72, 1966.

[Hil80]		J. A. Hillman. High-dimensional knot groups which are not 2-knot groups. *Australian Math. Soc.*, 21:449–462, 1980.

[Hil89]		J. A. Hillman. *2-knots and their groups.* Cambridge U. Press, 1989.

[HW85]		J. C. Hausmann and S. Weinberger. Charactéristiques d'Euler et groupes fondamentaux des variété de dimension 4. *Comment. Math. Helv.*, 60(1):139–144, 1985.

[Kea76]		C. Kearton. An algebraic classification of some even-dimensional knots. *Topology*, 15:363–373, 1976.

[Ker65a]	M. Kervaire. Les noeuds de dimension supérieures. *Bull. Soc. de France*, 93:225–271, 1965.

[Ker65b]	M. Kervaire. On higher-dimensional knots. In S. Cairn, editor, *Differential and combinatorial topology*, pages 105–120. Princeton Univ. Press, 1965.

[KM63]		M. Kervaire and J. Milnor. Groups of homotopy spheres I. *Annals of Math.*, 77:504–537, 1963.

[KN77]		L. H. Kauffman and W. D. Neumann. Products of knots, branched fibrations and sums of singularities. *Topology*, 16:369–393, 1977.

[Ko87]		K. H. Ko. Seifert matrices and boundary link cobordism. *Transactions AMS*, 299:657–681, 1987.

[Koj79]		S. Kojima. Classification of simple knots and Levine pairings. *Comment. Math. Helv.*, 54:356–367, 1979.

[KW77]		M. Kervaire and C. Weber. A survey of multidimensional knots. In *Proc. Knot Theory Conf.,Plans-sur-Bex, Lecture Notes in Math.* 685, pages 61–134. Springer, 1977.

[Let95]		C. Letsche. *Eta invariants and the knot slice problem.* Ph.D. thesis, Indiana University, Bloomington, 1995.

[Lev63]		J. Levine. On differential embeddings of simply-connected manifolds. *Annals of Math.*, 69:806–9, 1963.

[Lev65a]	J. Levine. A classification of differentiable knots. *Annals of Math.*, 82:15–50, 1965.

[Lev65b]	J. Levine. Unknotting spheres in codimension two. *Topology*, 4:9–16, 1965.

[Lev66]		J. Levine. Polynomial invariants of knots of codimension two. *Annals of Math.*, 84:537–554, 1966.

[Lev68]		N. Levitt. On the structure of Poincaré duality spaces. *Topology*, 7:369–388, 1968.

[Lev69a]	J. Levine. Invariants of knot cobordism. *Invent. Math.*, 8:98–110, 1969.

[Lev69b]	J. Levine. Knot cobordism in codimension two. *Comment. Math Helv.*, 44:229–244, 1969.

[Lev70]		J. Levine. An algebraic classification of knots some knots of codimension two. *Comment. Math. Helv.*, 45:185–198, 1970.

[Lev77a]	J. Levine. Knot modules I. *Trans. AMS*, 229:1–50, 1977.

[Lev77b]	J. Levine. Some results on higher dimensional knot theory. In *Proc. Knot Theory Conf., Plan-sur-Bex, Lecture Notes in Math.* 685, pages 243–70. Springer, 1977.

[Lev89]		J. Levine. Link concordance and algebraic closure of groups, II. *Invent. Math.*, 96:571–592, 1989.

[Lev94]		J. Levine. Link invariants via the eta invariant. *Comment. Math. Helv.*, 69:82–119, 1994.

[Lia77]		C. C. Liang. An algebraic classification of some links of codimension two. *Proc. Amer. Math. Soc.*, 67:147–151, 1977.

[LMO93]	J. Levine, W. Mio, and K. E. Orr. Links with vanishing homotopy invariant. *Comm. Pure and Applied Math.*, 66:213–220, 1993.

[LS69]     R. Lashof and J. L Shaneson. Classification of knots in codimension two. *Bull. AMS*, 75:171–175, 1969.

[Mat73]    Y. Matsumoto. Knot cobordism groups and surgery in codimension two. *J. Fac. Sci. Tokyo*, 20:253–317, 1973.

[Mil69]    J. Milnor. Isometries of inner product spaces. *Invent. Math.*, 8:83–97, 1969.

[Mil94]    D. Miller. Noncharacteristic embeddings of the $n$-dimensional torus in the $(n + 2)$-dimensional torus. *Trans. AMS*, 342(1):215–240, 1994.

[Mio87]    W. Mio. On boundary link cobordism. *Math. Proc. Camb. Phil. Soc.*, 101:259–266, 1987.

[Mio92]    W. Mio. On the geometry of homotopy invariants of links. *Proc. Camb. Phil. Soc.*, 111(2):291–98, 1992.

[Orr87]    K. E. Orr. New link invariants and applications. *Comment. Math. Helv.*, 62:542–560, 1987.

[Orr89]    K. E. Orr. Homotopy invariants of links. *Invent. Math.*, 95:379–394, 1989.

[Qui75]    F. Quinn. Semifree group actions and surgery on PL homology manifolds. In *Geometric Topology, Proc. Conf., Park City, Utah, Lecture Notes in Math.* 438, pages 395–419. Springer, 1975.

[Ran81]    A. Ranicki. *Exact sequences in the algebraic theory of surgery*. Princeton University Press, 1981.

[Ran98]    A. Ranicki. *High-dimensional knot theory*. Springer, 1998.

[Rol75]    D. Rolfsen. A surgical view of Alexander's polynomial. In L. C. Glaser and T. B. Rushing, editors, *Geometric Topology, Lecture Notes in Math.* 438, pages 415–423. Springer, 1975.

[Rub83]    D. Ruberman. Doubly slice knots and the Casson-Gordon invariants. *Trans. AMS*, 279:569–88, 1983.

[Sat84]    N. Sato. Cobordism of semi-boundary links. *Topol. and its Applic.*, 18:97–108, 1984.

[Smo86]    L. Smolinsky. Doubly slice knots which are not the double of a disk. *Trans. AMS*, 298(2):723–32, 1986.

[Smy66]    N. Smythe. Boundary links. In R. H. Bing, editor, *Annals of Math. Studies*, volume 60, pages 69–72, 1966.

[Sta63]    J. Stallings. On topologically unknotted spheres. *Annals of Math.*, 77:490–503, 1963.

[Sto77]    N. Stoltzfus. Unraveling the knot concordance group. In *Memoirs AMS*, volume 192. Amer. Math. Soc., 1977.

[Suc92]    A. Suciu. Inequivalent framed-spun knots with the same complement. *Comment. Math. Helv.*, 67:47–63, 1992.

[TW79]     L. Taylor and B. Williams. Local surgery: foundations and applications. In *Algebraic Topology, Aarhus 1978, Lecture Notes in Math.* 763, pages 673–695. Springer, 1979.

[Wal65]    C. T. C. Wall. Unknotting tori in codimension one and spheres in codimension two. *Proc. Camb. Phil. Soc.*, 61:659–664, 1965.

[Wal70]    C. T. C. Wall. *Surgery on compact manifolds*. Academic Press, 1970.

[Wei94]    S. Weinberger. *The topological classification of stratified spaces*. University of Chicago Press, 1994.

[Zee60]    E. C. Zeeman. Unknotting spheres. *Annals of Math.*, 72:350–361, 1960.

DEPARTMENT OF MATHEMATICS
BRANDEIS UNIVERSITY
WALTHAM, MA 02254-9110
*E-mail address*: `levine@max.math.brandeis.edu`

DEPARTMENT OF MATHEMATICS
INDIANA UNIVERSITY
BLOOMINGTON, IN 47405
*E-mail address*: `korr@ucs.indiana.edu`

# Surgery and $C^*$-algebras

## John Roe

### 1. INTRODUCTION

A $C^*$-*algebra* is a complex[1] Banach algebra $A$ with an involution $*$, which satisfies the identity

$$\|x^*x\| = \|x\|^2 \quad \forall x \in A.$$

Key examples are

- The algebra $C(X)$ of continuous complex-valued functions on a compact Hausdorff space $X$.
- The algebra $\mathfrak{B}(H)$ of bounded linear operators on a Hilbert space $H$.

The theory of $C^*$-algebras began when Gelfand and Naimark proved that any commutative $C^*$-algebra with unit is of the form $C(X)$, and a little while later that any $C^*$-algebra (commutative or not) is isomorphic to a subalgebra of some $\mathfrak{B}(H)$.

A simple consequence of Gelfand and Naimark's characterisation of commutative $C^*$-algebras is the so-called *spectral theorem*. Let $A$ be a $C^*$-algebra with unit, and let $x \in A$ be *normal*, that is $xx^* = x^*x$. Then $x$ generates a commutative unital $C^*$-subalgebra of $A$ which (by Gelfand-Naimark) must be of the form $C(X)$ for some $X$. In fact we can identify $X$ as the *spectrum*

$$X = \sigma(x) = \{\lambda \in \mathbb{C} : x - \lambda 1 \text{ has no inverse}\}$$

with $x$ itself corresponding to the canonical inclusion $X \to \mathbb{C}$. We thus obtain the *Spectral Theorem*: for any $\varphi \in C(\sigma(x))$ we can define $\varphi(x) \in A$ so that the assignment $\varphi \mapsto \varphi(x)$ is a ring homomorphism. If $x$ is *self adjoint* ($x = x^*$), then $\sigma(x) \subseteq \mathbb{R}$, since the canonical inclusion map $\sigma(x) \to \mathbb{C}$ must be invariant under complex conjugation.

As a consequence of Atiyah and Bott's elementary proof of the periodicity theorem, it was realised in the early sixties that it is possible and useful to consider topological $K$-theory for Banach algebras [34]. In particular one can define $K$-theory groups for $C^*$-algebras. For $A$ unital

- $K_0(A)$ = the Grothendieck group of f.g. projective $A$-modules (in other words, the ordinary algebraic $K_0$ group)

---

[1] There is also a theory of *real* $C^*$-algebras, but we will not consider it here.

- $K_1(A) = \pi_0 GL_\infty(A)$

and there is a simple modification which extends the definition to non-unital $A$. For an introduction one may consult [33].

For any integer $i$ define $K_i = K_{i\pm 2}$. Then to any short exact sequence of $C^*$-algebras

$$0 \to J \to A \to A/J \to 0$$

there is a long exact $K$-theory sequence

$$\ldots K_i(J) \to K_i(A) \to K_i(A/J) \to K_{i-1}(J) \ldots .$$

(The map $\partial\colon K_i(A/J) \to K_{i-1}(J)$ is known as the *connecting map*.) The 2-periodicity is a version of the Bott periodicity theorem. Of course, *algebraic* $K$-theory does not satisfy Bott periodicity; analysis is essential here.

Here is an important example. An operator $T$ on a Hilbert space $H$ is called *Fredholm* if it has finite-dimensional kernel and cokernel. Then the *index* of $T$ is the difference of the dimensions of the kernel and cokernel.

**(1.1)** DEFINITION: *The algebra of* compact operators, $\mathfrak{K}(H)$, *is the $C^*$-algebra generated by the operators with finite-dimensional range.*

One then has *Atkinson's theorem*:

**(1.2)** PROPOSITION: $T \in \mathfrak{B}(H)$ *is Fredholm if and only if its image in $\mathfrak{B}(H)/\mathfrak{K}(H)$ is invertible.*

Consider the short exact sequence of $C^*$-algebras

$$0 \to \mathfrak{K}(H) \to \mathfrak{B}(H) \to \mathfrak{B}(H)/\mathfrak{K}(H) \to 0.$$

A Fredholm operator $T \in \mathfrak{B}(H)$ maps to an invertible in the quotient $\mathfrak{B}(H)/\mathfrak{K}(H)$, by Atkinson's theorem, and hence defines a class $[T]$ in $K_1(\mathfrak{B}/\mathfrak{K})$. Under the connecting map $\partial$ this passes to $\partial[T] \in K_0(\mathfrak{K})$. However, $K_0(\mathfrak{K})$ is easily identified with $\mathbb{Z}$, and, under this identification, $\partial[T]$ corresponds to the index. Thus the Fredholm index may be regarded as a special case of the $K$-theory connecting map.

## 2. SURGERY AND TOPOLOGICAL $K$-THEORY

It seems that Gelfand and Mishchenko [12] were the first to point out that topological $K$-theory could be relevant in studying the surgery obstruction groups. They considered the following situation. Let $\Gamma$ be a discrete countable abelian group. The surgery obstruction group $L(\mathbb{Z}\Gamma)$ is constructed out of hermitian forms on finitely generated free $\mathbb{Z}\Gamma$-modules. Now by Pontrjagin duality there is a dual group $\widehat{\Gamma}$, a compact abelian topological group, such that every element of the complex group ring $\mathbb{C}\Gamma$ corresponds to a continuous complex-valued function on $\widehat{\Gamma}$. Thus a nonsingular hermitian form on a f.g. free $\mathbb{Z}\Gamma$-module gives rise to a map from $\widehat{\Gamma}$ to the space of all nonsingular hermitian matrices (suitably stabilized). This latter space, however, is easily seen to be homotopy equivalent to a classifying space $\mathbb{Z} \times BU$ for $K$-theory (with the integer component given by the signature) and so we get a class in $K(\widehat{\Gamma})$. We have thus got a map

$$\Sigma \colon L(\mathbb{Z}\Gamma) \to K(\widehat{\Gamma}).$$

When $\Gamma$ is a finite group, this is essentially the multisignature of Wall [32, §13A]. When $\Gamma = \mathbb{Z}^i$ is free abelian, $\widehat{\Gamma}$ is a torus, and Gelfand and Mischenko pointed out that in this case $\sigma$ is a rational isomorphism; the surgery groups on the left had been computed by Novikov. Since topological $K$-theory seems rather more malleable than algebraic $L$-theory, the existence of this isomorphism in the fundamental case of a free abelian group was suggestive and encouraging.

In order to extend the idea to non-abelian groups we note that the group $K(\widehat{\Gamma})$ can be expressed as the (topological) $K$-theory of the $C^*$-algebra of continuous complex-valued functions on $\widehat{\Gamma}$. This algebra is nothing other than a certain *completion* of the complex group algebra $\mathbb{C}\Gamma$; in fact it is the *group $C^*$-algebra* in the sense of the following definition.

**(2.1)** DEFINITION: *Let $\Gamma$ be a discrete (countable) group. The* reduced group $C^*$-algebra, $C^*_r\Gamma$, *is the completion of $\mathbb{C}\Gamma$ in the norm of bounded operators on the Hilbert space $\ell^2\Gamma$.*

Now $K$-theory can be defined for any $C^*$-algebra, commutative or not, and we will see that an extension of the Gelfand-Mischenko argument gives a map

$$\Sigma \colon L_*(\mathbb{Z}\Gamma) \to K_*(C^*_r\Gamma)$$

for *any* countable discrete group $\Gamma$. It is of vital importance that we are using $C^*$-algebras here. In fact, let $A$ be a unital $C^*$-algebra, and consider a nondegenerate hermitian form over $A$ — for simplicity, just think of a selfadjoint $x \in A$. The spectrum $\sigma(x)$ of $x$ is then a closed subset of $\mathbb{R}$, which does not contain 0 (because of nondegeneracy). The function $f_+$ which is equal to $+1$ on the positive reals and 0 on the negative reals is therefore continuous on $\sigma(x)$, as is the function $f_- = 1 - f_+$. Thus, using the spectral theorem, we find that

$$p_+ = f_+(x), \quad p_- = f_-(x)$$

are well defined self-adjoint projections in $A$ and so their difference $[p_+] - [p_-]$ gives an element of $K_0(A)$. It can be shown [29] that this procedure gives rise to an isomophism $L_0^p(A) \to K_0(A)$. Our map $\Sigma$ may now easily be defined as the composite

$$L_0(\mathbb{Z}\Gamma) \to L_0(C^*_r\Gamma) \to K_0(C^*_r\Gamma)$$

and there is also a definition in odd dimensions.

In Chapter 9 of [32], Wall gave an alternative definition of the surgery obstruction groups, as cobordism groups of a certain kind. It is helpful to consider the map from surgery groups to $C^*$-algebra $K$-theory also from this perspective. The key to doing so is the *signature operator*. This is a certain first order elliptic operator — we will denote it by $D$ — defined on differential forms on an oriented manifold $M$. If $M$ is compact and its dimension is a multiple of 4, then the index of the signature operator is just the Hirzebruch signature of $M$, as follows from an easy argument using Hodge theory. By applying the index theorem to this operator, Atiyah and Singer [2, 3] were able to rederive Hirzebruch's signature theorem.

Surgery obstructions can be thought of as sophisticated signatures. Analogously, on the analytic side, there is a more sophisticated index theory for manifolds with non-trivial fundamental group. Given such a manifold $M$, with $\pi_1 M = \Gamma$, let $\widetilde{M}$ be its universal cover. The signature operator $D$ lifts to a $\Gamma$-equivariant operator on the non-compact manifold $\widetilde{M}$. Now just as an ordinary elliptic operator like $D$ is invertible modulo the ideal of smoothing operators on $M$, so $\widetilde{D}$ is invertible modulo the ideal of *invariant* smoothing operators on $\widetilde{M}$. This ideal can be completed to a $C^*$-algebra, which is isomorphic[2] to $C_r^*(\Gamma) \otimes \mathfrak{K}$ where $\mathfrak{K}$ denotes the algebra of compact operators. The connecting homomorphism in $C^*$-algebra $K$-theory now produces an index in $K_*(C_r^*\Gamma \otimes \mathfrak{K}) = K_*(C_r^*\Gamma)$ for the signature operator, and this is the 'analytic' higher signature that we require.

Recall that in Wall's Chapter 9 a geometric group[3] $L_*^{geom}(\Gamma)$ is defined, equipped with a natural map $L_*^{geom}(\Gamma) \to L_*^{alg}(\mathbb{Z}\Gamma)$, defined by surgery obstructions, which subsequently turns out to be an isomorphism. Generalizing somewhat the discussion above[4] one can associate a 'higher analytic signature' in $K_*(C_r^*\Gamma)$ to every cycle for $L_*^{geom}(\Gamma)$ and thus fill in the diagonal arrow in the diagram

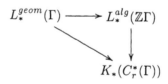

where the horizontal arrow is Wall's map and the vertical arrow is the generalized Gelfand-Mischenko construction described above.

---

[2] This is not hard to see. Essentially, the compact operators act on a fundamental domain, and the $C_r^*\Gamma$ handles the combinatorics of how the fundamental domains fit together.

[3] Actually two such groups, but I will conflate them for the purposes of this exposition.

[4] The cycles for Wall's geometric group are made up of certain Poincaré spaces and maps between them. Recall now that the ordinary signature can be defined for any Poincaré space, even one which is not a manifold. We need the 'higher, analytic' generalization of this; the $C^*$-algebraic higher signature can be defined for an arbitrary (compact) Poincaré space. Details of the construction will appear in [14].

The $C^*$-algebraic higher indices were described in [17, 1, 23] using slightly different language to that above. However, in the special case of free abelian $\Gamma$ they were first used by Lusztig [21] in an analytic approach to Novikov's conjecture on the homotopy invariance of the higher signatures. We now turn to this topic.

## 3. ANALYSIS AND THE NOVIKOV CONJECTURE

We recall the original statement of Novikov's conjecture (see the final pages of Wall's book). Let $M$ be a compact manifold with fundamental group $\Gamma$. Let $\varphi: M \to B\Gamma$ be a map classifying the universal cover. The *higher signature* of $M$ is $\varphi_*([M] \frown L(M)) \in H_*(B\Gamma; \mathbb{Q})$; Novikov conjectured that it is an (oriented) homotopy invariant. At the time Wall wrote the conjecture was known for poly-$\mathbb{Z}$ groups, and it is now proved for a much wider range of geometrically tractable groups.

There is a natural map, now called the assembly map, $A: H_*(B\Gamma; \mathbb{Q}) \to L_*(\mathbb{Z}\Gamma) \otimes \mathbb{Q}$. The image of the higher signature under the assembly map is the Mischenko-Ranicki symmetric signature of $M$, which is a homotopy invariant by construction. Thus to prove the Novikov conjecture it suffices to prove the injectivity of $A$; indeed, conjectures about the injectivity of assembly maps in a variety of contexts are now known as 'Novikov conjectures'.

**(3.1)** PROPOSITION: *The image of the symmetric signature under the natural map $L_*(\mathbb{Z}\Gamma) \to K_*(C_r^*\Gamma)$ is the higher index of the signature operator on the universal cover $\widetilde{M}$.*

It follows from this proposition that the $K_*(C_r^*\Gamma)$-index of the signature operator is an invariant of homotopy type, and proofs of this fact not depending on explicit appeal to the theory of the symmetric signature were given by Lusztig (in the case of interest to him), Kasparov, and Kaminker-Miller [16]. To prove the Novikov conjecture it therefore suffices to show that the 'analytic' assembly map

$$A: H_*(B\Gamma; \mathbb{Q}) \to K_*(C_r^*\Gamma) \otimes \mathbb{Q}$$

is injective. Lusztig did this for the case $\Gamma = \mathbb{Z}^n$, where $B\Gamma$ is a torus, and $C_r^*\Gamma$ is (by Fourier analysis) the algebra of continuous functions on a torus; the assembly map is just the inverse of the Chern character. This analytic approach has since been developed in two significantly different directions.

The first, especially associated with the work of Connes [7, 9, 10], is to develop a 'noncommutative de Rham theory' and use this to construct a faithful pairing between $K_*(C_r^*\Gamma)$ and the *cohomology* of $B\Gamma$. The relevant theory is Connes' cyclic homology. In the argument there is one crux at which the geometry of $\Gamma$ must be used; it is necessary to pass to a *subalgebra* of $C_r^*\Gamma$, sufficiently large that its $K$-theory agrees with the $K$-theory of the full $C^*$-algebra, but sufficiently 'regular' that cyclic homology classes are defined for it[5]; and it seems that there is no universal construction for such an algebra. (The significance if not the intractability of this point was anticipated by Gelfand and Mischenko; in their original paper, they wrote "Distinct completions of the group ring lead to distinct subrings [of the $C^*$-algebra]; however this probably influences only slightly the homotopy invariants determined by those subrings.")

The second approach is to consider an *analytic* assembly map, running from the $K$-homology $H_*(B\Gamma; \mathbb{K})$ of $B\Gamma$ (i.e. the generalized homology theory with coefficients in the periodic $K$-spectrum) to the $K$-theory of $C_r^*\Gamma$. This map is a refinement of the rationalized assembly map described above, analogous to the assembly map

$$A\colon H_*(X; \mathbb{L}(e)) \to L_*(\pi_1 X)$$

arising in a modern formulation of the surgery exact sequence for topological manifolds [25]. One can then conjecture that under suitable circumstances (e.g. when $\Gamma$ is torsion-free) this map is integrally an isomorphism. Kasparov [19] has followed this road, using sophisticated versions of bivariant $K$-theory. More recently methods of 'controlled functional analysis' have been used to approach the same set of results [13]. There is still much work to be done on the connection between $KK$-theory and controlled topology, but it is clear that these latter developments mark a reconvergence of topological and analytic approaches to the Novikov conjecture, after a period during which the methods employed seemed rather distinct.

The Novikov conjecture has been comprehensively surveyed elsewhere [11]; we will therefore not extend this section any further. However the existence of an analytic counterpart to the assembly map raises some further natural questions: is there an analytic counterpart to the whole surgery exact sequence? And, if so, what is the geometric meaning of the structure set term?

### 4. $C^*$-COUNTERPARTS OF THE EXACT SURGERY SEQUENCE

In this section I want to describe a construction, due to Higson and myself [15, 14, 26], of an analytic version of the surgery exact sequence. We begin with a compact space $X$. For simplicity of notation we will assume that $X$ is a manifold, and we equip it with some Riemannian metric; let $\Gamma = \pi_1(X)$. I will describe two $C^*$-algebras, $C_\Gamma^*(X)$ and $D_\Gamma^*(X)$, which will be algebras of operators on the Hilbert space $H = L^2(\widetilde{X})$.

Let $T$ be a bounded linear operator on $H$. It will be called *properly supported* if $Tu$ is compactly supported whenever $u$ is compactly supported.

---

[5]Recall that cyclic homology is modeled on de Rham cohomology. De Rham theory fits well into the context of the ring $C^\infty(M)$ of *smooth* functions, but not the ring $C(M)$ of *continuous* functions.

For a bounded continuous function $f$ on $M$ let $M_f$ denote the corresponding multiplication operator on $\widetilde{X}$. Then $T$ will be called *locally compact* if $TM_f$ and $M_f T$ are compact operators for all compactly supported functions on $\widetilde{X}$. Finally, $T$ will be called *pseudolocal* if $M_f T M_g$ is compact for all functions $f$ and $g$ with disjoint supports.

EXAMPLE: Any properly supported (and globally $L^2$-bounded) pseudo-differential operator of order $\leq 0$ is pseudolocal, and any such operator of order $< 0$ is compact.

As an important special case, the 'normalized signature operator' $F = D(1 + D^2)^{-\frac{1}{2}}$ on $\widetilde{X}$ is a pseudolocal operator[6], and ellipticity shows that it is invertible modulo locally compact operators.

REMARK: Pseudolocality is, of course, a standard property of pseudodifferential operators [31]. However, from the point of view of topology, it is helpful to think of pseudolocality as a condition of 'continuous control at infinity in the spectrum'. Roughly, one regards the Hilbert space $H$ as 'controlled' by a decomposition given by an orthonormal basis; pseudolocality suggests that only finitely many basis elements 'propagate' over distance greater than some fixed $\varepsilon$. For more on this see [27].

Notice that the properly supported pseudolocal operators form an algebra in which the locally compact operators form an ideal. We define $C^*_\Gamma(X)$ to be the $C^*$-algebra generated by the $\Gamma$-equivariant properly supported locally compact operators on $\widetilde{X}$, and $D^*_\Gamma(X)$ to be the $C^*$-algebra generated by the $\Gamma$-equivariant properly supported pseudolocal operators. Thus we have a short exact sequence

$$0 \longrightarrow C^*_\Gamma(X) \longrightarrow D^*_\Gamma(X) \longrightarrow D^*_\Gamma(X)/C^*_\Gamma(X) \longrightarrow 0$$

of $C^*$-algebras.

**(4.1)** DEFINITION: *The* analytic surgery exact sequence *for $X$ is the long exact sequence of $K$-theory groups associated to the short exact sequence of $C^*$-algebras above.*

By Bott periodicity this long exact sequence is 2-periodic (or 8-periodic if we refine matters by using real $C^*$-algebras and real $K$-theory). This contrasts with the 4-periodicity of surgery theory.

Of course the description 'surgery exact sequence' needs justification. Here are some significant facts.

**(4.2)** PROPOSITION: *The algebra $C^*_\Gamma(X)$ is Morita equivalent to (and hence has the same $K$-theory as) the group $C^*$-algebra $C^*_r\Gamma$.*

---

[6]It need not be properly supported, but standard techniques show that it can be abitrarily well approximated by properly supported operators.

In fact, $C_\Gamma^*(X)$ just is the appropriate 'error ideal' with respect to which $\Gamma$-invariant elliptic operators are invertible.

**(4.3)** PROPOSITION:   *The $K$-theory of the quotient $C^*$-algebra*

$$D_\Gamma^*(X)/C_\Gamma^*(X)$$

*is isomorphic to the $K$-homology of $X$:  $K_{i+1}(D_\Gamma^*(X)/C_\Gamma^*(X)) = H_i(X; \mathbb{K})$.*

This approach to $K$-homology arises from the work of Kasparov [17]; Paschke and Higson realized that it could be reformulated in the language of 'dual algebras'. A direct proof of this proposition using techniques more conventional in algebraic topology would begin by showing that the left hand side is a homology theory (a homotopy invariant and excisive functor) and would then identify the coefficient spectrum.

The connecting map in the long exact $K$-theory sequence is therefore a homomorphism

$$A\colon H_i(X; \mathbb{K}) \to K_i(C_r^*\Gamma)$$

and it is not too hard to identify it with other formulations of the 'analytic assembly' map. We therefore have obtained a long exact sequence into which the assembly map fits, so it seems reasonable to call it the analytic surgery exact sequence and to call $K_*(D_\Gamma^*(X))$ the *analytic structure set*. Since $K_1$ of an algebra is generated by invertibles (in matrix algebras) it is appropriate to think of the homology term $K_*(D_\Gamma^*(X)/C_\Gamma^*(X))$ as generated by 'elliptic' operators — pseudodifferential-like operators which are invertible modulo locally compact error. (In particular the (normalized) signature operator defines a $K$-homology class; this is the analytic counterpart to the canonical $L$-theory orientation of a manifold.) In the same spirit one should think of the structure set terms as generated by 'invertible elliptic' operators — elliptic operators 'without homology'. This should be compared with Ranicki's algebraic definition of the structure set, as the cobordism group of complexes which are locally Poincaré and globally contractible.

With Higson I have reinforced this analogy by constructing [14] a natural transformation from the (DIFF) surgery exact sequence to the analytic surgery exact sequence after tensoring with $\mathbb{Z}[\frac{1}{2}]$. In particular there is a map from the manifold structure set to the 'analytic structure set'. The construction uses large scale geometry to associate an analytic signature to the mapping cylinder of a homotopy equivalence between manifolds; this mapping cylinder has the structure of a Poincaré space and therefore a signature. From this signature one obtains the necessary 'structure invariant' by various geometric and analytic reductions. To see that equivalent structures give the same analytic invariant one remarks that the mapping cylinder of a diffeomorphism is actually a manifold, so that its signature is the index of an elliptic operator, i.e. belongs to the image of an assembly map. From a version of the analytic surgery exact sequence it follows that the analytic structure invariant of a diffeomorphism is zero.

The need to invert 2 arises from the difference between the Dirac and signature operators. In fact, a fundamental fact in working with $K$-homology is that if $W$ is a manifold with boundary $\partial W$, then the boundary map in the $K$-homology long exact sequence,

$$K_{i+1}(W, \partial W) \to K_i(\partial W)$$

takes the cycle defined by the Dirac operator on the interior of $W$ to the cycle defined by the Dirac operator on the boundary. Now surgery groups are cobordism groups, but the map from surgery to analysis is defined by the signature operator, not the Dirac operator. In the set-up above the boundary of the *signature* operator on $W$ is either 1 or 2 times the signature operator on $\partial W$ (depending on the parity of the dimension) and so we get certain extra factors of 2; these can be normalized away by multiplying by a suitable power of $\frac{1}{2}$ (depending on the dimension), but to do this we need to have tensored by $\mathbb{Z}[\frac{1}{2}]$.

## 5. THE ANALYTIC STRUCTURE SET

What sorts of 'structures' does the analytic structure set describe? It seems to be a universal analytic receptacle for index maps from various kinds of geometrical structure sets, of which the classic structure set of surgery theory is only one. For example, there is a map from the 'positive scalar curvature structure set' — the space of concordance classes of positive scalar curvature metrics on a given spin manifold $M$ — to $K_*(D_\Gamma^* M)$. We see this by considering the Dirac operator, rather than the signature operator. Indeed, the Dirac operator on a positive scalar curvature manifold is invertible, by the Weitzenbock formula, so its homology class is the image of a 'structure class' in $K_*(D_\Gamma^* M)$. We have simply formalized in homological language the argument of Lichnerowicz' vanishing theorem [20]; it ought to be possible to treat other vanishing theorems, such as Kodaira vanishing on Kähler manifolds, similarly, obtaining maps from other moduli spaces to the analytic structure set.

It is more difficult to see the map from the usual structure set to $K_*(D_\Gamma^* M)$. One attempt at a construction would begin from a model of the structure set described by Carlsson-Pedersen [4]. They show that the (TOP) structure set of $X$ can be described as the $L$-theory of a certain continuously controlled category

$$\mathcal{B}(\widetilde{X} \times [0,1), \widetilde{X} \times 1; \mathbb{Z})^\Gamma$$

whose objects are geometric $\mathbb{Z}$-modules over $\widetilde{X} \times [0,1)$ and whose morphisms are $\Gamma$ equivariant and continuously controlled at 1. If we now replace $\mathbb{Z}$ by $\mathbb{C}$ we obtain a category whose objects are infinite dimensional based complex vector spaces — so can naturally be completed to Hilbert spaces. Moreover, if we let continuous functions on $\tilde{X}$ act on these Hilbert spaces by multiplication along the first coordinate, the morphisms in the category are pseudolocal; in fact the continuous control condition shows that $M_f T M_g$ has finite rank whenever $T$ is a morphism and $f$ and $g$ have disjoint supports. The stage therefore seems to be set for a mapping from $L$-theory to $K$-theory as in section 2, but there is a problem; we have no reason to anticipate that the morphisms will be *bounded* operators on the relevant Hilbert spaces. This problem is linked the fact that we have tried to use techniques based on differential operators to study topological manifolds; some torus trickery is called for, and was worked out in a slightly different context in [24].

A fundamental result in the theory of topological manifolds is the $\alpha$-approximation theorem of Chapman and Ferry [5] which states that a homotopy equivalence $f: M' \to M$ of topological manifolds, such that the tracks of the homotopy are sufficiently small when measured in $M$, is homotopic to a homeomorphism: 'a sufficiently well controlled topological manifold structure is trivial'. In the context of the analytic structure set, Guoliang Yu [35] established an analogous result, which he then put to good use in the study of the Novikov conjecture.

**(5.1)** DEFINITION:   *An operator $T \in D_\Gamma^*(X)$ is $\varepsilon$-controlled if*

$$\|M_f T M_g\| < \varepsilon \sup |f| \sup |g|$$

*whenever the distance between the supports of $f$ and $g$ is greater than $\varepsilon$.*

Yu's result[7] is

**(5.2)** PROPOSITION:   *For any compact manifold (or even a finite complex) $X$ there is a constant $\varepsilon_X$ such that any $\varepsilon_X$-controlled analytic structure on $X$ is trivial (i.e., represents zero in $K_*(D_\Gamma^*(X)))$.*

Yu applied his result (actually a generalization to certain non-compact spaces) to prove the Novikov conjecture for groups of 'finite asymptotic dimension'. The relevant point in this proof is the scaling of the control constant when the space is expanded by some fixed factor. One might also, however, try to make use of Yu's result to get an analytic handle on some other classic results of manifold topology. For example, consider the topological invariance of the rational Pontrjagin classes. It is well known that this would follow from a similar invariance property for the $K$-homology class of the signature operator. But now, we might argue, consider a homeomorphism $h: M \to M'$ of compact smooth manifolds. Using $D$ for signature operators, we know from the analytic surgery

---

[7]Strictly speaking, Yu works with an alternative definition of the 'analytic structure set'; but doubtless his results extend to the present context also.

sequence that

$$h_*[D_M] - [D_{M'}] \in H_*(M'; \mathbb{K})$$

is the image of a class in the analytic structure set defined by $h$. But $h$ is arbitrarily well controlled, hence (?) by Yu's theorem it defines the zero class in $K_*(D_\Gamma^* M')$, and hence $h_*[D_M] = [D_{M'}]$ as required.

This argument is flawed. It turns out that in order that a structure should be well controlled in the analytic sense required by Yu's theorem, one needs *more* than just that the corresponding homotopy equivalence should be well controlled in the natural geometric sense. Essentially, one needs some extra information about derivatives. The point is exactly the same as the one about bouded operators in the Carlsson-Pedersen category, above; and the effect is that we can (probably) get the *Lipschitz* invariance of the Pontrjagin classes from an argument like this, but not the much deeper *topological* invariance. It is of course tempting to circumvent the whole issue by appealing to Sullivan's theory of Lipschitz structures on topological manifolds [30], but this is not wholly satisfactory; to my mind it remains a challenge to incorporate the deep facts about the topology of Euclidean space, used in Sullivan's theory or in the topological $\alpha$-approximation theorem, into an analytic framework such as that sketched above.

## 6. NOTES AND COMMENTS

In this article I have sketched out one approach to the connection between surgery theory and $C^*$-algebraic index theory. For a more detailed and rounded presentation the reader may wish to look at my CBMS lectures, [26]. These also discuss $C^*$-counterparts to 'bounded' and 'controlled' surgery theory, using index theory on non-compact manifolds.

The work of Kasparov provides other powerful tools for studying assembly maps in the context of analysis [17, 18, 19]. Bivariant $K$-theory is a $C^*$-algebraic tool which Kasparov invented and used extensively, and to which there does not presently seem to be a counterpart in surgery ('algebraic $LL$-theory'?) — although see [28] for some suggestions. Kasparov's work also yields results about the *surjectivity* of the assembly map — the so-called Baum-Connes-Kasparov conjecture. This seems to be a very delicate area, even more so than the corresponding surjectivity on the topological side[8], involving hard questions of analysis and representation theory.

A very careful account of the relation between $K$-theory and $L$-theory for $C^*$-algebras, using Kasparov's machinery, is in [22].

For Connes' approach to the Novikov conjecture, using cyclic homology as well as $C^*$-algebras, see the papers [6, 10, 8], among others.

---

[8] i.e. the Borel conjecture

## REFERENCES

[1] M.F. Atiyah. Elliptic operators, discrete groups and von Neumann algebras. *Astérisque*, 32:43–72, 1976.

[2] M.F. Atiyah and I.M. Singer. The index of elliptic operators I. *Annals of Mathematics*, 87:484–530, 1968.

[3] M.F. Atiyah and I.M. Singer. The index of elliptic operators III. *Annals of Mathematics*, 87:546–604, 1968.

[4] G. Carlsson and E. K. Pedersen. Controlled algebra and the Novikov conjectures for $K$ and $L$ theory. *Topology*, 34:731–758, 1995.

[5] T.A. Chapman and S. Ferry. Approximating homotopy equivalences by homeomorphisms. *American Journal of Mathematics*, 101:583–607, 1979.

[6] A. Connes. Cyclic cohomology and the transverse fundamental class of a foliation. In *Geometric methods in operator algebras*, pages 52–144. Pitman, 1986. Research Notes in Mathematics 123.

[7] A. Connes. *Non-Commutative Geometry*. Academic Press, 1995.

[8] A. Connes, M. Gromov, and H. Moscovici. Group cohomology with Lipschitz control and higher signatures. *Geometric and Functional Analysis*, 3:1–78, 1993.

[9] A. Connes and H. Moscovici. Conjecture de Novikov et groupes hyperboliques. *Comptes Rendus de l'Académie des Sciences de Paris*, 307:475–479, 1988.

[10] A. Connes and H. Moscovici. Cyclic cohomology, the Novikov conjecture, and hyperbolic groups. *Topology*, 29:345–388, 1990.

[11] S.C. Ferry, A. Ranicki, and J. Rosenberg. A history and survey of the Novikov conjecture. In S. Ferry, A. Ranicki, and J. Rosenberg, editors, *Proceedings of the 1993 Oberwolfach Conference on the Novikov Conjecture*, volume 226 of *LMS Lecture Notes*, pages 7–66, 1995.

[12] I.M. Gelfand and A.S. Mishchenko. Quadratic forms over commutative group rings and $K$-theory. *Functional Analysis and its Applications*, 3:277–281, 1969.

[13] N. Higson, E.K. Pedersen, and J. Roe. $C^*$-algebras and controlled topology. *K-Theory*, 11:209–239, 1996.

[14] N. Higson and J. Roe. Mapping surgery to analysis. In preparation.

[15] N. Higson and J. Roe. The Baum-Connes conjecture in coarse geometry. In S. Ferry, A. Ranicki, and J. Rosenberg, editors, *Proceedings of the 1993 Oberwolfach Conference on the Novikov Conjecture*, volume 227 of *LMS Lecture Notes*, pages 227–254, 1995.

[16] J. Kaminker and J.G. Miller. Homotopy invariance of the index of signature operators over $C^*$-algebras. *Journal of Operator Theory*, 14:113–127, 1985.

[17] G.G. Kasparov. Topological invariants of elliptic operators I: $K$-homology. *Mathematics of the USSR — Izvestija*, 9:751–792, 1975.

[18] G.G. Kasparov. The operator $K$-functor and extensions of $C^*$-algebras. *Mathematics of the USSR — Izvestija*, 16:513–572, 1981.

[19] G.G. Kasparov. Equivariant $KK$-theory and the Novikov conjecture. *Invent. Math.*, 91:147–201, 1988.

[20] A. Lichnerowicz. Spineurs harmoniques. *C. R. Acad Sci. Paris*, 257:7–9, 1963.

[21] G. Lusztig. Novikov's higher signature and families of elliptic operators. *J. Differential Geometry*, 7:229–256, 1972.

[22] J. Miller. Signature operators and surgery groups over $C^*$-algebras. *K-Theory*, 13:363–402, 1998.

[23] A.S. Mischenko and A.T. Fomenko. The index of elliptic operators over $C^*$-algebras. *Mathematics of the USSR — Izvestija*, 15:87–112, 1980.

[24] E.K. Pedersen, J. Roe, and S. Weinberger. On the homotopy invariance of the boundedly controlled analytic signature of a manifold over an open cone. In

S. Ferry, A. Ranicki, and J. Rosenberg, editors, *Proceedings of the 1993 Oberwolfach Conference on the Novikov Conjecture*, volume 227 of *LMS Lecture Notes*, pages 285–300, 1995.

[25] A. Ranicki. *Algebraic L-Theory and Topological Manifolds*. Cambridge, 1992.

[26] J. Roe. *Index theory, coarse geometry, and the topology of manifolds*, volume 90 of *CBMS Conference Proceedings*. American Mathematical Society, 1996.

[27] J. Roe. An example of dual control. *Rocky Mountain J. Math.*, 27:1215–1221, 1997.

[28] J. Rosenberg. *K* and *KK*: Topology and operator algebras. In W.B. Arveson and R.G. Douglas, editors, *Operator Theory/Operator Algebras and Applications*, volume 51 of *Proceedings of Symposia in Pure Mathematics*, pages 445–480. American Mathematical Society, 1990.

[29] J. Rosenberg. Analytic Novikov for topologists. In S. Ferry, A. Ranicki, and J. Rosenberg, editors, *Proceedings of the 1993 Oberwolfach Conference on the Novikov Conjecture*, volume 226 of *LMS Lecture Notes*, pages 338–372, 1995.

[30] D. Sullivan. Hyperbolic geometry and homeomorphisms. In J.C. Cantrell, editor, *Geometric Topology*, pages 543–555. Academic Press, 1979.

[31] M. Taylor. *Pseudodifferential Operators*. Princeton, 1982.

[32] C.T.C. Wall. *Surgery on Compact Manifolds*. Academic Press, 1970.

[33] N.E. Wegge-Olsen. *K-theory and C\*-algebras*. Oxford University Press, 1993.

[34] R. Wood. Banach algebras and Bott periodicity. *Topology*, 4:371–389, 1966.

[35] G. Yu. The Novikov conjecture for groups with finite asymptotic dimension. *Ann. of Math.* (2) 147:325–355, 1998.

DEPARTMENT OF MATHEMATICS
THE PENNSYLVANIA STATE UNIVERSITY
228 MCALLISTER BUILDING
UNIVERSITY PARK, PA 16802
*E-mail address*: roe@math.psu.edu

# The classification of Aloff-Wallach manifolds and their generalizations

R. James Milgram

## Introduction

The surgery program for classifying manifolds starts with the study of the homotopy type of the manifold and then applies the surgery exact sequence to determine the $h$ or $s$ cobordism classes of manifolds within the homotopy type. There are different sequences depending on whether we work with piecewise linear or differential classification. In this note we apply the surgery program to study the classification of the set of free, isometric $S^1$-actions on the Lie group $SU(3)$. Examples of these kinds originally occurred in surgery theory through the work of Kreck and Stoltz, [KS1], [KS2], motivated by results of Witten on possible models for unified field theories in physics. The cases studied in [KS1] and [KS2] provide the first examples of homeomorphic but non-diffeomorphic symmetric spaces. Also, recent work by differential geometers has concentrated on the natural metrics on these spaces. It is clear that they provide wonderful examples for studying all kinds of structure on manifolds*.

Up to a possible single $\mathbf{Z}/2$ indeterminacy, we obtain a complete classification in the piecewise linear case. These techniques can be extended to complete the classification up to the same indeterminacy in the differential case as well, and results along these lines have been obtained by Kruggel, [K1], [K2].

The set of distinct metric preserving differentiable actions of $S^1$ on $SU(2)$ is indexed by integer 4-tuples $(p_1, p_2, p_3, p_4)$ subject to the constraint that

$$p_1 \equiv p_2 \equiv p_3 \equiv p_4 \mod (3).$$

Research partially supported by a grant from the N.S.F.
* Ib Madsen has obtained similar results independently.

The set of relations below defines an equivalence relation on these 4-tuples and equivalent tuples give equivalent actions.

$$(p_1, p_2, p_3, p_4) \equiv (p_3, p_4, p_1, p_2) \equiv (p_2, p_1, p_3, p_4)$$
$$\equiv (p_2, -(p_1 + p_2), p_3, p_4).$$

The explicit action is given in 2.4. Associated to these actions we define certain semi-invariants:

$$\sigma_2 = \frac{1}{9}(p_1^2 + p_1 p_2 + p_2^2)$$

$$\sigma_2' = \frac{1}{9}(p_3^2 + p_3 p_4 + p_4^2)$$

$$L_2 = |\sigma_2 - \sigma_2'|$$

$$L_3 = \frac{1}{27}(p_3 - p_1)(p_4 - p_1)(p_1 + p_3 + p_4)$$

$$r = p_1 \bmod (3)$$

where $L_2$ and $r$ are actual invariants while $L_2$, $L_3$, $3\sigma_2$, and $3\sigma_2'$ are integers. Then we have

THEOREM: *The action is free if and only if $L_2$ and $L_3$ are relatively prime. If this is the case, then the homotopy type of the quotient $S^1 \backslash SU(3)$ is completely determined by $L_2$, $L_3 \bmod (L_2)$, $r$, and $3\sigma_2' \bmod (2)$. Moreover, the homotopy type of $S^1 \backslash SU(3)$ together with $6\sigma_2 \bmod (L_2)$ determine the piecewise linear homeomorphism type of the quotient up to at most two possibilities.*

REMARK: There is a single identification among these invariants explained in the second paragraph of §7, where we point out that the critical quadratic form invariant $b^2 \in \mathbf{Z}/L_2$ is only defined up to sign since the orientation of the quotient is not a homotopy invariant. This is reflected in the fact that the quotient for $(-p_1, \ldots, -p_4)$ is the same as the quotient for $(p_1, \ldots, p_4)$ but $L_3$ and $r$ in the list above change signs.

REMARK: The "at most two possibilities" in the theorem refers to the image of the Kervaire invariant in dimension 2 in the surgery exact sequence of §10, which is not determined by our analysis for $L_2$ odd or an associated element in dimension 4 when $L_2$ is even (see 10.1). This could affect the piecewise linear homeomorphism classification, but it might not. This is a question I don't know how to handle, but I suspect that understanding this point is a basic step in completely clarifying the surgery exact sequence.

Special cases of this classification have been considered previously. For example, the quotients with $p_3 = p_4 = 0$ are the Aloff-Wallach manifolds, studied in [AW], [W], [KS1], [KS2], for which complete diffeomorphism invariants have been obtained in the last two references, though there are some points of confusion (see the discussion in [AMP]). Partial results in the general case were first obtained by Eschenburg, [E1], [E2], though his objectives were different. In the slightly more general case where $p_3 = p_4$, these manifolds are obtained via the usual action of $U(3)$ on $SU(3)$ when $SU(3)$ is regarded as the Stiefel manifold $V_{2,1}$ of 2-frames in $\mathbf{C}^3$. In this case the author, and independently but somewhat later, [K1], [K2], determined the homotopy classification in terms of the Chern classes in $H^*(\mathbf{CP}^\infty; \mathbf{Z})$ associated to the inclusion $S^1 \hookrightarrow U(3)$. The theorem above specializes to give these results.

The work in [KS1], [KS2] depends on the fact that the action of $S^1$ when $p_3 = p_4 = 0$ extends to a free action of the maximal torus, $(S^1)^2$, and hence the quotient $S^1 \backslash SU(3)$ is a circle bundle over $(S^1)^2 \backslash SU(3)$. In particular it is the boundary of a manifold $M^8$ which we can understand pretty well. Eschenburg, [E1], [E2], [E3], found a second family of $S^1$-actions which come from a free $(S^1)^2$-action on $SU(3)$ distinct from the above. This is studied in [AMP] where, again, the fact that these $S^1$-quotients are boundaries of known 8-manifolds enables a complete diffeomorphism classification, given the homotopy classification above. However, it is also shown in [AMP] that there are no further examples of this kind. The best that one can hope for is that the $S^1$-action extends to an $(S^1)^2$-action with only isolated fixed points. Indeed, an argument of Stolz, (unpublished), shows that this can always be achieved.

Suppose now that a metric preserving action $(S^1)^2 \times SU(3) \rightarrow SU(3)$ with only isolated fixed points is given. It is defined by a sequence of 8 integers $(r_1, r_2, s_1, s_2, t_1, t_2, w_1, w_2)$ and, in terms of these integers we define two integral polynomials $A_2 x^2 + B_2 x + C_2 = f_1(x)$ and $A_3 x^3 + B_3 x^2 + C_3 x + D_3 = f_2(x)$, where the coefficients are integral combinations of monomials in the defining integers $r_1, \ldots, w_2$. The detailed definitions are given in 6.2, and we prove

THEOREM: Let $R(f_2(x), f_3(x))$ be the resultant of the two polynomials. Then, the number of fixed points for the action of $(S^1)^2$ is $\leq 6$, and we have

$$R(f_2(x), f_3(x)) = \sum_{x_i} |I(x_i)|$$

where $I(x_i)$ is the isotropy group for the fixed point $x_i$. Moreover, each $I(x_i)$ is cyclic.

In particular, if $S^1 \subset (S^1)^2$ acts freely on $SU(3)$ then the same construction as before, but deleting small neighborhoods of the fixed point singularities in the mapping cone of the projection $S^1\backslash SU(3) \to (S^1)^2\backslash SU(3)$ gives an 8-manifold with boundary consisting of $S^1\backslash SU(3)$ and a disjoint union of linear lens spaces. Thus, a procedure exists for resolving the remaining questions of determining the diffeomorphism classification of these $S^1$-quotients. We do not pursue this further in this note, however.

The original motivation for these results was work of C. Boyer, K. Galicki, and B. Mann on the metric structures of some of these quotients. They asked me for the structure of the homotopy types and wondered if it would be possible to get further information on their diffeomorphism and homeomorphism classification. During my attempts to answer these questions I became convinced that these quotients provide excellent examples for exploring the classification program, and I hope the partial results expounded here lead others to carry the program further.

In particular, hiding behind most of the results above is the cohomology of the classifying space $B_{SU(3)*SU(3)}$, where $SU(3) * SU(3)$, the central product of two copies of $SU(3)$ is the group of isometries of $SU(3)$. Fortunately, for the results above it was not necessary to write down these cohomology groups in the ramified case of $\mathbf{F}_3$-coefficients, but Eilenberg-Moore spectral sequence techniques and the examination of the structure of the maximal elementary subgroups $(\mathbf{Z}/3)^k \subset SU(3) * SU(3)$ do make this possible. In this case, these results arise in other contexts as well. B. Oliver pointed out to me that $SU(3) * SU(3)$ is a $\mathbf{Z}/3$-centralizer in $E_8$, for example. Moreover, further cases, e.g., $SU(2^n) * SU(2^n)$, seem to play important roles in other areas of topology.

## §1. Biquotients of Lie Groups

The set of left invariant metrics on a compact, connected, simple, and simply connected Lie group $G$ is identified with the set of metrics at the origin via left translation. Under the adjoint action of $G$ on the metrics at the origin the isotropy group, $K$, of the metric, $\gamma$, can be identified with the subgroup of $G$ which also leaves the metric invariant under right translations. Thus, as is well known, the component of the identity in the group of isometries of $G$ with the metric $\{\gamma\}$ is

$$G * K = \{(g,k) \in G \times K \mid (g,k) \sim (dg, dk) \text{ for } d \in Z(G) \cap K\}$$

the central product of $G$ with $K$. In particular, for the bi-invariant metric the component of the identity is $G * G$. Of course, the action is given by $\{g,k\}(h) = ghk^{-1}$.

EXAMPLE 1.1: In the case of $SU(n)$ we have that $Z(SU(n)) = \mathbf{Z}/n$ and we consequently have the central extension

1.2 $$\mathbf{Z}/n \xrightarrow{in} SU(n) \times SU(n) \xrightarrow{\pi} SU(n) * SU(n)$$

which defines the quotient, where the $\mathbf{Z}/n$ embeds *diagonally* into the product.

Perhaps a better way to look at $G * K$ is as follows. First there is a projection $p_2 \colon G * K \to K/(Z(G) \cap K)$ which realizes $G * K$ as the total space of a fibration

$$G \to G * K \to K/(Z(G) \cap K)$$

where the (normal) subgroup $G = \{g, 1\} \subset G * K$. But this fibration has a lift:

$$l \colon K/(Z(G) \cap K) \to G * K$$

defined by $l(\{k\}) = \{k, k\} \in G * K$ and we see that $G * K$ is the semi-direct product

$$G \times_\alpha (K/(Z(K) \cap K))$$

where the action $\alpha$ is given by the usual inclusion $K/(Z(K) \cap K) \hookrightarrow Inn(G) = G/(Z(G))$. The universal group of this type is clearly

$$G * G = G \times_\alpha Inn(G) \subset G \times_\alpha Aut(G),$$

the holomorph of $G$.

Note the following lemma pointed out at least by Eschenburg.

LEMMA 1.3: *Let $H \subset G * G$ be a subgroup and $\tilde{H} \subset G \times G$ the associated central extension. Then a necessary and sufficient condition that $H$ act freely on $G$ is that for no element $\{h_1, h_2\} \in \tilde{H}$ is $h_1$ conjugate to $h_2$ in $G$.*

PROOF: If $g \in G$ is a fixed point for $H$ then there is an element $\{h_1, h_2\} \in H$ and $g \in G$ so that $h_1 g h_2^{-1} = g$ or $h_1 = g h_2 g^{-1}$. Conversely, if this equation is satisfied for $g$, then $g$ is a fixed point of $H$. ∎

Let $H \subset G * G$ be any subgroup which acts freely on $G$. We are interested in the structure of the orbit space $H \backslash G$ which is a closed compact manifold when $G$ is a compact Lie group.

DEFINITION (Eschenburg) 1.4: *Let $H \subset G * G$ act freely on $G$, then the quotient $H \backslash G$ is called a biquotient.*

EXAMPLE 1.5: The following example is quite important in our applications. Let $\tilde{U}(3)$ be the three fold covering of $U(3)$, which we may regard as embedded in $U(3) \times S^1$ as the set of pairs

$$\tilde{U}(3) = \{(g, v) \in U(3) \times S^1 \mid \operatorname{Det}(g) = v^{-3}\}.$$

The covering transformations are given by $(g, v) \mapsto (g, \xi_3^j v)$. Then the map

$$\tilde{\mu}: \tilde{U}(3) \longrightarrow SU(3) \times SU(3), \qquad (g, v) \mapsto \left( gv, \begin{pmatrix} v & 0 & 0 \\ 0 & v & 0 \\ 0 & 0 & v^{-2} \end{pmatrix} \right),$$

preserves $\mathbf{Z}/3$-actions and induces an inclusion

1.6                          $\mu: U(3) \hookrightarrow SU(3) * SU(3).$

Note that if $A \in G * G$ is any element, then the orbit space of $H^A$, the conjugate of $H$ by $A$, is diffeomorphic to $H \backslash G$, so for this reason we only need to consider conjugacy classes of such $H \subset G * G$ as long as we are only interested in the diffeomorphism, topological, or piecewise linear classifications of biquotients. Similarly, $g \mapsto g^{-1}$ takes the action of $H$ to the action of $T(H)$ where $T: G * G \rightarrow G * G$ exchanges the factors, and so the quotients by $H$ and $T(H)$ are diffeomorphic as well.

Examples of biquotients are ordinary symmetric spaces (where $H$ is contained in one or the other copy of $G$) and many other types of manifolds, including some exotic spheres, [GM].

In particular the Aloff-Wallach manifolds and some of their generalizations extensively studied by Aloff-Wallach, Asti, Micha, and Pastor, Kreck-Stolz, and Witten, [AW], [AMP], [KS1], [KS2], [W], are very important examples. These are all given as biquotients of $SU(3)$ by $H = S^1$, and here is the description of the set of all biquotients of $SU(3)$ by $S^1$.

## §2. $S^1$-biquotients of $SU(3)$

We can regard $SU(3)$ as $V_{2,1}$, the Stiefel manifold of 2-frames in $\mathbf{C}^3$, i.e., pairs of vectors

2.1
$$\begin{pmatrix} z_1 & w_1 \\ z_2 & w_2 \\ z_3 & w_3 \end{pmatrix}, \quad \begin{cases} \sum_1^3 z_i \bar{z}_i = 1 \\ \sum_1^3 w_i \bar{w}_i = 1 \\ \sum z_i \bar{w}_i = 0. \end{cases}$$

Note that if $L \subset U(3) \times U(3)$ is the set of pairs of elements $(A, B)$ so that $\mathrm{Det}(A) = \mathrm{Det}(B)$, then $L$ is a subgroup containing the diagonal image of the center of $U(3)$, and the quotient of $L$ by this diagonal image is $SU(3) * SU(3)$. Let $(A, B) \in L$ lie in the central product of the two maximal tori,

$$T_i = \left\{ \begin{pmatrix} \lambda_1 & 0 & 0 \\ 0 & \lambda_2 & 0 \\ 0 & 0 & \lambda_3 \end{pmatrix} \, \Big| \, \prod_1^3 \lambda_i = 1, |\lambda_i| = 1, \ 1 \le i \le 3 \right\}$$

Then the action of $(A, B)$ on $V_{2,1}$ is given by

2.2
$$\begin{pmatrix} \lambda_1 & 0 & 0 \\ 0 & \lambda_2 & 0 \\ 0 & 0 & \lambda_3 \end{pmatrix} \begin{pmatrix} z_1 & w_1 \\ z_2 & w_2 \\ z_3 & w_3 \end{pmatrix} \begin{pmatrix} \tau_1 & 0 & 0 \\ 0 & \tau_2 & 0 \\ 0 & 0 & \tau_3 \end{pmatrix} = \begin{pmatrix} \lambda_1 \bar{\tau}_1 z_1 & \lambda_1 \bar{\tau}_2 w_1 \\ \lambda_2 \bar{\tau}_1 z_2 & \lambda_2 \bar{\tau}_2 w_2 \\ \lambda_3 \bar{\tau}_1 z_3 & \lambda_3 \bar{\tau}_2 w_3 \end{pmatrix}.$$

The factorization through the central product is clear. Consequently, we can assume that $A, B \in SU(3)$, so the fact that $\lambda_3 = (\lambda_1 \lambda_2)^{-1}$ shows that this action is really described by the equivalence class of the four elements $(\lambda_1, \lambda_2, \tau_1, \tau_2)$ in the quotient

$$T_4 = (S^1)^4/(\mathbf{Z}/3).$$

Given four integers $(p_1, p_2, p_3, p_4)$ satisfying the constraint

2.3
$$p_1 \equiv p_2 \equiv p_3 \equiv p_4 \bmod (3)$$

we have an $S^1$-action on $V_{2,1}$ given by

2.4
$$\lambda \begin{pmatrix} v_1 & w_1 \\ v_2 & w_2 \\ v_3 & w_3 \end{pmatrix} = \begin{pmatrix} \lambda^{\frac{p_1-p_3}{3}} v_1 & \lambda^{\frac{p_1-p_4}{3}} w_1 \\ \lambda^{\frac{p_2-p_3}{3}} v_2 & \lambda^{\frac{p_2-p_4}{3}} w_2 \\ \lambda^{\frac{-(p_1+p_2+p_3)}{3}} v_3 & \lambda^{\frac{-(p_1+p_2+p_4)}{3}} w_3 \end{pmatrix}$$

as $\lambda = e^{2\pi i \theta}$ runs over the points of $S^1$.

EXAMPLE 2.5: The quotients where $p_3 = p_4 = 0$ are the Aloff-Wallach manifolds, $W(p_1, p_2)$ which are subject only to the constraint that $p_1$ and $p_2$ are relatively prime. These are homogeneous spaces and are the examples studied by Aloff-Wallach, Witten, and Kreck-Stolz. One of the more important properties of these examples is that the free action of $S^1$ extends to a free action of the entire torus, $(S^1)^2$, and so the mapping

$$S^1 \backslash SU(3) \rightarrow (S^1)^2 \backslash SU(3)$$

realizes $W(p_1, p_2)$ as the total space of an $S^1$ bundle over $(S^1)^2 \backslash SU(3)$, or, by passing to mapping cylinders, as the boundary of a $D^2$-bundle over $(S^1)^2 \backslash SU(3)$.

DEFINITION 2.6: Let $(p_1, p_2, p_3, p_4)$ satisfy the constraints of 2.3 and suppose that the resulting action in 2.4 is free. Then the quotient $S^1 \backslash SU(3)$ is denoted $X^7(p_1, p_2, p_3, p_4)$ or simply $X^7$ when the context is clear.

EXAMPLE 2.7: Suppose that $(\theta_1, \theta_2, \theta_3)$ is any triple of integers. If we set

$$\begin{aligned} p_1 &= 2\theta_1 - \theta_2 - \theta_3 \\ p_2 &= -\theta_1 + 2\theta_2 - \theta_3 \\ p_3 = p_4 &= -(\theta_1 + \theta_2 + \theta_3) \end{aligned}$$

then $(p_1, p_2, p_3, p_4)$ satisfies 2.3 and the resulting action on $V_{2,1} = SU(3)$ is given by the formula

$$\begin{pmatrix} v_1 & w_1 \\ v_2 & w_2 \\ v_3 & w_3 \end{pmatrix} \mapsto \begin{pmatrix} \lambda^{\theta_1} v_1 & \lambda^{\theta_1} w_1 \\ \lambda^{\theta_2} v_2 & \lambda^{\theta_2} w_2 \\ \lambda^{\theta_3} v_3 & \lambda^{\theta_3} w_3 \end{pmatrix}.$$

The constraint on the three integers $(\theta_1, \theta_2, \theta_3)$ in order for the action to be free is simply that they be pairwise relatively prime. Indeed, given any element $(\vec{v}, \vec{w}) \in V_{2,1}$, it must have two independent rows, though the third might be zero. Hence, in order that we have a fixed point, we must have two of the three terms $(\lambda^{\theta_1}, \lambda^{\theta_2}, \lambda^{\theta_3})$ equal to one. But this will only happen for $\lambda \neq 1$ if two of the $\theta_i$ have a non-trivial greatest common divisor.

REMARK 2.8: In fact the examples of 2.7 all factor through the inclusion $\mu \colon U(3) \hookrightarrow SU(3) * SU(3)$ of 1.5 since one sees directly from the definition of the action on $V_{2,1}$ that

$$\mu(g)(\vec{v}_1, \vec{v}_2) = (g(\vec{v}_1), g(\vec{v}_2))$$

for $(\vec{v}_1, \vec{v}_2) \in V_{2,1}$.

If the free action of $S^1$ extends to a free action of $(S^1)^2$ then

$$S^1 \backslash SU(3) \to (S^1)^2 \backslash SU(3)$$

is an $S^1$-fibering, hence $X^7(p_1, p_2, p_3, p_4)$ is the boundary of the 8-dimensional manifold given as the associated disk bundle $D^2 \to E \to (S^1)^2 \backslash SU(3)$. This is the case for the Aloff-Wallach spaces from 2.5 and is exploited in [KS1], [KS2], to enable one to give a complete diffeomorphism classification of the Aloff-Wallach spaces, though the homotopy classification is somewhat sketchy.

Eschenburg, [E1], [E2], [E2], also discovered a second free $(S^1)^2$ action on $SU(3)$ given by the subtorus, $T_2 \subset SU(3) * SU(3)$ defined by its action

$$2.9 \qquad (z, w) \begin{pmatrix} v_1 & w_1 \\ v_2 & w_2 \\ v_3 & w_3 \end{pmatrix} = \begin{pmatrix} z^{-1} v_1 & z w_1 \\ z^{-2} w v_2 & w w_2 \\ z^{-1} w^{-1} v_3 & z w^{-1} w_3 \end{pmatrix}$$

and the embedded circles in this second $(S^1)^2$ all have actions of the form

$$2.10 \qquad \begin{pmatrix} \lambda^{-(l+m)} v_1 & \lambda^{l+m} w_1 \\ \lambda^{-l-2m} v_2 & \lambda^l w_2 \\ \lambda^{-2l-m} v_3 & \lambda^m w_3 \end{pmatrix}.$$

Moreover, in [AMP], 2.3, it is shown that every 2-dimensional torus in $T_4$ which acts freely on $SU(3)$ is conjugate to this one or to the one associated with the torus $T_2 \times 1$ or with the corresponding tori obtained by exchanging the two sides of $SU(3) * SU(3)$. Again, using the fact that the quotient by $S^1$ is the boundary of the associated disk bundle over $T_2 \backslash SU(3)$, explicit invariants are obtained which determine the diffeomorphism types of the resulting manifolds.

## §3. The fixed points of $X^7(p_1, p_2, p_3, p_4)$

We begin by introducing two assumptions on the set $(p_1, \ldots, p_4)$ of 2.3 which will remain in force for the remainder of this article.

(3.1) The gcd of the four integers $(p_1, p_2, p_3, p_4)$ is either 1 or 3. This is equivalent to the assumption that the induced map of the circle to the torus $T_4$ is an embedding.

(3.2) The intersection of the sets $(p_1, p_2, -(p_1 + p_2))$, $(p_3, p_4, -(p_3 + p_4))$ consists of at most one element. If this is not the case, then the action is conjugate to conjugation by the elements of the lifted circle $\tilde{S}^1 = (\lambda^{p_1/3}, \lambda^{p_2/3}, \lambda^{-(p_1+p_2)/3})$ in $SU(3)$. In particular the entire torus $T_2 \subset SU(3)$ is fixed under the action.

In fact, under assumptions (3.1) and (3.2), all the isotropy groups of the action are finite – hence, cyclic subgroups $S^1$.

LEMMA 3.3: *Let $X^7(p_1, p_2, p_3, p_4)$ be given with $(p_1, p_2, p_3, p_4)$ satisfying assumptions (3.1) and (3.2) above. Then the fixed point set is either empty if the action is free or consists of a union of disjoint components where each component is a copy of $S^1$, a copy of $SU(2)/(\mathbf{Z}/k)$ where $\mathbf{Z}/k \subset U(2)$ is the subgroup of elements having the form $\begin{pmatrix} \xi_k^j & 0 \\ 0 & 1 \end{pmatrix}$ and the action is by conjugation or a copy of $U(2)/S^1$ where the action is determined by three integers, $p_1 \equiv p_3 \bmod (3)$, $k$, with $p_1 \neq p_3$, and is given as the action of the cube root of*

$$\left( \begin{pmatrix} \lambda^{p_1} & 0 \\ 0 & \lambda^{p_1 + k(p_1 - p_3)} \end{pmatrix}, \begin{pmatrix} \lambda^{p_3} & 0 \\ 0 & \lambda^{2p_1 - p_3 + k(p_1 - p_3)} \end{pmatrix} \right).$$

PROOF: Let $(A, B) \in T_4$ be an element with non-trivial fixed point set. Then $A$ and $B$ have the same elements. There are two cases. The first is when the three diagonal elements of $A$ are all distinct. In this case the centralizer of $A$ is the torus $T_2 \subset SU(3)$, and the fixed point set of $(A, B)$ is the quotient $S^1 \backslash T_2 \cong S^1$. Otherwise, $A$ has two equal elements, and the centralizer is $U(2)$ embedded in $SU(3)$ as matrices of the form $\begin{pmatrix} g & 0 \\ 0 & \mathrm{Det}(g)^{-1} \end{pmatrix}$. We may assume that the action on $\mathrm{Det}(g)^{-1}$ is multiplication by $\lambda^{(p_3+p_4-p_1-p_2)/3}$. If this is identically one then $p_1 \neq p_3$ or $p_4$ and $p_1 \equiv p_2 \bmod (p_1 - p_3)$ and the action is completely specified by three integers, $p_1 \equiv p_3 \bmod (3)$ and $k$ so that $p_2 = p_1 + k(p_1 - p_3)$.

Hence, the remaining case is the situation where $(p_3+p_4-p_1-p_2) \neq 0$. In this case the action does change the determinant, except of course when $\lambda$ satisfies $\lambda^{(p_3+p_4-p_1-p_2)/3} = 1$ so the quotient is identified with the quotient of $SU(2)$ by the cyclic group above. ∎

## §4. The Borel construction associated to an action of $S^1$

Let $\mu: G \times X \rightarrow X$ be a continuous and proper action of a topological group $G$ on a space $X$. Then the Borel construction associated to $\mu$ is the space $E_G \times_G X$ which is the total space of a fibration

4.1 $$X \longrightarrow E_G \times_G X \longrightarrow B_G,$$

where $E_G$ is a free contractible $G$-space and $B_G$ is the classifying space of $G$. If the action is free then $E_G \times_G X$ has the homotopy type of $G \backslash X$. But, in any case, there is a projection

4.2 $$p: E_G \times_G X \longrightarrow G \backslash X$$

which has the key property that $p^{-1}(\{x\}) = B_{I(x)}$ where $I(x) \subseteq G$ is the isotropy group of $x \in X$. In particular, in the case of the isometric $S^1$-actions on $SU(3)$ we've been studying we have

LEMMA 4.3: *Let $S^1$ act on $SU(3)$ via the action associated to the 4-tuple $(p_1, p_2, p_3, p_4)$ satisfying assumptions (3.1) and (3.2). Then $E_{S^1} \times_{S^1} SU(3)$ has the homotopy type of a CW-complex constructed as the disjoint union of spaces of the form*

$$S^1 \times B_{\mathbf{Z}/k}, \quad E_{S^1} \times_{S^1} U(2)$$

*with a finite number of cells having dimensions $\leq 7$.*

(This is clear, the inverse images of the free points are contractible, and 3.3 shows that the part associated to the fixed point sets have the structure described.)

COROLLARY 4.4: *Let* $X^7(p_1, p_2, p_3, p_4)$ *be the quotient* $S^1 \backslash SU(3)$ *associated to the action above satisfying assumptions (3.1) and (3.2). Then the action is free if and only if* $H^i(E_{S^1} \times_{S^1} SU(3); \mathbf{Z}) = 0$ *for all* $i \geq 8$.

PROOF: This condition is clearly satisfied if the action is free, since then $E_{S^1} \times_{S^1} SU(3) \simeq X^7$. But conversely, suppose the action is not free. Note that associated to an $S^1$ component of the fixed point set the inverse image of this component is

$$E_{S^1} \times_{S^1} (S^1)^2 \simeq K(\mathbf{Z}/k, 1) \times S^1$$

since it is the total space of a fibering $(S^1)^2 \rightarrow E_{S^1} \times_{S^1} (S^1)^2 \rightarrow K(\mathbf{Z}, 2)$, and from the homotopy exact sequence of the fibering we see that if $\mathbf{Z}/k$ is the isotropy group of $x$, then the exact sequence has the form

$$0 \longrightarrow \pi_2(K(\mathbf{Z}, 2)) \overset{\partial}{\longrightarrow} \pi_1((S^1)^2) \longrightarrow \pi_1(E_{S^1} \times_{S^1} (S^1)^2) \longrightarrow 0$$

with all other groups being identically zero. Moreover, $\partial$ is injective with quotient $\mathbf{Z}/k \times \mathbf{Z}$. Thus, each circle leads to an inverse image with non-trivial homology in arbitrarily high dimensions.

Similarly, it is direct to see that the same property holds for $E_{S^1} \times_{S^1} U(2)$ in either the case where the action is non-trivial on the determinant or where it is not. ∎

## §5. The structure of the spaces $X^7(p_1, p_2, p_3, p_4)$

THEOREM 5.1: *Let* $(p_1, p_2, p_3, p_4)$ *satisfy assumptions (3.1) and (3.2), and write the map* $S^1 \rightarrow T_4$ *associated to* $(p_1, \ldots, p_4)$ *as* $\nu$. *Define integers*

$$L_2(\nu) = \frac{1}{9}(p_3^2 + p_4^2 + p_3 p_4 - p_1^2 - p_2^2 - p_1 p_2),$$

$$L_3(\nu) = \frac{1}{27}(p_3 - p_1)(p_4 - p_1)(p_1 + p_3 + p_4)$$

$$= \frac{1}{27}(p_3 p_4 + p_1^2 - p_1(p_4 + p_3))(p_1 + p_3 + p_4).$$

*Then we have*

*(1) $\nu$ is free on $SU(3)$ if and only if $L_2(\nu)$ and $L_3(\nu)$ are relatively prime.*

*(2) If the action is free, the cohomology of the quotient is given as follows:*

| Dimension | 0 | 1 | 2 | 3 | 4 | 5 | 6 | 7 |
|---|---|---|---|---|---|---|---|---|
| $H^i(S^1\backslash SU(3); \mathbf{Z})$ | $\mathbf{Z}$ | 0 | $\mathbf{Z}$ | 0 | $\mathbf{Z}/(L_2(\nu))$ | $\mathbf{Z}$ | 0 | $\mathbf{Z}$ |

*with $H^i(S^1\backslash SU(3); \mathbf{Z}) = 0$ otherwise. Here, let $b$ be a generator for $H^2(S^1\backslash SU(3); \mathbf{Z})$ and $f$ generate $H^5(S^1\backslash SU(3); \mathbf{Z})$, then $b^2$ generates $H^4$ and $b \cup f$ generates $H^7$.*

*(3) The element $p_1 b \in H^2(\mathbf{CP}^\infty; \mathbf{F}_3)$ is a characteristic class for the action.*

REMARK: The cohomology calculation in 5.1(2) was first obtained by Eschenburg in [E1].

PROOF: The torus of $SU(3) \times SU(3)$ three fold covers the torus of $SU(3) * SU(3)$ and the map of classifying spaces is induced from the following map

$$(z_1, z_2, (z_1 z_2)^{-1}) \times (w_1, w_2, (w_1 w_2)^{-1}) \mapsto \{z_1, z_2, w_1, w_2\}$$

where the equivalence class is given by the following relation

$$(z_1, z_2, w_1, w_2) \sim (\lambda z_1, \lambda z_2, \lambda w_1, \lambda w_2),$$

whenever $\lambda = e^{2pik/3}$ is a third root of 1.

We now consider the fibration

5.2 $$SU(3) \longrightarrow E_{S^1} \times_{S^1} SU(3) \overset{\pi}{\longrightarrow} B_{S^1}$$

associated to the action. Consider the Serre spectral sequence associated to the fibration. 5.1 will follow as a direct corollary of the following lemma.

LEMMA 5.3: *In the Serre spectral sequence for 5.2, $E_2 = E_4 = \mathbf{Z}[b] \otimes E(e_3, e_5)$, while $d_4(e_3) = L_2 b^2$. Consequently, $E_5 = \mathbf{Z}[b]/(L_2 b^2) \otimes E(e_5)$, and $d_6(e_5) = L_3 b^3$. Moreover, $E_7 = E_\infty$.*

PROOF: Let $SU(3) \rightarrow E \rightarrow B_{T_4}$ be the fibration induced from the universal fibration

$$SU(3) \rightarrow E \rightarrow B_{SU(3)*SU(3)}$$

by the inclusion of the maximal torus $(S^1)^4 \hookrightarrow SU(3) * SU(3)$. We have the commutative diagram of fibrations

$$
\begin{array}{ccccc}
SU(3) & \longrightarrow & E_{S^1} \times_{S^1} SU(3) & \overset{\pi}{\longrightarrow} & B_{S^1} \\
\Big\downarrow{\scriptstyle =} & & \Big\downarrow{\scriptstyle E_\nu \times id} & & \Big\downarrow{\scriptstyle B_\nu} \\
SU(3) & \longrightarrow & E_{T_4} \times_{T_4} SU(3) & \overset{\pi}{\longrightarrow} & B_{T_4}
\end{array}
$$

where $\nu S^1 \to T_4$ is the homomorphism associated to $(p_1, p_2, p_3, p_4)$.

Let us write the coordinates in $T_4$ as $(R, S, T, W)$ and we write the dual elements in $\mathrm{Hom}(T_4, S^1)$ as $r$, $s$, $t$, and $w$. Then the induced map

$$
B_{S^1} \overset{\nu}{\longrightarrow} B_{T_4}
$$

gives rise to the following map in cohomology:

5.4
$$
\begin{aligned}
r &\mapsto p_1 b \\
s &\mapsto \frac{(p_2 - p_1)}{3} b \\
t &\mapsto \frac{(p_3 - p_1)}{3} b \\
w &\mapsto \frac{(p_4 - p_1)}{3} b
\end{aligned}
$$

Hence, using naturality for the Serre spectral sequence it remains to determine $d_4(e_3)$ and $d_6(e_5)$ in the spectral sequence for $T_4$.

To check the first differential consider the diagram

$$
\begin{array}{ccccc}
SU(3) & \longrightarrow & E' & \overset{\pi}{\longrightarrow} & B_{SU(3) \times SU(3)} \\
\Big\downarrow{\scriptstyle =} & & \Big\downarrow & & \Big\downarrow{\scriptstyle B_\pi} \\
SU(3) & \longrightarrow & E & \overset{\pi}{\longrightarrow} & B_{SU(3) * SU(3)}
\end{array}
$$

where the top fibration is induced from the lower one via the map of classifying spaces induced from the three fold covering $SU(3) \times SU(3) \to SU(3) * SU(3)$. The action of the first copy of $SU(3)$ on the fiber is just multiplication on the left while the action of the second copy is inversion and then multiplication on the right in the upper fibration.

Looking just at the left action the space is contractible and similarly on the right. Thus, by naturality, in the Serre spectral sequence of the upper fibration $d^4(e_3) = c_2 - c'_2$, the minus sign due to the inversion needed in the multiplication on the right. Similarly $d^4(e_5) = c_3 - c'_3$.

We now restrict to the induced fibrations over the maximal tori. In the fibration for the torus $(S^1)^4 \subset SU(3) \times SU(3)$ we have $d_4(e_3) = \sigma_2 - \sigma'_2$ where $\sigma_2$ and $\sigma'_2$ are the second symmetric polynomials in the variables $(z_1, z_2, -(z_1 + z + 2))$ and $(w_1, w_2, -(w_1 + w_2))$ respectively. Similarly $d_6(e_5) = \sigma_3 - \sigma'_3$ and we need merely make these explicit and use naturality to determine the differentials $d_4(e_3)$ and $d_6(e_5)$ in the spectral sequence over $B_{T_4}$.

In the map $B_{(S^1)^4} \to B_{T_4}$ the induced cohomology map with coefficients in $\mathbf{Z}(\frac{1}{3})$ is an isomorphism, and has the form $\frac{1}{3}r \mapsto z_1$, $\frac{1}{3}r + s \mapsto z_2$, $\frac{1}{3}r + t \mapsto w_1$, $\frac{1}{3}r + w \mapsto w_2$, and from this it is direct that

$$d_4(e_3) = -(r(t + w - s) + t^2 + tw + w^2 - s^2).$$

However, when we plug in to determine $d_6(e_5)$ we get

$$\frac{1}{3}(r^2(s - t - w) + r(s^2 - t^2 - 4tw - w^2)) - tw(t + w)$$

for the element which maps to $\sigma_3 - \sigma'_3$. This element is not in the integral cohomology of $B_{T_4}$, but the terms multiplied by $1/3$ are in the ideal generated by the image of $d_4$ except for the term $\frac{1}{3}3rtw$. Factoring this ideal out we get $-tw(t + w + r)$ which is in the integral cohomology and is hence the image of $d_6(e_5)$.

It remains to prove that $r$ is in the image from $H^*(B_{SU(3)*SU(3)}; \mathbf{F}_3)$. To see this consider the Serre spectral sequence of the fibration

$$B_{SU(3) \times SU(3)} \to B_{SU(3)*SU(3)} \to K(\mathbf{Z}/3, 2),$$

where $K(\mathbf{Z}/3, 2)$ is the Eilenberg-Maclane space. Since the fiber is 3-connected, the fundamental class $\iota \in H^2(K(\mathbf{Z}/3, 2); \mathbf{F}_3)$ survives to generate $H^2(B_{SU(3)*SU(3)}; \mathbf{F}_3)$. Now, the fibration restricted to the classifying spaces of the Tori is not trivial so the image of the fundamental class in $H^*(B_{T_4}; \mathbf{F}_3)$ must be non-trivial as well. ∎

## §6. Isometric torus actions on $SU(3)$

Let $\psi\colon (S^1)^2 \to T_4 \subset SU(3) * SU(3)$ induce an isometric action of $(S^1)^2$ on $SU(3)$. Let $B_{(S^1)^2} = B_{S^1} \times B_{S^1} = \mathbf{CP}^\infty \times \mathbf{CP}^\infty$ and let $b_i \in H^2(\mathbf{CP}^\infty; \mathbf{Z})$, $i = 1, 2$, denote respective generators. Write

$$
\begin{aligned}
B_\psi^*(r) &= r_1 b_1 + r_2 b_2 \\
B_\psi^*(s) &= s_1 b_1 + s_2 b_2 \\
B_\psi^*(t) &= t_1 b_1 + t_2 b_2 \\
B_\psi^*(w) &= w_1 b_1 + w_2 b_2
\end{aligned}
$$

6.1

Next, set

$$
\begin{aligned}
A_2 &= r_1(t_1 + w_1 - s_1) + t_1^2 + t_1 w_1 + w_1^2 - s_1^2 \\
B_2 &= r_1(t_2 + w_2 - s_2) + r_2(t_1 + w_1 - s_1) + 2(t_1 t_2 + w_1 w_2 - s_1 s_2) \\
&\quad + t_1 w_2 + t_2 w_1 \\
C_2 &= r_2(t_2 + w_2 - s_2) + t_2^2 + t_2 w_2 + w_2^2 - s_2^2 \\
A_3 &= t_1 w_1(t_1 + w_1 + r_1) \\
B_3 &= t_1 w_1(t_2 + w_2 + r_2) + (t_1 w_2 + t_2 w_1)(t_1 + w_1 + r_1) \\
C_3 &= (t_1 w_2 + t_2 w_1)(t_2 + w_2 + r_2) + t_2 w_2(t_1 + w_1 + r_1) \\
D_3 &= t_2 w_2(t_2 + w_2 + r_2),
\end{aligned}
$$

6.2

and define the two polynomials $A_2 z^2 + B_2 z + C_2 = f_1(z)$, $A_3 z^3 + B_3 z^2 + C_3 z + D_3 = f_2(z)$.

Recall that the resultant of the two polynomials $f_1$ and $f_2$, $R(f_1, f_2)$, [L], pp. 199-204, is the determinant

$$
\operatorname{Det}
\begin{pmatrix}
A_2 & B_2 & C_2 & 0 & 0 \\
0 & A_2 & B_2 & C_2 & 0 \\
0 & 0 & A_2 & B_2 & C_2 \\
A_3 & B_3 & C_3 & D_3 & 0 \\
0 & A_3 & B_3 & C_3 & D_3
\end{pmatrix}
$$

which can also be given as $A_2^3 A_3^2 \prod(f_i - g_j)$ where $f_1, f_2$ are the roots of $f_1(x)$, $g_1, g_2, g_3$ are the roots of $f_2(x)$.

We now have

THEOREM 6.3: *Assume that the action of $(S^1)^2$ on $SU(3)$ induced by $\psi$ has only isolated fixed points and that there is an $S^1 \subset (S^1)^2$ so that the restricted action of $S^1$ on $SU(3)$ is free. Then we have that $R(f_1, f_2) \neq 0$ and*

(1) *The action of $(S^1)^2$ on $SU(3)$ is free if and only if $R(f_1, f_2)$ is $\pm 1$.*

(2) *More generally, there are at most six fixed points for the action, each with a finite, cyclic, isotropy group $I_x(\psi)$ and we have*

$$|R(f_1(z), f_2(z))| = \prod_{j=1}^{6} |I_j(\psi)|.$$

PROOF: Consider the Borel construction $E_{(S^1)^2} \times_{(S^1)^2} SU(3)$ of 4.1. The fixed point sets for the general action will occur for points $(A, B) \in \psi((S^1)^2)$ where we may assume $A = B$, and both are diagonal. If all three entries of $\psi(A)$ are equal then the action is not faithful. If two of the entries are equal, then the fixed point set has the form $(S^1)^2 \backslash U(2)$, which is not an isolated point. Consequently, it follows that the three entries are distinct, the quotient $SU(3)/(S^1)^2$ is a six dimensional complex and, using the projection in 4.2, the Borel construction has the homotopy type of a six dimensional complex attached to a disjoint union of classifying spaces $B_{I(x)}$. Thus

$$H^i(SU(3) \times_\psi E_{(S^1)^2}; \mathbf{Z})$$
$$= \sum H^i(B_{I_x(\psi)}; \mathbf{Z})$$
$$= \begin{cases} \sum (\mathbf{Z}/|I_x(\psi)|) & \text{if } i \text{ is even and } I_x(\psi) \text{ is cyclic,} \\ 0 & \text{for } i \text{ odd if } I_x(\psi) \text{ is cyclic} \end{cases}$$

provided that $i \geq 7$.

On the other hand, we have a direct calculation of $H^*(E_{(S^1)^2} \times_\psi SU(3); \mathbf{Z})$ from the Serre spectral sequence of the fibration. We have $d^4(e_3) = y^2 f_1(x/y)$, and modulo the ideal generated by $d^4(e_3)$ we know that $d^6(e_5) = y^3 f_3(x/y)$. Thus, we have that $H^8(SU(3) \times_\psi E_{(S^1)^2}; \mathbf{Z}) = H^8(B_{(S^1)^2}; \mathbf{Z})/(d^4(e_3), d^6(e_5))$. This ideal in dimension eight is precisely the lattice

$$\mathcal{L} = \langle x^2 d^4(e_3), xy d^4(e_3), y^2 d^4(e_3), x d^6(e_5), y d^6(d_5) \rangle.$$

Moreover, $H^8(SU(3) \times_\psi E_{(S^1)^2}; \mathbf{Z}) = H^8(B_{(S^1)^2}; \mathbf{Z})/\mathcal{L}$ as an easy verification shows. But the order of this quotient is just the absolute value of the determinant for any basis of $\mathcal{L}$. Hence, this order is exactly the absolute value of the determinant above, but this determinant is, by definition, the resultant of $f_1$ and $f_2$.

Now we verify that there are at most six fixed points, each with a cyclic isotropy group. To do this we assume that the image of $\psi$ is contained in the maximal torus $T_4$, so a typical element in the image has the form

$$(D(\lambda^{i_1}\tau^{j_1}, \lambda^{i_2}\tau^{j_2}, \lambda^{i_3}\tau^{j_3}), D(\lambda^{i_4}\tau^{j_4}, \lambda^{i_5}\tau^{j_5}, \lambda^{i_6}\tau^{j_6}))$$

and the point is a fixed point if and only if the entries in the second matrix are a permutation of the entries in the first matrix. Moreover if $\alpha \in S_3$ is the permutation matrix associated to this point, then $\{\mathrm{Det}(\alpha)\alpha\}$ in $(S^1)^2\backslash SU(3)$ is the fixed point. Hence, there are at most 6 fixed points and they form a subset of $S_3$.

Finally, we check that the isotropy groups are all cyclic. To do this choose a new basis for $(S^1)^2$ so that the first copy of $S^1$ gives the free action on $SU(3)$. Now, on this quotient, consider the action of the second $S^1$. Each isotropy group here is clearly a finite subgroup of $S^1$, hence cyclic, and the result follows.                                              ∎

## §7. Two homotopy invariants for the spaces $X^7$

Consider the pairing

7.1     $H^7(X^7; \mathbf{Q}/\mathbf{Z}) \otimes H_7(X^7; \mathbf{Z}) \to \mathbf{Q}/\mathbf{Z}, \qquad \theta \otimes (n[X^7]) \to \langle \theta, n[X^7] \rangle.$

There is a quadratic map $\psi: H^3(X^7; \mathbf{Q}/\mathbf{Z}) = \mathbf{Z}/L_2 \to \mathbf{Q}/\mathbf{Z}$ defined by $v \mapsto \langle v\beta(v), [X^7] \rangle$ where $\beta: H^3(X^7; \mathbf{Q}/\mathbf{Z}) \to H^4(X^7; \mathbf{Z})$ is the universal Bockstein. Incidentally, note that $\beta$ is an isomorphism in this case.

As we have pointed out in 5.1, there is a canonical generator $b^2 \in H^4(X^7; \mathbf{Z})$, hence a unique element $v_0 \in H^3(X^7; \mathbf{Z})$ with $\beta(v_0) = b^2$. Consequently, $\pm\psi(v_0) \in \mathbf{Z}/L_2 \subset \mathbf{Q}/\mathbf{Z}$ is a homotopy invariant of $X^7$. (The plus or minus indeterminacy is due to the fact that there is no canonical choice for the generator $[X^7] \in H_7(X^7; \mathbf{Z}) = \mathbf{Z}$). We will call $\pm\psi(v_0) \in \mathbf{Z}/\sigma_2$ the *link characteristic* of $X^7$.

Here is a second homotopy invariant for the space $X^7$.

Given $X^7$ there is a well defined map (up to sign) $\pi: X^7 \rightarrow \mathbf{P}^\infty$ with $\pi^*(b) = b$. It's fiber is $V_{2,1}$. Moreover if $g: X^7 \rightarrow \bar{X}^7$ is any homotopy equivalence, then there is a homotopy commutative diagram

$$
\begin{array}{ccc}
X^7 & \xrightarrow{\ g\ } & \bar{X}^7 \\
\downarrow{\scriptstyle \pi} & & \downarrow{\scriptstyle \pi'} \\
\mathbf{P}^\infty & \xrightarrow{\ \pm 1\ } & \mathbf{P}^\infty
\end{array}
$$

Thus, a homotopy invariant of $X^7$ is the mapping cone of $\pi$, $M(X^7)$. Moreover, if we have any multiplicative cohomology theory, $h^*$, then the action of $h^*(\mathbf{P}^\infty)$ on $h^*(M(X^7))$ is also intrinsic.

More generally, if $f: E \rightarrow B$ is a fibering with fiber $V_{2,1}$ we may take the mapping cone of $f$ as an invariant of the fibration. This mapping cone will not generally be a homotopy invariant of $E$, of course. However, as we will see the mapping cone behaves very much like the Thom complex of a spherical fibration and has good naturality properties which allow us to extract homotopy type information about $X^7$.

LEMMA 7.2: *Let $M(f)$ be the mapping cone of the fibration $V_{2,1} \xrightarrow{} E \xrightarrow{f} B$, and suppose that the group of $f$ is contained in $U(3)$ or $T_4$, or that the coefficients are $\mathbf{A} = \mathbf{Z}(\frac{1}{3})$ and the group is contained in $SU(3) * SU(3)$. Suppose, also that the element $d_4(e_3) \in H^4(B; \mathbf{A})$ is regular. Then there are natural classes $U_4$, $U_6$ in $H^*(M(f); \mathbf{Z})$ and a long exact sequence*

$$
\cdots \xrightarrow{\ \delta\ } H^{*-10}(B) \xrightarrow{\cup v} H^{*-4}(B) \otimes U_4 \oplus H^{*-6}(B) \otimes U_6 \longrightarrow \tilde{H}^*(M(f)) \xrightarrow{\ \delta\ } \cdots
$$

*where $v = c_3(f)U_4 - c_2(f)U_6$.*

PROOF: In the cases above, for $U(3)$, $T_4$, or $SU(3) * SU(3)$ with $\mathbf{Z}(\frac{1}{3})$ as coefficients, there are the three Serre spectral sequences with $E_2$-term $H^*(B_G; H^*(V_{2,1}; \mathbf{A}))$ converging to $H^*(E; \mathbf{A})$. In each case the spectral sequence is totally trangressive, with $(d_4(e_3), d_6(e_5))$ forming a regular sequence in $H^*(B; \mathbf{A})$. Consequently, $E_7 = E_\infty$ is concentrated on the bottom row and is given as the quotient

$$
H^*(B; \mathbf{A})/(d_4(e_3), d_6(e_5)).
$$

Consequently, in these particular examples the cohomology of $M(f)$ can be identified with the ideal $(d_4(e_3), d_6(e_5))$ itself.

Now, for the general case. The assumption that $d_4(e_3)$ is regular implies that

$$E_5 = H^*(B; \mathbf{A})/(d_4(e_3)) \bigoplus H^*(B; \mathbf{A})/(d_4(e_3))e_5$$

with only the differential $d_6(e_5)$ remaining. Now, use naturality and the truth of the result in the universal case of the $V_{2,1}$ fibration over $B_{T_4}$ or $B_{SU(3)*SU(3)}$ to obtain the classes $U_4$ and $U_6$.                       ∎

EXAMPLE 7.3: Consider the fibration $V_{2,1} \to X^7 \xrightarrow{f} \mathbf{P}^\infty$. This satisfies the conditions of the lemma. Hence there is an exact sequence

$$0 \longrightarrow \mathbf{Z}[b]x_{10} \xrightarrow{h} \mathbf{Z}[b](U_4, U_6) \longrightarrow H^*(M(f); \mathbf{Z}) \longrightarrow 0$$

with $h(x_{10}) = b^2(L_3bU_4 - L_2U_6)$.

EXAMPLE 7.4: In low dimensions the universal fibration over $B_{SU(3)*SU(3)}$ has the property that $H^*(M(f); \mathbf{Z}(\frac{1}{3}))$ is torsion free and has generators

$$\begin{cases} U_4 & \text{dimension 4,} \\ U_6 & \text{dimension 6,} \\ \sigma_2 U_4, \sigma_2' U_4 & \text{dimension 8.} \end{cases}$$

Note also that $Sq^2(U_4) = U_6$, and $U_4^2 = (\sigma_2 - \sigma_2')U_4$.

REMARK 7.5:   The situation for the universal mapping cone over $B_{SU(3)*SU(3)}$ with $\mathbf{Z}$ or $\mathbf{F}_3$ coefficients is quite a bit more complex due to an interior $d_3$-differential on $U_6$, $(d_3(e_5) = re_3)$. But when we take that into account, $H^*(M(f); \mathbf{F}_3)$ in the universal case turns out to have low dimensional generators described as follows:

7.6     $$\begin{cases} U_4 & \text{dimension 4,} \\ rU_4 & \text{dimension 6,} \\ r^2U_4, (\sigma_2 - \sigma_2')U_4, P^1(U_4) & \text{dimension 8,} \end{cases}$$

where $P^1(U_4)$ is the image of the Steenrod $p^{th}$-power operation on $U_4$.

## §8. The determination of the homotopy type of $X^7$

Here is our main result: the determination of the homotopy types of the $X^7$'s in terms of the map $\nu: S^1 \to T_4$ of 5.1.

THEOREM 8.1: *Let $\nu$, as in 5.1, determine $X^7$, and $L_2(\nu)$, $L_3(\nu)$ be the associated integers.*

(1) *The link characteristic of $X^7$ is given as $L_3(\nu)^{-1}/L_2(\nu)$.*

(2) *$L_2(\nu)$ and the link characteristic of $X^7$ determine the homotopy type of $X^7$ up to an indeterminacy of at most four. More precisely, given a quotient $X^7$ where $L_2(\nu)$ and the link invariant are fixed, then it belongs to one of at most four distinct homotopy types.*

(3) *These four types are, in turn, distinguished by the mod(2) value of $\nu^*(\sigma_2')$ and the value of $M(\nu)^*((r^2 + s(r + s))U_4 - U_4^2 + rU_6)$ mod (3).*

REMARKS 8.2: The indeterminacy of at most 4 in 8.1(2) is due to a close analysis of the Postnikov system for $X^7$. It turns out that the fact that $V_{2,1}/S^1 = X^7$, $L_2$ and the link invariant determine the first 2 Postnikov invariants completely, but we can only determine part of the next invariant from this data. Looking at the variation we will see, after 8.4, that there are at most 4 possibilities. Then the remainder of the proof is concerned with showing that the invariants of 8.1(3) determine the rest of this $k$-invariant.

For 8.1(3), it turns out that only thing which matters is whether $r$ is 0 or non-zero mod(3). Indeed, $U_6 \in H^6(M(\nu); \mathbf{F}_3)$ is only determined up to sign and the addition of $\epsilon b U_4$. Also, the term $(r^2 + s(r + s))U_4 - U_4^2 + rU_6$ is simply $P^1(U_4)$ in the mapping cone for $B_{T_4}$. Thus, the indeterminacy of $U_6$ allows us to assume that $P^1(U_4) = rU_6$ in $H^8(M(\nu); \mathbf{F}_3)$.

PROOF: The proof comprises three distinct steps. First we use the Postnikov system of $X^7$ through dimension 7 (which completely determines the homotopy type of $X^7$) to prove (1) and (2). Then we study the homotopy type of $M(\nu)$ in dimension 8 to distinguish the remaining four cases.

*The Postnikov system for $X^7$*

The first eight homotopy groups of $SU(3)$ are given by the following table, [Mi], p. 970:

8.3

| dim | 1 | 2 | 3 | 4 | 5 | 6 | 7 | 8 |
|---|---|---|---|---|---|---|---|---|
| $\pi_i(SU(3))$ | 0 | 0 | $\mathbf{Z}$ | 0 | $\mathbf{Z}$ | $\mathbf{Z}/6$ | 0 | $\mathbf{Z}/12$ |

We will use this to determine the first few stages of a Postnikov resolution for $X^7$, Consider the commutative diagram which relates the first four stages of the Postnikov systems for $X^7$ and the first three stages of the system for $V_{2,1}$.

8.4

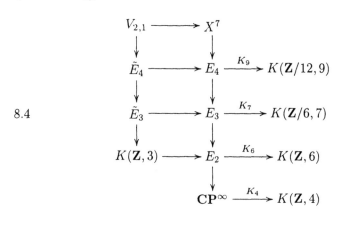

The first $k$-invariant for $V_{2,1}$ is $\beta Sq^2(\iota_3) \in H^6(K(\mathbf{Z},4);\mathbf{Z}) = \mathbf{Z}/2$. On the other hand the space $E_2$ is determined by the invariant $K_4 = L_2 b^2$, and it is direct to check that $H^6(E_2;\mathbf{Z}) = \mathbf{Z}/2 \oplus \mathbf{Z}/L_2$, where the $\mathbf{Z}/2$ is generated by an element which restricts to $\beta Sq^2(\iota_3)$ in $H^6(K(\mathbf{Z},3);\mathbf{Z}) = \mathbf{Z}/2$, and the $\mathbf{Z}/L_2$ summand is generated by $b^3$. Clearly, the invariant $K_6$ is thus the sum $\beta Sq^2(\iota_3) + L_3 b^3$.

Then a direct calculation shows that $H^7(\tilde{E}_3;\mathbf{Z}/6) = \mathbf{Z}/6$ is generated by $Sq^2(\iota_5)$ and $P^1(\iota_3)$, while $H^7(E_3;\mathbf{Z}/6) = \mathbf{Z}/6 \oplus \mathbf{Z}/6$, the first summand generated by the elements restricting to the fiber as described and the second generated by the class $L_3 b^2 \iota_3 - L_2 b\iota_5 = b(L_3\iota_3 - L_2\iota_5) = bf_5$

Hence, cupping with $b$ and dividing by $L_2$ we obtain that the generator is $-L_3 b^2 L_2$ so the link characteristic for $X^7$ must be $-L_3^{-1}/L_2$ at this stage, though there remains the analysis of the next $k$-invariant before we can claim this answer for $X^7$ itself.

We now consider the $k$-invariant in dimension 7. As we've seen, $H^7(E_3; \mathbf{Z}/6) = (\mathbf{Z}/6)^2$ with generators $[X^7]^*$, the reduction of the orientation class (this is imprecise – we will make it precise shortly however) and the class corresponding to $P^1(\iota_3)$ summed with the class corresponding to $Sq^2(\iota_5)$. We know that the $k$-invariant must involve these two classes in an essential way, so we may assume the $k$-invariant has the form $(w, 1, 1)$ where $w$ is the coefficient on $[X^7]^*$. The question now is the determination of $w$.

There is a homotopy equivalence of $E_3$ which changes the sign of $x_5$ but leaves $\iota_3$ and $b$ invariant. This changes the $k$-invariant $(w, 1, 1)$ to $(-w, 1, 1,)$, and hence $w$ may be assumed to be one of the four elements $0, 1, 2, 3 \in \mathbf{Z}/6$.

## The homotopy type of the mapping cone through dimension 8

Through dimension 7 the universal mapping cone has the homotopy type of the two cell complex $S^4 \cup_\eta e^6$, and $\pi_7$ of this space is $\mathbf{Z}/6 \oplus \mathbf{Z}$ where the $\mathbf{Z}$ is generated by the Hopf invariant class, $h$, the class of the Hopf fibration

$$S^3 \longrightarrow S^7 \longrightarrow S^4$$

and the $\mathbf{Z}/6$ is in the image of suspension from $\pi_6(S^3 \cup_\eta e^5)$. Also, $\pi_7(S^6) = \mathbf{Z}/2$ with generator $\eta S^6$, and $\pi_7((S^4 \cup_\eta e^6) \vee S^6)$ is the direct sum of these two groups, $\mathbf{Z} \oplus \mathbf{Z}/6 \oplus \mathbf{Z}/2$, since the first Whitehead product between the two pieces occurs in dimension 8. Thus, the homotopy type of the mapping cone for the fibration over $B_{T_4}$, $B_{U(3)}$ or $B_{S^1}$ through dimension 7 has the form

$$\left(S^4 \cup_\eta e^6\right) \vee \bigvee_1^s S^6$$

and $\pi_7$ of this space is $\mathbf{Z}/6 \oplus \mathbf{Z} \oplus (\mathbf{Z}/2)^s$ since there are no Whitehead products in this range. We will show that the cell dual to $\sigma_2' U_4$ attaches to $3v + h$ while the cell dual to $\sigma_2 U_4$ attaches to $h$. Moreover, the cell dual to $P^1(U_4)$ (with coefficients $\mathbf{F}_3$ attaches to $2v$ where $v$ generates the $\mathbf{Z}/6$. Thus, depending on the form of the maps to $h_*(M(\pi); \mathbf{F}_2)$ and $H_*(M(\pi); \mathbf{F}_3)$ we distinguish the four possible homotopy types with fixed $L_2$ and link invariant.

The $\mathbf{F}_3$-argument is direct. Since we may regard $U_4$ as identified with $r(t + w - s) + t^2 + tw + w^2 - s^2$, $U_6$ as $tw(t + w - r)$, we can directly calculate that

$$P^1(U_4) = r^3(t + w - s) + r(t^3 + w^3 - s^3) + t^3w + tw^3 + s^4 - t^4$$
$$= rU_5 + s(r + s)U_4 - U_4^2 + r^2U_4.$$

The $\mathbf{F}_2$-argument involves a check in the case of the mapping cone over $B_{U(3)}$, where $U(3)$ embeds in $SU(3) * SU(3)$ via the map $\mu$ of 1.5. For this example, because the action is just the usual left action by $U(3)$ on $V_{2,1}$, we see that the fibration $V_{2,1} \to E_{U(3)} \times_{U(3)} V_{2,1} \to B_{U(3)}$ is just the usual fibration

8.5                        $$V_{2,1} \longrightarrow B_{U(1)} \overset{\pi}{\longrightarrow} B_{U(3)}$$

induced from the usual inclusion $U(1) \hookrightarrow U(3)$.

The inclusion $B_{U(2)} \hookrightarrow B_{U(3)}$ induced by the usual inclusion induces a map of mapping cones, $B_{U(2)}/B_{U(1)} \hookrightarrow B_{U(3)}/B_{U(1)}$, which includes the Thom space of the universal $\mathbf{C}^2$-bundle over $B_{U(2)}$ into our mapping cone. This gives a map

8.6                   $$h: \Sigma^2\mathbf{P}^\infty \longrightarrow B_{U(2)}/B_{U(1)} \longrightarrow B_{U(3)}/B_{U(1)}$$

since $\Sigma^2\mathbf{P}^\infty$ is the Thom space of the $\mathbf{C}^2$-bundle $\xi + \epsilon$ over $\mathbf{P}^\infty$ where $\xi$ is the Hopf line bundle and $\epsilon$ is the trivial bundle. The cohomology map $h^*$ is given by $h^*(c_1^i U_4) = e^2 \otimes b^{i+1}$, and $h^*(w) = 0$ for any other monomial in $c_1, c_2, c_3$ cupped with $U_4$ or $U_6$.

LEMMA 8.7: *The attaching map, $\tau$ of $\Sigma^2 e^6 = e^8$ in $\Sigma^2\mathbf{P}^3 = (S^4 \cup_\eta e^6) \cup_\tau e^8$ has order 6 in $\pi_7(S^4 \cup_\eta e^6)$.*

PROOF: The reduced complex $K$-group for $\Sigma^2\mathbf{P}^3$, $\tilde{K}^U(\Sigma^2\mathbf{P}^3)$, is a copy of $\mathbf{Z}^3$ with generators $s \otimes x$, $s \otimes x^2$, $s \otimes x^3$ in $K^U(\Sigma^2\mathbf{P}^\infty) = s \otimes x\mathbf{Z}[x]$. The Adams $\psi^k$-operations are defined by the rules $\psi^k(a + b) = \psi^k(a) + \psi^k(b)$, $\psi^k(ab) = \psi^k(a)\psi^k(b)$, and $\psi^k(s) = ks$ where $s$ is a generator for $\tilde{K}^U(S^2) = \mathbf{Z}$. For $\tilde{K}^U(\mathbf{P}^3)$ we have $\psi^k(x) = kx + \binom{k}{2}x^2 + \binom{k}{3}x^3$. In particular, following work of J.F. Adams and Grant Walker, (see the discussion on pages 143-144 of [A]), the integral sub-lattice spanned by the eigenvectors for the $\psi^k$ operations in $\tilde{K}^U(\Sigma^2\mathbf{P}^3)$ is generated by the elements

$$w_1 = 6x - 3x^2 + 2x^3$$
$$w_2 = x^2 - x^3$$
$$w_3 = x^3$$

where $\psi^k(w_i) = k^i(w_i)$ for each integer $k \geq 1$, and each $i = 1, 2, 3$.

Now, let $j: S^8 \to \Sigma^2 \mathbf{P}^3$ be any map. We have $\tilde{K}^U(S^8) = \mathbf{Z}(s^4)$ with $\psi^k(s^4) = k^4 s^4$. Thus, since the $\psi^i$ are natural, we must have $j^*(s \otimes w_1) = j^*(s \otimes w_2) = 0$ and $j^*(s \otimes w_3) = \lambda s^4$ where $\lambda$ is the degree of $j$. Solving, we see that $j^*(x^2) = \lambda s^4$, so, plugging into the expression for $w_1$ we have $j^*(6x) = \lambda s^4$ as well and 6 must divide $\lambda$ as asserted. ∎

This, in turn, completes the proof of 8.1. ∎

## §9. The tangent bundle to $S^1 \backslash SU(3)$ where $S^1$ acts freely and isometrically

There is some discussion of the structure of $\tau(H \backslash G)$ in [S], at least for the case where $H$ is a product $H_1 \times H_2 \subset G * G$. In this case the result of [S] is

PROPOSITION 9.1: *The tangent bundle to $H_1 \backslash G / H_2$ fits into the following canonical exact sequence of vector bundles:*

$$0 \longrightarrow \alpha_{H_1}(Ad_{H_1}) \oplus \alpha_{H_2}(Ad_{H_2}) \longrightarrow \alpha_G(Ad_G) \longrightarrow \tau(H_1 \backslash G / H_2) \longrightarrow 0$$

*where $Ad_K$ is the adjoint action of $K$ on the Lie algebra of $K$ and*

$$
\begin{aligned}
\alpha_{H_1}(V) &= V \times_{H_1} G / H_2 \\
\alpha_{H_2}(W) &= H_1 \backslash G \times_{H_2} W \\
\alpha_G(U) &= (H_1 \backslash G \times G / H_2) \times_G U
\end{aligned}
$$

*and in the last $G$ acts as $(l, m)g = (lg, g^{-1}m)$.*

Most of the $S^1$-actions considered here are not immediately given in the form above though they all can be thought of as double quotients when we write $X$ as a double quotient of

$$L = \{(A, B) \in U(3) \times U(3) \mid \det A = \det B\}$$

as the referee kindly points out. In any case, the result below gives the explicit information that we will need about these bundles. Also, the discussion that we now give generalizes easily to isometric circle actions on $SU(n)$ as well as quotients by certain larger groups.

LEMMA 9.2: *In the notation of the previous section let the free isometric action of $S^1$ on $SU(3)$ be defined by the 4-tuple of integers $(p_1, p_2, p_3, p_4)$ as in 5.1. Then we have*

$$\tau(S^1 \backslash SU(3)) \oplus 3\epsilon_{\mathbf{R}} \oplus \xi_i^{\frac{p_3 - p_4}{3}} = \sum_{i=1}^{3} \sum_{j=1}^{2} \xi^{\frac{p_i - p_2 + j}{3}}.$$

PROOF: Embed $V_{2,1}$ into $S^{11} \subset \mathbf{C}^6$ by $(\vec{X}_1, \vec{X}_2) \mapsto \frac{1}{\sqrt{2}}(\vec{X}_1, \vec{X}_2)$. Then the normal bundle to $V_{2,1}$ in $S^{11}$, $\nu_{S^{11}}(V_{2,1})$, is $\epsilon_{\mathbf{R}} \oplus \epsilon_{\mathbf{C}}$. Indeed, if we define

$$\epsilon : S^{11} \longrightarrow \mathbf{R} \times \mathbf{C}$$

by

$$\vec{X} \mapsto (\sqrt{x_1 \bar{x}_1 + x_2 \bar{x}_2 + x_3 \bar{x}_3} - \frac{1}{\sqrt{2}}, x_1 \bar{x}_4 + x_2 \bar{x}_5 + x_3 \bar{x}_6)$$

then $V_{2,1}$ is the inverse image of $(0,0)$ and the map is regular there.

Next we note that we can extend our $S^1$-action to $S^{11}$ by the rule

$$(x_1, x_2, \ldots, x_6) \mapsto (\lambda^{\frac{p_1 - p_3}{3}} x_1, \lambda^{\frac{p_2 - p_3}{3}} x_2, \ldots, \lambda^{\frac{p_3 - p_4}{3}} x_6)$$

and this action extends to the normal bundle, trivially on the $\mathbf{R}$, but as multiplication by $\lambda^{\frac{p_3 - p_4}{3}}$ on the $\mathbf{C}$.

When we pass to quotients by the $S^1$ action note that the tangent bundle to $V_{2,1}$ is identified with $\tau(S^1 \backslash SU(3)) \oplus \epsilon_{\mathbf{R}}$, and the quotient of the tangent bundle to $S^{11}$, restricted to $V_{2,1}$ becomes

$$\tau(S^1 \backslash SU(3)) \oplus 2\epsilon_{\mathbf{R}} \oplus \xi^{p_3 - p_4}.$$

On the other hand, since $\tau(S^{11}) \oplus \epsilon_{\mathbf{R}} = 6\epsilon_{\mathbf{C}}$ with the action given above we obtain the expression in the lemma for $\tau(S^1 \backslash SU(3))$.   ∎

COROLLARY 9.3: *The Pontrjagin class $P_1$ for the normal bundle to $X^7(p_1, p_2, p_3, p_4)$ is given as*

$$\frac{2}{3}(p_1^2 + p_2^2 + p_1 p_2)b^2 \in H^4(X^7; \mathbf{Z}) = \mathbf{Z}/L_2.$$

PROOF: The tangent bundle is stably the pull-back of

$$\xi^{\frac{p_1 - p_3}{3}} \oplus \xi^{\frac{p_1 - p_4}{3}} \oplus \xi^{\frac{p_2 - p_3}{3}} \oplus \xi^{\frac{p_2 - p_4}{3}} \oplus \xi^{\frac{-(p_1 + p_2 + p_3)}{3}} \oplus \xi^{\frac{-(p_1 + p_2 + p_4)}{3}} \oplus -\xi^{\frac{p_3 - p_4}{3}}$$

from the canonical map $\pi \colon X^7 \to \mathbf{CP}^\infty$. Hence, $P_1$ for the normal bundle is the pull back of the class

$$\frac{1}{9}(4p_1^2 + 4p_1 p_2 + 4p_2^2 + 2p_3^2 + 2p_3 p_4 + 2p_4^2)b^2.$$

On the other hand, this is to be taken in $H^4(X^7; \mathbf{Z}) = \mathbf{Z}/L_2$, where $L_2$ is given in 5.1. Thus subtracting $2L_2(\nu)b^2$ gives the class of 9.3. ∎

## §10. The piecewise linear classification of $X^7$

The Pontrjagin class, $P_1$, is a homotopy invariant in $H^4(X^7; \mathbf{Z}/24)$ from [M], p. 249. Also, given two homotopy equivalent $S^1$-quotients, $X^7$ and $\bar{X}^7$, the difference of their Pontrjagin classes $P_1(\bar{X}^7) - P_1(X^7)$ is well defined and independent of the choice of homotopy equivalence since in cohomology, in dimensions $\leq 4$, the choices differ at most in the sign of $b$ in dimension 2, but the map on $b^2$ will, in either case be the identity.

The core of the surgery classification of manifolds in dimensions $\geq 5$ is the surgery exact sequence, [MM], pp. 40-44, which in our context, in view of the Browder-Novikov theorem and the fact that the dimension of $X^7$ is odd, takes the form

$$\mathcal{HT}(X^7) = [X^7, G/PL],$$

the set of homotopy classes of maps from $X^7$ to $G/PL$, where $\mathcal{HT}(X^7)$ is the set of homotopy triangulations of $X^7$. Recall that $\mathcal{HT}(X^7)$ surjects to the set of piecewise linear homeomorphism classes of manifolds homotopy equivalent to $X^7$.

LEMMA 10.1: *The set* $[X^7, G/PL]$ *contains* $2L_2$-*elements and is determined by the difference* $P_1(\bar{X}^7) - P^1(X^7)$ *and a mod(2) invariant in dimension 4 if* $L_2$ *is even, while it contains* $2L_2$ *elements and is determined up to an ambiguity of* $\mathbf{Z}/2$ *in dimension 2 by this difference if* $L_2$ *is odd.*

PROOF: Through dimension 7, $G/PL$ has the homotopy type of a product $K(\mathbf{Z}/2, 6) \times V$ where $V$ is the two stage Postnikov system

$$K(\mathbf{Z}, 4) \longrightarrow V \longrightarrow K(\mathbf{Z}/2, 2)$$

with $k$-invariant $2\beta(\iota_2^2)$. Consequently, there is an exact sequence

$$[X^7, K(\mathbf{Z}/2, 1)] \longrightarrow [X^7, K(\mathbf{Z}/2, 6) \times K(\mathbf{Z}, 4)]$$
$$\longrightarrow [X^7, G/PL] \longrightarrow [X^7, K(\mathbf{Z}/2, 2)] \longrightarrow 0$$

which becomes

$$0 \longrightarrow H^4(X^7; \mathbf{Z}) \longrightarrow [X^7, G/PL] \longrightarrow H^2(X^7; \mathbf{Z}/2) \longrightarrow 0$$

Hence, $|[X^7, G/PL]| = 2L_2$.

Now, $H^4(V; \mathbf{Z}) = \mathbf{Z}$ with generator $v$, restricting to $\iota^2$ mod (2) and restricting to the fiber as twice the generator $\iota_4$. Thus, if the map to $V$ factors through the inclusion of the fiber, the pull-back of this class must be an even multiple of $b^2$, and conversely, it is an odd multiple if it does not factor through the fiber. On the other hand, mod (2) the generator on the fiber corresponds to a sort of Stiefel-Whitney class, $\bar{w}_4$, which we cannot determine just from the Pontrjagin class. Consequently, with coefficents in $\mathbf{F}_2$, this class gives the remaining invariant. ∎

# References

[AW] R.S. Aloff, N.R. Wallach, An infinite family of distinct 7-manifolds admitting positively curved Riemannian structure, *Bull. Amer. Math. Soc.* **81** (1975), 93–97.

[AMP] L. Astey, E. Micha, G. Pastor, Homeomorphism and diffeomorphism types of Eschenburg spaces, *Diff. Geom. Appl.* **7** (1997), 41–50.

[A] M. Atiyah, *K-Theory*, Benjamin (1964).

[E1] J.-H. Eschenburg, New examples of manifolds with strictly positive curvature, *Invent. Math.* **66** (1982), 469–480.

[E2] J.-H. Eschenburg, Inhomogeneous spaces of positive curvature, *Diff. Geom. Appl.* **2** (1992), 123–132.

[E3] J.-H. Eschenburg, Cohomology of biquotients, *Manuscripta Math.* **75** (1992), #2, 151–166.

[GM] D. Gromoll, W. Meyer, An exotic sphere with non-negative sectional curvature, *Ann. of Math.* **100** (1974), 401–406.

[KS1] M. Kreck, S. Stolz, A diffeomorphism classification of 7-dimensional homogeneous Einstein manifolds with $SU(3) \times SU(2) \times U(1)$-symmetry, *Ann. of Math.* (2) **127** (1988), #2, 373–388.

[KS2] M. Kreck, S. Stolz, Some nondiffeomorphic homeomorphic homogeneous 7-manifolds with positive sectional curvature, *J. Diff. Geom.* **33** (1991), #2, 465–486.

[K1] B. Kruggel, A homotopy classification of certain 7-manifolds, *Trans. Amer. Math. Soc.* **349** (1997), 2827–2843.

[K2] B. Kruggel, Kreck–Stolz Invariants, normal invariants and the homotopy classification of generalised Wallach spaces, *Quart. J. Math. Oxford* (2) **49** (1998), no. 196, 469–485.

[L] S. Lang, *Algebra*, Third edition, Addison-Wesley (1993)

[MM] I. Madsen, R.J. Milgram, *The Classifying Spaces for Surgery and Cobordism of Manifolds*, Annals of Math Studies, #92, Princeton U. Press, (1979).

[M] R.J. Milgram, Some remarks on the Kirby-Siebenmann class, *Algebraic Topology and Transformation Groups*, LNM #1361, Springer Verlag, (1988), pp. 247–252.

[Mi] M. Mimura, Homotopy theory of Lie groups, *Handbook of Algebraic Topology*, ed. I.M. James, Elsevier, (1995), 953–991.

[S] W. Singhof, On the topology of double coset manifolds, *Math. Ann.* **297** (1993), 133–146.

[W] E. Witten, Search for a realistic Kaluza-Klein theory, *Nuclear Phys. B* **186** (1981), no. 3, 412–428.

Department of Mathematics
Stanford University
Stanford, CA 94305
e-mail: `milgram@math.stanford.edu`

# Elliptic cohomology

## Charles B. Thomas

## Introduction

From the algebraic point of view elliptic cohomology is a quotient of spin cobordism, and as such forms part of a chain

$$\Omega^*_{spin} \longrightarrow \cdots? \longrightarrow \mathcal{E}\ell\ell \longrightarrow KSpin^* \longrightarrow H^* \ ,$$

with each link corresponding to a 1-dimensional commutative formal group law. In the case of elliptic cohomology this was written down by Euler in the 18th century, and the validity of the Eilenberg–Steenrod axioms follows from properties of addition on a class of elliptic curves in characteristic $p$.

In 1988, G. Segal gave a talk in the Bourbaki Seminar [Se], in which he summarised what was known at the time under the headings

(a) $\mathcal{E}\ell\ell^*(X)$ is a cohomology theory [Lw1,2],

(b) the structural genus is rigid with respect to compact, connected group actions [Tb], and

(c) the completed localised ring $\mathcal{E}\ell\ell^*(BG)^\wedge_p$ is determined by 'elliptic characters' [HKR1].

He also explained how a genus related to the universal elliptic genus ought to be defined as the index of a Dirac operator in infinite dimensions, and suggested how a geometric model for $\mathcal{E}\ell\ell^*(X)$ might be constructed, using ideas from conformal field theory.

The purpose of the present survey is to bring Segal's up to date, and incidentally to provide a bibliography of some of the more important recent papers. The most important advance is that an extremely elegant algebraic theory is beginning to find geometric applications. Thus it is becoming clear that elliptic cohomology is the correct setting for the Moonshine phenomena associated with various sporadic simple groups, and that E. Witten's genus $\varphi_W$ carries information about Ricci curvature of a metric in much the same way as the $\widehat{A}$-genus does about scalar curvature. Both these applications are, in a roundabout way, taken from theoretical physics. At least in the opinion of the author, they illustrate a belief that interesting geometric problems are suggested by the models which we construct to explain the behaviour of the world around us. Through sheer ignorance there is nothing about physics in the sections that follow, although the reader may like to look at the collection of papers [AtRS], particularly that of David Olive with its hints about the role played by the exceptional Lie and sporadic simple finite groups. In this direction it may be worth noting that if there is a theory above elliptic cohomology in the chain, then it ought to be modelled by bundles over the double loop space $L^2X$ with 'paths' defined by finitely cusped 3-manifolds with a prescribed geometry.

The contents are as follows: in the first three sections we cover much the same ground as Segal. Section 4 is devoted to the definition of elliptic objects over $BG$, where $G$ is a finite group, and will appear in a greatly expanded version in a joint paper with A. Baker. Section 5 contains various results about the Mathieu groups — each of these is fascinating, but they have yet to receive a satisfactory common explanation. The last section contains introductions to the work of M. Kreck and S. Stolz [KS] showing that elliptic cohomology can be defined over $\mathbb{Z}$ rather then over $\mathbb{Z}[\frac{1}{2}]$, and to the still mysterious $K3$-theories. Again through ignorance there is little stable homotopy theory, although I would have liked to have included more on the spectrum $eo_2$, currently being studied by M. Hopkins and his collaborators.

This survey has its origins in a guide for the perplexed* prepared jointly with A. Baker at Mount Holyoke in the summer of 1994. I wish to thank him for his collaboration, J. Morava for several stimulating conversations and e-mail exchanges, and S. Cappell for his steady interest in my work. I wish to acknowledge financial support from the European Union's 'Human Capital and Mobility' programme, and also from the ETH in Zürich, where the final version was written.

It was a pleasure to be asked to contribute to a collection of papers dedicated to C.T.C. Wall. On several occasions he has asked "Elliptic Cohomology — very interesting — but what is it?" I hope that I have gone some way towards providing an answer.

---

* Moses Maimonides (born Córdoba 1135, died Cairo 1204), doctor and philosopher, is responsible for this turn of phrase.

## 1. Elliptic Genera

Let us continue and expand the discussion of cobordism started in [Ros]. Working away from the prime 2 we know that

$$R_\Omega := \Omega^*_{SO} \otimes \mathbb{Z}[\tfrac{1}{2}] \cong \Omega^*_{Spin} \otimes \mathbb{Z}[\tfrac{1}{2}], \quad \text{that rationally}$$
$$R_\Omega \otimes \mathbb{Q} = \mathbb{Q}[\mathbb{C}\mathbb{P}^2, \mathbb{C}\mathbb{P}^4, \dots, \mathbb{C}\mathbb{P}^{2n}, \dots],$$

and that the oriented cobordism class of an $m$-dimensional manifold ($m \equiv 0 \bmod 4$) is detected by its Pontryagin numbers.

We recall that, if $\xi$ is a real vector bundle over $X$ of fibre dimension equal to $k$, then the Pontryagin classes $p_i(\xi)$ satisfy

1. $p_i(\xi) \in H^{4i}(X, \mathbb{Z})$ and $p_\bullet(\xi) = 1 + p_1(\xi) + \cdots + p_{[k/2]}(\xi)$
2. $p_i(f^!\xi) = f^* p_i(\xi)$
3. $p_\bullet(\xi_1 \oplus \xi_2) = (p_\bullet \xi_1) \cdot (p_\bullet \xi_2)$ modulo 2 torsion, and
4. $p_\bullet(\eta_\mathbb{R}) = 1 + g^2$, where $\eta_\mathbb{R}$ is the real bundle underlying the Hopf bundle over complex projective space $\mathbb{C}\mathbb{P}^n$, and $g$ is the Poincaré dual to the homology class represented by $\mathbb{C}\mathbb{P}^{n-1}$.

Furthermore, if $m \equiv 0 \pmod 4$, and $(i_1 \cdots i_r)$ is a partition of $\frac{m}{4}$, then with $[X]$ equal to the orientation class of $X$ we can define a Pontryagin number by

$$\left( \prod_{j=1}^r p_{i_j}(TX) \right)[X].$$

Here, $TX$ denotes the tangent bundle of the manifold $X$. Half of R. Thom's determination of the ring $R_\Omega \otimes \mathbb{Q}$ consists in showing that $X_1$ and $X_2$ represent the same class iff their Pontryagin numbers coincide. It is then necessary to find a family of manifolds picking up all possible values in $\mathbb{Q}$. Since, as has already been discussed in [Ros], $\Omega^*_{SO}$ contains no odd torsion, in much of what follows we will work over $\mathbb{Z}[\tfrac{1}{2}]$ rather than over $\mathbb{Q}$.

**Definition.** Let $R$ be an integral domain over $\mathbb{Q}$. Then a *genus* is a ring homomorphism

$$\varphi : R_\Omega \longrightarrow R$$

with $\varphi(1) = 1$.

Examples are provided by the signature (or $L$-genus) and $\hat{A}$-genus (for Spin manifolds).

One way of constructing genera is to start with an even power series

$$Q(x) = 1 + a_2 x^2 + a_4 x^4 + \cdots ,$$

give the variable $x$ a weight equal to 2, and form the product

$$Q(x_1)\cdots Q(x_n) = 1 + a_2(x_1^2 + \cdots + x_n^2) + \cdots .$$

The term $K_r$ of weight $4r$ is a homogeneous polynomial in the elementary symmetric functions $p_j$ of the $x_i^2$; note that for $r > n$ we have $K_r = K_r(p_1, \ldots, p_n, 0, \ldots, 0)$, i.e., that we set an appropriate number of 'dummy' variables equal to zero.

**Definition.** The genus $\varphi_Q$ associated with a power series $Q$ is defined by

$$\varphi_Q(X) := K_m\big(p_1(TX), \ldots, p_m(TX)\big)[X] \in R$$

if the dimension of $X$ equals $4m$, and $\varphi_Q(X) = 0$ otherwise.

In fact every multiplicative genus arises in this way. To see this we define the *logarithm of $\varphi$* by

$$g'(y) = \log_\varphi'(y) = \sum_{n=0}^{\infty} \varphi(\mathbb{C}\mathbb{P}^{2n}) y^{2n} .$$

An exercise in the calculus of residues then shows that, if $f$ denotes the formal inverse function to $g$, and we write $Q(x) = x/f(x)$, then

$$\varphi(\mathbb{C}\mathbb{P}^{2n}) = \varphi_Q(\mathbb{C}\mathbb{P}^{2n}) .$$

The logarithm $g$ is not only associated with a genus, but also with a unique formal group law $G$ given by

$$g\big(G(y_1, y_2)\big) = g(y_1) + g(y_2) .$$

Pulling all this together we have

**Proposition 1.1.** The multiplicative genus $\varphi : R_\Omega \to R$ is uniquely determined by any one of the power series $Q(x)$, the logarithm $g(y)$ or formal group law $G(y_1, y_2)$.

**Examples.**

$$Q_L(x) = x/\tanh(x) ,$$
$$Q_{\widehat{A}}(x) = (x/2)/\sinh(x/2) .$$

The two genera $L$ and $\widehat{A}$ are the limiting values of the so-called *universal elliptic genus* $\Phi$ with logarithm

$$\log_\Phi(y) = \int_0^y \frac{dt}{\sqrt{1 - 2\delta t^2 + \varepsilon t^4}} \ .$$

The limiting values, associated with hyperbolic rather than elliptic functions, are obtained when $\varepsilon$ and $\delta$ are so chosen that the quartic equation $\varepsilon t^4 - 2\delta t^2 + 1 = 0$ has repeated roots. Thus for $L$ we take $\delta = \varepsilon = 1$ and for $\widehat{A}$, $\delta = -1/8$, $\varepsilon = 0$. More generally:

**Theorem 1.2.** *The genus $\varphi$ is elliptic if and only if $\varphi$ vanishes on the total space of all fibrations of the form $\mathbb{CP}(\xi)$, i.e., projective bundles associated with complex vector bundles $\xi$.*

For detailed proofs of this result see either [HBJ], Chapters 3-4, or the original paper of S. Ochanine [O]. However the main ideas are the following:

**1.** Choose new generators $H_{i,j}$ for the cobordism ring over $\mathbb{Q}$. These are the so-called Milnor manifolds $H_{i,j} \subset \mathbb{CP}^i \times \mathbb{CP}^j$ of double degree $(1,1)$ defined by the equation

$$x_0 y_0 + \cdots + x_k y_k = 0 \quad (k = \min\{i, j\}) \ .$$

That these $2(i + j - 1)$-dimensional manifolds do include generating sets follows from a calculation with characteristic numbers. The new geometric property is that $H_{i,j}$ fibres over $\mathbb{CP}^i$ with fibre $\mathbb{CP}^{j-1}$, so that with $j =$ even, $j \geq 4$, we obtain a manifold on which we are hoping to show that $\varphi$ vanishes.

**2.** In the cobordism ring let $J = $ ideal generated by the classes $\{[H_{3,2i}]i \geq 2\}$, and let $I$ be generated by $\mathbb{CP}^{2m-1}$ fibrations. Then $J \subseteq I$ by the discussion above.

**3.** If $\varphi$ is a multiplicative genus, then $\varphi(J) = 0$ if and only if $\varphi$ is elliptic. This is an exercise in elementary calculus, see [O] Proposition 3.

Note that one can use the formal group law for the elliptic genus (Euler's addition formula) to show that $\varphi(H_{2i+1,2j}) = 0$ for $j > i$.

If we write $h_{ij} = \varphi(H_{i,j})$, then direct calculation shows that

$$\sum_{i,j \geq 0} h_{ij} y_1^i y_2^j = G(y_1, y_2) g'(y_1) g'(y_2) \ .$$

Since

$$G(y_1, y_2) = \frac{y_1 \sqrt{P(y_2)} + y_2 \sqrt{P(y_1)}}{1 - \varepsilon \cdot y_1^2 y_2^2} \quad ,$$

where

$$P(y) = 1 - 2\delta y^2 + \varepsilon y^4 \quad , \quad g'(y) = \frac{1}{\sqrt{P(y)}} \quad ,$$

we have

$$\sum_{i,j \geq 0} h_{ij} y_1^i y_2^j = \left( \frac{y_1}{\sqrt{P(y_1)}} + \frac{y_2}{\sqrt{P(y_2)}} \right) \left( 1 + \varepsilon y_1^2 y_2^2 + \varepsilon^2 y_1^4 y_2^4 + \cdots \right) .$$

Now compare coefficients on the two sides. On the right there is no monomial of the form $y_1^{2i+1} y_2^{2j}$.

**4.** The ideals $I$ and $J$ coincide. This follows from the assertion that an elliptic genus $\varphi$ vanishes on $\mathbb{CP}^{2m-1}$-fibrations, and is proved by calculating $\varphi$ on the total space. More precisely one looks at the characteristic numbers of the bundle along the fibres, and exploits elementary properties of elliptic functions.

**Definition.** Let $F \to E \to B$ be a fibre bundle with $F$ a compact, oriented spin manifold and structural group $G$ a compact connected Lie group. We say that the genus $\varphi$ is *strongly multiplicative* if $\varphi(E) = \varphi(F)\varphi(B)$.

**Theorem 1.2.**[bis] *The genus $\varphi$ is strongly multiplicative if and only if $\varphi$ is elliptic.*

That this is at least plausible follows from the fact that for a $\mathbb{CP}^{2m-1}$-fibration an elliptic genus is strongly multiplicative, since $\mathbb{CP}^{2m-1}$ bounds the total space of a $D^3$-bundle over $\mathbb{HP}^{m-1}$, and therefore $\varphi(\mathbb{CP}^{2m-1}) = 0$.

The stronger version of Theorem 1.2 is due to C. Taubes [Tb], and expresses the *rigidity* of the elliptic genus. If the manifold $X$ admits an action by the compact topological group $G$, then any genus $\varphi$ can be thought of as taking values in the complex character ring $R(G) \otimes \mathbb{C}$, which in turn is isomorphic to the algebra of complex-valued class functions on $G$. (In a later section we will give an example of a family of genera, some being elliptic, parametrised by conjugacy classes in the discrete Mathieu group $M_{24}$.)

**Theorem 1.2.**[terce] *The genus $\varphi$ is elliptic if and only if for each smooth pair $(X, G)$ with $X$ a spin manifold and $G$ compact and connected, $\varphi_G(M)$ is constant as a function on $G$.*

**Proof.** See [Se] Theorem 3.7 *et seq.* In outline one reformulates the definition of $\varphi$ in terms of $K$-theory, applies it to the total space of a spin fibration and looks at the term coming from the 'bundle along the fibre'. Taking the augmentation gives $\varphi(F)$, which the assumed rigidity shows to be the whole fibre term. Conversely one reduces to the case of the circle group $S^1$, and then shows that strong multiplicativity forces any variation with respect to the group action to vanish.

After this brief introduction to rigidity let us return to the universal elliptic genus $\Phi$. Assume that the discriminant $\Delta = \varepsilon^2(\delta^2 - \varepsilon) \neq 0$, so that the integral defining the logarithm is the inverse function to an odd elliptic function $s$. This is characterised by the period lattice $L$ and by the fact that of the three points of order 2 in $\mathbb{C}/L$ one ($\omega$) is a zero and the others poles. If one replaces $(\delta, \varepsilon)$ by $(\lambda^2\delta, \lambda^4\varepsilon)$ the effect is to rescale $L$ to $\lambda^{-1}L$ and replace $\Phi(X)$ by $\lambda^{\frac{1}{2}(\dim X)}\Phi(X)$.

Let $H_1$ denote the upper half-plane, and normalise the period lattice $L$ as $L = 2\pi i\mathbb{Z} + 4\pi i\tau\mathbb{Z}$ with $\omega = 2\pi i\tau$ and $\text{Im}(\tau) > 0$. Denote the subgroup of matrices $A = \begin{pmatrix} a & b \\ c & d \end{pmatrix} \in SL_2(\mathbb{Z})$ such that $c \equiv 0 \pmod{2}$ by $\Gamma_0(2)$. Allowing $\tau$ to vary, we say that the holomorphic function

$$\Phi : H_1 \rightarrow \mathbb{C}$$

is a modular form of weight $2k$ with respect to the subgroup $\Gamma_0(2)$ provided that :

(i) $\Phi(\tau) = (c\tau + d)^{-2k}\Phi(A\tau)$, and

(ii) $\Phi$ has a non-negative Fourier expansion at the two "cusps" 0 and $i\infty$.

The previous paragraph implies that the genus $\Phi(X)$ becomes a modular form of weight $2k$ with invariance subgroup $\Gamma_0(2)$. In particular, as $\tau$ varies, $\delta$ and $\varepsilon$ become modular forms of weights 4 and 8 respectively, with Fourier expansions at zero given in Proposition 2.1 below. The geometric significance of $\Gamma_0(2)$ is that its natural action on $H_1$ preserves the half-period point $\omega$.

One further property of the universal genus, which may become clearer in our discussion of classifying spaces, is that it can be interpreted as the character of a virtual, projective, unitary representation of the group $\text{Diff}(S^1)$.

## 2. The Witten genus and Ricci curvature

In the previous subsection we showed that any elliptic genus is determined by a power series $Q(x)$. In the case of the universal genus $\Phi$ we have

**Proposition 2.1.** If $g'(y) = (1 - 2\delta y^2 + \varepsilon y^4)^{-1/2}$ and $Q(x) = x/f(x)$, where $f$ is the inverse power series to $g$, then

$$Q(x) = \frac{x/2}{\sinh x/2} \prod_{n=1}^{\infty} \left[ \frac{(1 - q^n)^2}{(1 - q^n e^x)(1 - q^n e^{-x})} \right]^{(-1)^n}$$

$$= \exp\left( \sum_{k>0,2|k} \widetilde{G}_k x^k \right),$$

where

$$\widetilde{G}_k(\tau) = -\frac{B_k}{2k} + \sum_{n \geq 1}\left( \sum_{d|n} (-1)^{n/d} d^{k-1} \right) q^n.$$

Furthermore

$$\delta = -\frac{1}{8} - 3\sum_{\substack{n \geq 1 \\ d|n \\ 2\nmid d}} \left( \sum d \right) q^n \quad \text{and} \quad \varepsilon = \sum_{n \geq 1}\left( \sum_{\substack{d|n \\ 2\nmid(n/d)}} d^3 \right) q^n.$$

**Proof.** (See [Za] pp. 218–219.) Consider meromorphic functions $\psi : \mathbb{C} \to \mathbb{C}$ which satisfy the conditions

$$(*) \qquad \begin{cases} \psi(x + 2\pi i) = -\psi(x) \,, \; \psi(x + 4\pi i\tau) = \psi(x) \,, \; \psi(-x) = -\psi(x) \,, \\ \text{the poles of } \psi \text{ all lie inside } L \,, \psi(x) = \dfrac{1}{x} + O(1) \text{ as } x \to 0 \,. \end{cases}$$

The lattice $L$ is as above. Such a function is unique, since if $\psi_1$ and $\psi_2$ were two such functions, $\psi_1 - \psi_2$ would be holomorphic and doubly periodic, hence constant, and hence equal to zero by oddness. The proposition is proved by showing that $\psi$ exists and equals $Q(x)/x$ and also $1/f(x)$, where $Q(x)$ equals the given power series. As usual $q = e^{2\pi i\tau}$.

First check that $Q(x)/x$ does indeed satisfy conditions $(*)$; call this $\psi_1$. Next for any $\psi$ satisfying $(*)$ $\psi(x)^2$ and $\psi'(x)^2$ are even, invariant under $L$-translation, and have poles at $x = 0$ with leading terms $x^{-2}$ and $x^{-4}$ as their only singularities (mod $L$). Therefore $\psi'^2$ must be a monic quadratic polynomial in $\psi^2$, i.e., $\psi'^2 = \psi^4 - 2\delta\psi^2 + \varepsilon$ for some $\delta, \varepsilon \in \mathbb{C}$. It follows that $\frac{1}{\psi(x)} = x + \cdots$ can be written as the inverse of a function

$g(y) = y + \cdots$ given by our original elliptic integral. The series in the second formula equals the Taylor series for $\log(Q(x))$, as can be seen from the first formula by taking the logarithm of both sides and substituting the calculable expansion for $\log((x/2)/\sinh x/2)$. The precise forms of $\delta$ and $\varepsilon$ now follow by a suitably sophisticated comparison of coefficients.

We now consider a variant of the power series in 2.1, due to E. Witten, and reverse the argument, obtaining a new genus $\varphi_W$. Witten interprets his formula, using ideas from Quantum Field Theory, as the equivariant index (Atiyah–Bott–Singer fixed point theorem) for a Dirac operator on the free loop space $LX$ of a smooth manifold $X$. To the best of this writer's knowledge we are still waiting for a mathematically rigorous definition of the operator; for a discussion of the circle of ideas involved see [Se]. We propose to show that it has at least one potentially important application in differential geometry, and the formula also suggests that $\varphi_W$ may be the restriction of a more general genus taking values in a ring of Siegel modular forms defined on a subset $H_2 \subseteq \mathbb{C}^3$ rather than $H_1 \subseteq \mathbb{C}$. We return to this latter point in our last section.

**Definition.** The Witten genus $\varphi_W$ is associated with the power series

$$Q_W(x) = \frac{x/2}{\sinh x/2} \prod_{n=1}^{\infty} \frac{(1-q^n)^2}{(1-q^n e^x)(1-q^n e^{-x})}$$

$$= \exp\left( \sum_{\substack{k>0 \\ 2|k}} \frac{2}{k!} G_k x^k \right)$$

where

$$G_k(\tau) = \frac{-B_k}{2k} + \sum_{n=1}^{\infty} \left( \sum_{d|n} d^{k-1} \right) q^n .$$

We only define $G_k$ when $k$ is non-zero and even. For $k \geq 4$ the close relation with Eisenstein series shows that $G_k$ is modular with respect to the whole group $SL_2(\mathbb{Z})$; for $k = 2$, $G_2 = -\frac{1}{24} + q + 3q^2 + \cdots$, and the modularity condition has a correction term

$$G_2\left(\frac{a\tau+b}{c\tau+d}\right) = (c\tau+d)^2 G_2(\tau) + \frac{i}{4\pi}c(c\tau+d) .$$

The factors $(-1)^{n/d}$ in the sum defining $\tilde{G}_k$ show that these functions are modular for all values of $k$ (including $k = 2$), since one has the relation

$$\tilde{G}_2(\tau) = -G_2(\tau) + 2G_2(2\tau) .$$

One way to avoid problems with the non-modularity of $G_2$ is to evaluate $\varphi_W$ only on oriented spin manifolds, satisfying the additional condition that $(1/2)p_1(X) = 0$. This amounts to beginning the series $\sum\limits_{k>0, 2|k} 2/k! G_k x^k$ above with $(1/12)G_4 x^4$ rather than $G_2 x^2$, and geometrically is the condition for a Spin structure on the loop manifold $LX$. Following S. Stolz [St] and calling such a manifold $X$ a $\widehat{Spin}$-manifold, we have :

**Theorem 2.2.** *There is a topological group $\widehat{Spin}$ and a continuous homomorphism*

$$\pi : \widehat{Spin} \to Spin$$

*such that the induced map of classifying spaces is the 7-connected covering of $BSpin$, i.e. $B\widehat{Spin} \cong BO\langle 8 \rangle$.*

**Proof:** See [St] Theorem 1.2. In view of the importance of $\widehat{Spin}$ structures in recent developments we will at least give the structure of the group. The classifying space $BPU(\mathcal{H})$ for projective unitary bundles fibred by an infinite dimensional separable Hilbert space $\mathcal{H}$ is a $K(\mathbb{Z}, 3)$; let $\mathcal{Q}$ be the principal bundle over $Spin$ corresponding to some generator of $H^3(Spin, \mathbb{Z}) \cong \mathbb{Z}$, and define $\widehat{Spin}$ as the subgroup of $Aut(\mathcal{Q})$ given by the short exact sequence

$$Gauge(\mathcal{Q}) \rightarrowtail \widehat{Spin} \overset{\pi}{\twoheadrightarrow} Spin .$$

Here, the gauge group of $\mathcal{Q}$ consists of those automorphisms which induce the identity on the base.

**Conjecture 2.3.** If $X$ is a $4k$-dimensional $\widehat{Spin}$ manifold, which admits a metric of positive definite Ricci curvature, then the Witten genus $\varphi_W(X) = 0$.

In the same paper that contains the full proof of Theorem 2.2, S. Stolz sketches a proof of the conjecture, assuming the existence of a Dirac operator $D(LX)$. Heuristically one expects the scalar curvature of a 'lifted' metric $Lg$ on $LX$ to be given in terms of the Ricci curvature of the metric $g$ on $X$. If $Ric(g)$ were positive definite, this would force $scalar(Lg) > 0$, and by an analogue of the finite-dimensional argument with the Weitzenböck formula the operator $D(LM)$ would have to be invertible. A consequence of this is the vanishing of the index.

**Examples. (i)** The total space $E$ of a bundle with fibre the homogeneous space $G/H$ and structure group $G$ (compact, semisimple), and base $B$

admitting a metric of positive definite Ricci curvature, also admits such a metric and satisfies $\varphi_W(E) = 0$.

**(ii)** By 'plumbing' (see [HBJ] Section 6.5) it is possible to show that $\varphi_W : \Omega^*_{\widetilde{Spin}} \otimes \mathbb{Q} \to \mathbb{Q}[G_4, G_6]$ is a surjective ring homomorphism. We can then construct a 24-dimensional simply connected $\widetilde{Spin}$-manifold $X$ which is such that

$$\widehat{A}(X) = 0 \quad \text{but} \quad \varphi_W(X) \neq 0 .$$

The manifold $X$ admits a metric of positive scalar curvature, but (modulo Conjecture 2.3) no metric of positive definite Ricci curvature. Such examples can also be constructed more directly by surgery, combined with facts about the spectrum $eo_2$, see [HM]. Their construction shows the continuing vitality of methods in the development of which C. T. C. Wall was so prominent in the 1960s.

## 3. Elliptic Cohomology

In order to define what is meant by a complex-oriented cohomology theory we start with a multiplicative theory $h^*(X)$, which satisfies the Eilenberg–Steenrod axioms and is such that $\frac{1}{2} \in h^0(\text{point}) = R$. The theory $h^*$ is defined on a category of spaces including finite $CW$-complexes. With $\mathbb{CP}^\infty = \bigcup_{n \geq 1} \mathbb{CP}^n$ we are interested in theories such that

$$h^*(\mathbb{CP}^n) = R[\xi]/(\xi^{n+1}) ,$$

where $\xi \in h^2(\mathbb{CP}^n)$ is an element which (i) maps to $-\xi$ under conjugation and (ii) restricts to the canonical generator $1 \in h^2(\mathbb{CP}^1)$ under the chain of maps

$$h^2(\mathbb{CP}^n) \xrightarrow[\text{res}]{} h^2(\mathbb{CP}^1) = h^2(S^2) \xleftarrow[\text{susp}]{\sim} h^0(\text{point}) = R .$$

The element $\xi$ is then said to define a complex orientation for $h^*$. The structure of $h^*(\mathbb{CP}^n)$ has numerous implications for the theory:

**(i)** Given a complex line bundle $E$ over $X$, the classifying map $f_E : X \to \mathbb{CP}^\infty$ induces a characteristic class $f^*\xi \in h^2(X)$. More generally for any complex vector bundle over $X$ there are $h^*$-valued Chern classes $c_i^h$ ($i = 1, \dots, \dim_\mathbb{C} E$), which satisfy the axioms corresponding to those for the Pontryagin classes in ordinary cohomology. The classes for the universal $n$-plane bundle generate $h^*(BU(n))$.

(ii) The orientation of a complex vector bundle $E$ determines a Thom isomorphism

$$h^i(X) \xrightarrow{\simeq} \widetilde{h}^{i+n}(E^+) \ .$$

(iii) Let $\mu : \mathbb{CP}^\infty \times \mathbb{CP}^\infty \to \mathbb{CP}^\infty$ be the map which classifies tensor products of line bundles. Then $\mu$ induces a map of $h^*$-algebras

$$\mu^* : h^*(\mathbb{CP}^\infty) \to h^*(\mathbb{CP}^\infty) \overset{\wedge}{\underset{h^*}{\otimes}} h^*(\mathbb{CP}^\infty) \ ,$$

determined by the element $\mu^*\xi = G^h(\xi \otimes 1, 1 \otimes \xi)$.

This amounts to using the first Chern class to define a commutative 1-dimensional *formal group law*

$$G^h(x, y) = \sum_{i,j} a_{ij} x^i y^j \ ,$$

which is odd $\left(-G^h(x, y) = G^h(-x, -y)\right)$ and graded $\left(a_{ij} \in R^{-2(i+j-1)}\right)$.

Proposition 1.1 together with the universality of the formal group law in cobordism $[Q]$ shows that (iii) is equivalent to giving a structural map (genus)

$$\varphi^h : R_\Omega \longrightarrow R = h^*(\text{point}) \ .$$

**Examples:**

(i) $G^H(x, y) = x + y$ gives real cohomology, and

(ii) $G^{KU}(x, y) = x + y + txy$ gives complex $K$-theory with coefficients $\mathbb{Z}[t, t^{-1}]$.

We are interested in the converse problem of using the genus to define the cohomology theory, i.e., defining

$$h^*(X) = \left(\Omega^*_{SO}(X) \otimes \mathbb{Z}[\tfrac{1}{2}]\right) \otimes_{R_\Omega} R \ .$$

In order to preserve exactness we therefore need the ring $R$ to satisfy certain flatness conditions over $R_\Omega$. These are encapsulated in the so-called Landweber conditions [Lw1], and hold for the universal elliptic genus. For a rigorous and elegant treatment, which works for all levels, not just two, see [F]. However, in order to avoid a digression into general quotients of localised cobordism theories and the formal group law of a generic elliptic curve, we use a more topological approach below. We emphasise that our argument implicitly uses much of the same elliptic input — this is hidden in Ochanine's Theorem 1.2.

**Theorem 3.1.** *Consider the ring* $R = \mathbb{Z}[\frac{1}{2}][\delta, \varepsilon]$ *and the diagram of localisations*

$$R[\Delta^{-1}]$$

$$\nearrow \qquad \nwarrow$$

$$R[\varepsilon^{-1}] \qquad\qquad\qquad R[(\delta^2 - \varepsilon)^{-1}]$$

$$\nwarrow \qquad \nearrow$$

$$R$$

*There are homology theories with each of these rings as coefficients. The corresponding cohomology theories are multiplicative and complex oriented. In each case the formal group law is the Euler law*

$$F(x,y) = \frac{x\sqrt{P(y)} + y\sqrt{P(x)}}{1 - \varepsilon\, y^2 x^2} \, ,$$

*where* $P(y) = 1 - 2\delta y^2 + \epsilon y^4$.

**Sketch proof.** It is possible to use bordism with singularities to construct a connective homology theory with coefficients equal to $R$. For this we recall from the general introduction the definition of the bordism groups $\Omega_n^{SO}(X)$ for the space $X$ in terms of equivalence classes of maps $f : M \to X$ with $M$ an oriented manifold. This gives us a homology theory with coefficients $R = \mathbb{Z}[\frac{1}{2}][x_4, x_8, x_{12}, \ldots]$, where, as in the proof of Theorem 1.2 we can take $x_4 = \mathbb{CP}^2$, $x_8 = H_{3,2}$ and the remaining generators so that the ideal $(x_{12}, x_{16}, \ldots)$ consists of all bordism classes killed by elliptic genera. Dividing out by this ideal to obtain $R = R_\Omega/(x_{12}, x_{16}, \ldots)$ corresponds to allowing the manifolds $M$ with target $X$ to have singularities of a prescribed form. Such singularities are compatible with the Eilenberg–Steenrod axioms — see [Ba] for a careful treatment of this. In this way we obtain the first homology theory. It is at least folklore that the obstructions to obtaining a product are 2-primary, and hence vanish in this case. Elliptic cohomology is obtained by passing to the dual cohomology theory and inverting one of the elements $\Delta$, $\varepsilon$ or $(\delta^2 - \varepsilon)$. Note that this last step introduces periodicity into the theory.

In Section 4 we will make a start in constructing a geometric model for the theory whose existence has just been algebraically proved. Motivation for the bundle-like objects needed is provided by a remarkable theorem of M. Hopkins, N. Kuhn and D. Ravenel [HKR]. They start from a result of M. Atiyah [At], which relates the complex representation ring $R(G)$ of a finite group to $K(BG)$. Note that $BG$ can be modelled as an infinite $CW$-complex with finite skeleta $BG^{(n)}$, and Atiyah defines $K(BG)$ to be $\varprojlim K\left(BG^{(n)}\right)$. He then proves that the completion of the natural map $\alpha : R(G) \to K(BG)$ with respect to the $I$-adic topology ($I$ = kernel of the augmentation map) is an isomorphism. The original method of proof was to start with $G = C_p$, a cyclic group of prime order, extend to solvable groups and finally use a version of Brauer induction for an arbitrary group. The first problem is therefore to describe $K(S^{2n-1}/C_p)$, the $K$-theory of the standard symmetric lens space $L^{2n-1}(p; 1, \ldots, 1)$. Using the fact that this fibres over $\mathbb{C}\mathbb{P}^{n-1}$ it is not hard to see that $K(S^{2n-1}/C_p) \cong R(C_p)/\lambda_{-1}(\rho)$ with $\rho$ equal to the defining representation. Passing to the limit gives the $p$-adic completion of the representation ring, which for groups of prime power order agrees topologically with the $I$-adic completion. Up to this point we have used little more than the 'flat bundle map' $\alpha$ and the existence of a complex orientation.

In order to generalise this argument to other cohomology theories without having an initial representation ring, we first $p$-localise and then take characters, obtaining an isomorphism

$$K(BG)_p \bigotimes_{\mathbb{Z}_p} \bar{\mathbb{Q}}_p \cong \mathrm{Map}(G_p^{(1)}, \bar{\mathbb{Q}}_p) \ .$$

Here, the suffix $p$ refers to $p$-adic completion, $\bar{\mathbb{Q}}_p$ denotes the completion of the algebraic closure of $\mathbb{Q}_p$ and $G_p^{(r)}$, $r = 1, 2, \ldots$, denotes conjugacy classes of commuting $r$-tuples of elements of $p$-power order in $G$.

**Theorem 3.2.** *Let $S = R[\varepsilon^{-1}] = \mathbb{Z}[\frac{1}{2}][\delta, \varepsilon, \varepsilon^{-1}]$ and $\bar{S}$ be the algebraic closure of the quotient field of the $p$-adic completion of $S$. There is an isomorphism*

$$\mathcal{E}\ell\ell^*(BG)_p \bigotimes_S \bar{S} \xrightarrow{\approx} \mathrm{Map}\,(G_p^{(2)}, \bar{S}) \ .$$

**Sketch proof.** There are two main ingredients. The first is a generalisation of the Artin induction theorem for representations, which, modulo a technical argument to circumvent the presence of quotients, reduces the problem first to abelian and then to cyclic groups. This can be thought of as a streamlined version of Atiyah's second and third steps. The burden of the proof is to construct the isomorphism above for $C_p = G$, starting from the description of $\mathcal{E}\ell\ell^*(BC_p)$ as $S[[\xi]]/[p]\xi$ coming from the complex orientation. The notation $[p]\xi$ is shorthand for $F\Big(\xi, \underbrace{F(\xi, \ldots F(\xi, \xi))}_{p} \ldots\Big)$,

where $F$ is the formal group law. After reduction modulo $p$, $F$ turns out to have height two — this is the key ingredient in the more 'arithmetic' proof of Theorem 2.1 — and application of a theorem of Lubin–Tate allows us to construct the required map into the class functions $\mathrm{Map}\,(G_p^{(2)}, \bar{S})$. The summary we have given of this step also makes it clear that the ring $S$ is taken to come equipped with a map into the ring of integers of some finite extension of $\mathbb{Q}_p$.

**Remark.** Full details of Theorem 3.2 have yet to appear in print. Several versions exist in preprint form. We have stated the theorem as it appears in [Se]. Useful references are [H1], [K] and [HKR2].

With applications to the elliptic objects associated with Moonshine phenomena in mind, we end this subsection with a brief description of theories of level higher than 2. The main references are [Bk] and [Br]. By way of notation let $\Gamma \subset \Gamma_0(2)$ be a congruence subgroup, and let $N$ be the smallest positive integer such that

$$\Gamma \supseteq \Gamma(N) = \mathrm{Ker}\left[SL_2(\mathbb{Z}) \to SL_2(\mathbb{Z}/N)\right].$$

With $\zeta_N = e^{2\pi i/N}$ let $S(\Gamma)$ be the $\mathbb{Z}[\zeta_N, \frac{1}{2}]$-algebra of meromorphic, even weight modular forms with invariance subgroup $\Gamma$ and Fourier coefficients at every cusp belonging to $\mathbb{Z}[\zeta_N, \frac{1}{2}]$. It is a deep result of P. Deligne and M. Rapoport [DR] that

(i) $S(\Gamma)$ is a finitely generated $\mathbb{Z}[\zeta_N, \frac{1}{2}]$-algebra, and

(ii) the $\mathbb{C}$-algebra of forms just described equals $S(\Gamma) \otimes_{\mathbb{Z}[\zeta_N, \frac{1}{2}]} \mathbb{C}$.

If $\Gamma_1 \subseteq \Gamma_2$ are principal congruence subgroups in $\Gamma_0(2)$, then $S(\Gamma_1)$ is faithfully flat both over $S(\Gamma_2)$ and $S(\Gamma_0(2))$. Hence up to inversion of the discriminant $\Delta$ we can define higher level theories by

$$\mathcal{E}\ell\ell_N^*(X) = \mathcal{E}\ell\ell^*(X) \bigotimes_R S(\Gamma).$$

This intuitively simple, but in detail sophisticated, approach is due to J. Brylinski. Starting from F. Hirzebruch's higher level elliptic genera, see [HBJ], Chapter 7 and A1.6, A. Baker has developed what appear to be similar theories. The technical tool here is the structure of the ring of level $N$ modular forms as a Galois extension of $\mathcal{E}\ell\ell[\frac{1}{N}]^*$, that is we obtain simpler proofs at the price of inverting $N$ in the coefficients.

What happens if we try to replace the elliptic by the Witten genus as the structural map for a cohomology theory? Put another way : is it possible to construct a spectrum $eo_2$ which bears the same relation to $MO\langle 8\rangle$ as the elliptic spectrum does to $MSpin$? Although the literature on the solution to this question is incomplete, I hope that the following summary based on [H2] and [HM] will convey the main ideas.

The first approach is abstract; one starts from the fact that the power series $Q$ associated to the genus $\varphi_Q$ determines a stable exponential characteristic class, which in turn is determined by its values on the Hopf line bundle $\eta$. For a genus defined on $MO\langle 8\rangle$-manifolds one is faced with the problem that the structural group of $\eta$ does not lift to $O\langle 8\rangle$ and one must work with the tensor product of three copies of $(\eta - 1)$. This leads Hopkins to study what he calls 'cubical structures' as a tool for the construction of maps

$$\sigma_h : MO\langle 8\rangle \to h \ .$$

Here the cohomology theory $h$ is such that $h^*(\text{point})$ is torsion free and concentrated in even dimensions (otherwise take $MU\langle 6\rangle$ as domain spectrum) and its formal group law is isomorphic to the formal completion of that of an elliptic curve $E$ over $h^0(\text{point})$. Provisionally let us describe such theories as being of "elliptic type".

Subject to the further technical condition of taking limits over families of elliptic curves which are (a) "étale" and (b) satisfy a higher associativity condition, the maps $\sigma_h$ above combine to give

$$\sigma : MO\langle 8\rangle \to \text{limit spectrum } eo_2 \ .$$

The spectrum $eo_2$ is no longer of elliptic type, but is such that $eo_2^*(pt)$, called the *ring of topological modular forms*, at least contains the natural target of the Witten genus $\varphi_W$. To explain this last claim recall the definition of the ring of modular forms over $\mathbb{Z}$

$$\widetilde{R}_1 \ = \ \mathbb{Z}[c_4, c_6, \Delta]/(c_4^3 - c_6^2 - 1728\Delta) \ ,$$

where $\widetilde{R}_1$ maps into the usual ring of modular forms with invariance subgroup $SL_2(\mathbb{Z})$ by mapping $c_4$ and $c_6$ to multiples of the forms $G_4$ and $G_6$ introduced in Section 2. The element $\Delta$ maps to the discriminant. More or less by definition $eo_2^*(pt)$ contains the subring $\widetilde{R}$ consisting of those forms known to be values of $\varphi_W$, in particular $\mathbb{Z}[c_4, c_6]$ and $\alpha(\binom{24}{i})\Delta^i$ where $\alpha(r)$ equals the product of the highest powers of 2 and 3 dividing $r$.

The second approach is constructive: build up a spectrum $\overline{X}$ such that $\widetilde{R}$ embeds in $\pi_*(\overline{X})$. The clue is provided by

**Proposition 3.3.** There exists a space $X$ such that for every $MO\langle 8\rangle$-manifold $M^{4m}$ there exists a manifold $N$ having the same Witten genus as $M$, and normal bundle classified by a composition

$$N \to \Omega X \to BO\langle 8\rangle \ .$$

The space $X$ is an $S^9$-fibration over $S^{13}$ with cross-sectional obstruction having order 12 in $\pi_{12}(S^9)$. The spectrum $\overline{X}$ is then the Thom spectrum of the bundle induced by a map from $\Omega X$ to $BO$ identifying 8-dimensional generators in homology.

The motivation for this construction comes from a table of pairs $(X, \varphi)$ for which 3.3 or an analogue holds:

| $X$ | $\varphi$ |
|---|---|
| $(S^1, S^5)$ | $\widehat{A}$ |
| $(S^5, S^9)$ 2-local | $\Phi_{ell}$ |
| $(S^9, S^{13})$ | $\varphi_W$ |

In order to embed $\widetilde{R}$ in $\pi_*(\overline{X})$ one makes detailed calculations in dimensions less than or equal to 24, and then uses the ring structure. (In this context recall the use of $\Delta$-periodicity plus calculations in low weight to obtain the structure of the ring of *arithmetic* modular forms.)

We have already referred in Section 2 to the use of $\overline{X}$ in the construction of a 24-dimensional manifold with $\widehat{A} = 0$, $\varphi_W \neq 0$.

## 4. Elliptic cohomology of classifying spaces of finite groups

The results in this section are suggested by Theorem 3.2 and by G. Segal's discussion of 'elliptic objects' in the last section of [Se]. Additional motivation comes from '2 variable Moonshine', which really dictates the conditions which must be imposed on infinite dimensional bundles over $LBG$. We start with a decomposition of the free loops in the classifying space $BG$, with $G$ finite.

**Proposition 4.1.** Let $h$ represent a conjugacy class $[h]$ of elements in $G$. The loop space $LBG$ is homotopy equivalent to the disjoint union $\bigsqcup_{[h]\subseteq G} BC_G(h)$, where $C_G(h)$ denotes the centraliser of $h$ in $G$.

**Proof.** Define a subspace in the space of all smooth paths in a universal free contractible $G$ space $EG$ by

$$L_h EG = \{p : \mathbb{R} \to EG \mid p(t+1) = hp(t) \ \forall t \in \mathbb{R}\} ,$$

and write

$$L_G EG = \coprod_{[h] \subseteq G} L_h EG .$$

There is a left action of $G$ on $L_G EG$ given by $(g \cdot p)(t) = gp(t)$, and this action maps $L_h EG$ to $L_{ghg^{-1}} EG$, so that $C_G(h)$ maps $L_h EG$ to itself. Furthermore (Exercise) each space $L_h EG$ is contractible, and we can construct an equivariant map from the space of paths into $EG \times G$. In the latter space $g$ sends the pair $(e, h)$ to $(ge, ghg^{-1})$, and passing to quotients shows that $LBG \simeq EG \times_G G_{(conj)}$, from which 4.1 follows.

Note that for any $G$-space $F$ the associated bundle $EG \times_G F$ pulls back to the bundle $L_h EG \times_{C_G(h)} F$ over $BC_G(h)$. All this depends on the choice of element $h$ representing the class $[h]$.

Proposition 4.1 implies that we can construct bundles over $LBG$, which are possibly infinite dimensional, by means of a family of graded bundles over the components $BC_G(h)$. Up to questions of completion a finite dimensional bundle is flat, i.e., of the form $EC_G(h) \times_{C_G(h)} V$ for some representation space $V$ for the centraliser. The compatibility conditions making the bundle over $LBG$ into an 'elliptic object' are expressed in terms of characters, and take the following form, due to J. Devoto [DV2].

Let $G$ be a group of odd order, and let $\Gamma_0(2) \times G$ act on the product of the set $G^{(2)}$ of commuting pairs and $H_1$ via

$$\left( \begin{pmatrix} a & b \\ c & d \end{pmatrix}, g \right) \times ((g_1, g_2), \tau) \xrightarrow{\rho} \left( g(g_1^d g_2^{-c}, g_1^{-b} g_2^a) g^{-1}, \frac{a\tau + b}{c\tau + d} \right) .$$

The action $\rho$ induces, for each $k \in \mathbb{Z}$, an action $\rho_k$ of $\Gamma_0(2) \times G$ on the ring of functions $\vartheta : G^{(2)} \times H_1 \to \mathbb{C}$. The action $\rho_k$ is defined by

$$\rho_k \left( \begin{pmatrix} a & b \\ c & d \end{pmatrix}, g \right) \vartheta((g_1, g_2), \tau) = (c\tau + d)^{-k} \vartheta \left( g(g_1^d g_2^{-c}, g_1^{-b} g_2^a) g^{-1}, \frac{a\tau + b}{c\tau + d} \right) .$$

**Definition.** The group $\mathcal{E}\ell\ell_G^{-2k}$ is the abelian group whose elements are the holomorphic functions $\vartheta : G^{(2)} \times H_1 \to \mathbb{C}$ that satisfy the following conditions:

(1) $\rho_k\left(\left(\begin{smallmatrix} a & b \\ c & d \end{smallmatrix}\right), g\right) \vartheta = \vartheta, \forall \left(\left(\begin{smallmatrix} a & b \\ c & d \end{smallmatrix}\right), g\right) \in \Gamma_0(2) \times G;$

(2) for each $(g_1, g_2) \in G^{(2)}$ the function $\vartheta((g_1, g_2), -) : H_1 \to \mathbb{C}$ has a power series expansion at both cusps of the form

$$\vartheta((g_1, g_2), \tau) = \sum_{n \geq K} a_n q^{\frac{n}{|g_1|}},$$

where $K \in \mathbb{Z}$, $q = \exp\{2\pi i \tau\}$, and $a_n \in \mathbb{Z}\left[\frac{1}{2}, \frac{1}{|G|}, \exp\left\{\frac{2\pi i}{|g_1 g_2|}\right\}\right]$.

(3) Let $C_{g_1}(G)$ be the centralizer of $g_1$ in $G$, and let

$$\psi = \exp\{2\pi i / |C_{g_1}(G)|\}.$$

If $n$ and $|C_{g_1}(G)|$ are coprime, and $\sigma_n$ is the ring automorphism of $\mathbb{Z}[\frac{1}{|G|}, \psi]$ defined by $\sigma_n(\psi) = \psi^n$, then

$$\sigma_n\big(a_m(g_1, g_2)\big) = a_m(g_1, g_2^n).$$

The group structure in $\mathcal{E}\ell\ell_G^*$ is induced by the sum of functions. Furthermore if $\vartheta \in \mathcal{E}\ell\ell_G^{-2k}$ and $\vartheta' \in \mathcal{E}\ell\ell_G^{-2k'}$, then $\vartheta\vartheta' \in \mathcal{E}\ell\ell_G^{-2(k+k')}$. Hence the direct sum $\mathcal{E}\ell\ell_G^* = \oplus_{k \in \mathbb{Z}} \mathcal{E}\ell\ell_G^{-2k}$ has a natural structure of a graded ring.

For each $g_1$ the $q$-expansion coefficient functions $a_n(g_1, -)$ are rational virtual characters on the centraliser $C_G(g_1)$, and hence lie in the rationalised representation ring $\mathbb{Q} \otimes R\big(C_G(g_1)\big)$. Indeed there exists a non-zero integer $N$ such that $Na_n(g_1, -) \in R\big(C_G(g_1)\big)$.

**Theorem 4.2.** *If $G$ is a finite group of odd order, then*

$$\mathcal{E}\ell\ell^*(BG) \otimes \mathbb{Z}\left[\frac{1}{|G|}\right] \cong \widehat{\mathcal{E}\ell\ell}_G^*,$$

*where the right hand side is completed with respect to $I_G = \mathrm{Ker}\big(\varepsilon : \mathcal{E}\ell\ell_G^* \to \mathcal{E}\ell\ell^*[\frac{1}{|G|}]\big)$ and $\varepsilon$ is the augmentation map corresponding to the inclusion of the trivial group into $G$.*

**Proof.** This is contained in [DV2] and has much in common with [HKR]. In particular, the basic step is to understand the isomorphism for the cyclic group $C_p$. Conjugation is trivial, and the action of $\Gamma_0(2)$ on $C_p^{(2)}$ has two orbits, represented by $(0,0)$ and $(1,0)$. Elements in $\mathcal{E}\ell\ell_{C_p}^*$ are determined by their values at 0 and 1 (identify $C_p$ with $\mathbb{Z}/p$), and we have a decomposition into the direct sum of $R = \mathcal{E}\ell\ell_0^* = $ non-equivariant coefficients and

the ring of modular forms whose power series expansions at the cusps have
the form

$$\sum_{n>K} a_n q^{\frac{n}{p}} , \ a_n \in \mathbb{Z}\left[\tfrac{1}{2}, e^{2\pi i/p}\right] ,$$

and whose invariance subgroup is $\Gamma_1(p) \cap \Gamma_0(2)$, with $\Gamma_1(p)$ consisting of
matrices $A \equiv \left(\begin{smallmatrix} 1 & 0 \\ * & 1 \end{smallmatrix}\right)$. In an earlier paper [DV1] Devoto shows that the
latter summand is isomorphic to $R[x]/\Psi(x)$ for a particular polynomial $\Psi$
generalising $1 + x + \cdots + x^{p-1}$. It turns out that $R[x]/\Psi(x)$ is isomorphic
to the augmentation ideal $I$, and taking powers of $I$, as in the case of $K$-
theory corresponds to calculating $\mathcal{E}\ell\ell_{C_p}^*(S^{2n-1})$, the elliptic cohomology of
a lens space. Allowing $n$ to tend to infinity proves the theorem for $C_p$. We
have gone into some detail of this example because of Devoto's assumption
that $G$ has odd order, and is hence solvable. Knowing the result for $C_p$,
and using the spectral sequence associated with the short exact sequence

$$1 \to G_1 \to G \to C_p \to 0$$

(care with filtrations!) allows us to prove 4.2 by induction.

**Remark.** It is plausible that Devoto's work can be extended to level-one
elliptic cohomology (where the coefficients are $SL_2(\mathbb{Z})$ rather than $\Gamma_0(2)$
invariant).

Having proved Theorem 4.2 it is now clear how to describe elliptic
objects. These should be vector bundles $E \to LX$ locally modelled on a
Hilbert space $\mathcal{H}$ over the field $\mathbb{C}$. Since $\mathcal{H}$ is infinite dimensional we must
ensure that we have partitions of unity — in the case of $LBG$ there will be
no problem by 4.1. We assume further that we have an associated principal
bundle $Q \to LX$ with structure group $\mathcal{G}$ acting on $\mathcal{H}$ by isometries.

**Definition.** An elliptic object over $BG$ is a bundle over $LBG$ as above
which satisfies the additional conditions

**(a)** On each component $BC_G(h)$ of $LBG$ there is a bundle decomposition

$$E|_{BC_G(h)} \cong \overset{\wedge}{\bigoplus_n} E_{[h],n}$$

where $n$ is greater than some integer $K$, and each $E_{[h],n}$ is the finite dimensional flat bundle associated to a finite dimensional representation space
$W_{h,n}^E$ of $C_G(h)$.

**(b)** The characters of each graded representation are equivariantly modular, that is they satisfy Devoto's conditions (1) – (3) above.

Note that $K$ provides a lower bound on the denominators of the exponents in the various power series expansions.

With this definition we can conclude that, at least for groups of odd order, up to completion and inversion of order, $\mathcal{E}\ell\ell^*(BG)$ is described by bundles.

The definition above is provisional and motivated by Devoto's definition of equivariant elliptic cohomology. There is a similar treatment in [GKV], which extends to compact Lie groups. For some groups $G$ there appears to be a version of 4.2 which holds without tensoring with an extension of $\mathbb{Z}$.

Together with A. Baker, the author has formulated a definition of an elliptic object which generalises both the one given above (for $BG$) and the one used by J. Brylinski in [Br] (for simply-connected manifolds). See [BT] for details, and [FLM] for the background.

## 5. *The mysterious role of the Mathieu group* $M_{24}$

The rings of modular forms which occur as coefficients in elliptic cohomology admit operations by the Hecke algebra. These operations can be thought of as being analogous to cohomology operations, are weight preserving, and satisfy the rules

$$T(m)T(n) = T(mn) \quad \text{if} \quad (m,n) = 1 \,,$$
$$T(p^{t+1}) = T(p)T(p^t) - p^k T(p^{t-1}) \,, \quad \text{for forms of weight } k \,.$$

It is possible to list those pairs ($k$ = weight, $N$ = level) for which the space of forms is one-dimensional, and hence generated by an eigenform of the Hecke algebra, see [R] for the basic results needed and [Ma1] for an outline of the calculation. By more than a coincidence the pairs $(k, N)$ are closely associated with the 'even' conjugacy classes in the sporadic simple group $M_{24}$ of order $2^{10} \cdot 3^3 \cdot 5 \cdot 7 \cdot 11 \cdot 23$.

**Definition.** The group $M_{24}$ is the automorphism group of the Steiner system $S(5,8,24)$ consisting of special octads from a 24 element set $\Omega$. Special means that each pentad from $\Omega$ is contained in a unique octad.

Note that $S(5,8,24)$ can be thought of as a 3-step extension of $S(2,5,21)$, a projective plane with automorphism group $PSL_3(\mathbb{F}_4)$. This construction can be used to prove the simplicity of the chain of automorphism groups $M_{22} \subset M_{23} \subset M_{24}$.

Since $M_{24}$ permutes the elements of $\Omega$, it admits a permutation representation in $S_{24}$ which splits as $(23) + (1)$, our notation being that of the "Atlas of Finite Groups" [BRB]. If $g$ represents a conjugacy class, we write $g$ as a product of disjoint cycles, $g = (1)^{j_1}(2)^{j_2} \cdots (r)^{j_r}$, $j_1 + 2j_2 + \cdots + rj_r = 24$, that is, the product contains $j_i$ cycles of length $i$ for $1 \leq i \leq r$. The link between Hecke eigenspaces and classes $g$ is then provided by the rule

$$\eta_g(\tau) = \eta(\tau)^{j_1}\eta(2\tau)^{j_2} \cdots \eta(r\tau)^{j_r} = \sum_n a_g(n)q^n .$$

Here, as in earlier sections, $\tau \in H_1$, $q = e^{2\pi i \tau}$ and

$$\eta(\tau) = q^{1/24} \prod_{m=1}^{\infty} (1 - q^m) .$$

Note that the form corresponding to the identity equals

$$\eta(\tau)^{24} = \Delta(\tau) ,$$

whose coefficients are the Ramanujan numbers.

Examples of the forms $\eta_g(\tau)$ are

| conjugacy class | Weight | Level | Fourier expansion |
|---|---|---|---|
| $(1\ 3\ 5\ 15)$ | 2 | 15 | $q - q^2 - q^3 - q^4 + q^5 + q^6 +$ |
| $(1^2\ 11^2)$ | 2 | 11 | $q - 2q^2 - q^3 + 2q^4 + q^5 + 2q^6 +$ |
| $(1\ 23)$ | 1 | 23 | $q - q^2 - q^3 \qquad\quad + q^6 + \ldots$ |

The weight $k(g)$ equals $(j_1 + j_2 + \cdots + j_r)/2$, and the level $N(g)$ equals the product of the maximum and minimum values of $j_n$. G. Mason has shown that the forms $\eta_g(\tau)$ combine to give the character of a graded virtual representation

$$\Theta(q) = \sum_{n=1}^{\infty} \theta_n q^n .$$

The associated flat bundles define an element in $K(BM_{24})[[q]]$, which is actually the restriction to the 'fixed point component' of a bundle over $LBM_{24}$ of the kind considered in the previous section. See [Ma2,3] for the more general construction.

**Remark.** In characteristic 2 the group $M_{24}$ admits an 11-dimensional representation, whose $\mathcal{E}\ell\ell^*$-Chern classes generate $\mathcal{E}\ell\ell^*(BM_{24})$. Note that we can omit consideration of the 2-Sylow structure, since $\frac{1}{2} \in \mathcal{E}\ell\ell^*(\text{point})$. At level one, when $\frac{1}{6}$ belongs to the coefficients, the calculation is even easier.

The construction outlined above can be pushed further. For each conjugacy class $g$ we have a formal $L$-series

$$L_g(s) = \sum_{n=1}^{\infty} a_g(n)n^{-s},$$

which, because of the Hecke invariance, has a product decomposition involving terms of the form

$$\left(1 - a_g(p)p^{-s} + p^{k-1-2s}\right)^{-1}.$$

Here, for the sake of simplicity we restrict to even values of $k$; for 'odd' conjugacy classes such as (1 23) it is necessary to twist $p^{k-1-2s}$ by a non-trivial Dirichlet character. When $k = 2$, i.e., for the conjugacy classes $(2\ 4\ 6\ 12)$, $(2^2\ 10^2)$, $(6^4)$, $(1\ 2\ 7\ 14)$, $(1\ 3\ 5\ 15)$ and $(1^2\ 11^2)$, we obtain the $L$-series of an elliptic curve defined over $\mathbb{Q}$. The formal group law for each of these curves will in turn define a genus $\varphi_g : \Omega^*_{SO} \otimes \mathbb{Z}[\frac{1}{2}] \to \mathbb{C}$. For the curves and the corresponding values of $\delta, \varepsilon$, see the forthcoming book [Th].

**Question.** Are the elliptic genera defined by conjugacy classes in $M_{24}$ of any geometric significance?

The group $M_{24}$ makes another surprising appearance in the structure of elliptic objects. In the final section we will show how a one-dimensional formal group law can be associated with a $K3$ surface, leading to the construction of a family of higher cohomology theories.

**Definition.** A compact complex surface $Y$ is called a $K3$ surface if $Y$ has a nowhere zero holomorphic 2-form $\omega$ and dim $H^1(Y, \mathcal{O}_Y) = 0$. An automorphism of $Y$ or an action of a group $G$ on $Y$ is said to be (complex) symplectic if it fixes $\omega$.

In [Mk] S. Mukai proves

**Theorem 5.1.** *For a finite group $G$ the following two conditions are equivalent:*

*(a) $G$ acts effectively and symplectically on a $K3$-surface $Y$.*

*(b) There is an embedding of G in the Mathieu group $M_{23}$, such that under the induced action on the 24 element Steiner set $\Omega$, there are at least 5 orbits.*

The subgroups $G$ can be determined by a careful reading of the tables in the "Atlas" [BRB], and each of them actually occurs as $\text{Aut}_\omega(Y)$ for some surface $Y$. The harder part of the theorem is to show that this group satisfies the Mathieu condition. Mukai does this by counting the number of fixed points of a periodic symplectic automorphism $h$. If the period is $N$, this number $\varepsilon(h) = 24(N\prod_{p|N}(1+p^{-1}))^{-1}$. Note that

$$\frac{24}{\varepsilon(h)} = [SL_2(\mathbb{Z}) : \Gamma_0(N)] ,$$

where $\Gamma_0(N)$ (generalising $\Gamma_0(2)$) consists of matrices $\left(\begin{smallmatrix} a & b \\ c & d \end{smallmatrix}\right)$ with $c \equiv 0 \pmod{N}$. This brings us back to the determination of one-dimensional Hecke eigenspaces above.

Here are Mukai's examples of $K3$ surfaces which embed in $\mathbb{CP}^3$ and their automorphism groups.

| Group | Order | $K3$ − surface |
|---|---|---|
| $L_2(7)$ | 168 | $X^3Y + Y^3Z + Z^3X + T^4 = 0$ |
| $M_{20} \cong (2^4) \rtimes A_5$ | 960 | $X^4 + \cdots + T^4 + 12XYZT = 0$ |
| $(4^2) \rtimes S_4$ | 384 | $X^4 + \cdots + T^4 = 0$ |
| $(Q_8 * Q_8) \rtimes S_3$ | 192 | $X^4 + \cdots + T^4 - \sqrt{-12}(X^2Y^2 + Z^2T^2) = 0$ . |

As usual our notation is close to that of [BRB], $Q_8 * Q_8$ denotes the central product of two copies of the quaternion group.

The first and third of these examples are probably the easiest to explain. The first goes back to F. Klein, who identified $L_2(7)$ with the complex automorphism group of the genus 3 curve defined by $X^3Y + Y^3Z + Z^3X = 0$. For the third, combine permutation of the coordinates with the automorphism

$$(X : Y : Z : T) \mapsto (i^a X, i^b Y, i^c Z, i^d T) \quad \text{with} \quad i = \sqrt{-1} ,$$

and note that $\text{Aut}_\omega(Y)$ has index 4 in this extension.

One (of many) possible future lines of research is to investigate the connections between these automorphism groups of Mathieu type for an arbitrary Kummer surface in $\mathbb{C}\mathbb{P}^3$ and the genera to be briefly described in the final section.

## 6. Variants and Generalisations

### (i) Elliptic cohomology at the prime 2

Starting from spin bordism, M. Kreck and S. Stolz [KS] have succeeded in constructing a homology theory, which when localised away from the prime 2, gives a homology theory dual to the cohomology theory constructed in Section 2 with coefficients in $R[\varepsilon^{-1}] = \mathbb{Z}[\frac{1}{2}][\delta, \varepsilon, \varepsilon^{-1}]$. Their integral definition is as follows:

$$\mathcal{E}\ell\ell_n^{\mathbb{H}}(X) = \bigoplus_{k \in \mathbb{Z}} \Omega_{n+8k}^{Spin}(X)/\sim$$

where the equivalence relation "$\sim$" is generated by identifying the class $[B, f] \in \Omega_n^{Spin}(X)$ with $[E, fp] \in \Omega_{n+8}^{Spin}(X)$ for every $\mathbb{H}\mathbb{P}^2$-bundle $p : E \to B$ having structural group $\Gamma = PSp_3$.

There is a related functor $ell_*(X)$ obtained by mapping to zero all bordism classes of the form $[E, fp]$ where $[B, f] = 0$ in $\Omega_{*-8}^{Spin}(X)$. Kreck and Stolz show that $ell_*(X)$ becomes a homology theory when localised at the prime 2 (though not at odd primes). The unlocalised theory fits into an exact sequence

$$\widetilde{\Omega}_*^{Spin}(\Sigma^8 B\Gamma \wedge X^+) \to \Omega_*^{Spin}(X) \to ell_*(X) \to 0 ,$$

and the problem is to construct a representing spectrum for $ell_*(\cdot)$ (after localising at 2). The first choice is the cofibre of the map of spectra

$$M\,Spin \wedge \Sigma^8 B\Gamma \to M\,Spin ,$$

which for technical reasons does not quite work. To circumvent this it is necessary to split $M\,Spin \wedge \Sigma^8 B\Gamma$ as $A \vee B$ and then construct a (2-local) homotopy equivalence $A \vee \bigvee \Sigma^{8k} ko \to M\,Spin$, where $ko$ is the spectrum for connective $KO$-theory. The final result is a (2-local) homotopy equivalence

$$ell \simeq \bigvee \Sigma^{8k} ko ,$$

but with a product structure other than that coming from multiplication in $ko$.

**Remark.** M. Hovey has shown that one obtains a cleaner version of theory by inverting $(\delta^2 - \varepsilon)$ rather than $\varepsilon$, see Theorem 3.1. Besides showing that elliptic homology is actually defined over $\mathbb{Z}$, the work of Kreck and Stolz is suggestive in another way.

**Question.** Do the groups

$$\mathcal{E}\ell\ell_n^{\mathbb{O}}(X) = \bigoplus_{k \in \mathbb{Z}} \Omega_{n+16k}^{\langle 8 \rangle}(X)/ \sim$$

satisfy the axioms for a homology theory, where this time "$\sim$" is generated by identifying the class of $[B, f]$ with $[E, fp]$, when $p : E \to B$ is an $\mathbb{O}\mathbb{P}^2$-bundle with structural group $F_4$?

Here, $\mathbb{O}$ denotes the Cayley numbers or Octonians, and $F_4$ is isomorphic to the isometry group of $\mathbb{O}\mathbb{P}^2$ with its natural metric.

One aim of the final subsection is to provide evidence for the existence of a cohomological dual to the theory proposed by the question, provided that we localise away from the primes 2 and 3. At these two primes the theory for $\mathbb{H}\mathbb{P}^2$-bundles suggests that we must look at the cofibre of a map

$$MO\langle 8 \rangle \wedge \Sigma^{16} BF_4 \to MO\langle 8 \rangle ,$$

for which the technical problems look formidable. Note in passing that the torsion in $\Omega_{\langle 8 \rangle}^*$ and $H^*(BF_4, \mathbb{Z})$ involves the primes 2 and 3 only, and that although there is also 5-torsion in $H^*(Sp_4(\mathbb{Z}), \mathbb{Z})$, this disappears on passing to an appropriate discrete subgroup.

Lack of space does not allow us to go more deeply into the stable homotopy theoretic arguments in [KS]. However, as has already been pointed out in [Ros], they provide vivid testimony to the power of a method first applied by Wall to $\mathrm{Tors}(\Omega_{SO}^*)$.

### (ii) K3-Cohomology

The bigraded cohomology of a $K3$-surface, and in particular of a quartic Kummer surface embedded in $\mathbb{C}\mathbb{P}^3$ is described by the so-called Hodge diamond

$$\begin{array}{ccccc}
 & & 1 & & \\
 & 0 & & 0 & \\
1 & & 20 & & 1 \\
 & 0 & & 0 & \\
 & & 1 & & \\
\end{array}$$

The construction of cohomological analogues of the formally completed Picard group $\widehat{Pic}(X)$, for $X$ an algebraic variety over some field $k$ is described in [AM]. However for our purposes the more 'ad hoc' account in

[Si] will be adequate. Associated with the field of definition we have two one-dimensional formal groups $\mathbb{G}_a^\wedge \left( (x_1, x_2) \mapsto x_1 + x_2 \right)$ and $\mathbb{G}_m^\wedge \left( (x_1, x_2) \mapsto x_1 + x_2 - x_1 x_2 \right)$, the latter of which by the usual process of sheafification defines a coefficient sheaf $\mathbb{G}_{m,X}^\wedge := \mathbb{G}_{m,\mathcal{O}_X}^\wedge$.

We then have the Artin–Mazur functors

$H^1(X, \mathbb{G}_{m,X}^\wedge) = $ formal Picard group and

$H^2(X, \mathbb{G}_{m,X}^\wedge) = $ formal Brauer group.

The latter turns out to be (pro)representable by an $h^{0,2}$-dimensional formal group, provided that $H^1(X, \mathcal{O}_X) = H^3(X, \mathcal{O}_X) = 0$, conditions which hold for our $K3$-surface. Furthermore, in a way explained to the author by J. Morava, the associated genus satisfies the Landweber exactness proposition — see the discussion before Theorem 3.1 — so that we obtain a family of cohomology theories.

**Example.** Consider the quartic surface $X^4 + Y^4 + Z^4 + T^4 = 0$. Then we have a formal group law over $\mathbb{Z}$ with logarithm

$$g(y) = \sum_{m \geq 0} \frac{(4m)!}{(m!)^4} \frac{y^{4m+1}}{4m+1} ,$$

which has a reassuring appearance, particularly as no powers with exponent $4m + 2$ occur in $g'(y)$. See [Si], page 919.

What does this family of genera have to do with the existence of 'higher' elliptic theory $\mathcal{E}\ell\ell_*^\mathbb{O}(X)$? We start with the Siegel space $H_2$ consisting of $2 \times 2$ complex symmetric matrices with positive definite imaginary part. Topologically this is an open convex subset in $\mathbb{C}^3$, and analytically the orbit space $Sp_4(\mathbb{Z})\backslash H_2$ is the moduli space for principally polarised abelian varieties of dimension 2 over $\mathbb{C}$. As in the case of the upper half-plane $H_1$ we can define holomorphic forms, for which the modularity condition is

$$\Phi(Z) := \det(CZ + D)^{-2k} \Phi(\alpha Z) ,$$

with $\alpha$ the $4 \times 4$ symplectic matrix $\begin{pmatrix} A & B \\ C & D \end{pmatrix}$ with

$$A^T D - C^T B = 1_2 , \quad A^T C = C^T A , \quad B^T D = D^T B .$$

Since the degree equals 2 the third (boundedness) condition is actually redundant, see [Fr] I.3.

The varieties parametrised by the points of the moduli space are either products of elliptic curves, or Jacobians $J(C)$ of curves of genus 2. The construction of the latter is beautifully described in the recently published volume of lecture notes [CF], where it is also shown that $J(C)$ double covers a Kummer surface $K$. Furthermore, since we work over an algebraically closed field, it is in principle possible (and in [CF] explained) how to reverse this process. Fact: a tangent plane to the surface $K$ meets $K$ in a curve birationally equivalent to $C$. We can now mimic the discussion at the end of Section 1 showing the modularity of the universal elliptic genus. Starting with (say) the quartic Fermat hypersurface we obtain a ring homomorphism from $\Omega^*_{(8)} \otimes \mathbb{Q}$ into $\mathbb{C}$, with prescribed generating manifolds mapping to complex images which we can denote $\delta, \varepsilon, \dots$. Rescaling $J(C)$ introduces modularity for $\delta, \varepsilon, \dots$, and varying the defining parameters inside $H_2$ converts them into modular forms, defined on at least some open subset. Thus the combination of the Artin–Mazur construction of the Brauer group, plus part of the theory of genus 2 curves, appears to allow the construction of $K3$-genera taking values in degree 2 modular forms. This is however very much work in progress . . . .

## References

The references starred below constitute a reading list for those wishing to find out more about the subject.

[AM] M. Artin, B. Mazur, Formal groups arising from algebraic varieties, Ann. Sci. École Norm. Sup. **10** (1977), 87–131.

[At] M. Atiyah, Characters and cohomology of finite groups, Publ. Math. IHES **9** (1961), 247–289.

[AtRS] M. Atiyah *et al.*, Physics and Mathematics of Strings, Royal Society Publications (London), 1989.

[Ba] N. Baas, On bordism theories of manifolds with singularities, Math. Scand. **33** (1973), 279–302.

*[Bk] A. Baker, Elliptic genera of level $N$ and elliptic cohomology, J. London Math. Soc. (2) **49** (1994), 581–593.

[BT] A. Baker, C. Thomas, Classifying spaces, Virasoro equivariant bundles, elliptic cohomology and Moonshine, ETH preprint (1996).

*[Br] J. Brylinski, Representations of loop groups, Dirac operators on loop spaces, and modular forms, Topology **29** (1990), 461–480.

[BRB] J. Conway *et al.*, An Atlas of Finite Groups, Clarendon Press (Oxford), 1985.

[CF] J. Cassels, E. Flynn, Prolegomena to a middlebrow arithmetic of curves of genus 2, London Math. Soc. Lecture Note Series **230**, Cambridge University Press (Cambridge), 1996.

[DV1] J. Devoto, An algebraic description of the elliptic cohomology of classifying spaces, J. Pure Appl. Algebra **130** (1998), 237–264.

*[DV2] J. Devoto, Equivariant elliptic cohomology and finite groups, Michigan Math. J. **43** (1996), 3–32.

[F] J. Franke, On the construction of elliptic cohomology, Math. Nachr. **158** (1992), 43–65.

[Fr] E. Freitag, Siegelsche Modulfunktionen, Grundlehren der Math. Wiss. **254**, Springer-Verlag (Berlin), 1983.

[FLM] I. Frenkel *et al.*, Vertex Operator Algebras and the Monster, Academic Press (Boston) 1988.

[GKV] V. Ginzburg *et al.*, Elliptic algebras and equivariant elliptic cohomology, preprint (1995).

*[HBJ] F. Hirzebruch *et al.*, Manifolds and Modular Forms, Vieweg (Braunschweig–Wiesbaden), 1988.

*[H1] M. Hopkins, Characters and elliptic cohomology, in Advances in Homotopy Theory (Cortona, 1988), London Math. Soc. Lecture Note Series **139**, Cambridge University Press (Cambridge), 1989, 87–104.

[H2] M. Hopkins, Topological modular forms, the Witten genus, and the theorem of the cube, Proc. of ICM (Zürich 1994), Birkhäuser (Basel) 1995.

[HM] M. Hopkins, M. Mahowald, The Witten genus in homotopy, MIT preprint (1996).

[HKR1] M. Hopkins *et al.*, Generalised group characters and complex oriented cohomology theories, to appear (?).

*[HKR2] M. Hopkins, N. J. Kuhn, D. C. Ravenel, Morava $K$-theories of classifying spaces and generalized characters for finite groups, in Algebraic Topology (San Feliu de Guíxols, 1990), Lecture Notes in Math. **1509**, Springer-Verlag (Berlin), 1992, 186–209.

*[KS]  M. Kreck, S. Stolz, $\mathbb{H}\mathbb{P}^2$-bundles and elliptic homology, Acta Math. **171** (1993), 231–261.

*[K]  N. Kuhn, Character rings in algebraic topology, in Advances in Homotopy Theory (Cortona, 1988), London Math. Soc. Lecture Note Series **139**, Cambridge University Press (Cambridge), 1989, 111–126.

*[Lw1]  P. Landweber, Elliptic cohomology and modular forms, in Elliptic Curves and Modular Forms in Algebraic Topology (Princeton, 1986), Lecture Notes in Math. **1326**, Springer-Verlag (Berlin), 1988, 55–68.

*[Lw2]  P. Landweber, Supersingular elliptic curves and congruences for Legendre polynomials, in Elliptic Curves and Modular Forms in Algebraic Topology (Princeton, 1986), Lecture Notes in Math. **1326**, Springer-Verlag (Berlin), 1988, 69–93.

[Ma1]  G. Mason, $M_{24}$ and certain automorphic forms, in Finite Groups— Coming of Age (Montreal, Que., 1982), Contemp. Math. **45**, Amer. Math. Soc. (Providence), 1985, 223–244.

[Ma2]  G. Mason, On a system of elliptic modular forms associated to the large Mathieu group, Nagoya Math. J. **118** (1990), 177–193.

[Ma3]  G. Mason, $G$-elliptic systems and the genus zero problem for $M_{24}$, Bull. Amer. Math. Soc. **25** (1991), 45–53.

[Mk]  S. Mukai, Finite groups of automorphisms of $K3$ surfaces and the Mathieu group, Invent. Math. **94** (1988), 183–221.

*[O]  S. Ochanine, Sur les genres multiplicatifs définis par les intégrales elliptiques, Topology **26** (1987), 143–151.

[Q]  D. Quillen, On the formal group laws of unoriented and complex cobordism theory, Bull. Amer. Math. Soc. **75** (1969), 1293–1298.

[R]  R. Rankin, Modular Forms and Functions, Cambridge University Press (Cambridge), 1977.

[Ros]  J. Rosenberg, Reflections on C. T. C. Wall's work on cobordism, volume 2 of this collection.

*[Se]  G. Segal, Elliptic cohomology, Astérisque 161–162 (1988), 187–201.

[Si]  J. Stienstra, Formal group laws arising from algebraic varieties, Amer. J. Math. **109** (1987), 907–925.

[St]  S. Stolz, A conjecture concerning positive Ricci curvature and the Witten genus, Math. Ann. **304** (1996), 785–800.

*[Tb]  C. Taubes, $S^1$-actions and elliptic genera, Comm. Math. Phys. **122** (1989), 455–526.

[Th]  C. Thomas, Elliptic Cohomology, Plenum/Kluwer (New York), 1999.

*[Za]  D. Zagier, Note on the Landweber–Stong elliptic genus, in Elliptic Curves and Modular Forms in Algebraic Topology (Princeton, 1986), Lecture Notes in Math. **1326**, Springer-Verlag (Berlin), 1988, 216–224.

Department of Pure Mathematics and Mathematical Statistics
University of Cambridge
Cambridge CB2 1SB
England, UK

email: c.b.thomas@pmms.cam.ac.uk

GPSR Authorized Representative: Easy Access System Europe - Mustamäe tee 50, 10621 Tallinn, Estonia, gpsr.requests@easproject.com

product-compliance

W, UK